QC 311   DEH

WITHDRAWN
FROM STOCK
QMUL LIBRARY

QMW Library

D0302922

104

17

05

06

16
2007

THERMODYNAMICS IN MATERIALS SCIENCE

McGraw-Hill Series in Materials Science and Engineering

Editorial Board

Ronald Gibala, University of Michigan
Matthew Tirrell, University of Minnesota
Charles A. Wert, University of Illinois

Brick, Pense, and Gordon: *Structure and Properties of Engineering Materials*
Courtney: *Mechanical Behavior of Materials*
DeHoff: *Thermodynamics in Materials Science*
Dieter: *Engineering Design: A Materials and Processing Approach*
Dieter: *Mechanical Metallurgy*
Flemings: *Solidification Processing*
Fontana: *Corrosion Engineering*
Geiger, Allen, and Strader: *VLSI Design Techniques for Analog and Digital Circuits*
Mielnik: *Metalworking Science and Engineering*
Seraphim, Lasky, and Li: *Principles of Electronic Packaging*
Smith: *Foundations of Materials Science and Engineering*
Smith: *Principles of Materials Science and Engineering*
Smith: *Structure and Properties of Engineering Alloys*
Vander Voort: *Metallography: Principles and Practice*
Wert and Thomson: *Physics of Solids*

THERMODYNAMICS IN MATERIALS SCIENCE

Robert T. DeHoff

University of Florida

McGraw-Hill, Inc.

New York   St. Louis   San Francisco   Auckland   Bogotá
Caracas   Lisbon   London   Madrid   Mexico   Milan   Montreal
New Delhi   Paris   San Juan   Singapore   Sydney   Tokyo   Toronto

This book was set in Times Roman by Electronic Technical Publishing Services.
The editors were B. J. Clark and John M. Morriss;
the production supervisor was Richard A. Ausburn.
The cover was designed by Carla Bauer.
Project supervision was done by Electronic Technical Publishing Services.
R. R. Donnelley & Sons Company was printer and binder.

QUEEN MARY & WESTFIELD
COLLEGE LIBRARY
(MILE END)

THERMODYNAMICS IN MATERIALS SCIENCE

Copyright ©1993 by McGraw-Hill, Inc. All rights reserved. Printed in the United States of
America. Except as permitted under the United States Copyright Act of 1976, no part of this
publication may be reproduced or distributed in any form or by any means, or stored in a data
base or retrieval system, without the prior written permission of the publisher.

1 2 3 4 5 6 7 8 9 0   DOC DOC   9 0 9 8 7 6 5 4 3

ISBN 0-07-016313-8

Library of Congress Cataloging-in-Publication Data

DeHoff, Robert T.
    Thermodynamics in materials science / R. T. DeHoff.
       p.   cm. — (McGraw-Hill series in materials science and
  engineering.)
    Includes index.
    ISBN 0-07-016313-8
    1. Materials science.   2. Thermodynamics.   I. Title.   II. Series.
  TA403.6.D44   1993
  620.1'1296—dc20   92-40109

ABOUT THE AUTHOR

Robert T. DeHoff arrived at the University of Florida with a fresh doctoral degree from Carnegie-Mellon University in 1959. He played a significant role in the development of the then fledgling Metals Research Laboratory into the present Department of Materials Science and Engineering, one of the largest and most successful in the world. In addition to his research in microstructures and how they evolve, he has been a major architect of the curriculum, and has developed and taught a variety of courses at the undergraduate and graduate levels, including courses in physical metallurgy, diffusion in solids, phase diagrams, quantitative stereology, and thermodynamics.

This textbook is the distillation of nearly three decades of teaching an undergraduate course in thermodynamics for materials scientists each year; throughout most of that time he also taught a graduate course in the same subject.

Professor DeHoff's reputation for making difficult and complex material clearly understandable has earned him three awards as an outstanding teacher at the University of Florida and was a component in his election as a Fellow of the American Society for Materials.

Dr. DeHoff has a son and daughter and one grandson; he lives with his wife of thirty-five years in Gainesville, Florida.

CONTENTS

PREFACE

In his classic paper in 1883 J. Willard Gibbs completed the apparatus called phenomenological thermodynamics that is used in engineering and science to describe and understand what determines how matter behaves. This work is all the more remarkable in the light of the enormous expansion of our knowledge in science and technology in the twentieth century. During the last century hundreds of books have been written on thermodynamics. In most cases these texts were directed at students in a particular field. Thermodynamics plays a key role in chemistry, physics, chemical engineering, mechanical engineering, engineering science, biology, and materials science and engineering. Each presentation offered its own slant for its intended audience. Several of these texts are classics that have endured for decades experiencing many revisions and many printings.

An author who undertakes an introductory text in thermodynamics in the face of this history had better be:

a. Sure of his or her subject, and

b. Have something unique to say.

After teaching introductory thermodynamics to materials scientists for nearly three decades at both the graduate and the undergraduate levels, I am convinced that the approach used in this text is both unique and in many ways better than those available elsewhere.

Thermodynamics in Materials Science is an introductory text, intended primarily for use in a first course in thermodynamics in materials science curricula. However, the treatment is sufficiently general so that the text has potential applications in chemistry, chemical engineering, and physics, as well as materials science. The treatment is sufficiently rigorous and content sufficiently broad to provide a basis for a second course either for the advanced undergraduate or at the graduate level.

Thermodynamics is a discipline that supplies science with a broad array of relationships between the properties that matter exhibits as it changes its condition. All of these relationships derive from a very few general and pervasive principles (the laws of thermodynamics) and the repetitive application of a very few very general strategies. It is not a collection of independent equations conjured out of misty vapors by an all-knowing mystic for each new application. There is a structure to thermodynamics that is both elegant and, once contemplated, reasonably simple.

The approach that undergirds the presentation in this text emphasizes the connections between the foundations and the working relations that permit the solution of practical problems. In this emphasis, and in its execution, it is unique among its competitors. This difference is crucial to the student seeing the subject for the first time.

Most texts spend a significant quantity of print and students' time in presenting the laws of thermodynamics, and in laying out arguments that justify the laws and lend intuitive interpretation to them. This presentation is based upon the recognition that such diversions are generally a significant waste of time and effort for the student, and, what is worse, are usually confusing to the uninitiated. Worse still, students may be left with an inadequate intuition that merely serves to mislead them when they attempt to apply it to complex systems. Thermodynamics is fundamentally a rational subject, rich with deductions and derivations. Intuition in thermodynamics is not for the uninitiated.

Thus, the laws are presented as *fait accompli*: "great accomplishments of the nineteenth century" that distilled a broad range of scientific observation and experience into succinct statements that reflect how the world works. Best at this beginning stage that these laws be presented with clear statements of their content, without the perpetual motion arguments, Carnot cycles, and other intuitive trappings.

The most significant departure of this text from its competitors lies in the treatment of the concept of *equilibrium* in complex physical systems, and in the presentation of a *general strategy for deriving conditions for equilibrium* in such systems. A *general criterion for equilibrium* is developed directly from the second law of thermodynamics. The mathematical procedure for deriving the equations that describe the internal condition of a system when it is at equilibrium is then presented with rigor. It is the central viewpoint of this text that, since all of the "working equations" of thermodynamics are mathematical statements of these internal conditions for equilibrium, establishment of the connection between these conditions and first principles is crucial to the development of a working understanding of thermodynamics. Indeed, the remainder of the text is a series of applications of this general strategy to the derivation of the conditions for equilibrium in systems of increasing complexity, together with strategies for applying these equations to solve problems of practical interest to the student. With each increment in the level of sophistication of the system being treated, new parts of the apparatus of thermodynamics are introduced and developed as they are needed. The general strategy for getting to the working equations is the same for all of these applications. Thus, the connection to the fundamental principles is visible for each new development. Further, this connection can be maintained without introducing any mathematical or conceptual shortcuts. Repetition builds confidence; rigor builds competence.

One early chapter introduces the concepts of statistical thermodynamics. This subject is treated as an algorithm for converting an atomic model for the behavior of a system, formulated as a list of the possible states that each atom may exhibit, into values of all of the thermodynamic properties of the system. The strategy for deriving conditions for equilibrium applies in this case to the derivation of the Boltzmann distribution function, which reports how the atoms are distributed over the energy levels when the system attains equilibrium. The algorithm is then illustrated for the ideal gas model and for the Einstein model of a crystal. Statistical thermodynamics is used very little in subsequent chapters because the classes of systems that are the domain of materials science tend to be too complex for tractable treatment, much less for presentation to first-time students of the subject.

Most chapters contain several illustrative examples designed to emphasize the strategies that connect principles to hard numerical answers. Each chapter ends with a summary that reviews the important concepts, strategies, and relationships that it contains. Each chapter also ends with a collection of homework problems. Many of these are designed so that they are best solved using a personal computer; the astute student will find it useful to write some more general programs that can be used repeatedly as the level of sophistication increases. Examples and homework problems will be drawn more or less uniformly from the major classes of materials: ceramics, metals, polymers, electronic materials, and composites. This approach serves to illustrate the power of the concepts, laws, and strategies of phenomenological thermodynamics by demonstrating that they can be applied to all states of matter.

The experience gained in twenty-five years of teaching an undergraduate course in thermodynamics in materials science, together with more than fifteen years of teaching a graduate course in the same area, has resulted in an approach to the topic which is unique. This approach accents rigor, generality, and structure in developing the concepts and strategies that make up thermodynamics. Because the connections between first principles and practical problem solutions are clearly and sharply illuminated, the first-time student can hope not only to apply thermodynamics to the sophisticated kinds of system that are the bread and butter of materials science, but to understand their application.

It is a pleasure to acknowledge the help of Heather Klugerman who provided advice in the more sophisticated aspects of word processing involved in putting together this text. Pamela Howell proofread the manuscript with remarkable skill before it was submitted to the publisher. David C. Martin, University of Michigan, and Monte Pool, University of Cincinnati, offered many helpful comments and suggestions while reviewing the manuscript. My thanks to the many students, both graduate and undergraduate, who for many years encouraged me to undertake this text. Finally, I am grateful to my wife, Marjorie, who sacrificed many evenings, weekends, and vacations as I disappeared into the den to work on the project.

Robert T. DeHoff

CHAPTER

1

INTRODUCTION

What determines how matter behaves?

This question has been at the core of scientific enquiry since man became curious about his environment. As the human experience unfolded, answers to this question were first shrouded in mysticism. Occasional flashes of insight flared in the mist, parting it, only to be engulfed again in the fog. Scientific understanding of this question began to emerge when Sir Francis Bacon suggested that we *examine* the behavior of matter in our attempt to explain its behavior, rather than accepting mystical explanations handed down from the ancients. Acceptance of the notion of experimental science required several centuries to digest, and clear demonstrations that this approach to the question in fact works by an ever increasing number of protagonists such as Copernicus, Kepler, Galileo, and ultimately Isaac Newton.

By the nineteenth century this *mechanical* description of the behavior of matter was well established, but it was also becoming evident that this view was incomplete. Mechanics recognized that the behavior of matter could be described in terms of two fundamental ideas, one associated with the *motion* of matter and the other associated with its *position* in a potential field. These rudimentary notions were formalized as *kinetic energy* and *potential energy*. However, experimental studies of the behavior of matter made it clear that the condition of matter in a system could be influenced by factors *besides* its motion and position.

The most obvious of these influences is formulated in the concept of *heat*. Once the idea of temperature appeared and was quantified, it became clear that the aspect of the system behavior indicated by temperature could be altered for matter at rest in its surroundings. Thus, some influence beside kinetic and potential energy could change the condition of matter.

It also became evident that a system could be made to expand and contract by supplying or removing heat. This motion is such that some of the parts enclosing a system could be put to use to do work previously done by men or beasts of burden, for example, to raise water from a well. The idea of *pressure* inside such a system served to quantify this aspect of the behavior of matter. These mechanical effects could, like heat, be made to occur without moving the center of mass of the system, and thus were influences not included in a picture restricted to kinetic and potential energy.

The realization that these various influences could be converted from one form to another set the groundwork for a general description of the behavior of matter. The heat from the fire in a rudimentary steam engine could be converted into the work of expanding the steam in a piston, which could in turn be converted into the work of raising a weight in a gravitational field. It was noted that the mechanical work done by a boring tool in making a cannon made the barrel hot. The observation that heat from the fire was produced by matter conversions, recognized to be chemical changes, added yet another class of influences capable of altering the condition of matter at rest. The notion evolved to an understanding that these influences were all different manifestations of the same thing.

In its origins, *thermodynamics* focused upon the most rudimentary effects beyond kinetic and potential energy: mechanical work derived from expansion and contraction of the system, and heat. In practical settings these effects were primarily viewed in terms of understanding how transfers of heat influenced matter. Since in mechanics, the description of the motion of matter was called dynamics, it was reasonable to refer to this developing field as *thermodynamics*.

As the field evolved it gradually grew in scope to encompass *all* of the influences that could affect the condition of matter, and all of the interrelations that could exist among these influences. Eventually, that expansion in scope came to embrace not only thermal, mechanical, and chemical effects, but also the original influences of mechanics including kinetic energy and the complete set of potential energies that physics enumerates: gravitational, electrical, magnetic, and body forces.

The development of the thermodynamic view of what determines how matter behaves reached a pinnacle in 1883 with the publication of J. Willard Gibbs' classic paper, "On the Equilibrium of Heterogeneous Substances." In this single publication Gibbs completed the apparatus of thermodynamics, providing the formalism that has come to be accepted as the basis for the description of all phenomena that influence the condition of matter. With the exception of atomic energy, which essentially adds a term to the apparatus, and was of course unknown to him, Gibbs set out the general principle for determining the equilibrium condition of matter of arbitrary complexity, defined all of the properties necessary for the description of the state of matter at rest and in equilibrium, and laid out the strategy for computing changes in the state of matter that occur when it is taken through processes of arbitrary complexity. During the past century many of the most widely used textbooks in thermodynamics are, as they are intended by the authors to be, elaborations of Gibbs' classic paper. This text is no exception.

What determines how matter behaves?

The answer to this question can be presented on a variety of levels of sophistication. The first is *phenomenological thermodynamics*, the primary subject of this text, which focuses upon the phenomena that matter can experience as exposed by experimental observation. This level of description of the behavior of matter seeks a complete enumeration of all the kinds of behavior that are possible, and all the observed relationships that exist among the various classes of behavior. It is unnecessary to know the nature of the constitution of matter in order to apply this level of description; it is only necessary to know what phenomena are possible. For example, in order to predict the change in volume that a system experiences when its temperature is raised, it is only necessary to measure the coefficient of thermal expansion for that substance. It is not necessary to explain why that particular substance has the value of the coefficient of expansion that it has in comparison with other systems; it is sufficient to have measured it experimentally.

The second level of sophistication in answering the fundamental question is *statistical thermodynamics*. This level attempts to explain why different substances have different values of their properties, and, indeed, to predict the values of their properties from a knowledge of the *structure* of matter. In its most rudimentary form, which is the kinetic theory of gases, this description begins with the assumption that matter is composed of atoms that are particles of a known mass and with a known distribution of velocities. As our breadth of experimental knowledge expanded, it became clear that these aspects of the behavior of matter required a more comprehensive view of its structure. The atom itself has an internal structure composed of a nucleus and surrounding electron cloud. The nucleus has a structure, as does the electron cloud. The atoms arrange themselves into molecules with a molecular structure that dictates much of their chemical behavior. In most solids, the atoms or molecules form a regular crystal structure. The crystal structure has defects that play a dominant role in some of its properties. The crystals fit together to form the *microstructure* of the system. Each level in this hierarchy of structures has an influence on the behavior of matter. It is the bold goal of the statistical thermodynamics of matter to explain, and ultimately predict, its properties and behavior from a knowledge of its structure.

A third, more sophisticated and more fundamental level of answer to the basic question, seeks to explain why the structure of atoms and molecules is as it is observed to be. *Quantum mechanics* formulates descriptions of isolated single atoms, as well as ensembles of atoms in molecules, liquids, and crystals, that yield predictions of the electronic structure. The spatial distribution of the electron cloud provides the basis for computing some subset of the properties of the system, which may then be tested more or less directly by experiment.

The practical everyday encounters with the question, "What determines how matter behaves?" are best handled with phenomenological thermodynamics. This apparatus is capable of describing the behavior of very complex systems that may experience very complex processes. The more fundamental understanding, embodied in statistical thermodynamics, provides the basis for generalizing our understanding of the behavior of matter. However, since the formalism is complicated, its application is limited to relatively simple systems. The computations of quantum mechanics are extremely involved and lengthy, severely restricting their application in practice. With the exception of Chap. 6, Statistical Thermodynamics, this text presents the practical apparatus of phenomenological thermodynamics formulated by Gibbs.

Because it explains what determines how matter behaves in the most complex kind of system, thermodynamics lies at the foundation of materials science. More than any other branch of science and engineering, materials science requires the full breadth of the thermodynamic apparatus. An understanding of how microstructures of materials evolve, a prerequisite for controlling the properties of materials, begins with the phase diagrams of these multicomponent, multiphase systems. Chemical reactions with the environment may limit the useful lifetime of a material at high temperatures, or in other situations may be used to protect a material during processing. Adsorption and capillarity effects are key factors in determining the development of microstructures. Electrochemical behavior may degrade a material by corrosion, or it can be used to protect or even purify it. Thus the full apparatus of phenomenological thermodynamics presented in this text finds application in materials science presented at the undergraduate level.

Ultimately, the most profound formulation of the question, "What determines how matter behaves?" enters the realms of philosophy, metaphysics or theology. Some of the brightest minds in human science have illuminated the sharp outlines and internal structure of the stuff of which the universe is made and the mechanism that operates it. The light of reason has penetrated deeply into the structure but the ultimate foundation remains shrouded in mysticism.

CHAPTER 2

THE
STRUCTURE OF
THERMODYNAMICS

Thermodynamics is rooted in logic and reason. At its foundation are a very few, very general, and therefore very powerful principles: the Laws of Thermodynamics. From these few principles can be deduced predictions about the behavior of matter in a very broad range of human experience. This structure can be visualized as an inverted pyramid with the laws at the apex and the consequences or deductions expanding upward and outward as the range of applications is developed. An understanding of how matter behaves in every situation rests directly upon these laws.

In their simplest and most general form the laws apply to the universe as a whole:

1. There exists a property of the universe, called its *energy*, which cannot change no matter what processes occur in the universe.
2. There exists a property of the universe, called its *entropy*, which can only change in one direction no matter what processes occur in the universe.
3. A universal absolute temperature scale exists and has a minimum value, defined to be *absolute zero*, and the entropy of all substances is the same at that temperature.

More precise mathematically formulated statements of the laws are developed in Chap. 3.

In practice the focus of thermodynamics is on a subset of the universe, called a *system*, Fig. 2.1. In order to apply thermodynamics, the first step is to identify the

5

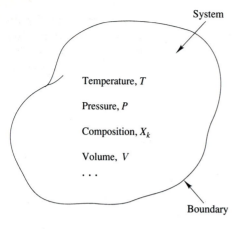

FIGURE 2.1
The subset of the universe in focus in a particular application of thermodynamics is usually called the *system*. At any given instant of observation, the condition of the system is described by an appropriate set of *properties*. Limitations on changes in these properties are set by the nature of its *boundary*.

subset of the universe that encompasses the problem at hand. It is necessary to be explicit about the nature of the contents of the system, and the specific location and character of its boundary.

The condition of the system at the time of observation is described in terms of its *properties*, quantities that report aspects of the condition of the system such as its temperature, T, its pressure, P, its volume, V, its chemical composition, and so on. As the system is caused to pass through a *process,* its properties experience changes, Fig. 2.2. A very common application of thermodynamics is the calculation of the changes in the properties of a specified system as it is taken through some specified process. Thus, an important aspect of the development of thermodynamics is the deduction of *relationships* between the properties of a system. Changes in some properties of interest, such as the entropy of the system can be computed from information given or determined about changes in other properties of the system such as temperature and pressure.

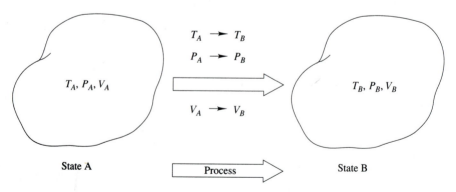

FIGURE 2.2
A *process* is a change in the condition or state of the system. Properties change from their values in some initial state A to some final state B.

An understanding of the structure of thermodynamics is greatly aided by deliberately organizing the presentation on the basis of a series of classifications. This organization compartmentalizes these elements of the field and thus permits a focus on the subset of the thermodynamic apparatus appropriate to a specific problem. Accordingly, this chapter presents classifications of:

1. Thermodynamic systems
2. Thermodynamic properties
3. Thermodynamic relationships

Mastery of the strategy for developing relationships among the properties of thermodynamic systems paves the way for deducing the working equations that are of most practical importance in thermodynamics: the *conditions for equilibrium*. A general criterion that can be used to determine when a system has attained thermodynamic equilibrium is introduced in Sec. 2.4. This concept is stated precisely and mathematically in Chap. 5, along with a general strategy for finding the conditions for equilibrium in the most complex kind of thermodynamic system.

2.1 A CLASSIFICATION OF THERMODYNAMIC SYSTEMS

The complete thermodynamic apparatus is capable of evaluating the behavior of the most complex kind of system. These systems are capable of experiencing the full range of influences that have been identified as possibly affecting its state. Most practical problems in thermodynamics do not require the application of the whole thermodynamic structure for their solution. In order to pinpoint the part of the apparatus that must be used to handle a given case, it is useful to devise a *classification of thermodynamic systems*. Use of such a classification at the beginning of any problem serves to focus attention on the specific set of influences that may operate, and, perhaps more importantly, those that may be excluded from consideration. This classification also serves as a basis for laying out the sequence of presentation in this text.

At the outset of consideration of any problem, classify the system under study according to each of the following five categories:

1. Unary versus multicomponent
2. Homogeneous versus heterogeneous
3. Closed versus open
4. Nonreacting versus reacting
5. Otherwise simple versus complex

Each of these descriptive words has explicit meaning in thermodynamics.

Category 1 identifies the complexity of the chemistry of the system. Systems with the simplest chemical composition are "unary," which means "one chemical component." If a system has more than one chemical component, that is, is multicomponent, additional apparatus must be invoked to describe its behavior.

The word "homogeneous" in category 2 specifically means "single phase." If a system is composed of more than one phase, (e.g., a mixture of water and ice), it is "heterogeneous." Treatment of heterogeneous systems adds to the thermodynamic apparatus.

In category 3, a "closed" system is one that makes no exchanges of matter with its surroundings for the processes considered. If matter is transferred the system is an "open system" and terms must be added to allow for influences transmitted by the addition of matter to the system.

Category 4 brings into consideration systems that can exhibit chemical reactions and focuses on the additional apparatus required to describe chemical reactions.

The last category lumps all other influences into a single listing. If a system is capable of exhibiting different kinds of energy exchange other than those arising from thermal, mechanical, or chemical changes, then it is a complex system. For example, if in the problem at hand there may be involved gravitational, electrical, magnetic, or surface influences, then it is classified as "complex" in this category. If none of these special kinds of influences can occur in the problem at hand, then it is an "otherwise simple" system.

Figure 2.3. is a cross section through a thin film device. From the point of view of thermodynamics the system consists of a large number of chemical components, some as impurities added to control the electronic properties and distributed through the several phases. Chemical reactions may occur at the gas/solid interface and between the solid phases. Thus, during processing, this system can be classified as a multicomponent, multiphase, closed, reacting, otherwise simple system.

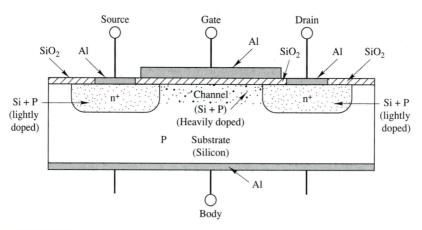

FIGURE 2.3

Cross section through a MOSFET (Metal Oxide Semiconductor Field Effect Transistor) thin film device shows it to be a multicomponent, multiphase system in which chemical reactions and the influence of an electrical field are important [2.1].

The most rudimentary kind of system that may be encountered is classified as a "unary, homogeneous, closed, nonreacting, and otherwise simple" system. This classification of simplest of systems is the focus of Chap. 4. Chapter 7 introduces unary *heterogeneous* systems. Chapter 8 presents the apparatus for handling *multicomponent*, homogeneous systems. *Multicomponent, heterogeneous systems* are dealt with in Chaps. 9 and 10. The apparatus for handling *reacting* systems is contained in Chap. 11. *Complex systems* are dealt with in Chaps. 12 through 15. The text progresses through a sequence of classes of systems of increasing complexity until, at the end, the tools for describing the behavior of matter in the most complicated kind of system, "multicomponent, heterogeneous, open, reacting, complex," are in hand.

2.2 CLASSIFICATION OF THERMODYNAMIC VARIABLES

The internal condition of a thermodynamic system, the changes in its condition, and the exchanges in matter and energy that it may experience, are quantified by assigning values to variables that have been defined for that purpose. These variables are the mathematical stuff of thermodynamics and their evaluation is the justification for inventing the apparatus in the first place. They fall into two major classes: *state functions* and *process variables*. Within these classes properties may also be defined that are *intensive* or *extensive*.

2.2.1 State Functions

A *state function* is a property of a system that has a value that depends on the current condition of the system and not on how the system arrived at that condition. The temperature of the air in the room has a certain value at the moment that does not depend on whether the room heated up to that temperature or cooled down to it. Other familiar properties that have this attribute are pressure, volume, and chemical composition. Figure 2.4 shows the *mathematical* nature of a state function.

One of the great accomplishments of thermodynamics is the identification of these and other properties of systems, perhaps not so familiar, which are also functions only of the current condition of the system. These include various measures of the *energy* of the system, its *entropy*, and a variety of properties associated with components in solutions. Complete definitions of these properties are developed in later chapters.

The fact that such *state functions* exist gives rise to one of the most important strategies for the thermodynamic analysis of the complicated processes likely to be encountered in the real world of science and technology. A process converts the condition of a system from some initial state, A, to some final state, B. Precisely because this class of properties, state functions, depend only on the state of the system, the change in any state function for any process is simply its value for the final state minus its value for the initial state. Thus, the change in any state function must be *identical* for *every* process that converts the system from the same initial state to the

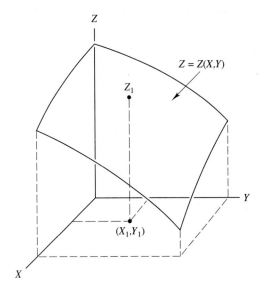

FIGURE 2.4
If the value of the variable Z depends only on the current values of the variables X and Y, then all three variables are *state functions*. The functional relationship among these variables, written $Z = Z(X, Y)$, is represented by a surface in (X, Y, Z) space. For any given values (X_1, Y_1) there is a corresponding value of $Z = Z_1$.

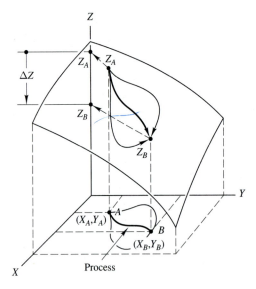

FIGURE 2.5
A process that changes the conditaion of the system from state A to state B may (if it is simple enough) be represented by a curve in the $(X$-$Y)$ plane; this represents the sequence of states through which the system passes in changing from state A to state B. Evidently, since Z is a state function, the change in Z, written $\Delta Z = Z_B - Z_A$, will be the *same* for all paths connecting A and B.

same final state. The value for the change in any state function is "independent of the path" or process by which the system is converted from state A to state B, Fig. 2.5.

As a consequence, the change in any state function accompanying a very complicated real-world process that alters the system from state A to state B may be computed by concocting or imagining the *simplest* process that connects the same two end states. A computation of the changes in state functions for this simple process yields the same result as would have been obtained for the very complex process. The importance of this strategy, its application, and other consequences of the existence of state functions are developed in Chap. 4.

2.2.2 Process Variables

In contrast to the notion of state functions, *process variables* are quantities that *only have meaning for changing systems*. Their values for a process depend explicitly upon the path, that is, the specific sequence of states traversed that takes the system from state A to state B. Change is inherent to the very concept of these quantities. There are two primary subcategories of process variables: *work* done on the system as it changes, and *heat* absorbed by the system.

The concept of work is developed in classical mechanics in physics. A force acts upon a body. If the point of application of the force moves, then the force does work. Let the vector F denote the force, and the vector dx denote an increment of its displacement, Fig. 2.6. The increment of work done by this displacement is defined to be

$$\delta w = F \cdot dx \qquad (2.1)$$

where the notation represents the dot product of the two vectors. For a finite process the force is moved along some path through space; the value of the force may change as it moves. The work done is defined to be

$$w = \oint F(x) \cdot dx \qquad (2.2)$$

where $F(x)$ describes how the force varies with position, and the integration is a line integral along the path traversed. The mathematical details are not important in the present context. It is important to note that the *displacement* of the force is an inherent component in the concept of work. Work cannot be associated with a system at rest, it is a *process variable*.

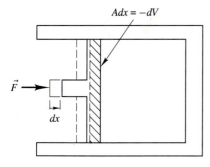

$Adx = -dV$

$\delta W = \vec{F} \cdot \vec{dx}$

$F = P_{ext} \cdot A$

$\delta W = P_{ext} \cdot A \cdot dx = -P_{ext}dV$

FIGURE 2.6
The generic concept of *work* is illustrated for mechanical work due to the external pressure on the system; the displacement of a force F through a distance dx does work.

It is possible for work to be done through a variety of influences that may act upon the system. Each of the forces that have been identified in physics:

1. the force exerted by the pressure on a system
2. the force due to gravity
3. body forces in a rotating system
4. the force acting on a charged particle in an electrical field
5. the force on a magnetic dipole in a magnetic field
6. the force associated with surface tension

may be displaced to do work. The early chapters in the text limit consideration primarily to work done by the mechanical force exerted by the external pressure on the system. The remaining forces listed are in the "complex system" category in the classification of systems and each has its own set of thermodynamic apparatus developed in Chaps. 12–15.

   If the boundary of a system is rigid and impermeable so that no matter can cross the boundry and no force acting upon the boundry may move it, the internal condition of that system can still be caused to change. Thus there exists a kind of influence that can alter the condition of a system that is not a form of work or matter transfer. During the past three centuries concepts and methods have gradually developed that quantified this *thermal* influence that begins with the development of the thermometer. A temperature scale was devised and evolved into a tool of general application. The *calorimeter* provided a means for determining the relative quantities of this thermal energy transferred in different processes. The quantity of *heat* that flows into or out of a system during a process can now be determined with accuracy, at least under carefully controlled experimental conditions. Heat is always associated with a change in the condition of a system; thus, it is also a process variable. Just as is true for work, it is meaningless to visualize a quantity of heat associated with a system that is not changing; the notion of the "heat content" of a system is meaningless. Flow and change are inherent aspects of the nature of heat.

2.2.3   Extensive and Intensive Properties

State functions may be further classified as *extensive* or *intensive* properties of the system.

   A property of.the system is *intensive* if it may be defined to have a value at a point in the system. For example, temperature T is an intensive thermodynamic property and the temperature has a value at each point in the system and may indeed vary from point to point. Pressure P may also be defined at each point in the system. In the earth's atmosphere, pressure varies with altitude as well as with geographical location. A variety of other intensive properties may be derived from the extensive properties of the system. Examples are developed following the introduction of the notion of an extensive thermodynamic property.

If the value of the property is reported for the system as a whole, then it is called an *extensive* property of the system. For example, the volume V of a system is an extensive property. The number of moles of a given chemical component n_k in the system is extensive, as are the internal energy U and entropy S, defined in Chap. 3. In general, extensive properties depend on the size or extent of the system. The most direct measure of size of a system is the *quantity of matter* that it contains. In a comparison of two systems that have the same intensive properties, doubling the quantity of matter doubles all of the extensive properties.

It is possible to derive an intensive property for each of the extensive properties defined in thermodynamics. Such a definition visualizes a limit of the ratio of two extensive properties. For example, the molar concentration c (moles of matter per liter) can be defined at a point by visualizing an infinitesimally small volume element neighboring the point (δV) and the number of moles (δn) in that element. The concentration is the limit of the ratio

$$c = \lim_{\delta V \to 0} \frac{\delta n}{\delta V} \qquad (2.3)$$

as δV approaches zero. This strategy can be used to define densities of internal energy, entropy, and any of the other extensive properties defined in Chap. 4. A rigorous development of these concepts is given in Chapter 14.

Similar definitions can be developed by reporting extensive properties per mole of matter in the system. Thus, the entropy per mole or volume per mole of the system can be defined for volume elements in the system and can vary from point to point. The most familiar and widely used example of molar properties is the mole fraction of component k, X_k, (see Chap. 8), used to describe chemical composition. The mole fraction is the ratio of the number of atoms (or molecules) of component k to the total number of atoms or molecules in the volume element. This measure of composition may vary from point to point in the system.

It may be confusing to find some intensive properties treated as if they were properties of the system as a whole. For example, a value for the temperature, pressure, or the atom fraction of CO_2 in a gas mixture can be reported for the system. However, this is possible only in the special case, frequently encountered in introductory thermodynamics, in which these intensive properties *do not vary with position* for the system under consideration. It is possible to quote a value of the temperature for a system if, and only if, the temperature is *uniform* throughout that system. Intensive properties, by concept, can be defined at each point in a system and have the potential to vary from point to point. If in simple systems intensive properties happen not to vary, then they have a single value that describes that property for the system. However, this does not convert them into extensive properties.

By their nature, intensive properties can depend only on the values of other intensive properties. It is clear that the value of a property that can be defined at a point in the system cannot depend on the value of another property that is a characteristic of the whole system. Extensive properties can be expressed as integrals of intensive properties over the extent of the system (see Chap. 14).

2.3 CLASSIFICATION OF RELATIONSHIPS

The apparatus of thermodynamics provides connections between the properties in a system. These connections are frequently unexpected and not intuitively evident. For example, how the entropy of a system varies with pressure is determined by the coefficient of thermal expansion of the system. This is demonstrated in Chap. 4. The apparatus also provides the equations for computing changes in these properties when a specified system is taken through a specified process. Also, thermodynamics introduces a variety of variables that are expressed in terms of previously defined quantities. As a result the formalism of thermodynamics generates a large number of relationships between the quantities with which it deals. The potential confusion that can result may be relieved to some extent by organizing the presentation of the relationships in thermodynamics.

This section introduces a classification of thermodynamic relationships so that this formidable array of equations can be subdivided into categories that reflect their origin.

1. *The Laws of Thermodynamics* are the fundamental equations that form the physical basis for all of these relations.

2. *Definitions* present new measures of the energy of systems expressed in terms of previously formulated variables. These defined quantities are generally introduced because they simplify the description of some particular class of system or process that may be commonly encountered. Another set of definitions introduces the quantities commonly measured in the laboratory and reported in thermodynamic tables or data bases.

3. *Coefficient relations* emerge from the description of changes in state functions during an infinitesimal step in a process. The differential relation

$$dZ = M dX + N dY + \cdots \tag{2.4}$$

describes, in a formal way, how the state function Z changes as a result of changes in the functions $(X, Y, \ldots)$ that describe the state of the system. The coefficients in this equation are related to specific partial derivatives of Z:

$$M = \left(\frac{\partial Z}{\partial X}\right)_{Y,\ldots} \quad \text{and} \quad N = \left(\frac{\partial Z}{\partial Y}\right)_{X,\ldots} \tag{2.5}$$

This differential equation is illustrated for two independent variables in Fig. 2.7. These coefficient relations frequently establish connections between variables that are not intuitively obvious.

4. *Maxwell relations* also derive from the mathematical properties of state functions. In order for Eq. (2.4) to be mathematically valid, it is necessary and sufficient that

$$\left(\frac{\partial M}{\partial Y}\right)_{X,\ldots} = \left(\frac{\partial N}{\partial X}\right)_{Y,\ldots} \tag{2.6}$$

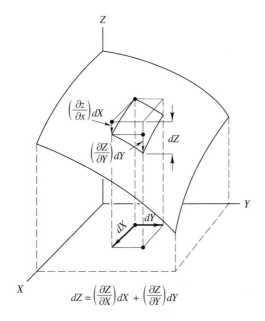

$$dZ = \left(\frac{\partial Z}{\partial X}\right)dX + \left(\frac{\partial Z}{\partial Y}\right)dY$$

FIGURE 2.7
Geometrically, the change dZ associated with changes dX and dY is given by the slope of the surface in the X direction times dX plus the slope in the Y direction times dY.

Since every state function can be expressed as functions of all of the possible combinations of the others, equations of the form Eq. (2.5) abound in thermodynamics. Maxwell relations hold for all of these equations.

5. *Conditions for equilibrium* are sets of equations that describe the relationships between state functions that must exist within a system when it attains its equilibrium state.

The laws are developed with precision and generality in Chap. 3. Definitions and the mathematical basis for the coefficient and Maxwell relations are presented in detail, together with examples for their application, in Chap. 4. Derivation of the conditions for equilibrium and the systematic application of these equations to systems of increasing complexity occupy most of the remainder of the text.

2.4 CRITERION FOR EQUILIBRIUM

A system placed in a given environment changes spontaneously until it has exhausted its capacity for change. When a system attains this final state of rest, it is described as being "in equilibrium" with itself and its surroundings. The prediction and description of this equilibrium state for any given system is a problem of central importance in thermodynamics. This description is expressed in terms of a set of equations, called the *conditions for equilibrium*, which are relationships among the internal properties that must be obtained in order for the system to be at rest. As a simple and familiar example of such equations, consider a unary, two phase, closed, nonreacting, otherwise simple system consisting of ice floating in water. The condition for thermal equilibrium in this system is evidently

$$T^{\text{ice}} = T^{\text{water}} \qquad (2.7)$$

Additional conditions must apply in order for the system to be in complete equilibrium.

The conditions for equilibrium are extremely important in thermodynamics, providing the basis for calculating phase diagrams, chemical equilibria in reacting systems, and the influence that factors like surface energy or electrical fields may have on the final resting state of a system. For the most complex kind of system these equations can be deduced from a single principle called the *general criterion for equilibrium*. This powerful principle, which follows directly from the second law of thermodynamics, is the subject of Chap. 5. The remainder of the text deals chiefly with the repeated application of this general criterion to the deduction of equations describing the equilibrium state and the application of the resulting equations to solving practical problems in thermodynamics. Each category in the hierarchy of the classification of thermodynamic systems yields such a set of equations that describes the equilibrium state.

2.5 SUMMARY OF CHAPTER 2

Thermodynamics provides the basis for describing the behavior of arbitrarily complex physical systems. This description flows from the laws of thermodynamics:

Energy is conserved
Entropy is created
Temperature has a zero

The part of the apparatus of thermodynamics required to handle a given problem can be brought into focus by classifying the system under consideration according to:

Its number of components
Its number of phases
The permeability of its boundary
Its ability to exhibit chemical reactions
Its interactions with nonmechanical forces

The strategy of presentation in this text is based upon a progressive development of systems of increasing complexity in this hierarchical classification.

Changes in the condition experienced by such systems are described in terms of:

State variables, with values that depend only on the current condition of the system

Process variables, which only have meaning for processes, that is, changes in state of the system

Changes in state functions are independent of the process by which the system passes from its initial to its final condition. Thus, these changes can be computed by visualizing the simplest process connecting the end states, and making the calculation for that process.

Properties of thermodynamic systems are either *intensive*, if in concept they are defined at each point in the system, or *extensive*, if they report an attribute of the system as a whole.

Changes in properties can be evaluated for any system taken through any process by applying the relationships of thermodynamics. Perception of this potentially bewildering array of relationships is aided by classifying them as

The laws of thermodynamics

Definitions

Coefficient relations

Maxwell relations

Conditions for equilibrium

A general criterion for equilibrium forms the basis for deducing these conditions for equilibrium. The conditions for equilibrium lead directly to the working equations that permit the prediction of the behavior of matter in the most complex kinds of systems.

PROBLEMS

2.1. Classify the following thermodynamic systems in the five categories defined in Sec. 2.1:
  (*a*) A solid bar of copper.
  (*b*) A glass of ice water.
  (*c*) A yttria stabilized zirconia furnace tube.
  (*d*) A styrofoam coffee cup.
  (*e*) A eutectic alloy turbine blade rotating at 20,000 rpm.
  You may find it necessary to qualify your answer by defining the system more precisely; state your assumptions.

2.2. It is not an overstatement to say that without state functions thermodynamics would be useless. Discuss this assertion.

2.3. Determine which of the following properties of a thermodynamic system are extensive properties and which are intensive.
  (*a*) The mass density.
  (*b*) The molar density.
  (*c*) The number of gram atoms of aluminum in a chunk of alumina.
  (*d*) The potential energy of a system in a gravitational field.
  (*e*) The molar concentration of NaCl in a salt solution.
  (*f*) The heat absorbed by the gas in a cylinder when it is compressed.

2.4. Why is *heat* a *process variable*?

2.5. Write the total differential of the function

$$z = 12u^3 v \cos(x)$$

(*a*) Identify the coefficients of the three differentials in this expression as appropriate partial derivatives.

(*b*) Show that three Maxwell relations hold among these coefficients.

2.6. Describe what the notion of *equilibrium* means to you. List as many attributes as you can think of that are exhibited by a system that has come to equilibrium. Why do you think these characteristics of a system in equilibrium are important in thermodynamics?

CHAPTER
3

THE
LAWS OF
THERMODYNAMICS

The laws of thermodynamics are highly condensed expressions of a broad body of experimental evidence. The formulation of these succinct empirical statements about the behavior of matter was essentially completed by the end of the last century and has not required significant alteration in the light of this century's scientific experience. This is a remarkable fact in view of the enormous scientific progress that has been achieved in the twentieth century. The laws of thermodynamics are thus soundly based and broad in their application.

The laws are empirical and no claim is made that they can be deduced from any fundamental philosophical principles.

The laws of thermodynamics have a status in science that is similar to Newton's laws of motion in mechanics and similarly, are subject to potential revision in light of new information. When it was found in physics that new evidence could be explained only by modifying Newton's laws with Einstein's relativistic concepts, the laws of motion were generalized. However, this generalization was in such a form that the new equations simplify to Newton's laws when the velocity of the system is not a significant fraction of the velocity of light. Newton's laws were not abandoned but were expanded to include new phenomena. Classical mechanics could then be viewed as a special case of the new relativistic mechanics. This strategy was necessary because classical mechanics successfully described a great body of scientific observations with plausibility and precision.

It is possible that new experimental evidence could require a reformulation of the laws of thermodynamics but up to now this has not been necessary, although the discovery of nuclear energy has had to be accommodated by expanding the framework established in the nineteenth century.

The laws of thermodynamics are:

1. There exists a property of the universe, called its *energy*, which *cannot change* no matter what processes occur.

2. There exists a property of the universe, called its *entropy*, which *always changes in the same direction* no matter what processes occur.

3. There exists a lower limit to the temperature that can be attained by matter, called the *absolute zero* of temperature, and the entropy of all substances is the same at that temperature.

A frequently cited "zeroth law of thermodynamics" acknowledges that a temperature scale exists for all substances in nature and provides an absolute measure of their tendencies to exchange heat.

This chapter presents each of the laws in turn, formulating the concepts involved with precision and stating the laws mathematically for the most complex class of thermodynamic system.

3.1 THE FIRST LAW OF THERMODYNAMICS

Energy has achieved the status of a household word and is in everyday use. It is thought to be a common sense concept with an attendant intuitive meaning. A careful development of the concept of energy in all of its diverse aspects reveals that such an intuition is superficial and potentially misleading. For example, try to visualize the intuitive meaning of kinetic energy, $mv^2/2$, a quantity that implicitly contains the "square of the rate of displacement" of the matter in a system. Fortunately, from a practical point of view, an intuitive understanding of the nature of energy is not crucial. Energy can be defined both physically and mathematically and can be measured with precision. The description of the state of a system and the prediction of its behavior rests on the mathematical and physical formulation of the concept, not on an intuitive understanding of it.

Three broad categories of energy have been identified in scientific experience:

1. *Kinetic energy*, which is associated with the *motion*, translation or rotation, of a particle or body and nothing else

2. *Potential energy*, which is associated with the *position* of a particle or body in a potential field and nothing else

3. *Internal energy*, which is associated with the *internal condition* of the body and does not otherwise depend upon its motion or position in space

Thermodynamics at first focuses on the influences that change the internal condition of a system at rest. The apparatus is then extended to include potential and kinetic energy as well as internal energy.

In its most pretentious form the first law can be stated for the behavior of the universe. In practical applications the focus of attention is on some small subset of the universe, called "the *system*." Changes that occur inside the system are almost always accompanied by changes in the condition of the matter in the vicinity of the system. The part of the universe that is external to the system but also affected by changes that are caused to occur in the system is called "the *surroundings*," Fig. 3.1. Thus, from the standpoint of any particular process that may occur in practice, the sum of the changes that occur in the system and its surroundings includes all the changes in the universe associated with that process.

By the first law the total energy of the universe cannot change for any process. Energy can be transported, or converted from one form to another but cannot be either created or destroyed. Since conversion of energy from one form to another does not change the total quantity of energy, the only way that the internal energy of a system can be changed is by *transferring energy across its boundary*. Thus, a mathematical statement of the first law for a system can be formulated from the statement that the change in internal energy of a system for a process must be equal to the sum of all energy transfers across the boundary of the system during the process. It is only necessary to enumerate all the possible kinds of energy transfers that can occur for the class of system considered and then set the sum of these energy transfers equal to the change in internal energy of the system for the process.

Let U be a thermodynamic state function called the *internal energy* of the system. For any process, define ΔU to be the increase in the internal energy of the system. Enumerate all the kinds of energy transfers that may occur:

Q is the quantity of *heat* that flows into the system during the process.

W is defined to be the *mechanical work* done on the system by the external pressure exerted by the surroundings.

W' is defined to be *all other kinds of work* done on the system during the process.

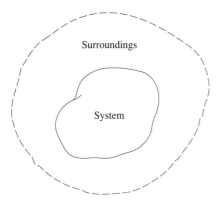

Surroundings

System

FIGURE 3.1
The part of the universe that is *outside* the system and experiences changes as a result of the changes that occur in the system is called the *surroundings*.

Note that in introducing these definitions, it is necessary to establish conventions for the sign of each quantity. By this choice of convention, if Q is positive, heat (and thus energy) flows into the system; if, for a given process, heat flows *out* of the system, Q is *negative*. Similarly, if W (or W') is positive, the surroundings does work on the system and energy flows into the system. If for the process under consideration, work is done by the system on the surroundings, W (or W') is negative and energy is transferred out of the system. The statement that the increase in the internal energy of the system during a process is the sum of the transfers across the boundary can be written:

$$\Delta U = Q + W + W' \qquad (3.1)$$

This is a mathematical statement of the first law of thermodynamics. In order to apply it in practice, it is necessary to devise an apparatus for evaluating the process variables Q, W and W' for each kind of process that may be encountered by each class of system.

In order to follow the course of a process in detail, it is useful to consider it as a succession of incremental steps producing infinitesimal changes in the internal condition of the system resulting from infinitesimal transfers of heat and work across the boundary. The first law is pervasive; it applies not only to the finite overall changes that a system may experience, but also to each incremental step along the way. Energy must be conserved during each small incremental step in a process, Fig. 3.2. Thus, for an infinitesimal change in the condition of the system, Eq. (3.1) can be written

$$dU = \delta Q + \delta W + \delta W' \qquad (3.2)$$

where dU is the change in the state function U for an infinitesimal step in the process, and δQ, δW, and $\delta W'$ are incremental quantities of heat and work transferred across the boundary of the system. It is widely accepted in thermodynamics that an infinitesimal quantity designated by a d, as in dU, represents the change in a state function and thus has the mathematical properties of a differential of the state function. Prefixing a quantity with δ, as in δQ above, denotes an infinitesimal quantity of heat or work. However, these infinitesimal quantities do not have the mathematical properties of a differential. There is no mathematical function Q, of which δQ is the differential, because Q is not a state function.

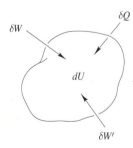

$$dU = \delta Q + \delta W + \delta W'$$

FIGURE 3.2
Illustration of the first law of thermodynamics for an infinitesimally small step in a process: the change in internal energy of a system must equal the sum of all of the energies transferred across its boundary.

Note that U is a state function, while Q, W, and W' are process variables. Thus, Q, W, and W' have values that depend explicitly upon the path, that is the sequence of thermodynamic states traversed by the system during the process; ΔU is independent of the path.

There are no restrictions involved in making the mathematical statement of the first law of thermodynamics contained in Eqs. (3.1) and (3.2). They apply to any system taken through any process. This very general form of statement is useful for emphasizing the concept that is the first law. However, it is not very useful for applications to practical systems and processes because it does not address the key question: "How can Q, W, and W' be evaluated?" This question is addressed in detail beginning in Chap. 4.

3.2   THE SECOND LAW OF THERMODYNAMICS

In human experience it is a common observation that processes observed in nature have a natural direction of change. If a motion picture or videotape is run backwards, it becomes immediately evident to any viewer that something is wrong with the motions. A pebble dropped into a pond produces a splash and a set of ripples that dissipate at the pond's edges. Viewed in reverse, ripples of waves spontaneously generate at the edges of the pond, converge to the middle, and amid an eruption of water, toss a pebble into the air. Energy is conserved in both processes, but the second process will never occur; there is something about it that is obviously contrary to experience.

When processes in nature are examined, either macroscopically or in detail, it is found that this principle of a proper direction for change is pervasive. A wilted flower does not revive to full bloom and then curl into a bud. Time flows in one direction. The second law of thermodynamics distills this aspect of experience and states it succinctly and quantitatively, albeit abstractly. It was a great accomplishment of the nineteenth century to identify a state function, for obscure reasons called the *entropy*, which when summed for both system and surroundings for any process, always changes in the same direction. By convention, this function is defined so that it always *increases*.[1]

Like the first law, the second law of thermodynamics is general and pervasive since no step in any process is exempt from its application. It applies microscopically to every volume element in a system experiencing change, as well as macroscopically to the system as a whole. It applies during each infinitesimal increment of time in the process, as well as to the entire process. Specifically stated, in every volume element of any system and surroundings that may be experiencing change, at every instant in

---

[1] In retrospect, this may have been a bad choice for the convention since the production of entropy is associated with dissipation or consumption of the capacity for spontaneous change when a process occurs. Every change that occurs for every process for every system uses up some of the finite capacity for change that the universe possesses. If the opposite sign had been chosen in this convention, then entropy would be destroyed rather than be produced in the universe for every real process that occurs.

time, the entropy production is *positive*. It will now be demonstrated that this does *not* imply that the entropy *of a system* can only increase.

Like energy, entropy changes arise from influences that operate to change the condition of a system. However, the relationship between the entropy changes and the influences that act is not the same as it is for energy. Like energy, entropy can be transferred across the boundary of a system in association with heat, work, and mass transfers. *Unlike energy*, the change in entropy of a system is not restricted to entropy transferred across its boundaries. There is an additional contribution; namely, the *production of entropy* inside the system. The first law is a conservation law but the second law is not.

The change in entropy experienced by a system undergoing change can always be decomposed into two parts:

ΔS_t the entropy transferred across the boundary during the process
ΔS_p the entropy production inside the system

The change in entropy for the system is the sum

$$\Delta S_{sys} = \Delta S_t + \Delta S_p \tag{3.3}$$

The transfer term can be positive or negative depending upon the nature of the process and the flows at the boundary of the system. The second law states that for all processes for all systems,

$$\Delta S_p \geq 0 \tag{3.4}$$

Whether ΔS_{sys} is positive or negative depends upon the sign of ΔS_t as well as its relative magnitude in comparison with ΔS_p. Thus, for a givin process, it is not unusual for the entropy of a system to decrease. Indeed, many of the practical processes that make technology work are explicitly set up to produce a negative change in entropy of a system, that is, to make a system change in a direction opposite to that which would occur "naturally." To accomplish this it is necessary to place the system in an environment, a *surroundings,* capable of causing the transfers necessary to produce the "unnatural" change.

Consider the entropy change experienced by the surroundings to the process and treat the surroundings as a system. Denote by a prime, the properties of the surroundings. Application of Eq. (3.3) gives:

$$\Delta S_{sur} = \Delta S_t' + \Delta S_p' \tag{3.5}$$

Now compute the change in entropy for the "universe," that is, the system plus surroundings:

$$\Delta S_{un} = \Delta S_{sys} + \Delta S_{sur} = \left[\Delta S_t' + \Delta S_p'\right] + \left[\Delta S_t + \Delta S_p\right]$$

The entropy transferred across the boundary into the system is the negative of the entropy transferred into the surroundings. Thus, these two terms cancel and

$$\Delta S_{un} = \Delta S_p' + \Delta S_p \tag{3.6}$$

The second law states that both of these production terms are positive. Thus, while the entropy of a system may increase or decrease during a process, the entropy of the universe, taken as system plus whatever surroundings are involved in producing the changes within the system, *can only increase.*

These statements about entropy have been presented for a finite change in the condition of a system. They can also be applied to an infinitesimal step in the course of this change, for which the entropy changes are infinitesimals:

$$dS_{sys} = dS_t + dS_p \qquad (3.7)$$

in which dS_t can be positive or negative but, by the second law, $dS_p \geq 0$. The most sophisticated formulation of this concept applies it locally, to an element of volume in an arbitrary system, and states that the rate of change of the local density of entropy is due to the divergence of the local entropy flows plus the local rate of entropy production, σ (reported as entropy per unit volume per unit time). In this local formulation (the details of which are beyond the scope of this text), the second law states that

$$\sigma \geq 0 \qquad (3.8)$$

See [1] and [2] for the rigorous formulation of the concept of the local rate of entropy production.

3.3 INTUITIVE MEANING OF ENTROPY PRODUCTION

It is a tautology to state that all changes that occur in the universe are "spontaneous." To state that a given system placed in a given surroundings experiences a specific process "spontaneously," merely acknowledges that for that system in those surroundings, that process happens and not its reverse, or some other process. Left in air iron oxidizes. Water placed in a 300°C oven vaporizes. A balloon kept in a refrigerator expands when taken out into the kitchen air. All of these changes can be reversed by placing the system in the proper surroundings. Iron oxide can be reduced to iron by putting it in the presence of carbon in the blast furnace. Steam condenses below 100°C. The balloon collapses if it is returned to the refrigerator. A given system, in a given surroundings, always changes in the same direction.

These ideas are inextricably intertwined with the notion of time. Time has a forward direction. If for a given familiar sequence of conditions that a system experiences, the time scale is reversed, it will be perceived that the changes observed are "improper."

It has been said that "entropy is time's arrow" [3]. Entropy is the property of the universe that monitors the pervasive experience that for a given system and surroundings, there is a direction of change that is proper or spontaneous. The total entropy of the system plus the surroundings always increases with increasing time; entropy is constantly produced. If a process or change is imagined for which entropy is destroyed, then the direction of time must be reversed for that process but this cannot happen.

Changes in the real world are always accompanied by friction, and something is *dissipated*. When the pebble is dropped into the pond, its kinetic energy at impact is dispersed by the wave motion throughout the pond and is dissipated in the absorbing water of the pond. In this dispersed state, it cannot be recovered. The entropy production in a system is a quantitative measure of this dissipation.

There exists a qualitative correlation between the rate of entropy production, or dissipation, occurring in a system during a process and two related aspects of the behavior of the system:

1. How far the process has to go before the system can achieve equilibrium.
2. The rate of the process.

Qualitatively, the further a system is from its equilibrium state, the faster it tends to change, the greater the frictional or dissipative effects, and the larger its rate of entropy production. *Quantitatively*, the correlation is very complex. These questions are developed in a sophisticated formalism for the description of rates of change of complex thermodynamic systems called the *thermodynamics of irreversible processes* and are beyond the scope of this text [1, 2]. To provide an idea of the level of detail required to compute the "dissipation function" and the rate of entropy production, an expression borrowed from that theory is presented in Table 3.1, which includes definitions of the factors involved.

Processes that are caused to occur slowly, by setting up conditions so that they are never very far removed from equilibrium, experience less dissipation and a

TABLE 3.1

Expression derived in the thermodynamics of irreversible processes for the local rate of entropy production [3.1]

$$\sigma = -\frac{1}{T^2} J_Q \operatorname{grad} T + \frac{1}{T} \sum_{k=1}^{c} J_k \left[K_Q - (\operatorname{grad} \mu_k)_T \right]$$

$$+ \frac{1}{T} \sum_{v=1}^{r} b_v A_v - \frac{1}{T} \sum_{i=1}^{3} \sum_{j=1}^{3} P_{ij} \left(\frac{\partial v_i}{\partial z_j} \right)$$

σ	is the local rate of energy production
T	is the absolute temperature
$\operatorname{grad} T$	is the temperature gradient
J_Q	is the heat flux
J_k	is the diffusional flux of component k
K_k	is the body force acting on the mass of component k
$\operatorname{grad} \mu_k$	is the gradient in chemical potential of component k
b_v	is the specific rate of the chemical reaction v
A_v	is the affinity for chemical reaction v
P_{ij}	is the ij component of the local pressure tensor
v_i	is the ith component of the local velocity vector
z_j	is the jth component of the local position vector

correspondingly small entropy production. It is possible to visualize a process that is carried out so slowly that its internal state is never more than infinitesimally removed from equilibrium with itself and its surroundings. For example, one may visualize adding heat to a system under conditions in which the temperature in the system is infinitesimally smaller than the temperature of the surrounding heat source. Similarly, work can be obtained from an expanding gas by reducing the external pressure applied to the system in very small increments, allowing the system to expand to meet the new level of pressure before the next incremental change. In the limiting case, in which the process is imagined to proceed infinitely slowly, and the incremental changes applied are infinitesimal, the dissipative effects approach zero. For such a limiting process, the entropy production is *zero*. The entropy change experienced by any system undergoing such a process is thus entirely due to entropy transferred across the boundary of the system.

Such processes, carried out with infinite slowness by infinitesimal changes in the influences that operate to produce change, are called *reversible* processes for reasons that will shortly become evident. All real processes have a finite rate in response to finite influences; such real processes are called *irreversible* to emphasize the contrast with imaginary "reversible processes." Irreversible processes suffer dissipations that result in the production of entropy and thus a permanent change in the universe.

Consider a piston containing a gas at some temperature and pressure, Fig. 3.3a. For a given temperature and number of molecules, the pressure in the gas when it is in equilibrium with its surroundings is determined by the pressure in the surrounding atmosphere and the weight of the piston, the pan, and its contents sitting on the top of the piston. Suppose the majority of the weight is supplied by the contents of the pan, which is lead powder.

Now, suppose the gas is allowed to expand through two different processes, which will be compared to illustrate the difference between *reversible* and *irreversible* processes. For process I, Fig. 3.3b, the pan and its contents are lifted off the top of the piston. The system experiences a sudden drop in the force applied to it at the top and expands rapidly. Eventually the piston comes to rest at some new position, and the gas again comes to equilibrium with its surroundings. Its temperature is that of its surroundings and its pressure is determined by the atmospheric pressure and the weight of the piston. It is likely that during this expansion, the piston may overshoot its final position, then drop below it, and after several oscillations settle down to its final position. Whether this actually happens in a given case depends upon the viscosity of the gas and the details of the process. During this process the internal condition of the gas is complex; gradients in pressure and temperature develop and a pattern of mass flow and heat flow emerges, develops, and ultimately dissipates as the system comes to its final uniform equilibrium state.

Now, the pan and its contents are then placed back on the piston, a second complex process ensues as the system ultimately returns to its initial state. The added force at the top immediately increases the gas pressure in that region, again producing gradients in pressure and temperature. Theseevolve and ultimately disappear as the

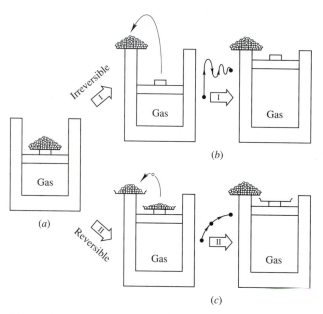

FIGURE 3.3

The pressure in the gas contained in the piston (a) is determined by the weight of lead powder in the pan. If the tray is removed, the gas expands suddenly and irreversibly and settles to a final state (b). If the gas expands slowly, by removing one particle of lead powder at a time, the process is reversible (c).

final resting state is attained for the system. This state is identical with that which originally existed for the system.

The key point in this demonstration is: when change in the external influence on the system (force derived from the pan of lead powder) is *reversed*, the complicated sequence of conditions experienced by the gas is *not simply the reverse* of that traversed when the pan was first removed. Indeed, it is not possible to devise a process that exactly retraces the sequence of conditions passed through during the expansion of the gas. The process is said to be *irreversible*.

Now consider a second process, II, Fig. 3.3c, in which the gas is caused to expand by removing the lead powder one particle at a time and letting the system come to rest before the next particle is removed. At each step during this expansion, the pressure in the gas is only very slightly different from that exerted by its surroundings. Gradients in pressure and temperature within the system are negligible. The description of the sequence of states that the system traverses is very simple since each state is specified by its current pressure and temperature, both of which are uniform in the system.

If the change in external influence is *reversed*, by individually replacing the particles of lead powder just removed, the sequence of states that the system traverses is identical to that which it experienced as those particles were being slowly removed, traversed in the opposite in direction. Thus, reversal of the change in external influence simply reverses the direction of trajectory of the system along the path connecting the initial and final states. This argument applies to any segment of the path in the process or to the whole path and the process is said to be carried out *reversibly*.

For a process carried out reversibly, there are no dissipations, no entropy production, and *no permanent changes* in the universe (system plus surroundings). Irreversible processes are accompanied by dissipation, finite entropy production, and *permanent changes* in the universe.

Calculation of the *process variables*, heat absorbed, and work done on the system is relatively easy for reversible processes, primarily because each intermediate state is described by just a few numbers, for example, temperature and pressure (T, P). In contrast the calculation of these process variables for irreversible processes is extremely complicated because the description of evolution of the system requires specification of the variation with time of the positional variation within the volume of the system of temperature and pressure $[T(x,y,z,t), P(x,y,z,t)]$.

On the other hand calculation of changes in *state functions* can be straightforward for both classes of processes because such changes depend only upon the initial and final states and are independent of the path connecting these two states. Thus, given a complex, irreversible process connecting states A and B, changes in state functions for this process can be computed by imagining the simplest process that connects the states A and B. This process is of course a reversible process. The changes in state functions for that simple process are identical to the changes in the state functions determined for the specified irreversible process. This strategy points up the importance of the fact that state functions have been identified in nature, and at the same time justifies the central role played by the analysis of reversible processes in thermodynamics even though such processes are, practically speaking, unrealistic.

3.4   RELATION BETWEEN ENTROPY TRANSFER AND HEAT

The previous section developed the qualitative relation between entropy production and dissipation occurring in irreversible processes. This section focuses on the other contribution to the change in entropy of a system, entropy flow across the boundary.

A quantitative treatment of entropy transfer for irreversible processes requires the formalism of the thermodynamics of irreversible processes and is beyond the scope of this text. A quantitative treatment of entropy transfer for *reversible* processes establishes a connection between the reversible heat flow across the boundary of the system and its change in entropy. The argument that demonstrates this connection is somewhat convoluted and is presented in Appendix K. The interested reader will find it useful to delay examining that argument until Chap. 4 has been studied since some of the relationships used are developed in that chapter. However, experience has shown that it is not essential to understand these arguments in order to apply their conclusion. Therefore, only an outline of the sequence of logic involved is presented here.

The argument demonstrates that if any system is taken through a process that is reversible, but otherwise may be completely general, then the line integral

$$\oint \frac{\delta Q_{rev}}{T} \tag{3.9}$$

along the path that defines the reversible process is a state function. This is demonstrated by considering a general *cyclic* process, that is, one that returns the system to its original state. Since the system ends up where it started, changes in all state functions for any cyclic process must be zero. The development, based upon the Carnot cycle, demonstrates that the function defined in expression (3.9) is zero for any reversible cyclic path and thus is a *state function*. It follows that the change in this function for any increment along the path is a state function, and ultimately, for each infinitesimal step along any path, $\delta Q_{rev}/T$ is a state function. This state function is then defined to be the entropy of the system. The practical form of this result is

$$\delta Q_{rev} = T dS \tag{3.10}$$

$$Q_{rev} = \oint T dS \tag{3.11}$$

which permits computation of the heat absorbed for any reversible process by integration of the combination of state functions, TdS. This is the most frequently used consequence of the argument.

One result of these developments is an important inequality that applies to *irreversible* (real) processes. For any reversible process that changes the condition of the system from state A to state B, the change in entropy is completely due to transfer of heat across the boundary:

$$\Delta S = \oint \frac{d Q_{rev}}{T} \tag{3.12}$$

Consider now any *irreversible* process that connects the same two end states, A and B. Because entropy is a state function, the change in entropy for this process is the same as for the reversible process:

$$\Delta S_{rev}[A \rightarrow B] = \Delta S_{irr}[A \rightarrow B]$$

However, the entropy change for the irreversible process must include some production of entropy inside the system. Therefore, the entropy transferred across the boundary will be smaller than in the reversible case:

$$\Delta S_{irr,t} + \Delta S_{irr,p} = \Delta S_{rev,t} + 0$$

$$\Delta S_{irr,t} = \Delta S_{rev,t} - \Delta S_{irr,p}$$

Insofar as the transfer of entropy is associated with heat transfer across the boundary,

$$\Delta S_{irr,t} = \oint \frac{\delta Q_{irr}}{T}$$

It follows that

$$\oint \frac{\delta Q_{irr}}{T} < \oint \frac{\delta Q_{rev}}{T} \tag{3.13}$$

For an isothermal process (one carried out in a system with temperature control so that T is held constant in the experiment), this result further simplifies to

$$[Q_{irr}]_T < [Q_{rev}]_T \tag{3.14}$$

In a comparison of all possible *isothermal* processes that convert a system from state A to state B, the heat absorbed by a reversible process is larger than for any irreversible process. Under these conditions, Q_{rev} is the *maximum* heat absorbed.

3.5 COMBINED STATEMENT OF THE FIRST AND SECOND LAWS

An expression for the mechanical work done during any arbitrary reversible process can also be derived. Recall the general definition of work, given by Eq. (2.1):

$$\delta W = F \cdot dx \tag{2.1}$$

The force acting in simple mechanical work is derived from the external pressure P_{ext} on the system as shown in Fig. 2.6. If this pressure acts over an area of the boundary A, the resulting force F is given by P_{ext}(force per unit area) $\times$ A(area). Work results from this applied force when it is displaced. In this case, the force is displaced if the boundary of the system on which it is acting moves some distance dx. Multiply and divide by A in Eq. (2.1):

$$\delta W = \frac{F}{A} \cdot A \cdot dx \tag{3.15}$$

(F/A) is P_{ext}, while $A \cdot dx$ is related to the change in volume of the system due to the motion of its boundary. (Note that the product Adx is equal to the volume swept through by the bounding area A when it moves through a distance dx.) The choice of sign convention established in formulating the first law defines W to be positive if work is done on the system. In order to be consistent with this convention, it is necessary to define dx to be positive when the force is displaced *toward* the system causing it to contract (because work is done on the system for such displacements) and negative when the displacement is outward and the system expands. With these choices the volume change of the system associated with a displacement dx is

$$A \cdot dx = -dV$$

and Eq. (3.15) becomes:

$$\delta W = P_{ext} \cdot (-dV) \tag{3.16}$$

The goal of this development is to obtain an expression for the mechanical work done in terms of functions of the *internal* state of the system. If the process is carried out reversibly, then the internal pressure in the system P, is essentially equal to the external pressure, $P = P_{ext}$, and Eq. (3.16) becomes:

$$\delta W_{rev} = -P \cdot dV \tag{3.17}$$

If the process is carried out irreversibly, Eq. (3.16) can still be used to evaluate the work done on the system; however, in this case $P_{ext} \neq P$, that is, the external pressure is not necessarily equal to the internal pressure.

Equations (3.10) and (3.17) permit evaluation of the process variables, heat, and mechanical work for any arbitrary reversible process. Substitution of these expressions into the general statement of the first law, Eq. (3.2), yields the most frequently used equation in the practical application of thermodynamics, called the *combined statement of the first and second laws*:

$$dU = T\,dS - P\,dV + \delta W'$$ (3.18)

The central role played by this equation in deriving relationships among the variables of thermodynamics is developed in detail in Chap. 4. It is also the basis of derivations of the conditions for equilibrium in thermodynamic systems, which are in turn the basis for most practical thermodynamic calculations in sophisticated systems.

3.6 THE THIRD LAW OF THERMODYNAMICS

Like energy, temperature is taken to be a familiar concept with daily practical application in household technology, as well as in science and industry. Weather reports give high, low, and current temperatures along with predictions about tomorrow's conditions. Thermostats control the temperature of the house, office, oven, or refrigerator. Temperature is understood to be a property of matter that provides a universal measure of the tendency of systems to exchange heat.

The earliest attempts to quantify this aspect of the behavior of matter date from Fahrenheit in the mid-seventeenth century. He monitored this property by recording changes in volume of a liquid in a tube, now familiar as a thermometer. In the Fahrenheit temperature scale that he devised, he chose his own benchmarks, defining a zero point and a point for 100°F. It is a familiar fact that water freezes at 32°F and boils at one atmosphere at 212°F on this scale. More than a century later, the scientific community agreed that pure water should be used to define the benchmarks on the scale and adopted the centigrade scale now called the Celsius scale, in which water freezes at 0°C and boils at 100°C.

As refrigeration devices came into use in laboratories, it became clear that matter could exist at temperatures significantly below the defined zero point for the Celsius scale. Near the turn of the century experimenters studying the behavior of matter at very low temperatures established that there is a lower limit to the temperature that matter can exhibit. In the light of this observation it made sense to define this lower limit to be the zero point on a new, *absolute* scale of temperature, now called the Kelvin scale. This zero point, written 0 K, has been shown to correspond to −273.150°C on the Celsius scale. The interval between temperatures was chosen to agree with the Celsius scale so that the temperature difference between the freezing and boiling points of water at one atmosphere pressure is also 100 units on the Kelvin scale. This observation, that there exists an ultimate lower limit to the temperature that matter can exhibit, is one component of the third law of thermodynamics.

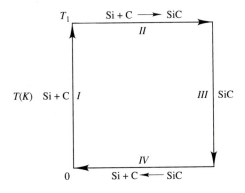

FIGURE 3.4
Cyclic process involving the reaction that forms silicon carbide from its components illustrates the third law of thermodynamics.

The same cryogenic studies that established the existence of the absolute zero in temperature explored other aspects of the behavior of matter in the range between absolute zero and room temperature. To illustrate the principle that evolved from these studies, consider the cyclic process shown in Fig. 3.4. A system consisting of one mole of pure silicon and one mole of carbon initially at absolute zero is heated from 0 K to 1500 K. The entropy change for this process (step I) can be computed from the heat capacities of silicon and carbon, which have been measured in this temperature range (see Chap. 4 and Appendix D). At 1500 K, the silicon and carbon are reacted to form the compound, silicon carbide, SiC. The entropy change for this reaction (step II) can be experimentally determined by measuring the heat absorbed in a calorimeter. The compound SiC is then cooled to absolute zero and the change in entropy for this process (step III) can be computed from measured values of the heat capacity of the compound SiC. The entropy difference between the compound SiC and pure Si and C at 0 K, which would be characteristic of reaction (step IV) in Fig. 3.4 if it could be made to occur, can be computed by invoking the fact that entropy is a state function.

If the system is taken through all four processes, it is returned to its original state, that is, the total process is a cycle. Since entropy is a state function, the change in entropy for the cycle must be zero:

$$\Delta S_{cyc} = \Delta S_I + \Delta S_{II} + \Delta S_{III} + \Delta S_{IV} = 0 \qquad (3.19)$$

However, when the individual entropy changes for steps I, II, and III are calculated from experimental measurements it is found that, within the experimental error,

$$\Delta S_I + \Delta S_{II} + \Delta S_{III} = 0 \qquad (3.20)$$

It is thus concluded that

$$\Delta S_{IV} = 0 \qquad (3.21)$$

That is, the entropy of the compound SiC is the same as that of the pure silicon and carbon at 0 K.

Repetition of this kind of computation for a large number of substances established that it is a pervasive observation. *The entropy of all substances is the same at 0 K.* (Where deviations from this observation have been found, they could be rationalized by establishing that the substances involved had not achieved total internal equilibrium at these low temperatures.) Since at absolute zero, all substances have the same value of entropy, it makes sense to choose this condition as a zero point for the definition and measurement of entropy. Thus, these empirical observations have established the third law of thermodynamics which follows:

> There exists a lower limit to the temperature that can be attained by matter, called the absolute zero of temperature, and the entropy of all substances is the same at that temperature.

The primary practical application of this law (in the context of this text) lies in the evaluation of entropy changes for chemical reactions. In the cyclic process just described the entropy change for the reaction (step II)

$$Si + C = SiC$$

can be evaluated without requiring that the reaction actually be experimentally studied. Rearrangement of Eq. (3.20) gives:

$$\Delta S_{II} = -(\Delta S_I + \Delta S_{III}) \tag{3.22}$$

The entropy changes on the right side of this equation, which are associated with heating the reactants from 0 K and cooling the products to 0 K, can be computed from heat capacity information about these substances. Thus, information about the chemical reaction is not required to compute this entropy change for step II. It is not necessary to set up and carry out an experiment to measure the change in entropy for the reaction. This entropy change can be computed as the absolute entropy of the products minus that of the reactants.

Computations of entropy changes for chemical reactions are made more convenient with the aid of tabulations of values of absolute entropy values of elements and compounds at room temperature, taken to be 25°C or 298 K (see Appendices D, G, H and I). To illustrate this application, consider the reaction that represents the formation of alumina (Al_2O_3) from aluminum and oxygen at 298 K. Refer to Appendices D and I to obtain the following values for the absolute entropies of these reactants at 298 K:

$$S^{\circ}_{Al,298K} = 28.32 \text{ J/mol-K}$$

$$S^{\circ}_{O_2,298K} = 205.03 \text{ J/mol-K}$$

$$S^{\circ}_{Al_2O_3,298K} = 51.00 \text{ J/mol-K}$$

A stoichiometrically balanced statement of the reaction is

$$2Al + \frac{3}{2}O_2 = Al_2O_3$$

The entropy change for this reaction at 298 K is

$$\Delta S^{\circ}_{298K} = S^{\circ}_{Al_2O_3,298K} - \left[2S^{\circ}_{Al,298K} + \frac{3}{2}S^{\circ}_{O_2,298K} \right]$$

$$\Delta S^{\circ}_{298K} = 51.00 - \left[2(28.32) + \frac{3}{2}(205.03) \right] = -313.18 \text{ J/mol-K}$$

3.7 SUMMARY OF CHAPTER 3

In all processes energy is conserved. Therefore the change in internal energy of any system in any process is the sum of the transfers across its boundary. Mathematically, for each infinitesimal step in any process,

$$dU = \delta Q + \delta W + \delta W' \qquad (3.2)$$

In every volume element at every instant in time, in all processes, entropy is created. This entropy production is largest for irreversible process that are occurring rapidly and far from equilibrium. It approaches zero in the limiting case of reversible processes for which each step is but infinitesimally removed from equilibrium.

Entropy is a state function. For reversible processes the change in entropy is related to the heat absorbed by the system at each step:

$$\delta Q_{rev} = T dS \qquad (3.10)$$

where T is the absolute temperature.

The most frequently used equation in thermodynamics is the combined statement of the first and second laws:

$$dU = T dS - P dV + \delta W' \qquad (3.18)$$

Both entropy and temperature have *absolute values* because both have an empirically observed zero point.

PROBLEMS

3.1. The laws of thermodynamics are "pervasive." Explain in detail the meaning of this important statement.

3.2. List the kinds of energy conversions involved in propelling an automobile.

3.3. List the kinds of energy conversions involved in operating a hand calculator.

3.4. List the kinds of energy conversions involved in using your right arm to turn the page in this text.

3.5. Suppose the convention were adopted that defines W and W' in the first law of thermodynamics to be the "work done **by** the system on the surroundings."
(*a*) Write the first law with this alternate convention.
(*b*) Why do the signs change?

3.6. Give five examples of the operation of the second law of thermodynamics in your daily experience; they must be different from those given in the text.

3.7. Biological systems, organelles, cells, organs, plants, and animals, are highly ordered, yet form spontaneously. Does the formation and growth of biological systems violate the second law of thermodynamics? Explain your answer.

3.8. "Irreversible" is an awkward adjective. Why is this term so appropriate in its application to the description of processes in thermodynamics? Suggest two or three alternate words or phrases that might be used to replace "irreversible" in these contexts.

3.9. Contrast the relative magnitudes of the entropy transfer versus entropy production in the following processes:

(a) A thermally insulated container has two compartments of equal size. Initially one side is filled with a gas and the other is evacuated. A valve is opened and the gas expands to fill both compartments.

(b) A gas contained in a steel cylinder is slowly expanded to twice its volume.

3.10. Consider an isolated system (no heat, matter or work may be exchanged with the surroundings) consisting of three internal compartments A, B and C, of equal volumes. The compartments are separated by partitions and each partition has a valve that may be opened remotely. Initially the central volume B is filled with a gas at 298 K (25°C) and the outer two are evacuated. Consider the following two processes:

(a) The valve to the A side is opened, the gas expands freely into the compartment A, and the system comes to equilibrium. Then the valve to the C side is opened, and the system again comes to equilibrium.

(b) Both valves are opened simultaneously, the gas expands freely into both compartments, and the system comes to its equilibrium.

Which of these processes produces more entropy?

3.11. It will be shown in Chap. 4 that the change in entropy ΔS associated with process (A) in Prob. 3.10 is 4.60 (J/mol-K) and that the initial and final states will be at the same temperature. Application of Eq. (3.10) suggests that the heat absorbed by the system during this process is

$$Q = T \Delta S = (298 \text{ K}) \times (4.60 \text{ J/mol-K}) = 13.71 \text{ J/mol}$$

Yet the description of the system says it is isolated from its surroundings so that $Q = 0$. Explain this apparent contradiction.

3.12. Give three examples of processes important in materials science that are thermodynamically "irreversible." Speculate briefly about the nature of the dissipations in these processes that contribute to the production of entropy.

3.13. The notion of a "reversible" process is a fiction in the real world. What makes this concept, which at first glance would appear to be only of academic interest, so useful in applying thermodynamics to real-world "irreversible" processes?

3.14. The combined statement of the first and second laws of thermodynamics, Eq. (3.15), evaluates the heat absorbed and mechanical work done on a system with relationships that are *valid only for reversible processes*. Since reversible processes do not occur in the real world, how is it possible for the combined statement to play an important role in the analysis of practical "irreversible" processes encountered in nature and in technology?

3.15. Describe the kinds of experimental observations that have been invoked to support the hypothesis that the entropy of all substances is the same at absolute zero.

3.16. Use the following values of absolute entropies of elements and compounds at 298 K to compute the entropy changes associated with as many chemical reactions as you can generate from this list.

Element	S°_{298K} (J/mol-K)	Compound	S°_{298K} (J/mol-K)
Al	11.5	CO	197.90
C(gr)	5.69	CO_2	213.64
Si	18.83	Al_2O_3	50.99
O_2	205.03	SiC	16.48
		SiO_2	42.09

REFERENCES

1. Haase, R., *Thermodynamics of Irreversible Processes*, Addison-Wesley, Reading, Mass. (1969) p. 243. Also reprinted by Dover Publications Inc., New York, N.Y. 1990.
2. DeGroot, S. R. and P. Mazur: *Non-Equilibrium Thermodynamics,* Amsterdam, 1962.
3. Eddington, Sir Arthur: *The Nature of the Physical World*, University of Michigan Press, Ann Arbor, Mich., 1958.

CHAPTER

4

THERMODYNAMIC
VARIABLES
AND
RELATIONS

The simplest class of problem that can be solved thermodynamically is that in which some change in the condition of a system is brought about by controlled or measured changes in the influences that operate on the system. A few representative examples illustrate some typical questions:

1. Hydrogen gas in a 20 liter steel cylinder at ambient temperature (18°C) is found to be at 10 atmospheres of pressure. Is the pressure significantly affected if the room temperature heats up to 25°C?

2. Calculate the change in entropy of 50 gm of nickel when it is heated from 0 to 500°C at one atmosphere pressure.

3. A 40 gram sample of the elastomer polyisoprene is stretched so that its length increases by 50 per cent. Estimate the change in temperature of the sample if it is initially at 20°C.

4. Estimate the amount of heat required to raise the temperature of an alumina (Al$_2$O$_3$) charge weighing 50 kg from room temperature to 1350°C at one atmosphere pressure.

Solutions to these problems are based on relationships between the properties given and required. The strategy for deriving such relationships is the subject of this chapter.

The first step in applying this strategy is the translation of the problem encountered in the real world (or in the text) from English into a statement of thermodynamics. This exercise in translation has the following elements:

1. Identify the properties of the system about which information is given. For example, in question 2 these properties are:

   Temperature (T) [Changes from 0 to $500°C$]

   Pressure (P) [Remains fixed at one atmosphere]

These variables are called the *independent variables* in the problem because in the context of the problem, they are subject to experimental control and thus may be changed independently. In simple systems there are two independent variables.

2. Identify the property in the system about which you are seeking information. In question 2 this property is the entropy of the system.

This property is called the *dependent variable* in the problem because its value is determined by the changes in the controlled (independent) variables.

3. Find or derive a relationships between the sought (dependent) variable and the given (independent) variables. In question 2, this relation has the generic form $S = S(T, P)$, read "entropy as a function of temperature and pressure."

This relationship necessarily contains quantities that are properties of the material comprising the system, such as heat capacity, thermal expansion coefficient, and the like.

4. Obtain values for these quantities, either from tabulations such as those reviewed in Appendices C and D, from the literature, from compiled data bases, or if necessary, by direct experimental evaluation.

5. Substitute these values into the relationship and carry out the mathematical operations necessary to obtain a numerical value for the dependent variable in the problem.

This strategy is applied repeatedly in this and subsequent chapters yielding numerical answers to practical problems.

The crucial application of the principles of thermodynamics in this strategy is evidently contained in step 3, which requires the derivation or deduction of the relationship between the dependent variable in the problem and the independent variables. A general procedure for deriving such relationships is the focus of this chapter.

4.1 CLASSIFICATION OF THERMODYNAMIC RELATIONSHIPS

Thermodynamics abounds with relationships. This potentially confusing aspect of the structure of thermodynamics exists because a variety of state functions are defined in terms of other state functions. This bewildering array of equations may be tamed by organizing them into the five classifications introduced in Sec. 2.3:

I. The laws of thermodynamics
II. Definitions in thermodynamics
III. Coefficient relations
IV. Maxwell relations
V. Conditions for equilibrium

The first two categories are self-explanatory; categories III and IV are derived from mathematical properties of state functions. Relationships in category V are derived from the general criterion for equilibrium developed in Chap. 5; much of the remainder of the text focuses upon these relationships.

4.1.1 The Laws of Thermodynamics

The relationships in this category have been presented in Chap. 3. Succinctly reviewed, they are:

The first law:

$$dU = \delta Q + \delta W + \delta W' \tag{4.1}$$

Reversible mechanical work:

$$\delta W_{\text{rev}} = -P dV \tag{4.2}$$

Reversible heat absorbed:

$$\delta Q_{\text{rev}} = T dS \tag{4.3}$$

Combined statement:

$$dU = T dS - P dV + \delta W' \tag{4.4}$$

Equation (4.4) is a centerpiece in the mathematical framework of thermodynamics and will be applied repeatedly in subsequent developments.

4.1.2 Definitions in Thermodynamics

Relationships presented in this section are simply *definitions*. The two subcategories in this classification are the *energy functions* and the *experimental variables*.

The *energy functions* are measures of the energy of a system that differ from the internal energy U originally introduced in the first law. It will be demonstrated that use of these functions simplifies the description of systems that may be subjected to certain classes of processes. They are simply more convenient measures of energy in such applications than is the function U.

THE ENTHALPY

$$H \equiv U + PV \tag{4.5}$$

is a state function, since it is defined in terms of other state functions. Consider a small change in the state of any system. In general, the change in enthalpy is obtained by taking the differential of its definition:

$$dH = dU + P\,dV + V\,dP$$

Use the combined statement of the first and second laws, Eq. (4.4), to substitute for dU in this equation:

$$dH = [T\,dS - P\,dV + \delta W'] + P\,dV + V\,dP$$

Simplifying

$$dH = T\,dS + V\,dP + \delta W' \tag{4.6}$$

This equation is an alternate form of the combined statement of the first and second laws of thermodynamics. It has the same level of generality and pervasiveness as does Eq. (4.4).

Historically, the enthalpy function was introduced because of its convenience in the description of heat engines taken through cycles. Only mechanical work is done in this case ($\delta W' = 0$) and cycles occur at atmospheric pressure ($dP = 0$). Thus, for such *isobaric* processes in simple systems,

$$dH_P = T\,dS_P = \delta Q_{\mathrm{rev},P}$$

For this and *only* this class of process, the state function, defined as the enthalpy, provides a direct measure of the reversible heat exchanges of the engine with its surroundings. For this reason the enthalpy is sometimes called the *heat content* of the system. This notion is potentially very misleading. A system at rest does not contain a "quantity of heat;" heat is a process variable and has a value only for a system that is changing its state. The *change* in the state function H is related to the process variable Q only for reversible, isobaric processes.

THE HELMHOLTZ FREE ENERGY

$$F \equiv U - TS \tag{4.7}$$

is also a state function. Again consider a small arbitrary change in the state of any system. The change in Helmholtz free energy is

$$dF = dU - T\,dS - S\,dT$$

Substitute for dU from Eq. (4.4):

$$dF = [T\,dS - P\,dV + \delta W'] - T\,dS - S\,dT$$

Simplifying

$$dF = -S\,dT - P\,dV + \delta W' \tag{4.8}$$

This is yet another form of the combined statement of the first and second laws, which can be used to relate property changes in any system taken through any process. This function was devised because it simplifies the description of systems subject to temperature control, which is a common experimental scenario in the laboratory. If the process is carried out at constant temperature (*isothermally*) then $dT = 0$ at each step along the way. For isothermal processes,

$$dF_T = -PdV_T + \delta W_T' = \delta W_T + \delta W_T' = \delta W_{T,\text{tot}}$$

That is, for isothermal processes, the Helmholtz free energy function reports the *total (reversible) work done on the system*. For this reason this property is sometimes called the *work function*. Indeed, in Helmholtz's original presentation, he used the symbol A for this function to stand for *arbeiten*, the German word for *work*. This expression, the "work function," may be misleading in the same way as the term "heat content" is misleading for the enthalpy function. A system at rest does not contain a given "quantity of work;" work can be associated only with a change in state along a specified path. The *change* in the state function F is related to the total work done on the system only for isothermal, reversible processes.

THE GIBBS FREE ENERGY

$$G = U + PV - TS = H - TS \qquad (4.9)$$

is evidently also a state function since it is also defined in terms of other state functions. Following the strategy used for the other two functions, for an arbitrary infinitesimal change in state, the change in Gibbs free energy is

$$dG = dU + PdV + VdP - TdS - SdT$$

Again substitute Eq. (4.4):

$$dG = [TdS - PdV + \delta W'] + PdV + VdP - TdS - SdT$$

Simplifying

$$dG = -SdT + VdP + \delta W' \qquad (4.10)$$

which is yet another equivalent but alternate statement of the combined first and second laws. The Gibbs free energy function was introduced because it simplifies the description of systems that are controlled in the laboratory so that *both* temperature and pressure remain constant. For processes carried out under these conditions [isothermal ($dT = 0$) *and* isobaric ($dP = 0$)]

$$dG_{T,P} = \delta W'_{T,P}$$

Thus, for such processes, which include, for example, phase transformations and chemical reactions, this function reports the total work done on the system *other than mechanical work*.

Four combined statements of the first and second laws have been presented in Eqs. (4.4), (4.6), (4.8), and (4.10). All represent valid ways of describing the thermodynamic changes for any system taken through any process. Each of the defined energy functions provides a simple measure of some process variable for specifically defined classes of processes. All of these equations are used frequently in subsequent developments. Mnemonic devices have been developed to aid in memorizing these equations. However, it is recommended that the *definitions* of H, F, and G be committed to memory. Then the simple two-line strategy presented here leads to these combined statements of the first and second laws.

A set of *experimental variables* provides the core of practical information about a specific material that is essential to solving thermodynamic problems involving that material. These quantities are commonly measured in the laboratory and published in extensive tables or data bases such as those contained in the Appendices *C*, and *D*, and in the list of references at the end of this chapter. The definitions of these experimental variables directly reflect the measurements involved.

The *coefficient of thermal expansion*, α, is obtained from a measurement of the volume change of the material when its temperature is increased, with the system constrained to a constant pressure. The definition:

$$\alpha = \frac{1}{V}\left(\frac{\partial V}{\partial T}\right)_P \quad (\text{K}^{-1}) \tag{4.11}$$

reports the fractional change in volume with temperature. The coefficient of expansion of any material substance varies with temperature, pressure, and composition of the system. Typical values of α are listed in Table 4.1 and in Appendix C.

The *coefficient of compressibility, β,* is determined by measuring the volume change of the substance as the pressure applied to it is increased while the temperature is held constant. Its definition,

$$\beta = -\frac{1}{V}\left(\frac{\partial V}{\partial P}\right)_T \quad (\text{atm})^{-1} \tag{4.12}$$

TABLE 4.1
Typical values of coefficients of thermal expansion α for common materials*

Material	$\alpha_L \times 10^6$ K^{-1}	$\alpha \times 10^6$ K^{-1}
Aluminum	23.5	70.5
Chromium	6.5	19.5
Copper	17.0	51.0
Lead	29.0	87.0
Potassium	83	250
Sodium	71	213
Alumina (Al$_2$O$_3$)	7.6	23
Silica (SiO$_2$)	22.2	66
Silicon Carbide (SiC)	4.6	14

* Reported values are for a linear expansion coefficient, α_L; for isotropic systems, the volume coefficient $\alpha = 3\alpha_L$

Source: Smithells Metals Reference Book, 6th edition, E. A. Brandes, ed., Butterworths, London, 1983.

is normalized to report the fractional change in volume with increasing pressure. The derivative in this definition is inherently negative since as P is increased, V decreases. Thus, inclusion of the minus sign guarantees that tabulated values of β are positive numbers. Note that β also varies with the temperature and pressure and is different for different materials. Table 4.2 and Appendix C list typical values for β.

Experimental information about the thermal behavior of substances is embodied in the concept of *heat capacity*. This quantity is determined experimentally by precisely measuring the rise in temperature when a measured quantity of heat is caused to flow into the system. Since heat is a process variable, it is necessary to specify the process path used in a particular heat capacity measurement. Virtually all heat capacity data are measured and compiled for either of two simple processes.

If the temperature rise dT is measured in a system with the pressure held constant and the heat δQ transferred reversibly, then the heat capacity at constant pressure C_P is obtained from the definition

$$\delta Q_{\text{rev},P} = C_P dT_P \qquad (4.13)$$

Tabulated heat capacities are normalized to obtain the value per mole of substance. Thus, the units of C_P are (J/mole-K). This experimental variable changes with temperature, pressure, and composition. Almost all heat capacity measurements are made at one atmosphere. The pressure dependence of this quantity can be computed theoretically but has not been widely investigated experimentally.

TABLE 4.2

Typical values of coefficients of compressibility β for common materials may be estimated from tabulated values of the modulus of elasticity, E^*

Material	β (atm)$^{-1} \times 10^7$
Aluminum	12
Carbon (Graphite)	340
Copper	6.6
Iron	5.9
Tungsten	2.9
Alumina	8.3
Boron Nitride	37
Silicon Carbide	6.5
Silica Glass	42

* For isotropic materials, $\beta = 3/E$ with properly converted units.

Source: C. L. Reynolds, Jr., K. A. Faugham and R. E. Barker, *J. Chem. Phys.*, vol. 59 (1973) p. 2934; W. D. Kingery, H. K. Bowen and D. R. Uhlman, *Introduction to Ceramics*, John Wiley & Sons, New York, (1976), p. 777.

Figure 4.1 illustrates the temperature dependence of C_P values for a variety of substances at one atmosphere pressure and above room temperature. The variation in the temperature range below room temperature is significantly more complicated. Above room temperature, where most practical applications of thermodynamics arise, it is found that the temperature variation of C_P follows the relatively simple empirical relation:

$$C_P(T) = a + bT + \frac{c}{T^2} \tag{4.14}$$

Tables of heat capacity data, such as Table 4.3 and Appendix D, provide values for a, b, and c for a variety of elements and chemical compounds.

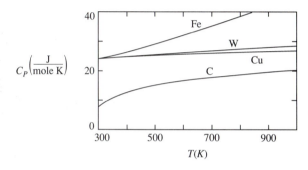

FIGURE 4.1
Variation of heat capacity with temperature for a variety of common materials.

TABLE 4.3
Experimental values of the variation of heat capacity at constant pressure (C_p) for common materials above room temperature*

Material	a	$b \times 10^3$	$c \times 10^{-5}$
Aluminum	20.6	12.4	—
Carbon	17.2	4.3	—
Copper	22.6	6.3	—
Gold	25.7	−0.7	3.8
Iron	14.1	30	—
Nickel	17.0	29	—
Tungsten	24.0	3.1	—
Alumina	21.9	3.7	—
Silica	15.6	11.4	—

* The empirical expression for the temperature dependence of the heat capacity is
$C_p = a + bT + \frac{c}{T^2}$
Values are reported in Joules per gram atom.

Source: Smithells Metals Reference Book, 6th edition, E. A. Brandes, ed., Butterworths, London, 1983.

Heat capacity can also be measured in a system contained in a rigid enclosure so that, as the heat flows into the system, the *volume* of the system is constrained to remain constant. Precise measurement of the temperature rise accompanying the influx of a known quantity of heat gives the *heat capacity at constant volume*, defined in the relation

$$\delta Q_{rev,V} \equiv C_V dT_V \qquad (4.15)$$

After normalization, the units for C_V are also (J/mole-K); C_V also varies with temperature, pressure, and composition.

For the system at constant pressure, absorption of heat results in both a temperature rise and an expansion of the volume of the system. In the case of the system constrained to constant volume, all of the heat acts to raise the temperature. Thus, more heat is required to raise the temperature of a substance one degree at constant pressure than at constant volume. It is concluded that, in general, $C_P > C_V$, and this is generally observed. It is demonstrated in a later section that these two experimental variables are related to each other [see Eq. (4.48)]. If given information about α and β for the system and if one heat capacity is known, the other can be computed.

It is be further demonstrated later in this chapter that if α, β, and C_P are known for any simple system, one for which $\delta W' = 0$, then changes in *all* the state functions can be computed for any arbitrary process through which that system may be taken. No additional information is required. Thus, these experimental variables are at the center of the strategy for solving practical problems requiring thermodynamics. The ranges of values that these quantities may have in solids, liquids, and gases are summarized in Tables 4.1–4.3.

4.1.3 Coefficient Relations

Coefficient relations are derived from a mathematical property of functions of several variables. Because the state variables of thermodynamics have been shown to satisfy the properties of ordinary mathematical functions, these coefficient relations apply to equations among all the state variables that have been identified and defined.

Suppose that X, Y and Z are state variables. There must exist a relationship among these variables such that, given values for any two of them, say X and Y, the third, Z, can be evaluated. Consequences of this kind of mathematical relationship, introduced in Sec. 2.3, are reviewed here. Mathematical shorthand for this relation is

$$Z = Z(X, Y) \qquad (4.16)$$

read "Z is a function of X and Y." In this statement of the relationship, X and Y are "independent variables" and Z is the "dependent variable" with a value that "depends" upon the values that may be "independently" assigned to X and Y. Figure 4.2 illustrates such a relation.

Assume that this function is well behaved, that is, it is smooth and continuous with continuous derivatives. An infinitesimal step in a process can be represented by

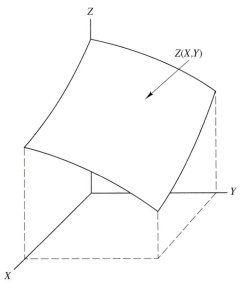

FIGURE 4.2
Geometric representation of the notion that "Z is a function of X and Y."

independent changes in X and Y. Because of the functional relation, Eq. (4.16), the corresponding change in Z can be written:

$$dZ = MdX + NdY \qquad (4.17)$$

The coefficients in this equation are certain explicit partial derivatives:

$$M = \left(\frac{\partial Z}{\partial X}\right)_Y \qquad \text{and} \qquad N = \left(\frac{\partial Z}{\partial Y}\right)_X \qquad (4.18)$$

The graphical meaning of these relations is evident in Fig. 4.2. This equation is the formal basis for all coefficient relations. Given any thermodynamic relation among differentials of state functions, it is possible, by inspection, to write down the coefficient relations. For example, the combined statement of the first and second laws, Eq. (4.4), applied to processes for which $\delta W' = 0$, becomes

$$dU = TdS - PdV \qquad (4.19)$$

which has the general form of Eq. (4.17). The coefficients in this equation are evidently the partial derivatives

$$T = \left(\frac{\partial U}{\partial S}\right)_V \qquad \text{and} \qquad -P = \left(\frac{\partial U}{\partial V}\right)_S \qquad (4.20)$$

Application of this strategy to the other three combined statements derived from the definitions of the energy functions, Eq. (4.6), (4.8), and (4.10), yields

$$T = \left(\frac{\partial H}{\partial S}\right)_P \qquad \text{and} \qquad V = \left(\frac{\partial H}{\partial P}\right)_S \qquad (4.21)$$

$$-S = \left(\frac{\partial F}{\partial T}\right)_V \qquad \text{and} \qquad -P = \left(\frac{\partial F}{\partial V}\right)_T \qquad (4.22)$$

$$-S = \left(\frac{\partial G}{\partial T}\right)_P \qquad \text{and} \qquad V = \left(\frac{\partial G}{\partial P}\right)_T \qquad (4.23)$$

In thermodynamic applications of this class of relations, it is important to specify explicitly not only the variables in the numerator and denominator of each partial derivative but also *the variable that is being held constant,* indicated in the subscript outside the parentheses of the partial derivative. The two derivatives

$$\left(\frac{\partial H}{\partial T}\right)_P \qquad \text{and} \qquad \left(\frac{\partial H}{\partial T}\right)_V$$

are not equal to each other. Indeed, it can be shown that

$$\left(\frac{\partial H}{\partial T}\right)_V = \left(\frac{\partial H}{\partial T}\right)_P + \left(\frac{\partial H}{\partial P}\right)_T \left(\frac{\partial P}{\partial T}\right)_V$$

It is absolutely essential to be explicit about all three variables required in the specification of a partial derivative in thermodynamics.

If the state of the system is a function of more than two variables $[Z = Z(X,Y,U,V,. . .)]$ the differential generalizes to

$$dZ = M\,dX + N\,dY + P\,dU + R\,dV + \cdots \qquad (4.24)$$

with each coefficient corresponding to a partial derivative of Z. This straightforward generalization is useful in later chapters where more complex systems are studied.

4.2   MAXWELL RELATIONS

This category of thermodynamic relationship is also derived from mathematical properties of state functions. Consider again the variables obeying the functional relationship in Eq. (4.16) with differential relationships given by Eq. (4.17) and coefficients given by Eq. (4.18). Consider the two partial derivatives

$$\left(\frac{\partial M}{\partial Y}\right)_X \qquad \text{and} \qquad \left(\frac{\partial N}{\partial X}\right)_Y$$

Substitute the coefficient relations for M and N given in Eq. (4.18).

$$\left(\frac{\partial M}{\partial Y}\right)_X = \left[\frac{\partial}{\partial Y}\left(\frac{\partial Z}{\partial X}\right)_Y\right]_X \qquad \text{and} \qquad \left(\frac{\partial N}{\partial X}\right)_Y = \left[\frac{\partial}{\partial X}\left(\frac{\partial Z}{\partial Y}\right)_X\right]_Y$$

Note that these two quantities are equal except for the order of differentiation. Thus,

$$\left(\frac{\partial M}{\partial Y}\right)_X = \left(\frac{\partial N}{\partial X}\right)_Y \qquad (4.25)$$

forms the mathematical basis for the Maxwell relations in thermodynamics.

In multivariable calculus, relations like this are called the Cauchy-Riemann conditions. They are the basis for determining whether a given differential form, *MdX*

+ *NdY*, for example, is in fact an *exact differential*, that is, whether there exists a function *Z(X,Y)* of which *(MdX + NdY)* is the differential. Evidently this is so if and only if Eq. (4.25) is satisfied. In thermodynamics, the establishment that *Z, X,* and *Y* are state functions guarantees that the form *(MdX + NdY)* is always an exact differential.

Maxwell relations for any thermodynamic differential form can be written by inspection. As examples, application of Eq. (4.25) to the combined statements derived earlier yields:

from $dU = TdS - PdV$,

$$-\left(\frac{\partial P}{\partial S}\right)_V = \left(\frac{\partial T}{\partial V}\right)_S$$

from $dH = TdS + VdP$,

$$\left(\frac{\partial T}{\partial P}\right)_S = \left(\frac{\partial V}{\partial S}\right)_P$$

from $dF = -SdT - PdV$,

$$-\left(\frac{\partial S}{\partial V}\right)_T = -\left(\frac{\partial P}{\partial T}\right)_V$$

from $dG = -SdT + VdP$,

$$-\left(\frac{\partial S}{\partial P}\right)_T = \left(\frac{\partial V}{\partial T}\right)_P \qquad (4.26)$$

These relations apply to any system at any condition. These are a few examples of Maxwell relations, not an exhaustive list. Equations of this form hold for every functional relationship that can be constructed relating any dependent state function to two independent state functions.

To illustrate the somewhat unexpected connections between properties that result from these relations, consider the last in the list, Eq. (4.26). From the definition of the coefficient of thermal expansion, Eq. (4.11), it follows that

$$\left(\frac{\partial V}{\partial T}\right)_P = V\alpha$$

Thus, from the last of the Maxwell relations, Eq. (4.26),

$$\left(\frac{\partial S}{\partial P}\right)_T = -V\alpha \qquad (4.27)$$

This result demonstrates that the pressure dependence of the entropy of a system is determined by its coefficient of thermal expansion, a result that is by no means intuitively evident.

The Maxwell relations are also readily generalized to systems that require more than two independent variables. Recall Eq. (4.24):

$$dZ = MdX + NdY + PdU + RdV + \cdots \qquad (4.24)$$

Maxwell relations must hold for each pair of terms in this equation, since each coefficient represents a differential of the state function, Z. Thus, for example,

$$\left(\frac{\partial R}{\partial X}\right)_{Y,U,V,\cdots} = \left(\frac{\partial M}{\partial V}\right)_{X,Y,U,\cdots} \qquad \left(\frac{\partial R}{\partial Y}\right)_{X,U,V,\cdots} = \left(\frac{\partial N}{\partial V}\right)_{X,Y,U,\cdots} \qquad \text{etc.}$$
(4.28)

Example 4.1. It is shown in Sec. 4.2 that the form of the equation relating entropy to temperature and pressure is

$$S = S(T, P) \qquad dS = \frac{C_p}{T} dT - V\alpha \, dP$$

Write the coefficient and Maxwell relations for this equation. The coefficients of dT and dP are respectively

$$\left(\frac{\partial S}{\partial T}\right)_P = \frac{C_P}{T} \qquad \left(\frac{\partial S}{\partial P}\right)_T = -V\alpha$$

The Maxwell relation is

$$\left(\frac{\partial (C_P/T)}{\partial P}\right)_T = \left(\frac{\partial(-V\alpha)}{\partial T}\right)_P$$

Evidently, the variation of heat capacity with pressure is related to the coefficient of thermal expansion for a system.

Two other relationships among partial derivatives that can be defined between the variables X, Y, and Z are derived with the aid of results presented in this section. These relationships are the *reciprocal relation*

$$\left(\frac{\partial Z}{\partial X}\right)_Y = \frac{1}{\left(\frac{\partial X}{\partial Z}\right)_y}$$
(4.29)

and the *ratio relation* (see Problem 4.2)

$$\left(\frac{\partial Z}{\partial X}\right)_Y \left(\frac{\partial X}{\partial Y}\right)_Z \left(\frac{\partial Y}{\partial Z}\right)_X = -1$$
(4.30)

In the process of developing relationships between thermodynamic state functions, these relations are useful in the manipulation of the partial derivatives.

As a further step in the development of the structure of thermodynamics, four classes of relationships between state functions have been presented. The laws of thermodynamics provide the physical basis for all these relationships. Definitions introduce energy functions that are convenient in specific applications and also a set of experimental variables that form the core of information needed to solve practical problems. Coefficient and Maxwell relations are consequences of the mathematical properties of differentials of state functions. The basis for deriving the fifth category of relations, conditions for equilibrium, is presented in Chap. 5.

4.3 GENERAL STRATEGY FOR DERIVING THERMODYNAMIC RELATIONS

The relationships presented and classified in the previous section provide the basis for developing completely general equations between arbitrarily selected thermodynamic state functions. Specifically, for a system that does no nonmechanical work, that is, for any system for which $\delta W' = 0$, it is possible to develop an equation relating any function of the state of that system as a dependent variable to any pair of other state functions as independent variables. As was pointed out at the beginning of this chapter, this is the key link in the chain of reasoning needed to translate and solve practical problems that require thermodynamics analysis.

This section presents a general procedure for deriving such relations. The procedure is simple, straightforward, and mathematically rigorous. Because there is no mathematical sleight-of-hand involved, and because the sequence of steps to be followed is clearly defined, it should become possible to solve the most general problem with rigor and confidence.

Extension of this procedure to systems that are more general, for which $\delta W'$ is not zero, is straightforward. Applications to this kind of system, which require the development of more of the apparatus of thermodynamics, are presented in later chapters.

An exhaustive list of the state functions thus far defined is given in Table 4.4. Others are introduced as more complex classes of systems are encountered and the thermodynamic apparatus expands.

The general procedure begins by choosing temperature and pressure as independent variables, since these variables are most frequently encountered in practical problems, and deriving relations for all the other state functions in Table 4.4 as functions of T and P. It is shown that once equations are derived for volume $[V = V(T,P)]$ and entropy $[S = S(T,P)]$, expressions for the four energy functions readily follow. Once all the functions have been expressed in terms of (T,P), conversion to other pairs of independent variables, for example, $H = H(V,S)$, requires no mathematics beyond algebra.

TABLE 4.4
State functions defined in thermodynamics

Equation of State Variables	
Temperature	T
Pressure	P
Volume	V
Energy Functions	
Internal Energy	U
Enthalpy	H
Helmholtz Free Energy	F
Gibbs Free Energy	G
Entropy	S

4.3.1   Entropy and Volume Relations to T and P

For every simple substance there exists a relationship between its volume, temperature, and pressure that can be written formally as

$$V = V(T, P) \tag{4.31}$$

since these are all state functions. The corresponding differential relation, analogous to the generic Eq. (4.17), is

$$dV = M dT + N dP$$

The coefficient relations for this equation give

$$dV = \left(\frac{\partial V}{\partial T}\right)_P dT + \left(\frac{\partial V}{\partial P}\right)_T dP$$

The partial derivatives in this equation can be expressed in terms of the experimental variables α and β defined in Eqs. (4.11) and (4.12).

$$\left(\frac{\partial V}{\partial T}\right)_P = V\alpha \quad \text{and} \quad \left(\frac{\partial V}{\partial P}\right)_T - -V\beta$$

Thus, without any restrictions on the nature of the substance that makes up the system, it can be stated that

$$V = V(T, P) \qquad dV = V\alpha dT - V\beta dP \tag{4.32}$$

This equation can be regarded as a differential form of an equation of state. If experimental information is available for the substance in the system for α and β as functions of temperature and pressure, this equation can be integrated (at least numerically) to obtain a complete description of the variation of the volume of the system with respect to temperature and pressure.

Consider now the analogous problem for the entropy function

$$S = S(T, P) \tag{4.33}$$

The corresponding differential relation is

$$dS = M dT + N dP$$

The coefficients in this equation can be written as partial derivatives so that

$$dS = \left(\frac{\partial S}{\partial T}\right)_P dT + \left(\frac{\partial S}{\partial P}\right)_T dP \tag{4.34}$$

The second coefficient, N, has already been evaluated in the consideration of Maxwell relations. Recall Eq. (4.26):

$$N = \left(\frac{\partial S}{\partial P}\right)_T = -\left(\frac{\partial V}{\partial T}\right)_P = -V\alpha \tag{4.35}$$

Evaluation of M requires a short thermodynamic argument.

For any reversible process, the second law connects the heat absorbed with the change in entropy:

$$\delta Q_{\text{rev}} = T dS \tag{3.10}$$

With equation (4.34)

$$\delta Q_{\text{rev}} = T \left[\left(\frac{\partial S}{\partial T} \right)_P dT + \left(\frac{\partial S}{\partial P} \right)_T dP \right]$$

The heat absorbed for a constant pressure process, for which $dP = 0$, can be written

$$\delta Q_{\text{rev},P} = T \left(\frac{\partial S}{\partial T} \right)_P dT_P \tag{4.36}$$

Measurements of the reversible heat absorbed for a constant pressure process are the basis for the experimental evaluation of C_P, the heat capacity at constant pressure,

$$\delta Q_{\text{rev},P} = C_P dT_P \tag{4.13}$$

Comparison of these two expressions for heat absorbed at constant pressure demonstrates that the coefficients of dT_P must be the same quantity in both equations. It is concluded that

$$C_P = T \left(\frac{\partial S}{\partial T} \right)_P$$

or,

$$\left(\frac{\partial S}{\partial T} \right)_P = \frac{C_P}{T} \tag{4.37}$$

Thus, a general differential relation for the variation of entropy with temperature and pressure for any system is

$$S = S(T, P) \qquad dS = \frac{C_P}{T} dT - V\alpha dP \tag{4.38}$$

If C_P and α are known as functions of temperature and pressure for the system under study, the entropy of the system can be computed by the integration of this equation as a function of temperature and pressure. There are no restrictions on the validity of this equation other than the limitations imposed by the simplification that $\delta W' = 0$ for all processes considered in this chapter.

Although Eqs. (4.32) and (4.38) are sufficient to implement the general procedure to be developed shortly, it is also useful to introduce relations between state functions and the experimental variable C_V. This brief development is introduced here because it parallels that just presented for C_P. Recall the definition of the heat capacity at constant volume, Eq. (4.15):

$$\delta Q_{\text{rev},V} = C_V dT_V \tag{4.15}$$

In the process used to determine C_V, the known variables are the temperature, which is measured, and the volume, which is held constant. The heat absorbed is related to the entropy changeof the system which suggests that the state function $S = S(T,V)$

should be examined. The differential form of this function is

$$dS = \left(\frac{\partial S}{\partial T}\right)_V dT + \left(\frac{\partial S}{\partial V}\right)_T dV$$

For a constant volume process, such as is used in determining C_V, $dV = 0$ and the change in entropy is

$$dS_V = \left(\frac{\partial S}{\partial T}\right)_V dT_V$$

Apply again the second law as embodied in Eq. (3.10):

$$\delta Q_{\mathrm{rev},V} = T dS_V = T \left(\frac{\partial S}{\partial T}\right)_V dT_V \tag{4.39}$$

Comparison of coefficients of Eqs. (4.39) and (4.15) gives

$$\left(\frac{\partial S}{\partial T}\right)_V = \frac{C_V}{T} \tag{4.40}$$

Thus, C_V can also be expressed in terms of state functions. Comparison of Eqs. (4.40) and (4.37) again demonstrates the necessity for specifying the variables that are held constant in taking each derivative. Clearly, since C_P and C_V are not equal to one another, their respective entropy derivatives must also be different.

4.3.2 Energy Functions Expressed in Terms of T and P

The four combined statements of the first and second laws derived in Sec. 4.1.2 follow directly from the definitions of each energy function and the original version of this statement expressed in terms of dU. The results are summarized here:

$$U = U(S, V) \qquad dU = T dS - P dV \tag{4.4}$$

$$H = (S, P) \qquad dH = T dS + V dP \tag{4.6}$$

$$F = F(T, V) \qquad dF = -S dT - P dV \tag{4.8}$$

$$G = G(T, P) \qquad dG = -S dT + V dP \tag{4.10}$$

Equations (4.32) and (4.38) provide general differential expressions for dV and dS for any system in terms of dT and dP. Conversion of these equations for the energy functions into expressions having T and P as independent variables merely requires substitution for dV or dS where they appear, followed by algebraic simplification. To illustrate, substitute for dS and dV in Eq. (4.4):

$$dU = T \left[\frac{C_P}{T} dT - V\alpha dP\right] - P[V\alpha dT - V\beta dP]$$

Collecting like terms

$$U = (T, P) \qquad dU = (C_P - PV\alpha)dT + V(P\beta - T\alpha)dP \tag{4.41}$$

To convert Eq. (4.6) for the enthalpy to a function of T and P, substitute for dS,

$$dH = T\left[\frac{C_P}{T}dT - V\alpha dP\right] + V dP$$

and collect terms

$$H = H(T, P) \qquad dH = C_P dT + V(1 - T\alpha)dP \qquad (4.42)$$

Similarly, Eq. (4.8) is easily converted to a function of T and P by substituting for dV:

$$dF = -SdT - P(V\alpha dT - V\beta dP)$$

Collecting terms

$$F = F(T, P) \qquad dF = -(S + PV\alpha)dT + PV\beta dP \qquad (4.43)$$

The Gibbs free energy function is already expressed in terms of the independent variables T and P in the version originally derived, Eq. (4.10).

This collection of all state functions expressed in terms of T and P just derived is compiled for quick reference in Table 4.5. Coefficient relations apply to each term in these equations, for example, Eq. (4.42),

$$\left(\frac{\partial H}{\partial T}\right)_P = C_P \quad \text{and} \quad \left(\frac{\partial H}{\partial P}\right)_T = V(1 - T\alpha) \qquad (4.44)$$

as do Maxwell relations, for example,

$$\left(\frac{\partial C_P}{\partial P}\right)_T = \left(\frac{\partial[V(1 - T\alpha)]}{\partial T}\right)_P \qquad (4.45)$$

The coefficients in these equations contain the following factors:

T, P which are the independent variables specified in any application

Experimental variables, α, β, C_P, assumed to be available in tables or the literature

S and V, which can be evaluated as functions of T and P, given α, β and C_P.

Thus, if the initial and final states for a process traversed by a system are specified by their temperature and pressure, knowledge of the values of α, β and C_P permits

TABLE 4.5
Thermodynamic state functions expressed in terms of the independent variables Temperature and Pressure

$V = V(T, P)$	$dV = V\alpha dT - V\beta dP$	(4.32)
$S = S(T, P)$	$dS = \left[\frac{C_P}{T}\right]dT - V\alpha dP$	(4.38)
$U = (T, P)$	$dU = (C_P - PV\alpha)dT + V(P\beta - T\alpha)dP$	(4.41)
$H = H(T, P)$	$dH = C_P dT + V(1 - T\alpha)dP$	(4.42)
$F = F(T, P)$	$dF = -(S + PV\alpha)dT - PV\beta dP$	(4.43)
$G = G(T, P)$	$dG = -SdT + V dP$	(4.11)

calculation of all the changes in state functions accompanying that process. The power of these equations is demonstrated by the fact that they apply to any system taken through any process.

4.3.3 The General Procedure

Translation of a statement of a practical problem into the language of thermodynamics requires the identification of the independent variables about which information is given, the dependent variable about which information is sought, and the development of the relationship that connects these variables. The derivation of such relationships is at the heart of the problem. The treatment presented in the previous section yields an exhaustive list of all such relationships for any problem in which the independent variables are temperature and pressure, Table 4.5. The procedure for solving the more general problem, in which any pair of state functions (X,Y) may be the independent variables and any third state function Z is the dependent variable, is laid out in this section. This part of the problem is considered solved (the relation derived) when the coefficients M and N, have been evaluated in terms of the experimental variables, α, β, C_P (or C_V) and the state functions, T, P, S and V, which can always be computed therefrom.

The step-by-step procedure for deriving such relationships is straightforward once all the relations in Table 4.5 have been obtained. It requires no higher mathematics than algebra and, indeed, no thermodynamic thinking is involved, since the manipulations are purely mathematical. There are seven steps in the procedure:

1. Identify the variables: $Z = Z(X,Y)$
2. Write the differential form: $dZ = MdX + NdY$
3. Use Table 4.5 to express dX and dY in terms of the variables dT and dP, (this is always possible, since Table 4.5 provides an exhaustive list for all the state functions):

$$dZ = M[X_T dT + X_P dP] + N[Y_T dT + Y_P dP]$$

   where the coefficients, X_T, X_P, Y_T, Y_P are the coefficients of dT and dP in the expressions for dX and dY in Table 4.5.
4. Collecting terms

$$dZ = [M \cdot X_T + N \cdot Y_T]dT + [M \cdot X_P + N \cdot Y_P]dP$$

5. Obtain $Z = Z(T,P)$ from Table 4.5:

$$dZ = Z_T dT + Z_P dP$$

6. The equations listed under steps 4 and 5 are alternate expressions for $Z = Z(T,P)$. Therefore, the coefficients of dT and dP in both equations are respectively

$$\left(\frac{\partial Z}{\partial T}\right)_P \quad \text{and} \quad \left(\frac{\partial Z}{\partial P}\right)_T$$

Equating like terms

$$M \cdot X_T + N \cdot Y_T = Z_T \tag{4.46a}$$

$$M \cdot X_P + N \cdot Y_P = Z_P \tag{4.46b}$$

7. This is a pair of linear simultaneous equations in M and N and can always be solved for these variables either by elimination of one variable then solving for the other, or, in the worst case, by determinants. The resulting expressions for M and N will certainly be expressed in terms of the experimental variables since all the terms in Table 4.5 (from which X_T, Y_T, Z_T, X_P, Y_P, and Z_P are obtained) are expressed in terms of these variables.

This approach to developing relationships among state variables has several advantages over alternate methods. It is mathematically rigorous so that it may be applied with confidence. It requires conscious identification of the independent and dependent variables so that the initial statement of the problem must be clear. The resulting equation has general applicability and can be used in any situation, avoiding confusion about which relations are special (such as being limited in validity to systems composed of ideal gases or to specific processes). Finally, the path from statement of the problem to its solution is straightforward, eliminating potential confusion about the manipulation of the quantities involved and further fostering confidence in the final result.

Since the mathematics involved in this methodology is only algebra, it is easy to make simple mistakes in computation in following the procedure. Once M and N are evaluated for a given problem, it is strongly recommended that a check of their *units* be carried out. Most algebraic errors are clearly exposed with this tactic since the units obtained for M and N in step 7 must agree with the units they are required to have in writing step 2. In this comparison, it is only necessary to represent each thermodynamic quantity in terms of *generic* units involving only combinations of P, V and T. It is found convenient to represent all measures of energy as having units (PV), then entropy has units (PV/T), heat capacities also have units (PV/T), α has units $(1/T)$, and β has units $(1/P)$.

In the application of the result to the solution of a numerical problem for a specified system, the actual units used are determined primarily by the units that happen to be quoted in the sources of the experimental information. Thus, energy functions may be expressed in Joules, calories, ergs, liter-atmospheres, Newton-meters, BTUs, and the like. Such alternate representations of energy can be interconverted at will using the conversion factors reported in Appendix A. Similarly, heat capacities have units of (energy/mole-K) and can be found carrying any of the energy units displayed above. The coefficients of expansion and compressibility can also appear in a variety of units (such as α as $1/°F$ and β as 1/psi or 1/megapascal). All correspond to the generic units listed in the previous paragraph.

The general procedure has been stated in the abstract; its application will now be made more concrete with a number of examples. The first set of examples (4.2–4.6) illustrates the step-by-step procedure to derive the differential expressions that relate a specific dependent variable to a pair of independent variables. Later examples apply this procedure as a module in the overall problem that begins with a verbal statement of the situation and ends with a numerical result.

Example 4.2. Relate the entropy of a system to its temperature and volume.

1. Identify the variables: $\quad\quad\quad\quad\quad S = S(T, V)$
2. Write the differential form: $\quad\quad\quad dS = M dT + N dV$
3. Convert dV, using Table 4.5: $\quad\quad dS = M dT + N(V\alpha dT - V\beta dP)$
4. Collecting terms $\quad\quad\quad\quad\quad dS = (M + NV\alpha)dT - NV\beta dP$
5. Obtain $S = S(T, P)$ from Table 4.5: $\; dS = [C_P/T]dT - V\alpha dP$
6. Compare coefficients $\quad\quad\quad M + NV\alpha = C_P/T \quad\quad -NV\beta = -V\alpha$
7. Solve this pair of equations for M and N:

$$N = \left(\frac{\partial S}{\partial V}\right)_T = \frac{\alpha}{\beta}$$

$$M = \left(\frac{\partial S}{\partial T}\right)_V = \frac{1}{T}\left(C_P - \frac{TV\alpha^2}{\beta}\right)$$

Check the units. From step 2, $M dT$ must have the same units as dS, which is (PV/T). Thus M has units (PV/T^2). This checks with the units for the value of M derived in step 7. In particular, for the second term in the brackets the units are $[TV(1/T)^2/(1/P) = PV/T]$ and multiplication by the factor $(1/T)$ outside the brackets yields the correct units. $N dP$ also has units of (PV/T). Thus, N has units (P/T). Substituting the units for α and β gives for N $[(1/T)/(1/P)] = P/T$.

Thus, entropy as a function of temperature and volume is given by

$$dS = \frac{1}{T}\left(C_P - \frac{TV\alpha^2}{\beta}\right)dT + \frac{\alpha}{\beta}dV \tag{4.47}$$

Incidentally, this result can be used to derive the general relation that exists between C_P and C_V mentioned in Sec. 4.1.2. By inspection, the coefficient of dT in this equation is

$$\left(\frac{\partial S}{\partial T}\right)_V = \frac{1}{T}\left(C_P - \frac{TV\alpha^2}{\beta}\right)$$

It was shown Eq. (4.40) that this coefficient is related to C_V

$$\left(\frac{\partial S}{\partial T}\right)_V = \frac{C_V}{T}$$

Thus,

$$C_V = C_P - \frac{TV\alpha^2}{\beta} \tag{4.48}$$

Accordingly, if either of the heat capacity coefficients has been measured for a system and α and β are known then the other may be calculated. Since all factors in the second term on the right-hand side are positive, C_V is always smaller than C_P.

Example 4.3. Relate the entropy of a system to its pressure and volume.

1. Identify the variables: $\quad\quad\quad\quad\quad\quad S = S(P, V)$
2. Write the differential form: $\quad\quad\quad dS = MdP + NdV$
3. Convert dV, using Table 4.5: $\quad\quad dS = MdP + N(V\alpha dT - V\beta dP)$
4. Collect terms: $\quad\quad\quad\quad\quad\quad\quad dS = NV\alpha dT + (M - NV\beta)dP$
5. Obtain $S = S(T,P)$ from Table 4.5: $\quad dS = [C_P/T]dT - V\alpha dP$
6. Compare coefficients $\quad\quad\quad\quad NV\alpha = C_P/T \quad\quad M - NV\beta = -V\alpha$
7. Solve this pair of equations for M and N:

$$N = \left(\frac{\partial S}{\partial V}\right)_P = \frac{C_P}{TV\alpha}$$

$$M = \left(\frac{\partial S}{\partial P}\right)_V = \left(\frac{C_P\beta}{T\alpha} - V\alpha\right)$$

Check the units. For M: $[V/T] = [(PV/T)(1/P)/(T/T) - V(1/T)]$
For N: $[P/T] = [(PV/T)/(TV/T)] = P/T$
The units check. Thus, the required relation is

$$dS = \left(\frac{C_P\beta}{T\alpha} - V\alpha\right)dP + \frac{C_P}{TV\alpha}dV \quad\quad\quad (4.49)$$

Example 4.4. Find the relationship needed to compute the change in Helmholtz free energy when the initial and final states are specified by their pressure and volume.

1. Identify the variables: $\quad\quad\quad\quad\quad\quad F = F(P, V)$
2. Write the differential form: $\quad\quad\quad dF = MdP + NdV$
3. Convert dV, using Table 4.5: $\quad\quad dF = MdP + N(V\alpha dT - V\beta dP)$
4. Collect terms $\quad\quad\quad\quad\quad\quad\quad dF = NV\alpha dT + (M - NV\beta)dP$
5. Obtain $F = F(T,P)$ from Table 4.5: $\quad dF = -(S + PV\alpha)dT + PV\beta dP$
6. Compare coefficients $\quad\quad\quad\quad NV\alpha = -(S + PV\beta) \quad M - NV\beta = PV\beta$
7. Solve this pair of equations for M and N:

$$N = \left(\frac{\partial F}{\partial V}\right)_P = -\left(\frac{S}{V\alpha} + P\right)$$

$$M = \left(\frac{\partial F}{\partial P}\right)_V = -\frac{S\beta}{\alpha}$$

Check the units. For M: $V = (PV/T)(1/P)/(1/T) = V$
For N: $P = [(PV/T)/(V/T) + P] = P$

The units check. Thus, the required relation is

$$dF = -\frac{S\beta}{\alpha}dP - \left(\frac{S}{V\alpha} + P\right)dV \tag{4.50}$$

Example 4.5. Express the change in enthalpy as a function of volume and entropy.

1. Identify the variables: $\qquad\qquad\qquad H = H(S, V)$
2. Write the differential form: $\qquad\qquad dH = MdS + NdV$
3. Convert both dS and dV, using Table 4.5:

$$dH = M[(C_P/T)dT - V\alpha dP] + N(V\alpha dT - V\beta dP)$$

4. Collecting terms:

$$dH = [M(C_P/T) + NV\alpha]dT - [MV\alpha + NV\beta]dP$$

5. Obtain $H = H(T, P)$ from Table 4.5:

$$dH = C_P dT + V(1 - T\alpha)dP$$

6. Comparing coefficients

$$MC_P/T + NV\alpha = C_P \qquad -MV\alpha - NV\beta = V(1 - T\alpha)$$

7. Solve this pair of equations for M and N:

Solution of this pair of equations is a nontrivial exercise in algebra but it is only algebra. Determinants or substitution might be used. The simplest approach in this case follows: Multiply the top equation by β and the bottom by α:

$$MC_P\beta/T + NV\alpha\beta = C_P\beta$$

$$-MV\alpha^2 - NV\alpha\beta = V\alpha(1 - T\alpha)$$

Add the equations: the term involving N drops out.

$$M[C_P\beta/T - V\alpha^2] = C_P\beta + V\alpha(1 - T\alpha)$$

This result can be simplified by applying Eq. (4.48). (It is usually true that if one of the independent variables is volume, expression of the coefficients in terms of C_V rather than C_P gives a result that is algebraically simpler.)

$$M \cdot \frac{\beta}{T}\left(C_P - \frac{TV\alpha^2}{\beta}\right) = \beta\left(C_P - \frac{TV\alpha^2}{\beta} + \frac{V\alpha}{\beta}\right)$$

or

$$M\frac{\beta}{T}C_V = \beta\left(C_V + \frac{V\alpha}{\beta}\right)$$

Thus,

$$M = \left(\frac{\partial H}{\partial S}\right)_V = T\left(1 + \frac{V\alpha}{\beta C_V}\right)$$

Substitute this result for M into the original version of the first equation

$$T\left(1 + \frac{V\alpha}{\beta C_V}\right) \cdot \frac{C_P}{T} + NV\alpha = C_P$$

Solve for N:

$$N = \left(\frac{\partial H}{\partial V}\right)_S = -\frac{C_P}{\beta C_V}$$

Check the units. For M: $T = T\{1 + V(1/T)/[(PV/T)(1/P)]\}$
For N: $P = (PV/T)/[(PV/T)(1/P)]$
Thus, the differential expression for $H = H(S,V)$ is

$$dH = T\left(1 + \frac{V\alpha}{C_V\beta}\right)dS - \frac{C_P}{C_V\beta}dV \qquad (4.51)$$

Example 4.6. Derive an expression for the increase in temperature for a process in which the volume of the system is changed at constant entropy. The variable for which information is sought is T; information is given for S and V.

1. Identify the variables: $T = T(S,V)$

   This case differs from those presented above because the dependent variable T is one of the variables we have chosen as an independent variable in developing the arsenal of relations contained in Table 4.5. In such cases rearrange the governing equation so that T is one of the *dependent* variables:

   $$S = S(T, V)$$

2. Find the differential relation:

   $$dS = MdT + NdV$$

3. Evaluate M and N and then solve the resulting equation for dT. The relation $S = S(T,V)$ was derived in Example 4.2. Combine Eq. (4.47) and (4.48):

   $$dS = \frac{C_V}{T}dT + \frac{\alpha}{\beta}dV \qquad (4.52)$$

4. Solve for dT:

   $$dT = \frac{T}{C_V}dS - \frac{T\alpha}{C_V\beta}dV \qquad (4.53)$$

From the statement of the problem for the process considered $dS = 0$. Thus, for this *isentropic* process,

$$dT_s = -\frac{T\alpha}{C_V\beta}dV_s$$

If the experimental variables are known for the system, integration of this equation from the initial to the final volume yields the change in temperature.

4.3.4   Application to an Ideal Gas

The gaseous state of matter is frequently encountered in systems of practical importance. Interaction with gases may lead to degradation through oxidation or hot corrosion. The vapor state can provide a medium for controlled addition of material to a system, as in nitriding or carburization in heat treatment or thin film deposition in processing microelectronic devices. Some chemical reactions may be more closely controlled in the vapor state. The apparatus for the description of gas mixtures, equilibrium between solids or liquids and their vapors and reactions in gases, is developed in later chapters. Applications in this section are limited to *unary gases*.

The *ideal gas model* provides a description of these gaseous systems that is adequate for most practical purposes. The *equation of state* for an ideal gas, that is, an algebraic equation that relates volume to temperature and pressure, was established by the early nineteenth century through experimental observation of the behavior of gases. This most important equation in the description of the behavior of gases is

$$PV = nRT \tag{4.54}$$

where P, V and T are as defined previously. R is the gas constant:

$$R = 8.314 \text{ (J/mol-K)} = 1.987 \text{ (cal/mol-K)}$$

$$= 0.08206 \text{ (liter-atm/mol-K)} = 82.06 \text{ (cc-atm/mol-K)} \tag{4.55}$$

and n is the number of moles of the gas in the system. If this equation is used to describe a system containing 1 mole of substance, then $n = 1$ and

$$PV = RT \tag{4.56}$$

where V can be interpreted as the **molar volume** of the gas (volume occupied by one mole).

The experimental variables can be computed for an ideal gas as follows. Apply the definition of the volume coefficient of thermal expansion, Eq. (4.11),

$$\alpha(\text{ideal gas}) = \frac{1}{V}\left(\frac{\partial V}{\partial T}\right)_P = \frac{1}{RT/P}\left[\frac{\partial}{\partial T}\left(\frac{RT}{P}\right)\right]_P = \left(\frac{P}{RT}\right)\left(\frac{R}{P}\right) = \frac{1}{T} \tag{4.57}$$

The coefficient of compressibility, Eq. (4.12), is

$$\beta(\text{ideal gas}) = -\frac{1}{V}\left(\frac{\partial V}{\partial P}\right)_T = -\frac{1}{RT/P}\left[\frac{\partial}{\partial P}\left(\frac{RT}{P}\right)\right]_T = -\frac{P}{RT}\left(-\frac{RT}{P^2}\right) = \frac{1}{P} \tag{4.58}$$

The kinetic theory of gases demonstrates that the heat capacity of an ideal gas is independent of temperature and pressure, but depends upon the number of atoms and their configuration in each gas molecule, (see Chap. 7). The relation between C_P and C_V expressed for a general system in Eq. (4.48) is particularly simple for an ideal gas since the factor

$$\frac{TV\alpha^2}{\beta} = T \cdot \left(\frac{RT}{P}\right) \cdot \left(\frac{1}{T}\right)^2 \cdot \left(\frac{1}{1/P}\right) = R$$

Thus, for an ideal gas,

$$C_V = C_P - R \tag{4.59}$$

For a *monatomic* gas (each gas molecule is just a single atom) it is found that

$$C_V \text{(monatomic ideal gas)} = \frac{3}{2}R \tag{4.60}$$

Note that C_V has the same units as does R. Insert this value into Eq. (4.59) to obtain the corresponding value for C_P:

$$C_P \text{(monatomic ideal gas)} = \frac{3}{2}R + R = \frac{5}{2}R \tag{4.61}$$

An important simplification in the thermodynamic behavior of an ideal gas can be obtained by considering the relation between internal energy and temperature and pressure, $U = U(T,P)$. From Table 4.5,

$$U = U(T, P) \qquad dU = (C_P - PV\alpha)dT + V(P\beta - T\alpha)dP \tag{4.41}$$

Apply Eqs. (4.57) and (4.58) to evaluate the coefficient of dP for an ideal gas:

$$V(P\beta - T\alpha) - V\left[P\left(\frac{1}{P}\right) - T\left(\frac{1}{T}\right)\right] = 0$$

The coefficient of dT can be simplified by evaluating α:

$$C_P - PV\alpha = C_P - PV\left(\frac{1}{T}\right) = C_P - R = C_V$$

Thus, for an ideal gas, Eq. (4.41) becomes

$$dU \text{(ideal gas)} = C_V dT + 0 \cdot dP = C_V dT$$

For any finite change in state from an initial condition given by (T_1, P_1) to a final state (T_2, P_2), the change in internal energy can be obtained by integrating this equation. This integration is simplified because C_V is a constant for an ideal gas. Thus, for any change in state of an ideal gas

$$\Delta U \text{(ideal gas)} = C_V(T_2 - T_1) \tag{4.62}$$

Thus, it is concluded that *the internal energy of an ideal gas is a function only of its temperature*. This results from the observation that the coefficient of dP in Eq. (4.41) is 0 but this is true only for ideal gases and for no other substances. The change in internal energy of an ideal gas for any process can, without exception, be computed from Eq. (4.62).

A similar result can be obtained for the enthalpy function. Recall Eq. (4.42) from Table 4.5.

$$H = H(T, P) \qquad dH = C_p dT + V(1 - T\alpha)dP \tag{4.42}$$

For an ideal gas, the coefficient of dP is found to be

$$V(1 - T\alpha) = V\left[1 - T\left(\frac{1}{T}\right)\right] = 0$$

and Eq. (4.42) becomes

$$dH(\text{ideal gas}) = C_P dT + 0 \cdot dP = C_P dT$$

Integration gives, for a finite change in state:

$$\Delta H(\text{ideal gas}) = C_P(T_2 - T_1) \tag{4.63}$$

from which it is seen that the change in enthalpy of an ideal gas for any process depends only upon the initial and final temperatures of the system and no other state function. This unusual result is valid only for an ideal gas because the value of the coefficient of dP is zero.

These results demonstrate that if an ideal gas is subject to any arbitrary process that takes it from state A to state B, the change in internal energy and enthalpy can be computed if the initial and final temperatures are either given or may be computed from the information given. This result holds even for irreversible processes since U and H are state functions.

The next series of examples (Examples 4.7–4.11) applies the general procedure and the experimental variables just evaluated to make thermodynamic calculations for processes described for a *monatomic ideal gas*.

Example 4.7. Compute the change in entropy when one mole of an ideal gas initially at 298 K and one atmosphere pressure is compressed isothermally to 1000 atm.

Identify the variables: $S = S(T, P)$ The differential form for this relationship is contained in Table 4.5 since the independent variables are T and P.

$$dS = \frac{C_p}{T} dT - V\alpha dP \tag{4.38}$$

For the problem at hand, $dT = 0$. The coefficient of dP can be evaluated by applying Eq. (4.57 and 4.58)

$$dS_T = -V\alpha dP_T = -\left(\frac{RT}{P}\right) \cdot \left(\frac{1}{T}\right) \cdot dP_T = -\frac{R}{P} dP_T$$

This is the change in entropy for each infinitesimal step in the process. The change in entropy for the whole process is the *integral* of dS_T:

$$\Delta S_T = \int_{P_1}^{P_2} \left(-\frac{R}{P}\right) dP = -R\ln P \Big]_{P_1}^{P_2} = -R[\ln P_2 - \ln P_1]$$

$$\Delta S_T(\text{ideal gas}) = -R\ln\frac{P_2}{P_1} \tag{4.64}$$

This expression gives the change in entropy when a monatomic ideal gas experiences a change in pressure at constant temperature. For the current example, $P_1 = 1$ atm and $P_2 = 1000$ atm, thus

$$\Delta S_T = -(8.314 \text{ J/mol-K}) \ln(1000/1) = -57.4 \text{ J/mol-K}$$

The next example introduces the notion of the *adiabatic* process. Imagine a system enclosed in a thermally insulating jacket. Then whatever processes occur in the system, no heat flows across its boundary. Processes that occur in such a system for which $\delta Q = 0$ are called *adiabatic*. If, in addition, the process considered is carried out reversibly then, since $\delta Q_{rev} = T dS$ and $\delta Q = 0$, it follows that $dS = 0$ for each infinitesimal step. Thus, *adiabatic, reversible processes are isentropic*, meaning that they occur at constant entropy. (This conclusion is not valid for an irreversible adiabatic process since entropy is produced within the system.)

Example 4.8. One mole of an ideal gas initially at temperature T_1 and occupying volume V_1 is compressed reversibly and adiabatically to a final volume V_2. Compute the final temperature, T_2 of the system.

Identify the variables: $T = T(S,V)$

This problem was treated previously for the general case in Example 4.6, yielding Eq. (4.52):

$$dT = \frac{T}{C_V}dS - \frac{T\alpha}{C_V\beta}dV \qquad (4.52)$$

Since the process under study is *isentropic*, $dS = 0$ and

$$dT_S = -\frac{T\alpha}{C_V\beta}dV_S \qquad (4.65)$$

For an ideal gas $\alpha = 1/T$ and $\beta = 1/P$:

$$dT_S = -\frac{T(1/T)}{C_V(1/P)}dV_S = \frac{P}{C_V}dV_S$$

Eliminate P through the equation of state:

$$dT_S = -\frac{(RT/V)}{C_V}dV_S$$

Separate the variables and integrate from the initial to the final state:

$$\frac{dT}{T} = -\frac{R}{C_V}\frac{dV}{V}$$

$$\ln\left(\frac{T_2}{T_1}\right) = -\frac{R}{C_V}\ln\left(\frac{V_2}{V_1}\right) = \ln\left(\frac{V_2}{V_1}\right)^{(-R/C_V)}$$

Two logarithms are equal if their arguments are equal:

$$\frac{T_2}{T_1} = \left(\frac{V_2}{V_1}\right)^{\left(-\frac{R}{C_V}\right)} = \left(\frac{V_1}{V_2}\right)^{(R/C_V)}$$

The final temperature at the end of the reversible adiabatic compression is

$$T_2 = T_1\left(\frac{V_1}{V_2}\right)^{(R/C_V)} \qquad (4.66)$$

Note that if V_2 is larger than V_1, T_2 is smaller than T_1, that is, if a gas expands adiabatically, its temperature drops. This phenomenon is the basis for some refrigeration cycles. Similarly, an adiabatic compression of the gas raises its temperature; thus a bicycle pump gets warm as the tire is inflated.

Equation (4.65) shows that this phenomenon is general, for all substances. The minus sign on the right-hand side of this equation implies that the temperature change and volume change are in the opposite direction for any reversible adiabatic process.

Equation (4.66), the adiabatic equation for an ideal gas, can also be expressed through relationships of T to P or P to V simply by applying the equation of state.

$$T_2 = T_1 \cdot \left[\frac{(RT_1/P_1)}{(RT_2/P_2)} \right]^{(R/C_V)}$$

Algebraic manipulation of this equation simplifies to:

$$T_2 = T_1 \left(\frac{P_2}{P_1} \right)^{(R/C_P)} \qquad (4.67)$$

which makes use of the previous result that $C_V = C_P - R$ for an ideal gas, Eq. (4.59). A similar substitution and manipulation yields the adiabatic relationship expressed in terms of P and V:

$$P_2 = P_1 \left(\frac{V_1}{V_2} \right)^{(C_P/C_V)} \qquad (4.68)$$

These three relationships all describe the same process; the reversible adiabatic compression of an ideal gas; they simply use different variables to describe the path.

The "free expansion" of an ideal gas is a favorite sample problem in every thermodynamics text. Picture a system, Fig. 4.3a, with an internal partition separating its volume into two equal parts. The partition contains a valve that can be opened as desired. The whole system is a rigid, thermally insulated box so that processes that occur within it are *adiabatic* and no work can be exchanged with the surroundings. Initially, the valve is closed and one side of the system is evacuated. The other contains one mole of an ideal gas at some temperature T_1. The valve is then opened and the gas expands freely into the evacuated side of the chamber, Fig. 4.3b.

Since the walls of the container do not move during this process, no work is done on the surroundings by the expanding gas, $\delta W = 0$ and since the system is adiabatic, no heat is exchanged with the surroundings, $\delta Q = 0$. During this process the system is *isolated* from its surroundings and no exchanges of any kind occur.

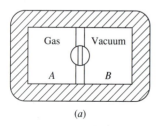

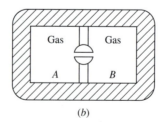

(a)   (b)

FIGURE 4.3
System illustrating the free expansion of an ideal gas. Initially the gas is contained in chamber A; chamber B is evacuated (a). The valve is opened and the gas expands irreversibly ("freely") to fill both chambers (b).

Since no transfers occur between system and its surroundings, the contribution to the entropy change due to exchanges with the surroundings is zero. Thus, whatever change in entropy occurs in the gas as it expands must be *entropy production* arising from the irreversible nature of the process that evidently occurs far from equilibrium. According to the second law this entropy production and thus the change in entropy of the system must be *positive*. This problem requires the computation of the entropy production for an irreversible process.

This irreversible process is complex. During the expansion, pressure and temperature gradients exist within the gas, heat and mass flows occur, and turbulent conditions may even arise. No information is supplied that would permit estimation of these complex details in the evolution of the system toward its final state. How then can the entropy produced be computed?

Recognize that entropy is a state function. Thus, the change in entropy for this complex irreversible process is exactly the same as that for the simplest reversible process that can be imagined that connects the initial and final states. The initial state is known and its volume and temperature (V_1, T_1) are specified. The volume of the final state is also known since the gas fills the whole container and since the two compartments have the same volume, $V_2 = 2V_1$. To complete the specification of the final state, its temperature must be determined.

Recall that the system is isolated during this process, $\delta Q = \delta W = 0$. Thus, according to the first law of thermodynamics, the internal energy of the system U cannot change during the process, and $\Delta U = 0$. Since the gas is ideal, the internal energy is determined by its temperature, Eq. (4.58). Since the internal energy of the initial and final states is the same, the final temperature must be equal to the initial temperature; the final temperature of the system is also T_1. Thus, the simplest process that can be visualized that connects the initial state (T_1, V_1) and final state $(T_1, V_2 = 2V_1)$ in this system is a reversible, isothermal expansion from V_1 to $2V_1$.

Example 4.9. Compute the change in entropy when one mole of an ideal gas expands freely to twice its volume.

Identify the variables: $S = S(T, V)$

Refer to Example 4.6, Eq. (4.52):

$$dS = \frac{C_V}{T} dT + \frac{\alpha}{\beta} dV \tag{4.52}$$

Set $dT = 0$ for the isothermal reversible process considered and evaluate α and β for an ideal gas:

$$dS_T = \frac{\alpha}{\beta} dV_T = \frac{1}{T} \frac{1}{(1/P)} dV_T = \frac{P}{T} dV_T = \frac{R}{V} dV_T$$

Integrate from the initial to the final state:

$$\Delta S_T = \int_{V_1}^{V_2} \frac{R}{V} dV = R \ln\left(\frac{V_2}{V_1}\right) = R \ln\left(\frac{2V_1}{V_1}\right)$$

$$\Delta S_T = R \ln 2 = 2.90 \text{ J/mole-k}$$

In this reversible, isothermal process, the entropy production is zero and the entropy change computed is completely due to heat transfer across the boundary of the system. In the irreversible process connecting the same two end states, illustrated in Fig. 4.3, the entropy change is completely due to entropy production in the system since there is no entropy transferred across the boundary. The change in entropy for the reversible isothermal process is identical to that for the irreversible free expansion in an isolated system because entropy is a state function and the initial and final states are the same. Thus, the entropy produced in the free expansion of an ideal gas to twice its initial volume is positive, as required by the second law, and is equal to +2.90 J/mole-K.

The heat absorbed and work done on a system can be computed for any reversible process by integrating $T dS$ or $-P dV$ along the path specified by the sequence of states that defines the process. In the general case this integration is a *line integral* along a curve that mathematically describes the process. This integration is simplified if one of the variables describing the state of the system is held constant. Example 4.10 illustrates this application of the general procedure.

Example 4.10. Derive expressions for the heat absorbed by the system for each of the following classes of reversible processes for one mole of an ideal gas:

Isothermal change in pressure
Isobaric change in volume
Isochoric (constant volume) change in temperature

Since the variables sought are process variables, the paths must be specified. Cases 1, 2, and 3 specify paths for three different processes. The heat absorbed in each case is evaluated through $\delta Q_{\text{rev}} = T dS$.

Case 1.

Relationship required: $S = S(T, P)$, Eq. (4.38)

$$dS = \frac{C_P}{T} dT - V \alpha dP$$

Process is isothermal: set $dT = 0$: $dS_T = -V \alpha dP_T$

$$dS_T = -\left(\frac{RT}{P}\right)\left(\frac{1}{T}\right) dP_T = -\frac{R}{P} dP_T$$

The heat absorbed:

$$\delta Q_{\text{rev},T} = T \, dS_T = -\frac{RT}{P} dP_T$$

Integrate:

$$Q_{\text{rev},T} = -RT \ln\left(\frac{P_2}{P_1}\right) \tag{4.69}$$

Case 2.

Relationship required: $S = S(P, V)$, Eq. (4.49)

$$dS = \left(\frac{C_P \beta}{T\alpha} - V\alpha\right) dP + \left(\frac{C_P}{TV\alpha}\right) dV \qquad (4.49)$$

Isobaric change; set $dP = 0$:

$$dS_P = \frac{C_P}{TV\alpha} dV_P$$

Ideal Gas: $(T\alpha = 1)$

$$dS_P = \frac{C_P}{V} dV_P$$

Heat absorbed:

$$\delta Q_{\text{rev},P} = T dS_P = T\frac{C_P}{V} dV_P$$

Substitute for $T = PV/R$:

$$\delta Q_{\text{rev},P} = \left(\frac{PV}{R}\right) \frac{C_P}{V} dV_P = P\frac{C_P}{R} dV_P$$

Integrate; (P is constant):

$$Q_{\text{rev},P} = P\frac{C_P}{R}(V_2 - V_1) \qquad (4.70)$$

Note that this result could also be expressed as

$$Q_{\text{rev},P} = C_P(T_2 - T_1)$$

since $PV_2/R = T_2$ and $PV_1/R = T_1$. This is consistent with the definition of heat capacity at constant pressure, where C_P is not a function of temperature.

Case 3.

Relationship required: $S = S(T,V)$, Eq. (4.52).

$$dS = \frac{C_V}{T} dT + \frac{\alpha}{\beta} dV \qquad (4.52)$$

Isochoric: set $dV = 0$:

$$dS_V = \frac{C_V}{T} dT_V$$

The heat absorbed:

$$\delta Q_{\text{rev},V} = T dS_V = T\frac{C_V}{T} dT_V = C_V dT_V$$

Integration gives:

$$\delta Q_{\text{rev},V} = C_V(T_2 - T_1)$$

which is consistent with the definition of heat capacity at constant volume, where C_V is not a function of temperature. This result could also be derived by recognizing that, since for an isochoric process $dV = 0$, and $\delta W = 0$, so that $\delta Q = dU_V$. Integration gives the result: $\Delta U_V = Q$. According to Eq. (4.62), since the internal energy of an ideal gas depends only upon its temperature, for *any* process, $\Delta U = C_V(T_2 - T_1)$.

Example 4.11. One mole of an ideal gas, initially at 273 K and one atmosphere, is contained in a chamber that permits programmed control of its state. Controlled quantities of heat and work are supplied to the system so that its pressure and volume change along a line given by the equation

$$V = 22.4 \left(\frac{\text{liter}}{\text{atm}} \right) P$$

Assume the process is carried out reversibly. Compute the heat required to be supplied to the system to take it to a final pressure of 0.5 atm. What is the final temperature of the gas? The independent variables are P and V; information is needed about S since the heat absorbed is $T dS$. Thus, the relation required is $S = S(P, V)$. From Example 4.2, Eq. (4.49),

$$dS = \left(\frac{C_P \beta}{T \alpha} - V \alpha \right) dP + \frac{C_P}{T V \alpha} dV \qquad (4.49)$$

For an ideal gas, evaluation of α and β yields

$$dS = \frac{C_P}{V} dV + \frac{C_V}{P} dP$$

In order to compute the change in entropy for an incremental step along the specified path, it is necessary to use the equation describing the path to express one independent variable in terms of the other. From the specified path,

$$V = 22.4 \cdot P \qquad \text{and} \qquad dV = 22.4 \cdot dP$$

Use these expressions for V and dV in the entropy relation just obtained to yield an expression for an incremental change in entropy along the controlled path,

$$dS(\text{this path}) = \frac{C_P}{(22.4 \cdot P)} (22.4 \cdot dP) + \frac{C_V}{P} dP$$

$$= (C_P + C_P) \frac{1}{P} dP$$

The heat absorbed for this step is

$$\delta Q_{\text{rev}} = T dS = T(C_P + C_V) \frac{1}{P} dP$$

In order to integrate this equation to obtain the total heat absorbed, it is necessary to relate T to P along this path. Evidently this requires combination of the equation of state with the equation for the path:

$$T = \frac{PV}{R} = \frac{P}{R} (22.4 \cdot P) = 22.4 \cdot \frac{P^2}{R}$$

Thus, for the given path the temperature of the system varies with the square of the pressure. Substitute:

$$\delta Q_{\text{rev}} = \left(22.4 \cdot \frac{P^2}{R}\right)(C_P + C_V)\frac{1}{P}dP$$

$$= 22.4 \cdot (C_P + C_V) \cdot \frac{P}{R}dP$$

Integration gives the heat absorbed

$$Q_{\text{rev}} = \int_{P_1}^{P_2} 22.4 \left(\frac{C_P + C_V}{R}\right) P\,dP$$

$$= 22.4 \left(\frac{C_P + C_V}{R}\right)\frac{P^2}{2}\Bigg]_{P_1}^{P_2}$$

Insert values for C_V and C_P for a monatomic gas, together with the initial and final pressures for the process,

$$Q_{\text{rev}} = -22.4\left(\frac{\frac{5}{2}R + \frac{3}{2}R}{R}\right) \cdot \frac{[(0.5)^2 - (1.0)^2]}{2} = 22.4 \cdot 4 \cdot (-0.75)$$

$$Q_{\text{rev}} = -67.2 \text{ liter-atm} = -6810 \text{ J}$$

The final temperature of the system is the value on the path that corresponds to 10 atmospheres of pressure:

$$T = (22.4 \text{liter/atm}) \cdot [(0.5\text{atm})^2/2]/[0.08206 \text{ liter-atm/mol-K}]$$

$$T = 34.1 \text{ K}$$

The final volume is $V = 22.4 \times 0.5 = 11.2$ liters. Evidently in order to lower *both* the pressure and the volume simultaneously, it is necessary to extract a significant quantity of heat from the system.

4.3.5   Applications to Solids and Liquids

Tables 4.1–4.3 survey the ranges of experimental variables values that are typical for solid and liquids. It is observed that, while α, β, and C_P are functions of temperature, pressure and composition in general, for a given material, these quantities are not *strong* functions of the state of the system. For many applications that do not involve the gas phase, particularly when only an *estimate* of the thermodynamic properties is sought, the assumption that α, β, and C_p are constants is a useful approximation. Where precise computations are required, the dependence of these quantities upon T and P must be obtained from the literature or experimentally determined. The examples presented in this section illustrate both precise and approximate calculations for selected thermodynamic state functions so that the magnitude of the errors introduced by these approximations can be assessed.

Example 4.12. One mole of nickel initially at 300 K and one atmosphere pressure is taken through two separate processes:

An isobaric change in temperature to 1000 K

An isothermal compression to 1000 atm

Compare the change in enthalpy of the nickel for these two processes. The experimental variables for nickel are

$$C_P = 16.99 + 2.95 \times 10^{-2}T \, (\text{J/mole-K}) \qquad V(3000 \text{ K, 1atm}) = 6.57 (\text{cc/mole})$$

$$\alpha = 40 \times 10^{-6} \left(\frac{1}{\text{K}} \right) \qquad \beta = 1.5 \times 10^{-6} \left(\frac{1}{\text{atm}} \right)$$

Identify the variables: $H = H(T,P)$

Obtain the relationship [Eq. (4.42) from Table 4.5]

$$dH = C_P dT + V(1 - T\alpha)dP \tag{4.52}$$

For *process 1*, dP = 0 and the change in enthalpy is

$$\Delta H = \int_{300K}^{1000K} (16.99 + 2.59 \times 10^{-2}T)dT$$

$$= \left[16.99 \cdot T + 2.95 \times 10^{-2} \frac{T^2}{2} \right]_{300K}^{1000K}$$

$$\Delta H = 11{,}893 + 13{,}423 = 25{,}316 (\text{J/mole})$$

The second term on the right-hand side of this equation arises from the temperature dependent contribution to the heat capacity. The first term represents the estimate of ΔH for the process if this temperature dependent term were ignored. Evidently a significant error would be introduced if the temperature dependent term were ignored.

For *Process 2, dT* = 0 and the change in enthalpy is

$$\Delta H = \int_{1}^{1000} V(1 - T\alpha)dP$$

If V and α are approximated as constants,

$$\Delta H = V(1 - T\alpha)(P_2 - P_1)$$

$$\Delta H = 6.57 \frac{\text{cc}}{\text{mole}} \cdot \left[1 - 40 \times 10^{-6} \frac{1}{\text{K}} \cdot 300 \text{ K} \right] \cdot [1000 - 1]\text{atm} \cdot \frac{8.314 \text{ J}}{82.06 \text{ cc-atm}}$$

$$\Delta H = 649.1 \text{ J/mole}$$

Since α typically has values in the range of $10^{-5}(1/\text{K})$, the product $T\alpha$ is in general small in comparison to 1. Thus, any second-order contributions to α, for example, due to its variation with pressure, is certainly negligible and the factor $[1 - T\alpha]$ may be treated as a constant without significant error. The pressure dependence of the molar volume can be expressed to a first approximation as

$$V(P) = V(P_1)[1 + \beta(P - P_1)]$$

Substitute this result for V in the equation for

$$dH_T = V(P_1)[1 + \beta(P - P_1)] \cdot [1 - T\alpha] \cdot dP_T$$

Integrate over the pressure range:

$$\Delta H = \int_{P_1}^{P_2} V(P_1)[1 + \beta(P - P_1)] \cdot (1 - T\alpha) \cdot dP$$

$$\Delta H = V(P_1)(1 - T\alpha) \cdot \left[P + \beta \frac{(P - P_1)^2}{2} \right]_{P_1}^{P_2}$$

$$\Delta H = V(P_1)[1 - T\alpha] \left[(P_2 - P_1) + \beta \frac{(P_2 - P_1)^2}{2} \right]$$

Insert the numerical values for the problem at hand:

$$\Delta H = (649.1 + 0.5) \text{ J/mole}$$

where the second term can be traced to the assumed pressure variation of the molar volume and the first term is identical to that computed approximately, earlier in this example. It will be noted that the second term is significant only if the pressure change in the system is extremely large, say 100,000 atm. Thus, for most processes, the assumption that V does not change with pressure yields an approximate result that is within the experimental error of measurement of the precise result.

   Comparison of the enthalpy changes for processes 1 and 2 in this example reveals another point that has broad application in the behavior of solids and liquids. Process 1 suggests a change of a factor of 3 in the absolute temperature; in process 2 the pressure is changed by a factor of 1000. The energy change for process 1, here reported as the enthalpy, is nonetheless about 25 times larger than for process 2. *Thus, for solids and liquids, energy changes associated with thermal influences tend to be much larger than those arising from mechanical influences.*

Example 4.13. Compute the change in Gibbs free energy of magnesia (MgO) when one mole is heated from 298 K (room temperature) to 1300 K at one atmosphere pressure.
   The properties of MgO obtained from reference [4.1] are:

$$C_P = 45.44 + 5.01 \times 10^{-3} T \quad \text{(J/mole-K)}$$

$$S_{298}^{\circ} = 26.8 \quad \text{(J/mole-k)}$$

where S_{298}° is the absolute entropy of MgO at 298 K, obtained by integrating heat capacity data from 0 K to 298 K. The variables in the problem:

$$G = G(T, P)$$

From Table 4.5:

$$dG = -SdT + VdP \tag{4.10}$$

In this problem,

$$dP = 0 \qquad dG_P = -SdT_P$$

The temperature dependence of S cannot be neglected in this case. Thus, to integrate this expression, it is first necessary to develop an expression for the variation of entropy with temperature at constant pressure. Recall again Eq. (4.38):

$$dS = \frac{C_P}{T} dT - V\alpha dP \tag{4.38}$$

The condition $dP = 0$ gives:

$$dP = 0 \qquad dS_P = \frac{C_P(T)}{T} dT_P$$

Insert the expression for the temperature dependence of the heat capacity and integrate from 298 K to a variable temperature T;

$$\int_{298K}^{T} dS = S(T) - S(298\ K) = \int_{298K}^{T} \frac{a + bT}{T} dT$$

$$S(T) = S(298) + \left[a \ln T + bT \right]_{298K}^{T}$$

$$S(T) = a \ln T + bT + C_1$$

where $C_1 = S_{298}^{\circ} - [a \ln(298) + b(298)]$ and a and b are the coefficients in the empirical formula for the heat capacity of MgO. Next, insert this result into the equation for Gibbs free energy and integrate again:

$$\int_{298K}^{T} dG = G(T) - G(298) = \int_{298K}^{T} -S(T) dT_P$$

$$G(T) - G(298) = - \int_{298K}^{T} [a \ln T + bT + C_1] dT$$

$$= \left[a(T \ln T - T) + \frac{b}{2} T^2 + C_1 T \right]_{298K}^{T}$$

$$G(T) - G(298) = A(T \ln T) + BT + CT^2 + D$$

$$A = a \qquad B = C_1 - a \qquad C = \frac{b}{2} \qquad \text{and}$$

$$D = a(298 \ln 298) + b \frac{(298)^2}{2} + C_1 298$$

For MgO, the values of these quantities are:

$$A = 45.44 \qquad B = -279.01 \qquad C = 2.51 \times 10^{-3} \qquad D = 7,764$$

Thus, the temperature dependence of the Gibbs free energy of MgO at one atmosphere pressure is given by:

$$G(T) - G(298) - 45.44(T \ln T) - 279.01T + 2.51 \times 10^{-3}T^2 + 7,764$$

This functional relationship is presented in Figure 4.4.

Example 4.14. One mole of copper initially at 700 K and one atmosphere is contained in a thermally insulated jacket. The system is compressed reversibly to a pressure of 10,000 atm. Compute the change in temperature for this reversible adiabatic process.
   Identify the variables: $T = T(S, P)$
   Rearrange (because the independent variable is T)

$$S = S(T, P)$$

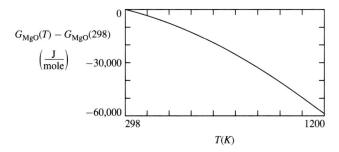

FIGURE 4.4
Variation of Gibbs free energy of MgO with temperature at one atmosphere pressure.

Recall this relation from Table 4.5:

$$dS = \frac{C_P}{T}dT - V\alpha dP \qquad (4.38)$$

Solve for dT:

$$dT = \frac{T}{C_P}dS + \frac{TV\alpha}{C_P}dP$$

Set $dS = 0$:

$$dT_S = \frac{TV\alpha}{C_P}dP_S$$

Assuming V, α, and C_p are constants in the problem, separate the variables:

$$\frac{dT}{T} = \frac{V\alpha}{C_P}dP$$

Integrate:

$$\ln\left(\frac{T_2}{T_1}\right) = \frac{V\alpha}{C_P}(P_2 - P_1)$$

Substitute the properties of copper and compute the result:

$$T_2 = 700 \text{ K}\left[\exp\left(-\frac{7.11 \text{ cc/mol} \cdot 49.3 \times 10^{-6}K^{-1}}{27.0 \text{ J/mol-K}} \cdot (10000 - 1) \cdot \left(\frac{8.314 \text{ J}}{82.06 \text{ cc-atm}}\right)\right)\right]$$

$$T_2 = 691 \text{ K}$$

Example 4.15. This example presents an analog to the illustration of the free expansion of an ideal gas presented in Ex. 4.9. One mole of pure iron is initially at 300 K and 100,000 atm and the system is contained in a thermally insulating jacket. The pressure is suddenly released and the iron expands irreversibly to a final pressure of one atmosphere. Compute the entropy produced in this irreversible process.

Since for an elastic solid the volume change is very small, even under extreme hydrostatic pressures such as that considered here, it can be assumed that the work done on the system during this expansion is negligible. Since the process is also adiabatic, by the first law the internal energy of the system is constant. Further, since there is essentially no exchange with the surroundings, the entropy flow across the boundary of the system is also negligible. Thus, the change in entropy for the process is virtually all due to entropy produced inside the system. By the second law, this entropy production must be positive.

The entropy production in this irreversible process can be computed by again applying the familiar strategy. Visualize the simplest reversible process that connects the same initial and final states. Since entropy is a state function, the change for that process is the same as that for the irreversible process in the problem. Thus, it is necessary to compute the change in entropy associated with a reversible decompression of one mole of iron at *constant internal energy*.

1. Identify the variables: $S = S(U,P)$
2. Write the differential form: $dS = MdU + NdP$
3. Use Eq. (4.41) in Table 4.5 to convert dU to a function of T and P:

$$dS = M[(C_P - PV\alpha)dT + V(P\beta - T\alpha)dP] + NdP$$

4. Collect terms:

$$dS = M(C_P - PV\alpha)dT + [MV(P\beta - T\alpha) + N]dP$$

5. Recall Eq. (4.38) from Table 4.5:

$$dS = \frac{C_P}{T}dT - V\alpha dP \qquad (4.38)$$

6. Compare coefficients:

$$M(C_P - PV\alpha) = \frac{C_P}{T}$$

$$MV(P\beta - T\alpha) + N = -V\alpha$$

7. Solve for M and N:

$$M = \left(\frac{\partial S}{\partial U}\right)_P = \frac{C_P}{T(C_P - PV\alpha)}$$

$$N = \left(\frac{\partial S}{\partial P}\right)_U = -\frac{PV(C_P\beta - VT\alpha^2)}{T(C_P - PV\alpha)}$$

The required relation is:

$$dS = \frac{C_P}{T(C_P - PV\alpha)}dU - \frac{PV(C_P\beta - TV\alpha^2)}{T(C_P - PV\alpha)}dP$$

In the problem at hand, $dU = 0$ and the change in entropy is

$$dS_U = -\frac{PV(C_P\beta - VT\alpha^2)}{T(C_P - PV\alpha)}dP_U$$

Integration of this equation requires a knowledge of the dependence upon pressure of all the experimental variables contained in the coefficient of dP_U. However, examination of the numerical values of these quantities shows that the term $PV\alpha$ is small in comparison with C_P, even at the start of the process with $P = 100,000$ atm. The factor in the numerator, $TV\alpha^2$, is even smaller. Thus, to a very good approximation,

$$dS_U = -\frac{V}{T}dP_U$$

Computation of the change in temperature [based upon $T = T(U,P)$] reveals that the temperature may decrease by about 10 K in this process. This change is small compared to the value of the temperature for the system. If it also is neglected, the entropy produced can be calculated approximately to be:

$$\Delta S = -\left[\left(7\frac{cc}{mol}\right)300 \text{ K}\right] \cdot (1 - 100,000)(atm)$$

$$\Delta S = +2300 \left(\frac{cc\text{-}atm}{mol\text{-}K}\right) = +220 \left(\frac{J}{mol\text{-}K}\right)$$

4.4   SUMMARY OF CHAPTER 4

Thermodynamic relationships may be usefully classified as:

1. The laws of thermodynamics
2. Definitions
3. Coefficient relations
4. Maxwell relations
5. Conditions for equilibrium

The laws of thermodynamics form the physical basis for solving practical problems. Additional energy functions are defined:

$$H \equiv U + PV \qquad F \equiv U - TS \qquad G \equiv U + PV - TS$$

because they provide measures of energy convenient for specific classes of processes. Alternate forms of the combined statements of the first and second laws may be derived for these energy functions:

$$dU = TdS - PdV + \delta W' \tag{4.4}$$

$$dH = TdS + VdP + \delta W' \tag{4.6}$$

$$dF = -SdT - PdV + \delta W' \tag{4.8}$$

$$dG = -SdT + VdP + \delta W' \tag{4.10}$$

A collection of experimental variables are defined:

$$\alpha \equiv \frac{1}{V}\left(\frac{\partial V}{\partial T}\right)_P \tag{4.11}$$

$$\beta \equiv -\frac{1}{V}\left(\frac{\partial V}{\partial P}\right)_T \tag{4.12}$$

$$\delta Q_{rev,P} \equiv C_P dT_P \tag{4.13}$$

$$\delta Q_{\text{rev},V} \equiv C_V dT_V \tag{4.15}$$

These variables provide complete information about the thermodynamic behavior of systems for which $\delta W' = 0$; no other empirical information is required.

Coefficient and Maxwell relations derive from the mathematical properties of differentials of functions of several variables. For any thermodynamic relation between differentials of state functions, these relations may be read by inspection of the equation:

$$Z = Z(X, Y) \tag{4.16}$$

$$dZ = MdX + NdY \tag{4.17}$$

$$M \left(\frac{\partial Z}{\partial X} \right)_Y \qquad N = \left(\frac{\partial Z}{\partial Y} \right)_X \tag{4.18}$$

$$\left(\frac{\partial M}{\partial Y} \right)_X = \left(\frac{\partial N}{\partial X} \right)_Y \tag{4.25}$$

A general, rigorous procedure provides an algorithm for deriving the thermodynamic relationship between any dependent variable Z and any pair of independent variables (X,Y). This procedure begins with the derivation of functional relationships for every other state function in terms of independent variables T and P. Expressions for the energy functions are easily derived once relationships for $V = V(T,P)$ and $S = S(T,P)$ are obtained:

$$dV = V\alpha dT - V\beta dP \tag{4.32}$$

$$dS = \frac{C_P}{T} dT - V\alpha dP \tag{4.38}$$

The dependence of the energy functions upon T and P may be derived by inserting these expressions for dS and dV into the combined statements of the first and second laws summarized at the beginning of this summary:

$$U = U(T, P) \qquad dU = (C_P - PV\alpha)dT + V(P\beta - T\alpha)dP \tag{4.41}$$

$$H = H(T, P) \qquad dH = C_P dT + V(1 - T\alpha)dP \tag{4.42}$$

$$F = F(T, P) \qquad dF = -(S + PV\alpha)dT - PV\beta dP \tag{4.43}$$

$$G = G(T, P) \qquad dG = -SdT + VdP \tag{4.11}$$

These key equations are summarized in Table 4.5. For any given problem with arbitrary variables $Z = Z(X,Y)$ with differential form $dZ = MdX + NdY,$ the general procedure uses these equations to convert dX and dY to differential forms involving dT and dP. Then dZ is written as a function of dT and dP. The resulting pair of equations for dZ are alternate expressions for $Z = Z(T,P)$. Corresponding coefficients must therefore be equal. Equating corresponding coefficients yields two simultaneous linear equations in M and N. Evaluation of M and N in terms of experimental variables completes the derivation of the required relationship.

To obtain a numerical result for a given practical problem that seeks the change in Z for initial and final states specified by values of X and Y, it is necessary to obtain experessions for the experimental variables in M and N and integrate, first over dX, then over dY. If the problem at hand requires the evaluation of a process variable involving dZ (either $\delta Q = T dS$ or $\delta W = -P dV$), it is necessary to carry out the integration along the path that must be described explicitly as a relationship between X and Y.

A variety of examples illustrate the procedure. It is demonstrated that its application is particularly simple for an ideal gas, defined by the equation of state:

$$PV = RT \tag{4.56}$$

and a fixed value of heat capacity:

$$C_V = \frac{3}{2}R \qquad C_P = C_V + R = \frac{5}{2}R$$

for a monatomic gas. These examples demonstrate that the internal energy and the enthalpy of an ideal gas are each a function only of temperature:

$$\Delta U = C_V(T_2 - T_1) \tag{4.62}$$

$$\Delta H = C_P(T_2 - T_1) \tag{4.63}$$

A final set of examples applied to condensed phases demonstrates the generality of the procedure.

PROBLEMS

4.1. Write out the combined statements of the first and second laws for the energy functions $U = U(S,V)$, $H = H(S,P)$, $F = F(T,V)$ and $G = G(T,P)$. Assume $\delta W'$ is zero.
(a) Write out all eight coefficient relations
(b) Derive all four Maxwell relations for these equations.

4.2. Derive the ratio relation, Eq. (4.30):

$$\left(\frac{\partial Z}{\partial X}\right)_Y \left(\frac{\partial X}{\partial Y}\right)_Z \left(\frac{\partial Y}{\partial Z}\right)_X = -1$$

[*Hint*: Begin with the differential form of $Z = Z(X,Y)$; solve for dX; write the differential form of $X = X(Y,Z)$ and compare coefficients.]

4.3. The molar volume of Al_2O_3 at 25°C and 1 atm is 25.715 cc/mole. Its coefficient of thermal expansion is 26×10^{-6} K^{-1} and the coefficient of compressibility is 8.0×10^{-6} atm^{-1}. Estimate the molar volume of Al_2O_3 at 400°C and 10 kilobars pressure (10×10^3 atm).

4.4. Compare the entropy changes for the following processes:
(a) One gram atom of nickel is heated at one atm from 300 K to 1300 K.
(b) One gram atom of nickel at 300 K is compressed isothermally from 1 atm to 100 kbars.
(c) One mole of zirconia is heated at one atmosphere from 300 K to 1300 K.
(d) One mole of zirconia at 300 K is isothermally compressed from 1 atm to 100 kbars.
(e) One mole of oxygen is heated at one atmosphere from 300 K to 1300 K.
(f) One mole of oxygen at 300 K is isothermally compressed from 1 atm to 100 kbars.

What general qualitative conclusions do you draw from these calculations?

4.5. Express the results obtained for (c) and (d) of Problem 4.4 in values *per gram atom* of ZrO_2. (Each mole of zirconia contains three gram atoms of its elements; one gram atom of zirconium and two of oxygen.) How does this result influence the conclusions you made in the comparisons in Prob. 4.4?

Use the *general procedure for deriving relationships in thermodynamics* to solve the remaining problems in this chapter. Begin by identifying explicitly the dependent and independent variables in each case.

4.6. Compute the change in internal energy when 12 liters of argon gas at 273 K and 1 atm is compressed to 6 liters with the final pressure equal to 10 atm. Solve this problem in two different ways.

(a) Apply the general procedure to evaluate $U = U(P,V)$ for an ideal gas and integrate from initial to final (P,V).

(b) Use the information given to compute the final temperature of the gas and apply the general relation $\delta U = C_V \Delta T$.

Assume $C_V = (3/2)R$ for this monatomic gas.

4.7. Compute and plot the surface that represents the relationship for the entropy of nitrogen gas as a function of temperature and pressure in the (P,T) range from (1 atm, 300 K) to (10 atm, 1000 K). Since nitrogen is diatomic, $C_P = (7/2)R$. Assume it obeys the ideal gas law in this domain.

4.8. Compute and plot the surfaces that represent the variation with pressure and volume of

(a) the internal energy of one mole of nitrogen gas.

(b) the enthalpy of one mole of nitrogen gas.

Cover the range in (P,V) space from (1 atm, 22.4 liters) to (10 atm, 8.2 liters). Plot constant energy lines on the surfaces.

4.9. Evaluate the partial derivative

$$\left(\frac{\partial H}{\partial G}\right)_S$$

in terms of experimental variables.

4.10. Derive the relationship that describes the dependence of Helmholtz free energy upon entropy and temperature. Design an experiment that would require this relationship in analyzing the results.

4.11. Estimate the pressure required to keep a sample of polyvinyl chloride from expanding as it is heated from room temperature to 100°C.

4.12. A system permits continuous programmed control of the pressure and volume of the gas that it contains. The system is filled with one gram atom of helium and brought to an initial condition of one atmosphere and 18 liters. It is then reversibly compressed to 12 liters along a programmed path given by the relationship

$$V = 2P^2 + 20$$

where P is in atmospheres and V is in liters. Compute:

(a) The initial and final temperature of the system.

(b) The heat absorbed by the system.

(c) The work done by the system.

(d) The changes in U, H, F, G and S.

4.13. Estimate the pressure increase required to impart one Joule of mechanical work in reversibly compressing one mole of titanium at room temperature. What pressure rise is required to impart one Joule of work to one mole of alumina at room temperature?

4.14. Compare the numerical values of the coefficient

$$\left(\frac{\partial F}{\partial S} \right)_V$$

for a monatomic ideal gas and for pure iron.

4.15. Compute and plot the Gibbs free energy of polyethylene in the temperature range from 298 K to 398 K.

4.16. Compute and plot the surface representing the Gibbs free energy of oxygen gas as a function of temperature and pressure in the range from (298 K, 10^{-10} atm) to (1000 K, 100 atm). Use (298 K, 1 atm) as the zero point for the calculation. The absolute entropy of O_2 at 298 K and 1 atm is 205 (J/mol-K); assume that $C_P = 34.6$ (J/mol-K) and is independent of P and T.

4.17. Use a mathematics applications program, spreadsheet, or computer language to program and plot the generic equations for computing the temperature dependence of enthalpy, entropy, and Gibbs free energy as a function of temperature at one atmosphere pressure. Assume as input the absolute entropy, of the substance at 298 K and values of a, b, and c in the empirical heat capacity expression

$$C_P = a + bT + \frac{c}{T^2}$$

Use the program to compute and plot H, S, and G (relative to their values at 298 K) as functions of temperature at one atmosphere for

(a) Argon
(b) Titanium
(c) TiO_2

4.18. In an attempt to produce barium tungstate ($BaWO_4$) by autocatalytic conversion, a mixture of one mole each of BaO_2 and WO_2 is heated to 1100 K, at which point ignition occurs and the reaction

$$BaO_2 + WO_2 = BaWO_4$$

goes spontaneously to completion. Assuming that the system is enclosed so that heat losses are negligible, make a rough estimate of the temperature rise that might accompany this reaction.

4.19. Using the bounding values in Tables 4.1 to 4.3, estimate the range of the magnitude of the difference between C_P and C_V for condensed phases.

REFERENCES

1. *Handbook of Physics and Chemistry*, CRC Press, Cleveland, Ohio (Revised annually).
2. *JANAF (Joint Army-Navy-Air Force) Thermochemical Tables*, D. R. Stull, dir., Clearinghouse for Technical and Scientific Information, Springfield, Va, 1965. Addenda, supplements and updates published periodically.
3. Hultgren, R., R. L. Orr, P. D. Anderson, and K. K. Kelley: *Selected Values of Thermodynamic Properties of Metals and Alloys*, John Wiley & Sons, New York, N.Y. 1963.
4. *Thermophysical Properties Research Center Data Book*, Thermophysical Properties Research Center, Purdue University, West Layfayette, Ind. Data Base.

5. *TRCTHERMO*, Thermodynamic Research Center, Texas Engineering Experiment Station, The Texas A & M University System, College Station, Texas. Data Base.

6. Bales, C. W., A. D. Pelton, and W. T. Thompson, developers: *FACT (Facility for the Analysis of Chemical Thermodynamics)*, McGill University, Montreal, Quebec, Canada. Data Base.

7. Kelley, K. K., *Contributions to the Data on Theoretical Metallurgy, vols. I–XV*, U.S. Bureau of Mines, 1932–1962. Valuable compilations are long out of print, but may be available in libraries.

CHAPTER

5

EQUILIBRIUM IN
THERMODYNAMIC
SYSTEMS

This chapter introduces a general principle that is the basis for determining the internal condition of the most complex kind of thermodynamic system at its equilibrium state. This *general criterion for equilibrium* has the same level of importance in the development of thermodynamics as do the laws themselves. It is the foundation that provides the basis for the calculation of phase diagrams, chemical equilibria, predominance diagrams without and with electrical effects, the role played by capillarity effects, the chemistry of defects in crystals, and the primary results of statistical thermodynamics. These applications of this criterion occupy most of the rest of this text.

Some intuitive notions of equilibrium are first presented. These ideas are then formalized providing a thermodynamic statement of the criterion that determines when a system isolated from its surroundings during its approach to equilibrium has attained its equilibrium state:

In an isolated system, the entropy is a maximum at equilibrium.

This thermodynamic *extremum principle* (the entropy has an extreme value) is then formulated mathematically. A set of equations, called the *conditions for equilibrium,* that describe the relationships the internal properties of the isolated system must have when it achieves the equilibrium state are derived from this extremum principle. It is then demonstrated that although these *conditions for equilibrium* are derived for an isolated system, they are valid for *any* system at equilibrium whether or not it was isolated during its approach to that final condition. The strategy for obtaining these conditions for equilibrium is illustrated in this chapter by applying it to the simplest case; equilibrium in a unary, two phase system.

This strategy is applied repeatedly in subsequent chapters to derive the conditions for equilibrium in systems of progressively increasing sophistication. These equations are the basis for making practical calculations about the final resting condition of complex systems.

The chapter ends with a collection of alternate statements of the criterion for equilibrium, each of which is useful in the description of systems with specific external constraints.

5.1   INTUITIVE NOTIONS OF EQUILIBRIUM

The idea that a system changes toward some final condition and, once there, remains in that condition unless acted upon by some external agent is a familiar one. Such a system is said to come to equilibrium or approach its equilibrium state.

A simple application of this idea is illustrated in Fig. 5.1. The system acted upon by the set of mechanical forces comes to equilibrium when it arrays itself with respect to the forces so that they are in balance. The mathematical statement that describes this condition is particularly simple for this mechanical system. The vector sum of the forces acting on the point P is zero:

$$\sum_i \vec{F_i} = 0 \qquad\qquad (5.1)$$

which implies:

$$\sum_i F_{ix} = 0 \qquad \sum_i F_{iy} = 0 \qquad \sum_i F_{iz} = 0 \qquad\qquad (5.2)$$

where the subscripts x, y, and z represent components of the force vectors resolved in some orthogonal coordinate system and the summation over the index i is taken

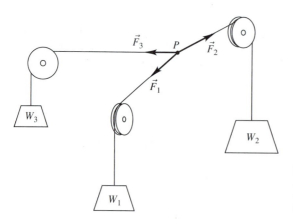

FIGURE 5.1
A simple mechanical system in equilibrium. The vectoral sum of the forces is zero.

over all the forces acting on the body. Once the system has found the position O that achieves this balance, it will remain fixed in that position until one of the weights is changed. Furthermore, if the system is displaced from its equilibrium position such that the body is moved to any position in the neighborhood of O, it will return to O.

In thermodynamics the influences that operate to modify the condition of a system are more general than mechanical forces. Nonetheless, the intuitive notion of an equilibrium condition also applies to such systems. This idea of an equilibrium state has two components:

1. It is a state of rest.
2. It is a state of balance.

The first component means that the condition of the system, no matter what it might be, is *time independent*. The system has achieved a *stationary state*. No changes can occur in a system that has come to equilibrium except by the action of influences that originate outside the system/surroundings complex. The second component to the concept means that if the system is perturbed from its equilibrium condition by some outside influence, it will return again to the same condition when it again comes to rest.

5.2 THERMODYNAMIC FORMULATION OF A GENERAL CRITERION FOR EQUILIBRIUM

Systems can experience two distinct classes of conditions that are time invariant, that is, two classes of stationary states. Either the system is in an *equilibrium state* or it has achieved a *steady state*.

A simple example of a steady state is shown in Fig. 5.2. A copper rod is surrounded by an insulating jacket except at its ends. One end is placed in contact with a furnace maintained at temperature T_1 and the other end contacts a water-cooled plate maintained at temperature T_2. Heat flows through the rod from left to right and the temperature profile evolves with time. Eventually, because the external conditions are fixed, the temperature profile achieves a distribution that no longer changes with time. To maintain this time invariant condition, it is necessary to supply heat continuously to the left end of the rod and extract heat from the right. The resulting condition is time independent, but it is not an equilibrium condition; it is a steady state.

To distinguish a steady state from an equilibrium state, apply a simple but general test. Isolate the system from the surroundings, that is, surround the system (at least in a thought experiment) with a rigid, impermeable, thermally insulating boundary. No heat, no work, no matter crosses such a boundary. If a system that is not changing with time is isolated from its surroundings and changes *then begin to occur* within the system, then the initial stationary condition is a *steady state*. This steady state is maintained by flows across the boundary that are cut off upon isolation. If the ends of the bar in Fig. 5.2 are isolated from the heat source and sink, the internal steady state temperature profile begins to change.

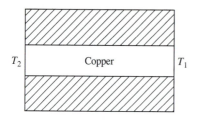

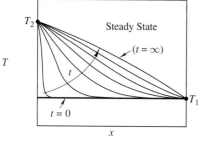

FIGURE 5.2
The thermally insulated copper rod achieves a time invariant (steady state) temperature distribution if the temperatures at its ends are maintained constant with time.

In contrast, if upon isolation no changes occur in the internal condition of the system, then the initial time invariant state is an equilibrium state.

Examine the result of this test for the equilibrium state closely, for it contains an important principle, first stated succinctly by J. Willard Gibbs [1, p. 62]. If a system is in equilibrium both internally and with its surroundings, then isolating it from its surroundings produces no change in the internal state of the system. Thus, the internal condition of any system that achieves equilibrium with its surroundings is identical with the condition of some other system that comes to the same final equilibrium state but is isolated from its surroundings during its approach to equilibrium. If a criterion can be devised that determines the final resting state for isolated systems, and a strategy can be devised for deriving the relationships that characterize the internal condition of that isolated system at equilibrium, then the resulting set of relationships are *general*. These relationships describe any system that comes to equilibrium, whether or not it is isolated from its surroundings during its approach to equilibrium.

This insight is potentially very useful because isolated systems are relatively easy to treat thermodynamically. Changes inside the system are not complicated by exchanges of heat, work, or matter with the surroundings. Focus upon the change in entropy of an isolated system. Since there are no exchanges of any kind with the surroundings, whatever processes occur within the system, no entropy is transferred across the boundary. Thus, the total entropy change experienced by an isolated system, no matter what processes produce that change, results entirely from entropy production within the system. The second law of thermodynamics guarantees that entropy production is always positive. It is concluded that:

For any real process that occurs within an isolated system, the total entropy of the system can only increase.

This conclusion identifies a thermodynamic property S, which monitors the direction of possible change in an isolated system. No matter how complicated the system or how irreversible the process, since the second law is pervasive this conclusion applies.

Thus, as an isolated system evolves through its process of spontaneous change, its entropy continually increases. When the system finally attains its state of rest, its equilibrium state, the entropy of the system must be the largest value that system can exhibit. If this were not true, meaning that there exists a state of that isolated system that has a higher entropy than the equilibrium state, then a change from that state toward the equilibrium state would be accompanied by a decrease in entropy. In an isolated system this violates the second law. It is thus concluded that:

In an isolated system the equilibrium state is the state that has the maximum value of entropy that the system can exhibit.

This extremum principle is referred to as the *criterion for equilibrium* in an isolated system.

Note that this statement does not imply that the equilibrium state for *any* thermodynamic system is the state that has the maximum entropy the system can exhibit. This statement applies only if the system were *isolated from its surroundings* during its approach to equilibrium. If, during its evolution, the system were not isolated from its surroundings, then entropy may be transferred across its boundary. The change in the total entropy of a system that is not isolated includes a contribution due to transfers that may be positive or negative. The entropy of such a system may thus increase or decrease and the total entropy of the system is not a valid monitor of the direction of spontaneous change. The entropy change is guaranteed by the second law to be positive only for systems that are isolated from their surroundings so that the total entropy change derives solely from entropy production.

Return now to the test proposed to determine whether a given system is in equilibrium both internally and with its surroundings. Imagine an arbitrary system exchanging heat, work, and matter with its surroundings as it evolves. No general statement can be made about the expected direction of change of any of its thermodynamic properties as it proceeds spontaneously toward its final state of equilibrium. Eventually the system attains its time-invariant state of balance called its equilibrium state. The internal condition of the system at this point is characterized by relationships (more explicitly, a set of equations) that must exist between its properties, called *conditions for equilibrium*. Now *isolate the system*. If the system is in equilibrium with itself and its surroundings, *no changes will occur in its internal condition*. Thus, the relationships that define the equilibrium state of a general system that approached equilibrium from an arbitrary direction are identical to the set of relations characterizing the conditions for equilibrium in a system that is isolated from its surroundings as it approaches that equilibrium state.

This argument, first advanced by Gibbs [1, p. 62], demonstrates that although the criterion for equilibrium (the extremum principle) that describes the equilibrium state for the isolated system is not valid for a general system, the equations between the internal properties of the system derived from that criterion (the conditions for equilibrium) are the same. It is concluded that:

> *The conditions for equilibrium that are derived for an isolated system are the same as those that hold for any system that achieves an equilibrium state, no matter what the history of the system may be.*

This set of equations, the *conditions* for equilibrium, is derived from the extremum principle, which provides the *criterion* for equilibrium in an isolated system, which holds that the entropy is a maximum. The strategy for deriving the conditions for equilibrium in an isolated system from the criterion for equilibrium is developed in the next section.

5.3 MATHEMATICAL FORMULATION OF THE GENERAL CONDITIONS FOR EQUILIBRIUM

At first glance the mathematical problem posed by the principles spelled out in the preceding section is familiar and straightforward. Given a function (in this case the entropy, S) of several variables, find the values of the set of variables that correspond to an extreme value (in this case a maximum) of the function. However, there are two complicating factors:

1. The entropy may be a function of a very large number of variables (see Chap. 6);
2. The set of variables that describe the change in entropy are *not independent* of each other in an isolated system.

More explicitly, the fact that the system is isolated implies that certain relationships exist among some of the variables. The extremum sought in this case is therefore a *constrained* maximum. Although this class of problems is treated in most standard calculus texts in those sections that deal with derivatives of functions of several variables, it is not widely emphasized in calculus courses.

The mathematical nature of the problem can be clarified by first considering a function z of two independent variables x and y.

$$z = z(x, y) \tag{5.3}$$

Figure 5.3 is a geometric illustration of this functional relationship. To find an extreme value of z, write its differential:

$$dz = \left(\frac{\partial z}{\partial x}\right)_y dx + \left(\frac{\partial z}{\partial y}\right)_x dy \tag{5.4}$$

The extreme value occurs at those points on the $z(x,y)$ surface for which both derivatives are simultaneously zero:

$$\left(\frac{\partial z}{\partial x}\right)_y = 0 \qquad \text{and} \qquad \left(\frac{\partial z}{\partial y}\right)_x = 0 \tag{5.5}$$

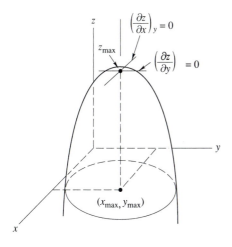

FIGURE 5.3

Illustration of a function z of the two variables, x and y, in which z shows an extremum that is a *maximum*. The maximum value of z, z_{max}, occurs at the combination of dependent variables (x_{max}, y_{max}).

This pair of equations characterizes the extremum; their simultaneous solution yields a pair of values for the variables (x_m, y_m), which are the coordinates in the xy plane that locate the extremum. In order to determine whether the extremum is a maximum, minimum or saddle point, it is necessary to examine the second derivatives of z with respect to x and y. If both second derivatives are negative, then the extremum is a maximum.

These tactics apply to a function z of an arbitrarily large number n of *independent* variables:

$$z = z(u, \ v, \ x, \ \ldots) \tag{5.6}$$

The differential of z has n terms in it:

$$dz = \left(\frac{\partial z}{\partial u}\right) du + \left(\frac{\partial z}{\partial v}\right) dv + \left(\frac{\partial z}{\partial x}\right) dx + \left(\frac{\partial z}{\partial y}\right) dy + \cdots \tag{5.7}$$

Evidently an extreme value of z occurs when *all* of the n coefficients in this equation simultaneously vanish, or when

$$\left(\frac{\partial z}{\partial u}\right) = 0 \qquad \left(\frac{\partial z}{\partial v}\right) = 0 \qquad \left(\frac{\partial z}{\partial x}\right) = 0 \qquad \left(\frac{\partial z}{\partial y}\right) = 0 \qquad \cdots \tag{5.8}$$

This set of n simultaneous equations in n unknowns can be solved (at least in principle) for the set of values of the n variables $(u_m, \ v_m, \ x_m, \ y_m, \ldots)$ that are the coordinates of the extremum in the function z.

The extrema described by the Eqs. (5.3) to (5.5) and (5.6) to (5.8) are unconstrained extrema because the set of variables that describe z in these two examples are explicitly assumed to be independent variables. In the thermodynamic application at hand, this assumption is not valid and the variables that most conveniently describe the change in entropy in an isolated system are not independent of each other.

If x and y are *not* independent variables in the first illustration, then there exists an equation relating these two variables; that is, in addition to the relation

$$z = z(x, y)$$

it will be true that

$$y = y(x) \tag{5.9}$$

This situation is shown in Fig. 5.4. Only those values of z that map above the curve describing the relation between the variables x and y are of interest in the problem. The relation $y = y(x)$ can be thought of as a *constraint* operating on the system that restricts the possible values of x and y. It is observed in Fig. 5.4 that the set of values that z takes on over this allowed domain of values of x and y has a maximum z_1 at the point (x_1, y_1) that is different from z_{max} at (x_{max}, y_{max}). This is a *constrained maximum* in z subject to the condition $y = y(x)$.

Figure 5.5 illustrates the proposition that a function $z(x,y)$ that is monotonic in x and y, meaning that it exhibits no unconstrained extrema, may nonetheless exhibit a constrained maximum z_{max} if the constraining relation $y = y(x)$ is appropriately complicated.

These notions generalize to the case in which z is a function of n variables. The system is represented by a *set* of equations, the first representing the variation of z with the n variables,

$$z = z(u, \ v, \ x, \ y, \ \ldots)$$

and an additional m equations ($m < n$) among the variables

$$u = u(x, \ y, \ldots) \tag{5.10}$$

$$v = v(x, \ y, \ldots)$$

Here z is the function whose extreme value is sought and the remaining equations are constraints placed upon the variables in z. An extreme value found for z with these auxiliary relations is a *constrained extremum*.

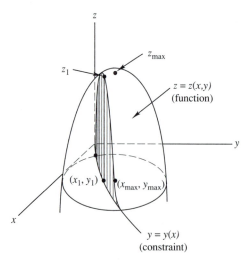

FIGURE 5.4
If the problem of interest involving the function in Figure 5.3 requires that the values of the variables x and y be constrained by a relationship $y = y(x)$, then, for this subset of (x, y) values, the maximum value of z, labelled z_1, is different from the absolute maximum, z_{max}.

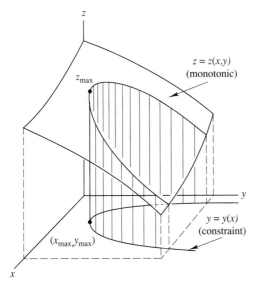

FIGURE 5.5
Even if the function $z(x, y)$ is monotonic, that is, it exhibits no extrema, a constrained maximum can exist if the constraining relation between x and y has an extremum.

The general approach to the solution of this class of problems makes use of the constraining Eq. (5.10) to eliminate dependent variables in the expression for z. The result expresses z in terms of $(n - m)$ variables that are now *independent* of each other. The extreme value then can be found by applying the strategy in Eqs. (5.6) through (5.8) for the case of independent variables, writing the differential of z in terms of a set of independent variables and setting all the coefficients equal to zero.

Example 5.1. Finding a constrained maximum by eliminating dependent variables by direct substitution.

Suppose z is expressed as a function of four variables,

$$z = xu + yv$$

but, in the problem at hand, u and v are related to x and y:

$$u = x + y + 12$$

$$v = x - y - 8$$

To find the extreme value of z subject to these constraints, eliminate the dependent variables u and v in the expression for z by substituting their relationships to the independent variables x and y:

$$z = x(x + y + 12) + y(x - y - 8)$$

Simplify:

$$z = x^2 + 12x + 2xy - 8y - y^2$$

The variable z is now expressed in terms of independent variables x and y. To find an extremum, form the differential

$$dz = (2x + 12 + 2y)dx + (2x - 8 - 2y)dy$$

where the coefficients are the partial derivatives of z with respect to x and y. Set the coefficients equal to zero:

$$2x + 2y + 12 = 0$$

$$2x - 2y - 8 = 0$$

Solving: $x = -1$, $y = -5$. These are the coordinates of the constrained extremum. Substitute these values into the expression for z to obtain the value of z at the extremum: $z = 14$.

Example 5.2. Finding a constrained extremum by eliminating the differentials of the dependent variables.

Consider again the equation for z given in Example 5.1 with its attendant constraints. An alternate strategy for finding the extremum in z subject to the constraining relations $u = u(x, y)$, $v = v(x, y)$ begins by taking the differentials of the function z and the constraining equations

$$dz = udx + xdu + vdy + ydv$$

$$du = dx + dy$$

$$dv = dx - dy$$

Eliminate the dependent differentials by substituting for du and dv in the expression for dz:

$$dz = udx + x(dx + dy) + vdy + y(dx - dy)$$

Collect terms:

$$dz = (u + x + y)dx + (v + x - y)dy$$

Set the coefficients equal to zero:

$$u + x + y = 0$$

$$v + x - y = 0$$

To solve for x and y, use the constraining equations to evaluate u and v:

$$(x + y + 12) + x + y = 0 \rightarrow 2x + 2y + 12 = 0$$

$$(x - y - 8) + x - y = 0 \rightarrow 2x - 2y - 8 = 0$$

Note that these conditions for the extremum are identical with those derived in Example 5.1 and thus yield the same solution.

The strategy outlined in Example 5.1 cannot be applied if the constraining equations are not explicit, which means they if they cannot be solved explicitly for u and v. For example, suppose the equation relating u to x and y has the form:

$$x^2 yu \cdot \ln u = 12$$

This equation cannot be solved explicitly to obtain an expression of the form $u = u(x, y)$, to be substituted wherever u appears in the expression for z. The strategy presented in Example 5.2 *can* be applied, even where the constraining relations are implicit in the dependent variables. For this reason, the strategy employed for finding constrained maxima in the entropy function to obtain conditions for equilibrium is that presented in Example 5.2.

Now that the mathematical nature of the problem has been presented in detail, it is possible to lay out a general strategy for finding the conditions for equilibrium in a thermodynamic system of arbitrary complexity.

1. Write a differential expression for the change in entropy that the system may experience when taken through an arbitrary process, explicitly including terms for all possible variations in the state of the system.

2. Write differential expressions that describe the constraints that apply to the system because it is considered to be isolated from its surroundings.

3. Use the isolation constraints to eliminate dependent variables in the description of the change in entropy for the system. The resulting differential expression for the change in entropy applies explicitly to the possible changes that may occur in an isolated system with all variables otherwise independent.

4. Set the coefficients of each of the differentials in the new expression for dS equal to zero. These equations are the *conditions for equilibrium* for the system.

This strategy for finding conditions for equilibrium is illustrated for a unary two phase system in the next section.

5.4 APPLICATION OF THE GENERAL STRATEGY FOR FINDING CONDITIONS FOR THERMODYNAMIC EQUILIBRIUM: THE UNARY TWO PHASE SYSTEM

Focus on a unary (one component), two phase (α and β), nonreacting otherwise simple system. The strategy for dealing with a two phase system is straightforward. Each of the separate phases can be treated as a one phase system with its own set of extensive and intensive properties. The apparatus for describing the behavior of such single phase systems was developed in Chap. 4. The description of a two phase system simply makes use of the fact that, for all the *extensive properties* defined, the value for a system is the sum of the values for each of its parts. Thus, for a two phase system, each of the properties S', V', U', H', F', and G' is the sum of the values of the corresponding quantities for the separate phases. (Extensive properties are designated with a prime (′) to distinguish them from their molar counterparts.)

Because the phase boundary separating two phases is a natural boundary, it is not possible to restrict the flow of matter between the phases. Accordingly, it is necessary to treat each phase as an *open system* in describing its behavior; that is, to allow for the possibility that the number of moles of each phase may change during an arbitrary process.

The first step in applying the general strategy requires an expression for the change in entropy experienced by the system when it is taken through an arbitrary process. The entropy for the system is the sum of the entropies of its parts and therefore, the change in entropy of the system is

$$dS'_{sys} = d(S'^{\alpha} + S'^{\beta}) = dS'^{\alpha} + dS'^{\beta} \tag{5.11}$$

where the superscript α denotes the properties of one phase, designated the α phase, and the superscript β designates properties of the other phase called the β phase. It is thus necessary to compute the changes in entropy experienced separately by the α and β parts of the system during an arbitrary process.

Focus first on the behavior of the α phase. Its internal energy U'^{α} is in general a function of its entropy S'^{α}, volume V'^{α}, and number of moles n^{α}:

$$U'^{\alpha} = U'^{\alpha}(S'^{\alpha}, \ V'^{\alpha}, \ n^{\alpha}) \tag{5.12}$$

From the combined statement of the first and second laws for the α phase

$$dU'^{\alpha} = T^{\alpha} dS'^{\alpha} - P^{\alpha} dV'^{\alpha} + \mu^{\alpha} dn^{\alpha} \tag{5.13}$$

This equation introduces the coefficient

$$\mu^{\alpha} = \left(\frac{\partial U'^{\alpha}}{\partial n^{\alpha}} \right)^{\alpha}_{S'}, \ V'^{\alpha} \tag{5.14}$$

which is called the *chemical potential* of the component that makes up the α phase. This quantity plays a central role in the description of the chemical aspects of the behavior of matter; the concept is developed in detail in Chap. 8 and beyond. For the present purpose, which is to illustrate the application of the general strategy for obtaining conditions for equilibrium, a definition, Eq. (5.14), is all that is required. Rearranging Eq. (5.13) gives an expression for the change in entropy of the α phase:

$$dS'^{\alpha} = \frac{1}{T^{\alpha}} dU'^{\alpha} + \frac{P^{\alpha}}{T^{\alpha}} dV'^{\alpha} - \frac{\mu^{\alpha}}{T^{\alpha}} dn^{\alpha} \tag{5.15}$$

Identical reasoning for the β phase produces the analogous expression for the change in entropy of the β phase:

$$dS'^{\beta} = \frac{1}{T^{\beta}} dU'^{\beta} + \frac{P^{\beta}}{T^{\beta}} dV'^{\beta} - \frac{\mu^{\beta}}{T^{\beta}} dn^{\beta} \tag{5.16}$$

Apply Eq. (5.11); the expression for the change in entropy of the system has six terms:

$$dS'_{sys} = dS'^{\alpha} + dS'^{\beta}$$

$$dS'_{sys} = \frac{1}{T^{\alpha}} dU'^{\alpha} + \frac{P^{\alpha}}{T^{\alpha}} dV'^{\alpha} - \frac{\mu^{\alpha}}{T^{\alpha}} dn^{\alpha}$$

$$+ \frac{1}{T^{\beta}} dU'^{\beta} + \frac{P^{\beta}}{T^{\beta}} dV'^{\beta} - \frac{\mu^{\beta}}{T^{\beta}} dn^{\beta} \tag{5.17}$$

The second step in the strategy requires evaluation of the constraints imposed by isolating the system. The boundary of an isolated system is rigid (so that no work can be done on or by the system), thermally insulating (so that no heat can be exchanged with the system), and impermeable (so that no matter crosses the boundary of the system). These characteristics place the following *constraints* upon the state functions that describe the two phase system:

Because there are no exchanges with the surroundings, then, by the first law, the internal energy of the system must remain constant. Thus, whatever processes occur within the system,

$$dU'_{sys} = d(U'^{\alpha} + U'^{\beta}) = dU'^{\alpha} + dU^{\beta} = 0 \qquad (5.18)$$

Because the boundary is rigid,

$$dV'_{sys} = d(V'^{\alpha} + V'^{\beta}) = dV'^{\alpha} + dV'^{\beta} = 0 \qquad (5.19)$$

Because the boundary is impermeable,

$$dn_{sys} = d(n^{\alpha} + n^{\beta}) = dn^{\alpha} + dn^{\beta} = 0 \qquad (5.20)$$

Thus, the differentials in the six terms that are required in Eq. (5.17) to describe the change in entropy of a unary two phase system when it is taken through an arbitrary process are not independent when the process is carried out in an isolated system. Three equations relate these six variables, Eqs. (5.18), (5.19), and (5.20). Use these relations to eliminate three of the six variables in Eq. (5.17):

$$dU'^{\alpha} = -dU'^{\beta} \qquad (5.21)$$

$$dV'^{\alpha} = -dV'^{\beta} \qquad (5.22)$$

$$dn^{\alpha} = -dn^{\beta} \qquad (5.23)$$

Internal energy, volume, and moles of material may be exchanged between the phases but are conserved for the isolated system. Substitute these expressions for dU'^{β}, dV'^{β}, and dn^{β} in Eq. (5.17):

$$dS'_{sys,iso} = dS'^{\alpha} + dS'^{\beta}$$

$$= \frac{1}{T^{\alpha}}dU'^{\alpha} + \frac{P^{\alpha}}{T^{\alpha}}dV'^{\alpha} - \frac{\mu^{\alpha}}{T^{\alpha}}dn^{\alpha}$$

$$+ \frac{1}{T^{\beta}}(-dU'^{\alpha}) + \frac{P^{\beta}}{T^{\beta}}(-dV'^{\alpha}) - \frac{\mu^{\beta}}{T^{\beta}}(-dn^{\alpha}) \qquad (5.24)$$

Collecting terms:

$$dS'_{sys,iso} = \left(\frac{1}{T^{\alpha}} - \frac{1}{T^{\beta}}\right)dU'^{\alpha} + \left(\frac{P^{\alpha}}{T^{\alpha}} - \frac{P^{\beta}}{T^{\beta}}\right)dV'^{\alpha} - \left(\frac{\mu^{\alpha}}{T^{\alpha}} - \frac{\mu^{\beta}}{T^{\beta}}\right)dn^{\alpha} \qquad (5.25)$$

This equation describes the change in entropy accompanying an arbitrary change in state of a unary two phase system when it is isolated from its surroundings. The function S' is now expressed in termsof three *independently variable* properties, U'^{α}, V'^{α}, and n^α. The conditions that describe the maximum in entropy for such an isolated system are obtained by setting the coefficients of each of the three differentials in Eq. (5.25) equal to zero

$$\frac{1}{T^\alpha} - \frac{1}{T^\beta} = 0 \rightarrow T^\alpha = T^\beta \qquad \text{(Thermal Equilibrium)} \qquad (5.26)$$

$$\frac{P^\alpha}{T^\alpha} - \frac{P^\beta}{T^\beta} = 0 \rightarrow P^\alpha = P^\beta \qquad \text{(Mechanical Equilibrium)} \qquad (5.27)$$

$$\frac{\mu^\alpha}{T^\alpha} - \frac{\mu^\beta}{T^\beta} = 0 \rightarrow \mu^\alpha = \mu^\beta \qquad \text{(Chemical Equilibrium)} \qquad (5.28)$$

When these three *conditions for equilibrium* are met within this two phase isolated system, its entropy is a maximum and the *criterion for equilibrium* is satisfied. (Strictly speaking, it is also necessary to examine the second derivatives in the function to determine that the extremum is a *maximum;* however, exploration of this aspect of the problem is beyond the scope of this presentation.) Further, it has been shown that the conditions for equilibrium derived for an isolated system are valid equations in an equilibrated system whether it is isolated or not. Thus, *Eqs. (5.26)–(5.28) are general conditions for thermal, mechanical and chemical equilibrium in a unary two phase system.*

Stated in words, a unary two phase system is in equilibrium when the temperatures, pressures, and chemical potentials of both phases are equal.

In Chap. 7 it is shown how these equations form the basis for computing phase diagrams for unary systems.

5.5 ALTERNATE FORMULATIONS OF THE CRITERION FOR EQUILIBRIUM

The criterion for equilibrium developed in this Chapter is chosen as the basis for presentation in the remainder of this text because it follows very directly from the second law of thermodynamics. Other criteria can be formulated, expressed in terms of extreme values of other state functions, combined with constraining relations that are appropriate to *their* application.

For example, consider a system that is constrained to processes in which the *entropy, volume, and number of moles of components* cannot change. For a system with these constraints, it can be demonstrated that the internal energy U' *can only decrease.* Thus, the internal energy function is a monitor of the direction of spontaneous change in such a system. As such a system proceeds toward equilibrium, its internal energy decreases until, at the equilibrium state, its internal energy is a minimum. Thus, an alternate criterion for equilibrium may be stated:

> *For a system constrained to constant entropy, volume, and quantity of each of its components, the internal energy is a minimum at equilibrium.*

There are two significant problems in formulating this alternate criterion:

1. Its derivation from the second law is less direct than the maximum entropy criterion.

2. While a system with the desired constraints of S', V', and n constant can be visualized mathematically, it may not be able to be realized practically.

Since the criterion is general, it must apply to systems that approach equilibrium irreversibly. In such cases entropy is produced; the constraint, "S' is a constant," can be achieved only if the entropy flow out of the system is remains equal the entropy production. The construction of a real system satisfying this constant entropy constraint would incorporate very sophisticated monitoring and control devices. In contrast, the constraints required for the maximum entropy criterion are those of an isolated system, which can be easily visualized.

It can be easily demonstrated that the minimum internal energy criterion (at constant S', V', and n) presented above, yields the same set of conditions for equilibrium as does the maximum entropy criterion (at constant U', V', and n). To find the constrained minimum in internal energy follow the same mathematical strategy. For each phase, the change in internal energy can be written

$$dU'^{\alpha} = T^{\alpha}dS'^{\alpha} - P^{\alpha}dV'^{\alpha} + \mu^{\alpha}dn^{\alpha} \tag{5.29}$$

and

$$dU'^{\beta} = T^{\beta}dS'^{\beta} - P^{\beta}dV'^{\beta} + \mu^{\beta}dn^{\beta} \tag{5.30}$$

For the system,

$$dU'_{\text{sys}} = dU'^{\alpha} + dU'^{\beta}$$

$$dU'_{\text{sys}} = T^{\alpha}dS'^{\alpha} - P^{\alpha}dV'^{\alpha} + \mu^{\alpha}dn^{\alpha} + T^{\beta}dS'^{\beta} - P^{\beta}dV'^{\beta} + \mu^{\beta}dn^{\beta} \tag{5.31}$$

The constraints for this criterion are

$$dS'_{\text{sys}} = dS'^{\alpha} + dS'^{\beta} = 0 \rightarrow dS'^{\alpha} = -dS'^{\beta} \tag{5.32}$$

$$dV'_{\text{sys}} = dV'^{\alpha} + dV'^{\beta} = 0 \rightarrow dV'^{\alpha} = -dV'^{\beta} \tag{5.33}$$

$$dn_{\text{sys}} = dn^{\alpha} + dn^{\beta} = 0 \rightarrow dn^{\alpha} = -dn^{\beta} \tag{5.34}$$

Use these constraints to eliminate dependent variables in Eq. (5.31)

$$dU'_{\text{sys}} = T^{\alpha}dS'^{\alpha} - P^{\alpha}dV'^{\alpha} + \mu^{\alpha}dn^{\alpha} + T^{\beta}(-dS'^{\alpha}) - P^{\beta}(-dV'^{\alpha}) + \mu(-dn^{\alpha})$$

Collecting terms

$$dU'_{\text{sys}} = (T^{\alpha} - T^{\beta})dS'^{\alpha} - (P^{\alpha} - P^{\beta})dV'^{\alpha} + (\mu^{\alpha} - \mu^{\beta})dn^{\alpha} \tag{5.35}$$

This expresses the internal energy of the system in terms of independent variables, S'^{α}, V'^{α}, and n^{α}. To find the extremum (minimum) in the internal energy, set all the coefficients equal to zero:

$$T^{\alpha} - T^{\beta} = 0 \rightarrow T^{\alpha} = T^{\beta} \qquad \text{(Thermal Equilibrium)} \tag{5.36}$$

$$P^{\alpha} - P^{\beta} = 0 \rightarrow P^{\alpha} = P^{\alpha} \qquad \text{(Mechanical Equilibrium)} \tag{5.37}$$

$$\mu^\alpha - \mu^\beta = 0 \rightarrow \mu^\alpha = \mu^\beta \qquad \text{(Chemical Equilibrium)} \qquad (5.38)$$

These *conditions for equilibrium* are identical with those obtained from the maximum entropy criterion.

In his classic work, Gibbs [1, p. 62–65] applied this minimum internal energy criterion for equilibrium. The resulting working equations are the same and the attendant mathematics are slightly simpler. However, the system constraints associated with this criterion are experimentally unrealistic. Accordingly, in this text, the criterion that is used consistently in arriving at conditions for equilibrium is the maximum entropy isolated system criterion. The strategy based upon this criterion provides the most direct connection between the laws of thermodynamics and the conditions for equilibrium.

Each of the other energy functions, *H*, *F*, and *G*, may be associated with the condition for spontaneous change, leading ultimately to a criterion for equilibrium, provided the system under consideration is subject to constraints that are appropriate for the given energy function:

1. In a system constrained to constant *entropy* and *pressure*, the direction of spontaneous change is monitored by a decrease in the *enthalpy* function; *H* is a *minimum* at equilibrium.

2. In a system constrained to constant *temperature* and *volume*, the *Helmholtz free energy function* decreases during every spontaneous change; *F* is a *minimum* at equilibrium.

3. In a system constrained to constant *temperature* and *pressure*, the *Gibbs free energy function* decreases for every spontaneous change; *G* is a *minimum* at equilibrium.

To illustrate the origin of these criteria, consider the last case for which it is assumed the system under consideration is experimentally controlled so that, whatever processes occur, its temperature and pressure remain constant. Consider two states, I and II, along the isothermal, isobaric path traversed by the system. From the definition of Gibbs free energy, for each of these states,

$$G'^{\mathrm{I}} = U'^{\mathrm{I}} + P^{\mathrm{I}} V'^{\mathrm{I}} - T^{\mathrm{I}} S'^{\mathrm{I}} \qquad (5.39)$$

$$G'^{\mathrm{II}} = U'^{\mathrm{II}} + P^{\mathrm{II}} V'^{\mathrm{II}} - T^{\mathrm{II}} S'^{\mathrm{II}} \qquad (5.40)$$

For the constraints imposed, $P^{\mathrm{I}} = P^{\mathrm{II}} = P$ and $T^{\mathrm{I}} = T^{\mathrm{II}} = T$. The difference in Gibbs free energy between the states can be written

$$G'^{\mathrm{II}} - G'^{\mathrm{I}} = (U'^{\mathrm{II}} - U'^{\mathrm{I}}) + P(V'^{\mathrm{II}} - V'^{\mathrm{I}}) - T(S'^{\mathrm{II}} - S'^{\mathrm{I}})$$

or

$$\Delta G'_{T,P} = \Delta U' + P \Delta V' - T \Delta S' \qquad (5.41)$$

The first law of thermodynamics in its most general form evaluates the change in internal energy for any real, irreversible process as

$$\Delta U' = Q + W + W' \qquad (5.42)$$

The other two terms in Eq. (5.41) can be expressed in terms of the process variables associated with an isothermal, isobaric *reversible* process connecting states I and II:

$$W_{\mathrm{rev},T,P} = \int_{V'_1}^{V'_2} P\,dV' = -P\int_{V'_1}^{V'_2} dV' = -P(V_2 - V_1) = -P\Delta V' \qquad (5.43)$$

The heat absorbed for such a reversible process is given by

$$Q_{\mathrm{rev},T,P} = \int_{S'_1}^{S'_2} T\,dS' = T\int_{S_1}^{S'_2} dS' = T(S_2 - S_1) = T\Delta S' \qquad (5.44)$$

Insert Eqs. (5.42–5.44) into Eq. (5.41):

$$\Delta G'_{T,P} = [Q + W + W'] - Q_{\mathrm{rev}} - W_{\mathrm{rev}}$$

or

$$\Delta G'_{T,P} = [Q - Q_{\mathrm{rev}}] + [W - W_{\mathrm{rev}}] + W' \qquad (5.45)$$

According to the second law, the heat absorbed during a reversible process connecting two end states is larger (algebraically) than for any irreversible process connecting those states: $Q_{\mathrm{rev}} > Q$. Further, the reversible mechanical work is also larger than that for any irreversible process: $W_{\mathrm{rev}} > W$. Finally, it can be argued that any spontaneous process can be arranged so that the system does nonmechanical work on the surroundings: $W' < 0$. When these conclusions are applied term by term to Eq. (5.45), it is seen that both terms in brackets are negative and W' is also negative. Thus, for any irreversible spontaneous change in an isothermal, isobaric system, $\Delta G'_{T,P} < 0$. In other words, the Gibbs free energy function *decreases* for every spontaneous change in a system constrained to constant temperature and pressure. Thus, G' may be used as a monitor of the direction of spontaneous change in systems so constrained (*but only in systems so constrained!*). It follows that, when a system controlled at constant temperature and pressure comes to rest at its equilibrium state, its Gibbs free energy will be a *minimum*.

Since temperature and pressure must be held constant in applying this criterion, it is not possible to deduce from this extremum principle, the conditions for thermal and mechanical equilibrium for a system. Only the conditions for chemical equilibrium can be derived from the principle of minimization of the Gibbs free energy. This is another point of argument that favors the maximum entropy-isolated system criterion as a basis for arriving at general conditions for equilibrium.

5.6 SUMMARY OF CHAPTER 5

Because the entropy of an isolated system can only increase, the final equilibrium state of an isolated system has the maximum value of entropy that the system can exhibit.

This extremum principle is the *criterion for equilibrium* in an isolated system.

Mathematical formulation of this constrained maximum leads to the *conditions for equilibrium* for this isolated system; these conditions are equations relating properties of the system that must be satisfied for the extremum to be achieved.

If a system that is not isolated from its surroundings comes to equilibrium with itself and its surroundings then, at that point, isolation of the system from its surroundings does not change the internal condition of the system.

Thus, the *conditions for equilibrium* derived for an isolated system are *general conditions for equilibrium.*

Application of this strategy to a unary, two-phase system shows that, at equilibrium,

1. the temperatures of the two phases must be the same
2. the pressures of the two phases must be the same
3. the chemical potentials of the two phases must be the same

Alternate conditions for spontaneous change and criteria for equilibrium may be derived in terms of other thermodynamic state functions:

1. For systems constrained to constant S' and V', U' decreases, and is a minimum at equilibrium.
2. For systems constrained to constant S' and P, H' decreases and is a minimum at equilibrium.
3. For systems constrained to constant T and V', F' decreases and is a minimum at equilibrium.
4. For systems constrained to constant T and P, G' decreases, and is a minimum at equilibrium.

Application of any of these criteria leads to the same set of conditions for equilibrium.

PROBLEMS

5.1. Discuss the meaning of the statement: "An equilibrium state is a state of balance."

5.2. Give three illustrative examples each of
   (*a*) An equilibrium state
   (*b*) A steady state

5.3. State in outline form the logical steps that lead to our *general criterion for equilibrium.*

5.4. Contrast the concepts presented in this chapter as
   (*a*) *Criterion* for equilibrium
   (*b*) *Conditions* for equilibrium

5.5. State in words and mathematically the set of relationships implied by the statement that a thermodynamic system is *isolated* from its surroundings during a process.

5.6. The development presented in this chapter describes the behavior of an *isolated system.* Systems of practical interest are rarely isolated from their surroundings in their approach toward equilibrium. How is it then possible to conclude that the results of this development are *general*, and not limited to an isolated system?

5.7. Find the extreme values of the function

$$z = (x - 2)^2 + (y - 2)^2 + 4$$

Find the constrained maximum of this function corresponding to the condition

$$x + y = 1$$

(a) Solve the problem first by eliminating y as a variable and finding the extreme value of the function $z = z[x, \ y(x)]$.

(b) Then solve the same problem by writing the differential forms of the two equations and substituting for dy in the expression for dz, then set the coefficient of dx equal to zero.

5.8. The steps in the strategy for finding conditions for equilibrium are:

(a) Write an expression for the change in entropy of the system when it is taken through an arbitrary process.

(b) Write the isolation constraints in differential form.

(c) Use the isolation constraints to eliminate dependent variables in the expression for the entropy.

(d) Collect terms.

(e) Set the coefficients of each differential equal to zero.

(f) Solve these equations for the conditions for equilibrium.

Use the example of a unary two phase system presented in Sec. 5.4 to write out each of these steps mathematically.

5.9. The combined statement of the first and second laws for the change in enthalpy of a unary, single-phase system may be written:

$$dH' = T\,dS' + V'dP + \mu\,dn$$

Use this result to write an expression for the change in enthalpy of a two phase $(\alpha + \beta)$ system. If the entropy, pressure, and total number of moles are constrained to be constant, then the criterion for equilibrium is that the enthalpy is a minimum. Paraphrase the strategy used to deduce the conditions for equilibrium in an isolated system to derive them for a system constrained to constant S', P, and n. What happens to the condition for mechanical equilibrium?

REFERENCES

1. Gibbs, J. Willard: *The Scientific Papers of J. Willard Gibbs, Volume 1, Thermodynamics*, Dover Publications, New York, 1962.

CHAPTER
6

STATISTICAL
THERMODYNAMICS

All of the concepts and relationships developed so far in this text visualize a thermodynamic system as consisting of some continuous medium or as a collection of separate media, each of which is continuous. A system is endowed with properties like heat capacity, coefficient of expansion, an equation of state and so on. Knowledge of the variation of these properties with the state of the system is sufficient to describe the macroscopic phenomena that the system may experience. This level of description of the behavior of matter is called *phenomenological* thermodynamics. No use has been made of the idea that the substance of this continuum is actually composed of atoms or molecules and that the behavior of the system is somehow related to the properties of the particles that compose it.

Development of the connection between the thermodynamic behavior of macroscopic systems and that of atoms on their submicroscopic scale is useful for a variety of reasons. A new level of understanding of how matter behaves emerges from this connection. At the phenomenological level of description, it is sufficient to know that experimental measurements show that the heat capacity of substance A is different from that of substance B, for example. At the atomic level of description, it is possible to predict not only that substances A and B have different heat capacities but also the expected magnitudes of those differences. Atomistic models provide a level of *explanation* for the behavior of a substance; phenomenological information merely provides a consistent *description* of how it behaves.The atomistic point of view provides a perspective that helps make sense out of the continuum description of matter. In addition, the atomistic description may provide insight into the physical meaning of some of the phenomenological concepts that are unfamiliar. Indeed, much of the atomistic approach developed in this chapter centers on the concept of entropy, raising the level of sophistication of the understanding of this elusive phenomenological property.

An atomistic description of the behavior of matter begins with the idea that each atom in the system can be assigned values of properties that describe its condition. For example, each atom in a gas has a position vector x and velocity vector v. The state of an atom in a condensed phase is usually described by its energy. Such a description has an obvious problem: one cubic centimeter of a typical condensed phase contains about 10^{22} atoms (about one tenth of a mole). Specification of the properties (e.g., energy level) of 10^{22} atoms is a hopeless task. Even a modern computer listing an entry every nanosecond (10^9 entries per second) would take over 3,000 *centuries* to make such a list. Specification of the thermodynamic state of a system with such an enormous table is called the *microstate* of the system.

The mathematical discipline that provides the tools necessary to analyze very large collections of numbers is *statistics*. The primary tool supplied for the purpose is the concept of the *distribution function*. Atoms that have similar values of properties are lumped together into classes (in this case, energy levels); the distribution function simply reports the number of particles in each class (energy level). In this way the quantity of information required to specify the condition of the system is greatly reduced. Description of the behavior of matter in terms of the distribution of particles over their allowable states is called *statistical thermodynamics*. Specification of the thermodynamic state of a system in terms of such a distribution function is called the *macrostate* for the system.

This chapter begins with a more detailed development of the concepts of microstate and macrostate just introduced. The number of distinct microstates corresponding to a given macrostate is computed by applying some combinatorial analysis adapted from statistics. A fundamental hypothesis of statistical thermodynamics, due to Boltzmann, is introduced. This hypothesis relates the entropy of the system to the number of microstates that correspond to a given macrostate. Then the general strategy for finding conditions for equilibrium developed in Chap. 5 is applied to this statistical description of the behavior of matter. In this case the equilibrium condition is described by a specific distribution of atoms over their allowable states, called the *Boltzmann distribution function*. The important distinguishing physical quantity contained in the description of this equilibrium distribution is found to be the *partition function* for the system. The partition function for a given system can be evaluated from a list of energy levels that the particles in the system can exhibit. Finally, it is shown that, given the partition function for a system, all of the macroscopic phenomenological properties of the system can be computed. This completes the connection between an atomistic model for the system, formulated in terms of the list of allowable energy states, and the experimentally measurable thermodynamic properties. Application of this algorithm is illustrated for the ideal gas model and the Einstein model for a crystalline solid.

6.1 MICROSTATES, MACROSTATES AND ENTROPY

A unary thermodynamic system is composed of a very large number of particles (atoms or molecules) that are all structurally identical. The condition of such a collection of particles at any instant in time could, in principle, be described by listing the condition of each particle in the array. The phrase, *in principle*, is included because, in practice, such a specification is not possible, as noted in the introduction to this chapter. Notwithstanding that such a description is impractical, the concept of such a specification of the state of the system may be contemplated. Such a description, specifying the condition or state of each particle that composes the system, is defined to be a *microstate* that the system may exhibit. If any entry in this list is altered, that is, if any particle changes its condition, the system is considered to be in a different microstate. Evidently, because the number of particles and conditions is large, the number of different microstates that a given system might exhibit in its evolution is enormous. This is illustrated in Table 6.1 for a system consisting of only four particles each of which may exhibit only two states.

Since the particles are assumed to be physically identical, the macroscopically observable behavior of the system is not dependent on *which* particles exist in a given state, but merely on *how many* particles are in that state. The macroscopic properties will be the same whether particles a and b are in state ϵ_2, or whether

TABLE 6.1
Microstates and macrostates for a simple system

Particles: a, b, c, d
States: ϵ_1, ϵ_2

List of Microstates (Number of microstates $= 2^4 = 16$)

State	ϵ_1	ϵ_2	State	ϵ_1	ϵ_2
A	abcd	—	I	bc	ad
B	abc	d	J	bd	ac
C	abd	c	K	cd	ab
D	acd	b	L	a	bcd
E	bcd	a	M	b	acd
F	ab	cd	N	c	abd
G	ac	bd	O	d	abc
H	ad	bc	P	—	abcd

List of Macrostates

State	No. of particles ϵ_1	ϵ_2	Corresponding microstates	Number	Probability
I	4	0	A	1	1/16
II	3	1	B,C,D,E	4	4/16
III	2	2	F,G,H.I.J.K	6	6/16
IV	1	3	L,M,N,O	4	4/16
V	0	4	P	1	1/16

particles c and d are. Thus the microstates listed as B, C, D and E in Table 6.1 give the same values for the macroscopic properties of the system. Each of these microstates corresponds to the condition "two particles are in state ϵ_1 and two particles are in state ϵ_2." This observation gives rise to a much more efficient and useful way of describing the state of a system at an atomic level, called the *macrostate* for the system.

To specify the macrostate of a system at a given instant in time, focus not upon the particles but upon the list of possible conditions or states that the individual atoms can exhibit. In Table 6.1, the number of states is two. More generally, suppose that there are r states that each atom can exhibit. In Table 6.1, specification of a given macrostate requires two numbers: the number in state ϵ_1 and the number in state ϵ_2. Specification of the macrostate labelled II requires two numbers: 3, 1. This pair of numbers constitutes a rudimentary distribution function: "3 particles in state ϵ_1, 1 particle in state ϵ_2." In the general case a macrostate is specified by assigning a number of particles to each of the r available atomic states. Let the list of possible states be $\epsilon_1, \epsilon_2, \epsilon_3, \ldots, \epsilon_i, \ldots, \epsilon_r$. Then a particular macrostate is given by

$$
\begin{array}{ccccccc}
\epsilon_1 & \epsilon_2 & \epsilon_3 & \cdots & \epsilon_i & \cdots & \epsilon_4 \\
n_1 & n_2 & n_3 & \cdots & n_i & \cdots & n_r
\end{array}
$$

where n_i is the number of particles that have energy ϵ_i. The set of numbers $(n_1, \ldots n_r)$ is a distribution function specifying how the atoms are distributed over the energy levels. This distribution describes the *macrostate* for the system.

For the example of four particles that each may exhibit two energy states, Table 6.1 shows that the system may exhibit 16 microstates. The number of possible macrostates is significantly smaller; there are five macrostates possible for the system. Evidently *a given macrostate may correspond to a number of different microstates.* These relationships are spelled out explicitly for the simple example in Table 6.1. As the number of particles and energy levels is increased, the number of microstates that may correspond to a given macrostate may become very large. For example, for a system consisting of just 10 particles distributed over 3 energy levels, the total number of microstates is $3^{10} = 59,049$. The number of distinguishable macrostates for this system is 60. Thus for this system, on the average, a macrostate corresponds to about 1000 microstates. Real thermodynamic systems may contain 10^{22} particles and 10^{15} energy levels. It can be appreciated that the number of microstates corresponding to the typical macrostate in these cases is almost incredibly large. The number of microstates that corresponds to a given macrostate is a central quantity in the development of statistical thermodynamics.

From the atomistic point of view a thermodynamic process, which is a change in the macroscopic state of the system, corresponds to a redistribution of the atoms over their allowable states, or to a collection of changes in the number of particles in each atomic energy state. As atoms experience changes in their condition, altering from one state to a neighboring one, the system as a whole evolves through a sequence of macrostates. The fraction of its lifetime that each particle spends in a given energy state can in the long run be expected to be the same for all particles, since they are identical. Accordingly, it can be argued that the time the system spends in any given *microstate* is the same for all microstates. The time the system spends in a particular *macrostate* is the sum of the times it spends in the various microstates that correspond to that macrostate. The fraction of time it spends in any given macrostate is thus the ratio of the number of microstates that correspond to that macrostate to the total number of microstates that the system is capable of exhibiting. This fraction may be plausibly interpreted as the probability that the system exhibits the given macrostate in any randomly selected instant in time.

The simple system analyzed in Table 6.1 serves to illustrate this principle. There are a total of 16 microstates for this system of four particles and two energy levels (2^4). If all are equally likely, the system spends 1/16 of its time in each microstate. There are five macrostates labelled with Roman numerals. As an example, focus on the macrostate labelled II. It occurs when any one of the four microstates, B, C, D, or E, exists. Since each of these microstates exists 1/16 of the time, macrostate II will occur 4/16 or 1/4 of the time. The probability that macrostate II will be observed at any instant in time is 1/4.

To apply these ideas to a thermodynamic system, it is necessary to generalize the computation of the total number of microstates a system can exhibit and the number of microstates that correspond to any given macrostate. Consider a system that contains a large number N_0 of particles. (For example, for one mole of system N_0 is 6.023×10^{23}.) Suppose that each of these particles can exist in any of a large number of conditions or states; let the number of such states be r. In Table 6.1, N_0 is four and r is two. The total number of microstates that a system may exhibit can be computed through the following exercise in induction. Particle a can be placed in any of the r states. Particle b may also be independently placed in any of the r states. If these were the only two particles in the system, the number of possible arrangements in the r states would be $r \cdot r$. If a third particle is added to the system, it can also be placed independently in any of the r states. The number of arrangements increases to $r \cdot r \cdot r$. Each particle that is added to the system may be added in any of the r states. Thus, the number of different ways the N_0 particles may be arranged in the r states is r^{N_0}. Even if r is not large, raising it to a power of the order of 10^{23} makes the total number of microstates enormous.

The number of microstates corresponding to each of the five macrostates available to the four-particle two-state system is listed in Table 6.1. For a general system composed of N_0 particles distributed over r states, the description of a particular macrostate has the form of a distribution function

$$(n_1, n_2, n_3, \ldots, n_i, \ldots, n_r)$$

where n_i is the number of particle in state ϵ_i. For example, for a twenty-particle system with seven energy states, this list might read

$$(1, \ 3, \ 4, \ 6, \ 2, \ 3, \ 1)$$

for a particular macrostate. The number of microstates that correspond to a given macrostate is a standard problem in that branch of the foundations of statistics known as *combinatorial analysis*. The statement of the problem, "How many microstates correspond to the macrostate $(n_1, n_2, \ldots, n_r)$?" can be restated: "How many different ways can N_0 balls be arranged in r boxes such that there are n_1 balls in the first box, n_2 in the second and so on to n_r balls in the rth box?" The answer to this question, familiar to those acquainted with combinatorial analysis, is

$$\Omega = \frac{N_0!}{n_1! \cdot n_2! \cdots n_r!} \tag{6.1}$$

Thus, Ω is the number of microstates that correspond to the macrostate given by the set of n_i values in the denominator. The notation $n!$, read "n factorial," represents the product of the sequence of numbers

$$n! \equiv n \cdot (n-1) \cdot (n-2) \cdots 3 \cdot 2 \cdot 1$$

Note that as n increases, $n!$ increases very rapidly.

Application of Eq. (6.1) to each of the five macrostates listed for the system considered in Table 6.1 demonstrates the validity of the formula. Recall that $0! = 1$.

Macrostate I: $\Omega_{\text{I}} = \dfrac{4!}{4!0!} = \dfrac{(4 \cdot 3 \cdot 2 \cdot 1)}{(4 \cdot 3 \cdot 2 \cdot 1) \cdot (1)} = 1$

Macrostate II: $\Omega_{\text{II}} = \dfrac{4!}{3!1!} = \dfrac{(4 \cdot 3 \cdot 2 \cdot 1)}{(3 \cdot 2 \cdot 1) \cdot (1)} = 4$

Macrostate III: $\Omega_{\text{III}} = \dfrac{4!}{2!2!} = \dfrac{(4 \cdot 3 \cdot 2 \cdot 1)}{(2 \cdot 1) \cdot (2 \cdot 1)} = 6$

Macrostate IV: $\Omega_{\text{IV}} = \dfrac{4!}{1!3!} = \dfrac{(4 \cdot 3 \cdot 2 \cdot 1)}{(1) \cdot (3 \cdot 2 \cdot 1)} = 4$

Macrostate V: $\Omega_{\text{V}} = \dfrac{4!}{0!4!} = \dfrac{(4 \cdot 3 \cdot 2 \cdot 1)}{(1) \cdot (4 \cdot 3 \cdot 2 \cdot 1)} = 1$

It has been argued that the probability that the system exists in a given macrostate, interpreted as the fraction of the time that it spends in microstates that correspond to that macrostate, is the ratio of the number of microstates Ω_J that correspond to the Jth macrostate to the total number of microstates the system may exhibit. This probability is thus [6.1]

$$P_J = \frac{\Omega_j}{r^{N_0}} = \frac{N_0!}{\displaystyle\prod_{i=1}^{r} n_i!} \cdot \frac{1}{r^{N_0}} \tag{6.2}$$

The notation $\Pi n_i!$ is shorthand for the product $n_1! n_2! \ldots n_r!$.

Macrostates for which Ω_J is large will exist in the system for a greater fraction of time than those for which Ω_J is small. For example, in Table 6.1, macrostate III is six times more probable than states I or V.

Of all the macrostates that can exist for a system, one contains more microstates than any other. This macrostate has the maximum value of Ω and the maximum probability of appearing at any instant in time. Examination of the behavior of the P_J function for a variety of macrostates demonstrates that, for systems with large numbers of particles, this function has an extremely sharp peak at the macrostate containing the maximum number of microstates, Fig. 6.1. Macrostates that are only slightly different from the maximum probability state have a much smaller probability of being observed. Macrostates that differ significantly from the maximum probability state have negligible probability of occurrence. Thus, the maximum probability state, or those very near to it, are observed almost all of the time. If this most likely state is interpreted as the macrostate that corresponds to the *equilibrium state* for the system, then this hypothesis forms the basis for connecting the statistical, atomistic description of the system with phenomenological thermodynamics.

In phenomenological thermodynamics the equilibrium state is also characterized by an extremum; the entropy of an isolated system is a maximum at equilibrium. This correspondence suggests a connection between entropy and Ω, the number of microstates corresponding to a given macrostate. If the functional relationship between these quantities is *monotonic*, that is, both either increase together or decrease together, then when one function maximizes, so does the other. Further, if one compares the values of these two quantities for the range of conditions a system might exhibit, it is evident that Ω varies over orders of magnitude of orders of magnitude, while values of entropy vary over, at most, one or two orders of magnitude. These considerations lead to the basic assumption that connects the atomistic and phenomenological descriptions known as the Boltzmann hypothesis:

$$S = k \ln \Omega \tag{6.3}$$

where k turns out to be a universal constant, called *Boltzmann's constant*. It is demonstrated later that the value of k is simply R/N_0, where R is the ideal gas constant and N_0 is Avogadro's number. Thus, k is simply the value of the gas constant per atom (or molecule) rather than per mole of gas. The consequences of this hypothesis will now be explored.

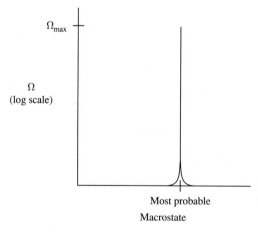

FIGURE 6.1
The probability distribution for macrostates has an extremely sharp peak for systems with large numbers of particles and macrostates.

6.2 CONDITIONS FOR EQUILIBRIUM IN STATISTICAL THERMODYNAMICS

In the atomistic description of the thermodynamic behavior of a system, the equilibrium state is that particular macrostate which maximizes the entropy of the system when it is isolated. Thus, the equilibrium state is a particular set of values of the numbers that define macrostates: $(n_1, n_2, \ldots, n_r)_{eq}$. To find this set of numbers and relate them to the original specification of the behavior of the system contained in the list of states the individual atoms may exhibit $(\epsilon_1, \epsilon_2, \ldots, \epsilon_r)$ it is only necessary to apply the general strategy for finding conditions for equilibrium developed in Chap. 5. This strategy involves the following steps.

1. Write an expression for the change in entropy of the system in terms of the variables that define its state. In statistical thermodynamics, these variables are $(n_1, n_2, \ldots, n_r)$.
2. Write expressions for the constraints on the variation of these variables imposed by the limitation of an isolated system.
3. Derive the set of equations that must be satisfied in order for the entropy function to be a maximum, subject to the isolation constraints.

These steps are carried out in this section.

6.2.1 Evaluation of Entropy

An explicit relation between the entropy of a macrostate and the set of numbers that specify its distribution is obtained by combining Eqs. (6.1) and (6.3):

$$S = k \ln \left(\frac{N_0!}{\displaystyle\prod_{i=1}^{r} n_i!} \right) \tag{6.4}$$

The computation of the factorials required to evaluate S is a very time consuming process, even for a high speed computer. Computation of 1000! requires the computation of 1000 products of numbers. Since the numerator in Eq. (6.4) is of the order of 10^{23}, its computation would be impractical. Fortunately, the values of factorials of large numbers can be estimated at very high precision by a simple formula known as *Stirling's approximation* ([1], pp. 472–474):

$$\ln x! \cong x \ln x - x \tag{6.5}$$

For example, for $x = 100$, $\ln x! = 364$, while $x \ln x - x$ gives 361, an error of less than one percent. The precision of the estimate increases as x increases and is excellent for very large x values, as are characteristic of the n_i quantities in statistical thermodynamics.

Use the properties of logarithms to simplify Eq. (6.4). First, the log of a ratio of numbers is the difference of their logs

$$S = k \left[\ln N_0! - \ln \left(\prod_{i=1}^{r} n_i! \right) \right]$$

In the second term, the log of a product of numbers is the sum of their logs

$$S = k \left[\ln N_0! - \sum_{i=1}^{r} \ln(n_i!) \right]$$

Third, apply Stirling's approximation to all $\ln x!$ terms in this equation

$$S = k \left[(N_0 \ln N_0 - N_0) - \sum_{i=1}^{r} (n_i \ln n_i - n_i) \right] \tag{6.6}$$

The summation can be expressed as the difference between two summations:

$$S = k \left[(N_0 \ln N_0 - N_0) - \sum_{i=1}^{r} n_i \ln n_i + \sum_{i=1}^{r} n_i \right]$$

Recognize that the total number of particles in the system is the sum of the numbers in each level

$$N_0 = \sum_{i=1}^{r} n_i \tag{6.7}$$

Thus,

$$S = k \left[N_0 \ln N_0 - N_0 - \sum_{i=1}^{r} n_i \ln n_i + N_0 \right]$$

or

$$S = k \left[N_0 \ln N_0 - \sum_{i=1}^{r} n_i \ln n_i \right]$$

Substitute equation (6.7) for the coefficient of the first term

$$S = k \left[\left(\sum_{i=1}^{r} n_i \right) \ln N_0 - \sum_{i=1}^{r} n_i \ln n_i \right]$$

Combine as a single summation

$$S = k \left[\sum_{i=1}^{r} n_i (\ln N_0 - \ln n_i) \right]$$

The difference of two logs is the log of their ratio

$$S = k \sum_{i=1}^{r} n_i \ln \left(\frac{N_0}{n_i} \right)$$

Finally, use the property that the log of an inverse of a ratio is equal to the negative of the log of the ratio:

$$S = -k \sum_{i=1}^{r} n_i \ln \left(\frac{n_i}{N_0} \right) \tag{6.8}$$

This development thus converts Eq. (6.4) from an intractable computation problem involving factorials to a much simpler problem involving logarithms of large numbers. Equation (6.8) can be used to compute the entropy of any macrostate that exists for any given atomistic model of the system.

In statistical thermodynamics a process, that is, a change in the thermodynamic state, is described by a change in the macrostate for the system. More explicitly, a process involves changes in the number of particles in some or all of the energy states, or a redistribution of the particles over the energy levels in the system. The number of particles in some energy levels increases while those in others decrease. Mathematically, a process can be represented by a collection of changes in the n_i terms:

$$(\Delta n_1, \ \Delta n_2, \ldots, \Delta n_i, \ldots, \Delta n_r)$$

For an infinitesimal change in the macroscopic state of the system, these finite changes Δn_i can be replaced with infinitesimal changes in the variables dn_i since the number of particles in each energy level is very large. Thus, for an infinitesimal change in the thermodynamic state of the system, the corresponding redistribution of the particles among the energy states can be described by a set of infinitesimal changes in the (large) numbers, n_i

$$(dn_1, \ dn_2, \ldots, dn_i, \ldots, dn_r)$$

Since Eq. (6.8) describes the entropy of any macroscopic state, the change in entropy for an arbitrary change in state can be obtained by differentiating this equation.

$$dS = -kd \left[\sum_{i=1}^{r} n_i \ln \left(\frac{n_i}{N_0} \right) \right]$$

$$dS = -k \sum_{i=1}^{r} d \left[n_i \ln \left(\frac{n_i}{N_0} \right) \right]$$

$$dS = -k \sum_{i=1}^{r} d[n_i \ln n_i - n_i \ln N_0]$$

$$dS = -k \sum_{i=1}^{r} \left[\ln n_i \cdot dn_i + n_i \left(\frac{1}{n_i} \right) dn_i - \ln N_0 \cdot dn_i - n_i \left(\frac{1}{N_0} \right) dN_0 \right]$$

$$dS = -k \left[\sum_{i=1}^{r} (\ln n_i - \ln N_0) dn_i + \sum_{i=1}^{r} dn_i - \sum_{i=1}^{r} \left(\frac{n_i}{N_0} \right) dN_0 \right]$$

From Eq. (6.7),

$$dN_0 = \sum_{i=1}^{r} dn_i \quad \text{and} \quad \sum_{i=1}^{r} \frac{n_i}{N_0} = 1$$

Thus,

$$dS = -k \left[\sum_{i=1}^{r} (\ln n_i - \ln N_0) dn_i + dN_0 - 1 \cdot dN_0 \right]$$

Finally,

$$dS = -k \sum_{i=1}^{r} \ln \left(\frac{n_i}{N_0} \right) \cdot dn_i \tag{6.9}$$

This expression describes the change in entropy for any process through which the system may be taken. No restrictions are imposed upon the nature of the redistribution of the particles over their energy levels.

6.2.2   Evaluation of the Isolation Constraints

Recall that the application of the criterion for equilibrium requires that the system be isolated from its surroundings. This constraint places some restrictions upon the interchanges that can occur during the redistribution of the particles over their energy levels. Isolation from the surroundings implies that, whatever processes occur inside the system, the total number of particles cannot change and the internal energy of the system cannot change. The total number of particles is related to the number in each energy level in Eq. (6.7). Computation of the internal energy of the system is straightforward. If ϵ_i is the energy of a particle in the ith state, then $n_i \epsilon_i$ is the energy of all of the particles in that state. The total energy of the whole collection is just the sum of the energies of the particles in each level

$$U = \sum_{i=1}^{r} \epsilon_i n_i \tag{6.10}$$

Since N_0 and U cannot change in an isolated system, redistributions of particles over the energy levels are subject to the constraints

$$dN_0 = \sum_{i=1}^{r} dn_i = 0 \tag{6.11}$$

and

$$dU = \sum_{i=1}^{r} \epsilon_i dn_i = 0 \tag{6.12}$$

A second term in the differentiation of the product $\epsilon_i n_i$ is not required because the energy levels over which the particles are distributed, the ϵ_i, do not change during a process. Processes are viewed to occur as a redistribution of the particles on a fixed set of energy levels.

6.2.3 The Constrained Maximum in the Entropy Function

The final step in applying the strategy for finding conditions for equilibrium involves determining the set of values of the variables that yield a maximum in entropy in a system subject to the constraints imposed by isolation. Equations (6.11) and (6.12) must be used to eliminate the dependent variables in the expression for the change in entropy, Eq. (6.9). In Eq. (6.9) the change in entropy is expressed as a function of r variables, with a value of dn_i for each of the r energy levels. In an isolated system only $(r-2)$ of these dn_i terms are independent because they are related by Eqs. (6.11) and (6.12). The simple substitution procedure used to eliminate dependent variables in the previous application of this strategy in Chap. 5 cannot be applied in this case since the large number of variables involved makes the algebra in the problem unwieldy.

Fortunately, a general procedure for solving this class of mathematical problems exists, and is known as the method of Lagrange multipliers. For the present purpose it is sufficient to outline the procedure and apply it to the problem at hand. The steps involved in finding a constrained extremum in a function of many variables (such as S) subject to constraining equations [such as Eq. (6.11) and (6.12)] are:

1. Multiply each of the differential forms of the constraining equations by an arbitrary constant.
2. Add these forms to the differential of the function whose extreme value is sought.
3. Collect like terms and set the resulting combination of differential forms equal to zero.
4. Set the coefficients of each of the differentials that appear in this equation equal to zero.
5. Solve the resulting set of equations, evaluating the Lagrange multipliers in the process.

This procedure solves the general mathematical problem of finding a constrained extremum. If the function whose extreme value is sought is the entropy function, Eq. (6.9), and the constraining equations are the isolation constraints, Eqs. (6.11) and (6.12), then this set of solutions derived from setting the coefficients equal to zero are the conditions for equilibrium for the system.

To apply this strategy to the problem at hand first multiply the constraint equations by arbitrary constants.

From Eq. (6.11):

$$\alpha \cdot dN_0 = \alpha \cdot \sum_{i=1}^{r} dn_i = 0$$

From Eq. (6.12):

$$\beta \cdot dU = \beta \cdot \sum_{i=1}^{r} \epsilon_i dn_i = 0$$

where α and β are the Lagrange multipliers. Add these to the expression for dS [Eq. (6.9)] and set the result equal to zero

$$dS + \alpha \cdot dN_0 + \beta \cdot dU + 0 \tag{6.13}$$

Substitute for dS, dN_0, and dU

$$-k \sum_{i=1}^{r} \ln \left(\frac{n_i}{N_0} \right) \cdot dn_i + \alpha \cdot \sum_{i=1}^{r} dn_i + \beta \cdot \sum_{i=1}^{r} \epsilon_i dn_i = 0$$

Collect terms

$$\sum_{i=1}^{r} \left[-k \ln \left(\frac{n_i}{N_0} \right) + \alpha + \beta \epsilon_i \right] dn_i = 0 \tag{6.14}$$

This equation contains r terms, each identical in form but each with its own value of ϵ_i and n_i. To find the maximum in S subject to the isolation constraints, set each of these coefficients equal to zero. This yields r equations of the form

$$-k \ln \left(\frac{n_i}{n_0} \right) + \alpha + \beta \epsilon_i = 0 \qquad (i = 1, 2, \ldots, r)$$

Algebraic manipulation of this result gives

$$\frac{n_i}{N_0} = e^{\alpha/k} \cdot e^{\beta \epsilon_i / k} \qquad (i = 1, 2, \ldots, r) \tag{6.15}$$

This set of r equations constitutes the conditions for equilibrium in the statistical description of thermodynamic behavior. The equilibrium macrostate is characterized by this relationship: the fraction of particles in the ith energy level (n_i/N_0) is exponentially related to the value of the energy for that level, ϵ_i.

It remains to evaluate the Lagrange multipliers α and β in terms of the thermodynamic properties of the system. The isolation constraints expressed in Eqs. (6.7) and (6.10) are used for this purpose. Substitute the result obtained in Eq. (6.15) into Eq. (6.7):

$$\sum_{i=1}^{r} \frac{n_i}{N_0} = 1 = \sum_{i=1}^{r} e^{\alpha/k} \cdot e^{\beta \epsilon_i / k} = e^{\alpha/k} \cdot \sum_{i=1}^{r} e^{\beta \epsilon_i / k}$$

Solve for the factor containing α

$$e^{\alpha/k} = \frac{1}{\displaystyle\sum_{i=1}^{r} e^{\beta \epsilon_i / k}} \tag{6.16}$$

The summation in the denominator on the right-hand side of this equation plays a central role in statistical thermodynamics. Define this quantity to be the *partition function*, $\mathcal{P}$, for the system.

$$\mathcal{P} = \sum_{i=1}^{r} e^{\beta\epsilon_i/k} \tag{6.17}$$

It is shown that all the thermodynamic properties of the system can be computed if the partition function is known.

Equation (6.16) permits elimination of the Lagrange multiplier α from the expression for the equilibrium distribution, Eq. (6.15)

$$\frac{n_i}{N_0} = \frac{1}{\mathcal{P}} \cdot e^{\beta\epsilon_i/k} \tag{6.18}$$

It remains to evaluate β.

The Lagrange multiplier associated with the energy constraint can be determined by comparing the expression for the change in entropy for a reversible process computed in statistical thermodynamics with that computed in phenomenological thermodynamics. The statistical thermodynamics expression for entropy change is given in Eq. (6.9). A reversible process is a sequence of equilibrium states. Thus, the ratio (n_i/N_0) that appears in Eq. (6.9) can be evaluated by applying Eq. (6.18)

$$dS = -k\sum_{i=1}^{r} \ln\left(\frac{n_i}{N_0}\right) \cdot dn_i = -k\sum_{i=1}^{r} \ln\left(\frac{1}{\mathcal{P}} \cdot e^{\beta\epsilon_i/k}\right) dn_i \tag{6.19}$$

Use the properties of logarithms, including $\ln(e^x) = x$

$$dS = -k\sum_{i=1}^{r} \left(\frac{\beta\epsilon_i}{k} - \ln\mathcal{P}\right) dn_i$$

$$dS = -\beta\sum_{i=1}^{r} \epsilon_i dn_i + k\ln\mathcal{P}\sum_{i=1}^{r} dn_i$$

Recognize the summations as related to dU and dN_0:

$$dS = -\beta dU + k\ln\mathcal{P}dN_0 \tag{6.20}$$

The combined statement of the first and second laws provides the analogous phenomenological expression. For an open system,

$$dU = TdS - PdV + \mu dN_0$$

(N_0 is the total number of atoms in the system and μ is thus the chemical potential per atom.) Solve for dS

$$dS = \frac{1}{T}dU + \frac{P}{T}dV - \frac{\mu}{T}dN_0 \tag{6.21}$$

The volume term in Eq. (6.21) has no counterpart in the statistical expression for the entropy, Eq. (6.20), because in this introductory development of statistical thermodynamics it is assumedthat the average volume occupied by an atom is the same for all energy levels. Compare Eqs. (6.20) and (6.21)

$$\beta = -\frac{1}{T} \tag{6.22}$$

$$\frac{\mu}{T} = k \ln \mathcal{P} \tag{6.23}$$

Thus, β is the negative reciprocal of the absolute temperature.

Substitution of this result into Eqs. (6.17) and (6.18) gives the complete expression for the equilibrium distribution of particles over the energy levels:

$$\frac{n_i}{N_0} = \frac{1}{\mathcal{P}} e^{-(\epsilon_i/kT)} \tag{6.24}$$

where $\mathcal{P}$, the partition function, is

$$\mathcal{P} \equiv \sum_{i=1}^{r} e^{-(\epsilon_i/kT)} \tag{6.25}$$

A list of the energy levels available to the particles in the system, the set of ϵ_i values, constitutes a model for the behavior of the atoms in the system. According to Eq. (6.25) the partition function for the model can be computed from such a list. Equation (6.24) then yields the number of particles in each energy level at equilibrium. This description of the behavior of the system is complete and detailed: all of the macroscopic thermodynamic properties of the system can be derived from these results.

6.2.4   Calculation of the Macroscopic Properties from the Partition Function

Equation (6.8) gives the entropy of a system for any distribution of particles over their energy levels. The value of the entropy for the equilibrium distribution is obtained by substituting Eq. (6.24) into Eq. (6.8):

$$S = -k \sum_{i=1}^{r} n_i \ln \left(\frac{n_i}{N_0} \right) = -k \sum_{i=1}^{r} n_i \ln \left[\frac{1}{\mathcal{P}} e^{-(\epsilon_i/kT)} \right]$$

$$= -k \sum_{i=1}^{r} n_i \left[-\frac{\epsilon_i}{kT} - \ln \mathcal{P} \right] = +\frac{k}{kT} \sum_{i=1}^{r} \epsilon_i n_i + k \ln \mathcal{P} \sum_{i=1}^{r} n_i$$

The summation in the second term is the total number of particles N_0 in the system; the summation in the first term is the internal energy U of the system.

$$S = \frac{1}{T} U + k \ln \mathcal{P} N_0 \tag{6.26}$$

Recall the definition of the Helmholtz free energy function, F, Eq. (4.7). Equation (6.26) can be written:

$$F \equiv U - TS = U - T \left[\frac{1}{T} U + N_0 k \ln \mathcal{P} \right]$$

$$F = -N_0 kT \ln \mathcal{P} \tag{6.27}$$

Thus, the Helmholtz free energy function can be computed if the partition function is known, and no other information is required.

Next, recall the combined statement of the first and second laws for the Helmholtz free energy function, Eq. (4.8)

$$dF = -SdT - PdV + \delta W' \tag{4.8}$$

The coefficient relationship corresponding to dT is

$$S = -\left(\frac{\partial F}{\partial T}\right)_V$$

Apply this relationship to Eq. (6.27) to compute the entropy of the system in terms of the partition function

$$S = -\left[\frac{\partial}{\partial T}(-N_0 kT \ln \mathcal{P})\right]_V$$

$$S = N_0 k \ln \mathcal{P} + N_0 kT \left(\frac{\partial \ln \mathcal{P}}{\partial T}\right)_V \tag{6.28}$$

The internal energy of the system can now be computed by rearranging the definitional relationship for F

$$U = F + TS = -N_0 kT \ln \mathcal{P} + T\left[N_0 k \ln \mathcal{P} + N_0 kT \left(\frac{\partial \ln \mathcal{P}}{\partial T}\right)_V\right]$$

$$U = N_0 kT^2 \left(\frac{\partial \ln \mathcal{P}}{\partial T}\right)_V \tag{6.29}$$

To assess the validity of an atomistic model, it is necessary to compare its predictions with experimental observations. Since the experimental information that is determined directly is the heat capacity, it is useful to obtain an expression for C_V in terms of the partition function. Equation (4.40) gives the relationship between heat capacity and the temperature dependence of the internal energy:

$$C_V = \left(\frac{\partial U}{\partial T}\right)_V = 2N_0 kT \left(\frac{\partial \ln \mathcal{P}}{\partial T}\right)_V + N_0 kT^2 \left(\frac{\partial^2 \ln \mathcal{P}}{\partial T^2}\right)_V \tag{6.30}$$

Computation of the remaining thermodynamic state functions, V, H, G, and C_p, requires a formulation that includes the volume dependence of the partition function, which is neglected in this introduction to statistical thermodynamics.

Equations (6.27) through (6.30) complete the algorithm. An atomistic model for the thermodynamic behavior of a system begins with a complete list of the energy levels that the particles in the system may exhibit. Given this list and no other information the partition function for the model can be computed. Given the partition function, all the macroscopic thermodynamic properties of the system can be computed. Of particular interest is the heat capacity since it provides the most direct test of the predictions of such an atomistic model. Thus, statistical thermodynamics provides an algorithm that converts an atomistic model for the behavior of a system as input into values of the macroscopic thermodynamic properties as output. Examples of the application of this algorithm are presented in the next section.

6.3   APPLICATIONS OF THE ALGORITHM

Three models for the available energy levels of a system are explored. The first visualizes a system that exhibits only two energy levels. This simple model illustrates the procedures involved in pursuing the algorithm. The second, the Einstein model for a crystal, is more realistic and more sophisticated. The third example, the ideal gas model, is mathematically advanced, though physically straightforward and it yields results that are plausible and familiar.

6.3.1   A Model with Two Energy Levels

Consider a system that has N_0 particles that can exist in either of only two energy states, ϵ_1 and ϵ_2. Suppose further that ϵ_2 has a value that is twice that of ϵ_1. For simplicity, let ϵ be the energy of a particle in level 1, then $\epsilon_2 = 2\epsilon$ in this model. This list, $(\epsilon, 2\epsilon)$, completes the description of the model. The equilibrium distribution of the particles between these states can now be computed, as well as its macroscopic thermodynamic properties.

The partition function for this model is given by

$$\mathcal{P} = \sum_{i=1}^{r} e^{-(\epsilon_1/kT)} = e^{-(\epsilon_1/kT)} + e^{-(\epsilon_2/kT)} = e^{-(\epsilon/kT)} + e^{-(2\epsilon/kT)}$$

$$= e^{-(\epsilon/kT)}\left[1 + e^{-(\epsilon/kT)}\right] \tag{6.31}$$

Apply Eq. (6.24) to find the equilibrium distribution of particles.

$$\frac{n_1}{N_0} = \frac{e^{-(\epsilon_1/kT)}}{\mathcal{P}} = \frac{e^{-(\epsilon/kT)}}{e^{-(\epsilon/kT)}\left[1 + e^{-(\epsilon/kT)}\right]} = \frac{1}{\left[1 + e^{-(\epsilon/kT)}\right]}$$

$$\frac{n_2}{N_0} = \frac{e^{-(\epsilon_2/kT)}}{\mathcal{P}} = \frac{e^{-(2\epsilon/kT)}}{e^{-(\epsilon/kT)}\left[1 + e^{-(\epsilon/kT)}\right]} = \frac{e^{-(\epsilon/kT)}}{\left[1 + e^{-(\epsilon/kT)}\right]}$$

The ratio of occupancy of the states is

$$\frac{n_2}{n_1} = e^{-(\epsilon/kT)} \tag{6.32}$$

The relative occupancy of the energy levels is determined by the magnitude of ϵ in comparison with thermal energy represented by kT. At very low temperatures, $\epsilon/kT \gg 1$, the ratio is small and most of the particles lie in level 1. At sufficiently high temperatures in this system, $\epsilon/kT \ll 1$, approaching zero and the particles tend to become evenly distributed between the two states.

Computation of the thermodynamic properties, applying Eqs. (6.27) through (6.30), requiresevaluation of the temperature derivative of the natural logarithm of the partition function. From Eq. (6.31), after some manipulation,

$$\ln \mathcal{P} = -\frac{\epsilon}{kT} + \ln \left(1 + e^{-(\epsilon/kT)}\right) \tag{6.33}$$

$$\left(\frac{\partial \ln \mathcal{P}}{\partial T}\right)_V = \frac{\epsilon}{kT^2} \frac{\left[1 + 2e^{-(\epsilon/kT)}\right]}{\left[1 + e^{-(\epsilon/kT)}\right]} \tag{6.34}$$

Substitution of these results into Eqs. (6.27), (6.28), and (6.29) yields, after some algebraic simplification,

$$F = N_0\epsilon - N_0 kT \ln \left[1 + e^{-(\epsilon/kT)}\right] \tag{6.35}$$

$$S = \frac{N_0\epsilon}{T} \cdot \frac{e^{-(\epsilon/kT)}}{\left[1 + e^{-(\epsilon/kT)}\right]} + N_0 k \ln \left[1 + e^{-(\epsilon/kT)}\right] \tag{6.36}$$

$$U = N_0\epsilon \left[\frac{1 + 2e^{-(\epsilon/kT)}}{1 + e^{-(\epsilon/kT)}}\right] \tag{6.37}$$

The heat capacity can be evaluated from the coefficient relation,

$$C_V = \frac{N_0\epsilon^2}{kT^2} \cdot \frac{e^{-(\epsilon/kT)}}{\left[1 + e^{-(\epsilon/kT)}\right]^2} \tag{6.38}$$

The properties of this atomistic model are determined by a single parameter ϵ.

Example 6.1. To illustrate the calculation of property changes for a simple process, choose a value for this parameter; let $\epsilon = 2 \cdot 10^{-20}$ J/atom. Compute the change in entropy for this system when one mole is heated from 300 K to 1000 K. From Eq. (6.36) the entropy at 300 K is computed to be 0.31 J/mole; at 1000 K, the value is 3.92 J/mole. The change in entropy for the process is thus

$$\Delta S = S_{1000} - S_{300} = 3.92 - 0.31 = 3.61 \left(\frac{\text{J}}{\text{mol-K}}\right)$$

It should be pointed out that this simple model is not realistic. For example, at low temperatures C_V approaches infinity while at high temperatures C_V approaches zero. It is presented here only to illustrate the application of statistical thermodynamics to the computation of macroscopic thermodynamic properties from a proposed atomistic model for the behavior of the system.

6.3.2 Einstein's Model of a Crystal

Most solids are crystalline. The atoms arrange themselves on a lattice that repeats a simple structural unit in three dimensions. Einstein developed a conceptually simple model as a first attempt to understand the thermodynamic behavior of crystalline solids. Although the model proved inadequate in quantitative comparisons to the behavior of real crystals, it successfully predicts the qualitative form of important aspects of behavior. This theory provided the basis for subsequent descriptions that were more successful and are still being perfected. It is also very useful as an illustration of the application of the statistical thermodynamics algorithm.

To make the model concrete, suppose that the atoms arrange themselves in a simple cubic structure, where the unit cell is a cube with an atom at each corner. Each atom has six nearest neighbors. The energy of the crystal is assumed to reside entirely in the bonds between near neighbor pairs and the total energy of the crystal is viewed as the sum of the energies of the bonds it contains. The bonding between pairs of atoms is modeled as a simple spring with a spring constant that represents the strength of the bond. The atoms vibrate about their equilibrium positions, each connected by springs to its nearest neighbors, Fig. 6.2. The energy of the crystal is the kinetic energy of the atoms as they vibrate about their equilibrium positions. It is shown in physics that the kinetic energy of motion of particles connected by a spring is proportional to this vibrational frequency, v, of the spring.

A cubic crystal consisting of N_0 atoms contains $3N_0$ bonds and each of the six bonds associated with a specific atom is shared between two neighbors. Since all the springs connecting pairs of atoms are coupled together in the crystal, analysis of coupled oscillators shows that only certain discrete vibrational frequencies can occur in such a system. Since each frequency is associated with a corresponding energy value, the bond energies in the system can exhibit only discrete values. The bond energies are thus *quantized*. Einstein showed that the list of allowable energies for bonds could be described by the expression

$$\epsilon_i = (i + \tfrac{1}{2}) \cdot \hbar v \tag{6.39}$$

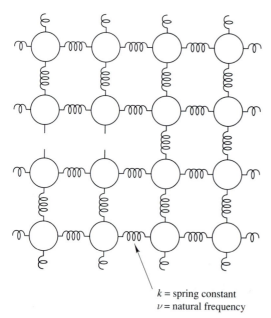

k = spring constant
v = natural frequency

FIGURE 6.2
The Einstein model for a simple cubic crystal. Atoms vibrate about their equilibrium positions harmonically, as if joined by simple, ideal springs.

where i is an integer, $\hbar$ is Planck's constant (a universal constant equal to 6.624×10^{-27} erg-sec/atom, and common in quantum theory), and v is a *characteristic* vibrational frequency from which all the frequencies of the bonds can be computed. The characteristic frequency v is related to the spring constant and hence the strength of the binding energy in the crystal. This model has a single adjustable parameter v, that can be used in an attempt to explain observed differences between the behavior of the elements. It is not surprising that a one parameter model is not sufficient for this purpose.

The list of energy levels in the crystal given in Eq. (6.39) is a complete description of the model. Application of the statistical thermodynamics algorithm permits prediction of the thermodynamic properties of such a system. First, evaluate the partition function

$$\mathcal{P} = \sum_{i=0}^{r} e^{-(\epsilon_i/kT)} = \sum_{i=0}^{r} e^{-[(i+1/2)\hbar v/kT]} \tag{6.40}$$

$$\mathcal{P} = \sum_{i=0}^{r} e^{-i(\hbar v/kT)} \cdot e^{-(1/2)(\hbar v/kT)} = e^{-(1/2)(\hbar v/kT)} \sum_{i=0}^{r} e^{-i(\hbar v/kT)}$$

Since the contribution from high energy levels (large i) is small this sum is well approximated by an infinite sum. Let $x = \exp(-(\hbar v/kT))$. Then the infinite series in this expression can be recognized as the familiar *geometric series* with a sum that converges to

$$\sum_{i=0}^{\infty} \left(e^{-(\hbar v/kT)}\right)^i = \frac{1}{1 - e^{-(\hbar v/kT)}}$$

The partition function can be evaluated;

$$\mathcal{P} = \frac{e^{-1/2(\hbar v/kT)}}{\left[1 - e^{-(\hbar v/kT)}\right]} \tag{6.41}$$

Thus,

$$\ln \mathcal{P} = -\frac{1}{2}\frac{\hbar v}{kT} - \ln\left[1 - e^{-(\hbar v/kT)}\right]$$

Given the partition function, the thermodynamic properties of the system can be computed by applying Eqs. (6.27) through (6.30). In this calculation the "particles" that exhibit the energy values in Eq. (6.29) are the *bonds* in the system. Thus, for a simple cubic crystal containing N_0 atoms, there are $3N_0$ "particles."

$$F = -3N_0 kT \ln \mathcal{P} = \frac{3}{2}N_0 \hbar v + 3N_0 kT \ln\left[1 - e^{-(\hbar v/kT)}\right] \tag{6.42}$$

The entropy is the temperature derivative of F:

$$S = -\left(\frac{\partial F}{\partial T}\right)_V = 3\frac{N_0 \hbar v}{T}\left[\frac{e^{-(\hbar v/kT)}}{1 - e^{-(\hbar v/kT)}}\right] - 3N_0 k \ln\left(1 - e^{-(\hbar v/kT)}\right) \tag{6.43}$$

The internal energy can be computed from the definition, $U = F + TS$:

$$U = F + TS = \frac{3}{2}N_0\hbar v \left[\frac{1 + e^{-(\hbar v/kT)}}{1 - e^{-(\hbar v/kT)}}\right]$$ (6.44)

The heat capacity is the temperature derivative of U:

$$C_V = \left(\frac{\partial U}{\partial T}\right)_V = 3N_0k \left(\frac{\hbar v}{kT}\right)^2 \cdot \frac{e^{-(\hbar v/kT)}}{\left(1 - e^{-(\hbar v/kT)}\right)^2}$$ (6.45)

The validity of this model for the description of the thermodynamic behavior of crystals can be assessed by comparing experimental measurements of the heat capacity as a function of temperature with the curve computed from Eq. (6.45). The characteristic frequency v, the only adjustable parameter in the theory, can be statistically selected to provide the best fit. Such a comparison is shown in Fig. 6.3. The qualitative form of the computed curve agrees well with the experimentally observed dependence of heat capacity upon temperature. However, the agreement is not quantitative. More sophisticated theories, incorporating a distribution of values of v and taking account of contributions to the energy of the crystal other than the kinetic energy of lattice vibrations, have been devised and applied with better success. Nonetheless, this simple, one parameter model provides a plausible basis upon which to build more sophisticated rationalizations of the thermodynamic behavior of crystals.

6.3.3   Monatomic Gas Model

Consider a gas composed of identical particles, each of which is a single atom. Rare gases like argon and neon satisfy this description. The thermodynamic properties of such a gas can be computed from an atomistic model that is based on the assumption that the energy contributed by each particle is simply the kinetic energy associated

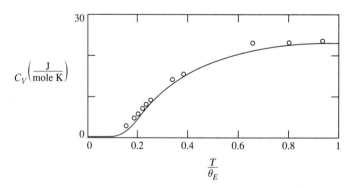

FIGURE 6.3

Comparison of experimentally measured heat capacity for diamond with that computed from the Einstein one parameter model. $\theta_E = 1320$ K. Figure from C. Kittel, *Introductin to Solid State Physics*, 4th ed. John Wiley, New York, NY (1971)

with its translation through space. The state of a particle is completely determined by its mass m, velocity $\mathbf{v}$ and its position $\mathbf{x}$ in space. Since $\mathbf{v}$ and $\mathbf{x}$ are vectors, each with three independent components, it is necessary to specify the values of six variables to specify the state of a gas atom in the system. These are its position (x, y, z) and its velocity components (v_x, v_y, v_z). The positional variables range over the set of positions that correspond to the volume of the system that contains the gas. In principle, the velocity components can vary from $-\infty$ to $+\infty$.

In order to relate macroscopic properties to this atomistic model, it is first necessary to evaluate the partition function for the model. In its definition the partition function is a summation, taken over all the states that the particles can exhibit, of the values of the quantity $\exp(-\epsilon_i/kT)$ corresponding to each state [see Eq. (6.25)]. It is assumed that the energy exhibited by a particle in this model is the kinetic energy associated with its motion through space

$$\epsilon = \frac{1}{2}mv^2 = \frac{1}{2}m\left(v_x^2 + v_y^2 + v_z^2\right) \tag{6.46}$$

Imagine that the gas is enclosed in a box with dimensions of l_x, l_y, l_z. In this model the variables are not quantized and the position of a particle may vary continuously within the domain of the enclosing box. In principle, the velocity components vary continuously between $-\infty$ and $+\infty$. Because the variables involved are *continuous*, the summation over all the available states that constitute the partition function must be replaced by an integral over all states. Since the state of a particle is a function of six variables, this is a sextuple integral:

$$\mathcal{P} = \int_0^{l_z}\int_0^{l_y}\int_0^{l_x}\int_{-\infty}^{\infty}\int_{-\infty}^{\infty}\int_{-\infty}^{\infty} e^{-(\epsilon/kT)}\,dv_x dv_y dv_z dx dy dz \tag{6.47}$$

Substitute the expression for the energy in this model, Eq. (6.46):

$$\mathcal{P} = \int_0^{l_z}\int_0^{l_y}\int_0^{l_x}\int_{-\infty}^{\infty}\int_{-\infty}^{\infty}\int_{-\infty}^{\infty} e^{-1/(2kT)m(v_x^2+v_y^2+v_z^2)}\,dv_x dv_y dv_z dx dy dz \tag{6.48}$$

The exponential in the integrand can be written as the product of three exponentials:

$$\mathcal{P} = \int_0^{l_z}\int_0^{l_y}\int_0^{l_x}\int_{-\infty}^{\infty}\int_{-\infty}^{\infty}\int_{-\infty}^{\infty} e^{-(mv_x^2/2kt)} \cdot e^{-(mv_y^2/2kT)} \cdot e^{-(mv_z^2/2kT)}\,dv_x dv_y dv_z dx dy dz \tag{6.49}$$

The position and velocity coordinates are independent of each other. As a consequence this sextuple integral can be written as the product of six independent integrals, each of which is a function of only one variable.

$$\mathcal{P} = \int_0^{l_z}\int_0^{l_y}\int_0^{l_x} dx dy dz \int_{-\infty}^{\infty} e^{-(mv_x^2/2kT)}\,dv_x \int_{-\infty}^{\infty} e^{-(mv_y^2/2kT)}\,dv_y \int_{-\infty}^{\infty} e^{-(mv_z^2/2kT)}\,dv_z \tag{6.50}$$

Evaluation of these integrals is straightforward. The first three integrate to the volume V of the enclosing container, no matter what shape it might have:

$$V = \int_0^{l_z}\int_0^{l_y}\int_0^{l_x} dx dy dz \tag{6.51}$$

The velocity integrals are all identical and have a form which is familiar in mathematics:

$$\int_{-\infty}^{\infty} e^{-a^2 x^2} dx = \frac{\sqrt{\pi}}{a}$$

where $a^2 = m/2kT$. Thus, for each of these integrals

$$\int_{-\infty}^{\infty} e^{-(m/2kT)v^2} dv = \sqrt{\frac{2\pi kT}{m}}$$

The partition function for a monatomic ideal gas is

$$\mathcal{P} = V \left[\frac{2\pi kT}{m} \right]^{3/2} \tag{6.52}$$

In order to evaluate the thermodynamic properties of the gas, the natural logarithm of the partition function

$$\ln \mathcal{P} = \ln \left[V \left(\frac{2\pi kT}{m} \right)^{3/2} \right] = \ln V + \frac{3}{2} \ln \left(\frac{2\pi k}{m} \right) + \frac{3}{2} \ln T \tag{6.53}$$

and its temperature derivative

$$\left(\frac{\partial \ln \mathcal{P}}{\partial T} \right)_V = \frac{3}{2} \cdot \frac{1}{T} \tag{6.54}$$

are required. Apply Eqs. (6.27) through (6.30) to compute the properties:

$$F = -N_0 kT \ln \mathcal{P} = -N_0 kT \ln \left[V \left(\frac{2\pi kT}{m} \right)^{3/2} \right] \tag{6.55}$$

$$S = N_0 k \ln \mathcal{P} + N_0 kT \left(\frac{\partial \ln \mathcal{P}}{\partial T} \right)_V = N_0 k \ln \left[V \left(\frac{2\pi kT}{m} \right)^{3/2} \right] + N_0 kT \left(\frac{3}{2T} \right)$$

$$S = N_0 k \ln \left[V \left(\frac{2\pi kT}{m} \right)^{3/2} \right] + \frac{3}{2} N_0 k \tag{6.56}$$

$$U = N_0 kT^2 \left(\frac{\partial \ln \mathcal{P}}{\partial T} \right)_V = N_0 kT^2 \left(\frac{3}{2} \cdot \frac{1}{T} \right) = \frac{3}{2} N_0 kT \tag{6.57}$$

$$C_V = \left(\frac{\partial U}{\partial T} \right)_V = \frac{3}{2} N_0 k \tag{6.58}$$

Note that it is deduced from the model that the internal energy of an ideal gas is a function only of its temperature; the heat capacity is found to be independent of temperature. Further, recognizing that, for one mole of the gas, $N_0 k = R$, the gas constant, the value of the heat capacity agrees well with experimental values for the *monatomic* gases in Table 6.2.

TABLE 6.2
Comparison of heat capacities of monotonic and diatomic gases with values predicted from the principle of equipartition of energy

Monatomic gas	C_P (J/mole-K)	Diatomic gas	C_P (J/mole-K)
Ideal Gas	$5/2R = 20.79$	Ideal Gas	$7/2 = 29.10$
Argon	20.72	Chlorine (Cl_2)	33.82
Krypton	20.69	Fluorine (F_2)	31.32
Neon	20.76	Hydrogen (H_2)	28.76
Radon	20.81	Oxygen (O_2)	29.32
Xenon	20.76	Nitrogen (N_2)	29.18

Finally, recall again the combined statements of the first and second laws for the Helmholtz free energy

$$dF = -SdT - PdV + \delta W' \tag{4.8}$$

The coefficient relation for the second term is

$$\left(\frac{\partial F}{\partial V}\right)_T = -P \tag{6.59}$$

Applying this relation to the ideal gas model expression for the Helmholtz free energy, Eq. (6.55):

$$P = -\left(\frac{\partial F}{\partial V}\right)_T = -\left(\frac{\partial(-N_0 kT \ln \mathcal{P})}{\partial V}\right)_T = N_0 kT \left(\frac{\partial \ln \mathcal{P}}{\partial V}\right)_T$$

$$= N_0 kT \left[\frac{\partial}{\partial V}\left(\ln V + \frac{3}{2}\ln\left(\frac{2\pi k}{m}\right) + \frac{3}{2}\ln T\right)\right]_T$$

Carry out the operation (the second and third terms are constant for this derivative)

$$P = N_0 kT \cdot \frac{1}{V}$$

Rearranging:

$$PV = N_0 kT = RT \tag{6.60}$$

which is the familiar equation of state developed experimentally for the behavior of simple gases. The coefficients of expansion and compressibility for the gas may be derived directly by applying this result to their definitional relations. Thus, all of the properties of monatomic gases can be explained on the basis of a simple atomistic model in which it is assumed that the energy of each atom is contained in its kinetic energy of translation through space.

If the gas under consideration is composed of molecules containing two or more atoms (such as H_2, CO_2, CH_4, etc.) then there are contributions to the kinetic energy of the molecule due to motions in addition to the translation of its center of mass. Specifically, a kinetic energy is associated with the rotation of the molecule and with the vibrational displacements of its atoms relative to its center of mass. A contribution may also arise from the motion of the electrons within the molecule. In order to treat polyatomic molecules, it is necessary to develop models for the energy states associated with these added kinds of motion and to compute the corresponding contributions for the partition function for the system. The partition function could then be used to compute the heat capacity of a system for a variety of assumed configurations for the molecules that compose it. Comparison with experimental measurements of heat capacities then provides insight into the molecular structure and the behavior of matter at the atomic level.

An interesting simplification arises in the treatment of molecular gases. Kinetic energies of rotation and vibration have the same mathematical form as does the translation energy, though the physical factors involved are different. For example, in describing the kinetic energy of rotation, mass is replaced by the moment of inertia of the molecule and translational velocity is replaced by angular velocity. The total kinetic energy of the molecule has the form

$$\epsilon = \sum_{j=i}^{n} b_j v_j^2 \tag{6.61}$$

There is a term for *each independent component of motion* that the molecule can exhibit. Thus the upper limit on the sum n is the number of independent motion components for the molecule. Clearly, the value of n depends on the details of the structure of the molecule. There is one rotational term if the molecule has an axis of symmetry; two if it does not. There is a vibrational term for each bond in the molecular structure.

The evaluation of the partition function for this case appears complex because it involves integration over three positional variables and n velocity variables.

$$\mathcal{P} = \int_0^{l_z} \int_0^{l_y} \int_0^{l_x} dx\,dy\,dz \int_{-\infty}^{\infty} \int_{-\infty}^{\infty} \cdots \int_{-\infty}^{\infty} e^{-(1/kT)\sum_{j=1}^{n} b_j v_j^2} dv_1 dv_2 \ldots dv_n \tag{6.62}$$

The integrand can be rewritten as a product of exponentials

$$\mathcal{P} = V \cdot \int_{-\infty}^{\infty} \int_{-\infty}^{\infty} \cdots \int_{-\infty}^{\infty} \prod_{j=1}^{n} \left[e^{(-b_j/kT)v_j^2} \right]_d v_1 dv_2 \ldots dv_n$$

Since all these factors are independent (each contains only one variable, v_j) this multiple integral can be written as a product of n integrals, *all mathematically identical in form:*

$$\mathcal{P} = V \cdot \prod_{j=1}^{n} \left[\int_{-\infty}^{\infty} e^{-(b_j/kT)v_j^2} dv_j \right]$$

This standard form of integral was evaluated for the monatomic case

$$\int_{-\infty}^{\infty} e^{-a^2 x^2} dx = \frac{\sqrt{\pi}}{a}$$

where the parameter $a^2 = b_j/kT$. Thus, for each factor in the product of integrals,

$$\int_{-\infty}^{\infty} e^{-(b_j/kT)v_j^2} dv_j = \left[\frac{\pi kT}{b_j}\right]^{1/2}$$

The partition function may be written as the product

$$\mathcal{P} = V \cdot \prod_{j=1}^{n} \left[\frac{\pi kT}{b_j}\right]^{1/2} \tag{6.63}$$

The natural logarithm of this function is

$$\ln \mathcal{P} = \ln V + \sum_{j=1}^{n} \ln \left(\frac{\pi kT}{b_j}\right)^{1/2}$$

which can be written

$$\ln \mathcal{P} = \ln V + \sum_{j=1}^{n} \frac{1}{2} \ln \left(\frac{\pi k}{b_j}\right) + \sum_{j=1}^{n} \frac{1}{2} \ln T$$

The derivative with respect to temperature at constant volume, required in the evaluation of the internal energy and heat capacity for the gas, is particularly simple

$$\left(\frac{\partial \ln \mathcal{P}}{\partial T}\right)_V = 0 + 0 + \sum_{j=1}^{n} \frac{1}{2} \cdot \frac{1}{T} = \frac{n}{2T}$$

The only parameter in the model that remains in this derivative is n, the number of independent components of motion that the system can exhibit. The internal energy of a molecular gas is obtained by inserting this value into Eq. (6.29)

$$U = N_0 kT^2 \left(\frac{\partial \ln \mathcal{P}}{\partial T}\right)_V = N_0 kT^2 \cdot \left(\frac{n}{2T}\right)$$

$$U = n \cdot \frac{1}{2} N_0 kT \tag{6.64}$$

The heat capacity is the temperature derivative of the internal energy

$$C_V = n \cdot \frac{1}{2} N_0 k \tag{6.65}$$

Thus, the heat capacity of a molecular gas is predicted to depend only upon *the number of independent components of motion that the molecule can display*. Table 6.2 presents some examples that test this prediction.

The principle derived from this application of the statistical thermodynamical algorithm, namely that each independent component of motion of the molecules in the gas contributes the *same quantity*, $\frac{1}{2}kT$, to the internal energy of the gas, is called the *principle of equipartition of energy*. Modern analytical tools provide many avenues for the direct exploration of the structure of molecules. Application of this principle provided this kind of insight before these tools developed.

Statistical thermodynamics provides the link between phenomenological information supplied by experiments in classical thermodynamics and the structure and behavior of matter at an atomic level. An explanation for an observation that the heat capacity for a system composed of molecules of gas A is larger than that for a system composed of molecules of gas B is put forward and tested. Patterns of behavior emerge and the intricacies of the behavior of matter are brought to a new level of understanding.

6.4   ALTERNATE STATISTICAL FORMULATIONS

Over the years experience has accumulated in attempts to apply the algorithm of statistical thermodynamics to an increasing range of classes of systems. It became evident that the specific formulation that computes the number of microstates corresponding to a given macrostate, Eq. (6.1), is not adequate to describe all aspects of the behavior of matter. For example, in the formulation presented here no restrictions are placed upon the number of particles that can exist in any energy state in the system. In the treatment of the behavior of *electrons*, it has been found necessary to invoke the Pauli exclusion principle, which states that no more than two electrons can exhibit the same energy at any instant in time. In distributing electrons in their allowable energy levels, this restriction must be strictly obeyed. With this restriction it is clear that the number of possible microstates corresponding to a given macrostate is not given by Eq. (6.1); a modified combinatorial analysis must be devised. Since this relationship is at the foundation of the development of statistical thermodynamics, it leads to different relationships for the equilibrium distribution of electrons over energy levels and to different forms in the predictions of macroscopic thermodynamic properties of electrons.

Three different forms of the relationship between microstates and macrostates have been prominent in the development of statistical thermodynamics. Each is labelled with the names of the original authors of the formulation. The development presented in this chapter is called *Maxwell-Boltzmann statistics*. The other developments are referred to as *Bose-Einstein* and *Fermi-Dirac* statistics. These six pioneers who lend their names to these three developments in the field of thermodynamics are giants who contribute greatly to our understanding of how the world works.

The relationship between entropy and the number of microstates per macrostate remains the same in all three developments, from Boltzmann's hypothesis, Eq. (6.3). By following the general strategy for finding the conditions for equilibrium, or more specifically, for determining the distribution of particles on energy levels that yields the maximum entropy in an isolated system, the equilibrium distribution can be obtained in each case.

Maxwell-Boltzmann Distribution

$$\frac{n_i}{N} = \frac{1}{\mathcal{P}_{\mathrm{MB}} e^{\epsilon_i/kT}} \tag{6.66}$$

Bose-Einstein Distribution

$$\frac{n_i}{N} = \frac{1}{\mathcal{P}_{\mathrm{BE}} \left(e^{\epsilon_i/kT} - 1 \right)} \tag{6.67}$$

Fermi-Dirac Distribution

$$\frac{n_i}{N} = \frac{1}{\mathcal{P}_{\text{FD}}\left(e^{\epsilon_i/kT} + 1\right)} \tag{6.68}$$

For precise definitions of the quantities in these equations consult reference 6. The remainder of the algorithm, which computes the macroscopic thermodynamic properties from these distribution functions, is formally identical to that presented in this chapter. Application of these formulations, each in the circumstances for which the underlying statistical assumptions are appropriate, has greatly broadened the scope of application of statistical thermodynamics.

6.5   SUMMARY OF CHAPTER 6

A *microstate* for a system is a specification of the state or condition of each particle in the system. The number of different microstates a given system can exhibit is enormous.

A *macrostate* for a system is a list of the number of particles existing in each condition that they may exhibit. The number of microstates that correspond to a given macrostate can be computed from combinatorial analysis:

$$\Omega = \frac{N_0!}{n_1! \cdot n_2! \cdots n_r!} \tag{6.1}$$

Boltzmann hypothesized that the entropy of a macrostate is logarithmically related to the number of microstates it contains

$$S = k \ln \Omega \tag{6.3}$$

Application of the general strategy for finding conditions for equilibrium yields, in this case, the Boltzmann distribution function

$$\frac{n_1}{N_0} = \frac{1}{\mathcal{P}} \cdot e^{-(\epsilon_i/kT)} \tag{6.24}$$

where $\mathcal{P}$ is the partition function for the system, given by

$$\mathcal{P} = \sum_{i=0}^{r} e^{-(\epsilon_i/kT)} \tag{6.25}$$

All the macroscopic thermodynamic properties of the system can be computed from the partition function.

Application of the algorithm that converts a list of energy levels into predictions of thermodynamic properties provides a rudimentary model for crystals and a satisfactory model for the behavior of gases at ordinary conditions.

Maxwell-Boltzmann statistics are inadequate to explain some aspects of the behavior of matter. Alternate formulations, namely Bose-Einstein and Fermi-Dirac statistics, significantly broaden the range of application of the statistical thermodynamics algorithm.

PROBLEMS

6.1. Explain in words your understanding of the difference between *phenomenological* thermodynamics and *statistical* thermodynamics.

6.2. Show, by listing all the macrostates, that the number of macrostates possible for a system composed of ten particles that can occupy three energy levels is 60.

6.3. Consider a system with two particles A and B that may each exhibit any of four energy levels, $\epsilon_1, \epsilon_2, \epsilon_3,$ and ϵ_4.

(*a*) How many microstates may this system exhibit?

(*b*) Enumerate its microstates.

(*c*) Use the list of microstates to generate a list of macrostates for this system.

(*d*) Identify the microstates corresponding to each macrostate.

6.4. Calculate the number of microstates corresponding to each of the following combinations:

(*a*) A system with three particles and 4 energy levels.

(*b*) A system with 15 particles and 4 energy levels.

(*c*) A system with 4 particles and 15 energy levels.

(*d*) A cluster of 50 particles each of which may reside in any of 30 energy levels.

(*e*) 1000 particles that may reside in 100 energy levels.

6.5. Consider a model in which the available energy levels are linearly spaced along the energy axis

$$\epsilon_n = \left(n + \frac{1}{2} \right) \epsilon_0 \qquad (n = 0, 1, 2, \ldots, 9)$$

The system contains ten particles. Consider two macrostates:

$$\text{State I} \quad \{0, 0, 1, 2, 4, 2, 1, 0, 0, 0\}$$

$$\text{State II} \quad \{0, 1, 1, 2, 2, 2, 1, 1, 0, 0\}$$

(*a*) Which macrostate has the higher energy?

(*b*) Which macrostate has the higher entropy?

(*c*) Which macrostate is more likely to be observed?

6.6. Compute the factorials of the following numbers: 10; 30; 60,

(*a*) Directly.

(*b*) Using Stirling's approximation.

(*c*) In each case, compute the error in $\ln x!$ that results from the approximation.

6.7. A system containing 500 particles and 15 energy levels is in the following macrostate:

$$\{14, 18, 27, 38, 51, 78, 67, 54, 32, 27, 23, 20, 19, 17, 15\}$$

This system experiences a process in which the number of particles in each energy level changes by the following amounts:

$$\{0, 0, -1, -1, -2, 0, +1, +1, +2, +2, +1, 0, -1, -1, -1\}$$

Estimate the change in entropy for this process.

6.8. Using a convenient computer applications package, calculate and plot the partition function for the Einstein model as a function of (T/θ_E). Calculate and plot the heat capacity at constant volume of a simple cubic Einstein crystal as a function of temperature in the range $0 \leq T \leq 1000$ K. Repeat the calculation for each of the following values of the

Einstein temperature, θ_E:

$$100K; \qquad 200K; \qquad 300K; \qquad 500K.$$

6.9. Use the Einstein model to compute the change in internal energy of a crystal when it is heated reversibly at one atmosphere pressure from 90 K to 210 K. Assume $\theta_E = 250$ K.

6.10. Compute the change in entropy when one mole of a monatomic ideal gas is compressed from an initial condition at 273 K and 1 atm to a final condition of 500 K at 3.5 atm.

   (a) Make the calculation using familiar phenomenological thermodynamics.

   (b) Repeat the calculation using the results of statistical thermodynamics. (*Hint*: first calculate the initial and final volumes.)

6.11. At ordinary temperatures and pressures the heat capacity of ammonia (NH_3) is 37 J/mole. Apply the principle of equipartition of energy to speculate on the spatial arrangement of the atoms in the ammonia molecule.

REFERENCES

1. Boas, M. L.: *Mathematical Methods in the Physical Sciences*, John Wiley & Sons, New York, pp. 699–706, 1983.
2. Lee, J. F., F. W. Sears, and P. L. Turcotte: *Statistical Thermodymics,* Addison Wesley Pub. Co. Inc., Reading, Mass., pp. 148–165, 1963.

CHAPTER
7

UNARY HETEROGENEOUS SYSTEMS

A system is considered to be *unary* if, for the range of states under study, it consists of a single chemical component. Each of the elements forms a unary system over its full range of existence. Molecular compounds such as CO_2 or H_2O may be treated as unary systems over most of the range of temperatures and pressures normally encountered in the laboratory. Under conditions in which they may decompose to form significant quantities of other molecules, they cannot be treated as unary systems.

From a thermodynamic point of view a system is *homogeneous* if it consists of a single phase. More specifically, a system consists of a single phase if its intensive properties are uniform or, at most, vary continuously throughout the system. A *heterogeneous* system consists of more than one phase. Some of its intensive properties exhibit discontinuities at the boundaries between the phases in a heterogeneous system.

All the elements may exist in at least three distinct states of matter or phase forms: solid, liquid, gas. In addition, many of the elements exhibit more than one phase form in the solid state. For example, at one atmosphere pressure, when pure iron is equilibrated below 910°C, it exists in a body centered cubic (BCC) crystal structure. Between 910°C and 1455°C the equilibrium state for iron is the face centered cubic (FCC) crystal structure. Between 1455°C and its melting point at 1537°C the stable form of iron is again body centered cubic. Different solid phase forms of the same element are called *allotropes*; the BCC and FCC structures are allotropic forms of iron.

132

When a piece of pure iron is heated reversibly from room temperature it remains in the BCC structure until the temperature of 910°C is attained. Continued input of heat into the system does not raise the temperature above 910°C. Instead small crystals of FCC iron nucleate within the BCC structure and begin to grow, consuming the BCC structure. The system requires the input of heat to make this transformation occur. Eventually all of the BCC structure is replaced by the FCC phase and only then does the continued input of heat begin to raise the temperature of the remade structure. This phase change from BCC to FCC iron at 910°C and one atmosphere pressure is called an *allotropic transformation*. Below 910°C, BCC is the *stable phase form*, above 910°C, FCC iron is stable. Thus, at one atmosphere, 910°C is the *limit of stability* of both phase forms.

If during the transformation from BCC to FCC iron at 910°C, the flow of heat is brought to zero, then the mixture of BCC and FCC iron can exist at equilibrium for an indefinite period. This two phase system is a *heterogeneous* system in the thermodynamic usage of the word. The part of the system that is BCC has the intensive properties of that crystal structure, such as molar volume, entropy, internal energy, and the like. The part of the structure that is FCC has the properties of FCC iron. At boundaries where these two crystal structures meet these intensive properties exhibit a discontinuity.

The allotropic transformation is a special case of a more general concept, the *phase transformation*, which applies to any change in the phase form of a system. *Melting* or *fusion*, which is a change from a solid to a liquid phase, and *boiling* or *vaporization*, which is the change from a liquid to the vapor state, are familiar examples of phase transformations. The study of all classes of phase transformations in unary as well as multicomponent systems is extremely important in materials science because these processes control the microstructure of materials. Since many properties that materials exhibit are very sensitive to the microstructure of the material, control of microstructure is tantamount to control of properties.

The set of thermodynamic conditions under which a given phase form is stable for a particular element may be succinctly summarized in a graph on temperature-pressure coordinates called a *phase diagram*. Figures 7.1 and 7.2 represent examples of phase diagrams for elements and other systems that may be treated as unary. The lines on these diagrams are called *phase boundaries* and represent the limits of stability of each of the phase forms that the system exhibits. Thus, these lines represent conditions, that is the combinations of temperature and pressure, at which transformations between the impinging phase forms occur. A system that exists at a combination of P and T lying on a phase boundary consists, at equilibrium, of a mixture of the two phases that the line bounds on the diagram. If the system is taken through a reversible process represented by any path on the (P, T) diagram, the change in temperature and pressure will be arrested while the phase transformation occurs, which is when the path intersects a phase boundary. When the change in structure is complete, the change along the (P, T) path will resume.

The limits of stability of the single phase regions in Figs. 7.1. and 7.2 are defined by the conditions under which pairs of phases may coexist at equilibrium. The equations that define the conditions for equilibrium in a unary two phase system were

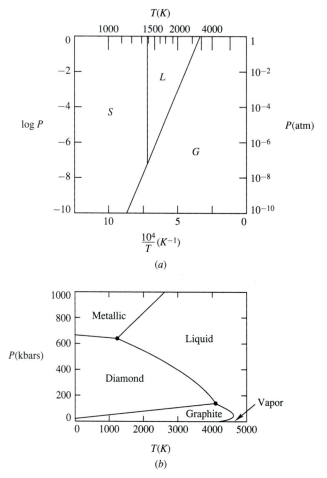

FIGURE 7.1
Typical phase diagrams for elements: (*a*) copper; (*b*) carbon.

derived in Chap. 5 as an illustration of the application of the general strategy for finding conditions for equilibrium. In this chapter it is shown how application of these conditions for equilibrium lead to the rules for constructing unary phase diagrams in (P, T) space. Combination of the three equations that describe the conditions for equilibrium into a single relationship yields the working equation, called the Clausius-Clapeyron equation, which is used for calculating such unary phase diagrams. Application of this result to the computation of phase boundaries for real systems and the estimation of phase boundaries where information is incomplete, are presented for sublimation, vaporization, melting, and solid-solid transformations. Inverting the strategy permits the estimation of some thermodynamic properties from phase diagram information.

The simplest representation of regions of stability of the phases in a unary system is obtained when the phase diagram is plotted in (P, T) coordinates. Alternate

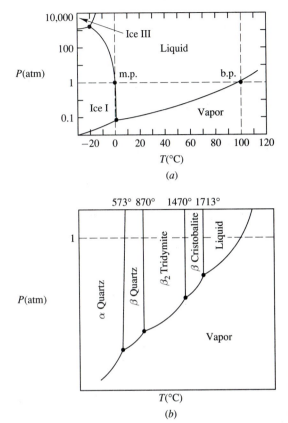

FIGURE 7.2
Pressure - temperature phase diagrams for compounds that may be treated s unary systems: (*a*) water; (*b*) SiO_2.

representations can be obtained by choosing other combinations of properties as the axes for the plot. For some purposes plotting the phase diagram in (P, V) or (S, V) coordinates may be useful. All such plots share the same underlying thermodynamic basis. However, each representation provides unique information about the relationships of the phases. Also, the rules for constructing the phase diagram are different for different choices of variables. Examples of these alternate constructions are presented in Sec. 7.6.

7.1 STRUCTURE OF UNARY PHASE DIAGRAMS IN (P, T) SPACE

As shown in Figs. 7.1 and 7.2, the representation of the limits of stability of phases on a unary phase diagram in the pressure-temperature plane has the following characteristics:

1. The domain of stability of each single phase is represented by an *area*.
2. The domain of stability for two phases coexisting in equilibrium is a *line*.

3. The domain of stability for three phases existing simultaneously in equilibrium is a *triple point*, at which three single phase areas and three two phase lines meet.
4. There are no regions where more then three phases may coexist at equilibrium.

These characteristics of unary phase diagrams are direct consequences of the conditions for equilibrium in two and three phase structures and the choice of variables (P, T) to represent the state of the system. It is shown in Chap. 9 that these characteristics can be deduced from a general principle, called the *Gibbs phase rule*, which gives rules of construction for phase diagrams with any number of components and phases. In this illustration of that principle for a unary system, it is possible to demonstrate the origins of these construction rules graphically.

7.1.1 Chemical Potential and the Gibbs Free Energy

In order to develop this graphical illustration, it is first necessary to establish the connection between the chemical potential, introduced in Chap. 5, Eqs. (5.13) and (5.14), and the molar Gibbs free energy. Fortunately, this connection is straightforward for a unary system. The term involving the chemical potential is added to the apparatus of thermodynamics in order to describe systems capable of exchanging matter with their surroundings during a process, called *open systems*. Recall that, in the treatment of open systems, it is necessary to distinguish explicitly the extensive properties of the system, designated by a prime ('), from their intensive counterparts, left unprimed in this notation. Thus, U' is the internal energy of the system, no matter how many moles it may contain and U is the corresponding molar internal energy.

If it is recognized that the number of moles of the component in a unary system may be varied during a process, then the combined statements of the first and second laws must have an additional term describing this possible change in state for the system, Eq. (5.13)

$$dU' = T dS' - P dV' + \mu dn \qquad (5.13)$$

where

$$\mu = \left(\frac{\partial U'}{\partial n} \right)_{S', V'} \qquad (5.14)$$

is the chemical potential of the component in the system. Recall the definition of the Gibbs free energy function. For a system composed of an arbitrary number of moles of substance,

$$G' = U' + PV' - TS' \qquad (4.9)$$

Differentiating

$$dG' = dU' + P dV' + V' dP - T dS' - S' dT$$

For an open, unary system, substitute for dU' with Eq. (5.13)

$$dG' = (T dS' - P dV' + \mu dn) + P dV' + V' dP - T dS' - S' dT$$

Simplifying

$$dG' = -S'dT + V'dP + \mu dn \tag{7.1}$$

Applying the coefficient relation to the third term

$$\mu = \left(\frac{\partial G'}{\partial n}\right)_{T,P} \tag{7.2}$$

In this equation G' is the Gibbs free energy of the system. Let G be the corresponding value of the Gibbs free energy *per mole of system*, i.e., the *molar Gibbs free energy*. Then $G' = nG$ and

$$\mu = \left(\frac{\partial G'}{\partial n}\right)_{T,P} = \left[\frac{\partial (nG)}{\partial n}\right]_{T,P} = G\left(\frac{\partial n}{\partial n}\right)_{T,P} = G \tag{7.3}$$

Thus, in a unary system, *the chemical potential of the component in any state is identical with the molar Gibbs free energy for that state.* This simple relationship is invoked frequently in this and subsequent chapters.

7.1.2 Chemical Potential Surfaces and the Structure of Unary Phase Diagrams

Because $\mu = G$ in a unary system, the dependence of the chemical potential upon temperature and pressure is identical to the variation of molar Gibbs free energy.

$$d\mu = dG = -SdT + VdP \tag{7.4}$$

where the coefficients are the molar entropy and molar volume for the phase form of the substance under study. If, for example, the α phase is taken through an arbitrary change in state,

$$d\mu^{\alpha} = -S^{\alpha}dT^{\alpha} + V^{\alpha}dP^{\alpha} \tag{7.5}$$

where the superscript α indicates a property that is evaluated for the α phase. The molar entropy and molar volume of the α phase can be computed as functions of temperature and pressure from heat capacity, expansion, and compressibility data as spelled out in Chap. 4. Equation (7.5) can be integrated, at least in principle, to yield a function

$$\mu^{\alpha} = \mu^{\alpha}(T^{\alpha}, P^{\alpha}) \tag{7.6}$$

This functional relationship can be visualized graphically as the surface shown in Fig. 7.3*a*. A similar argument can be applied to each phase that may exist in the system. Thus, for the liquid phase there exists a chemical potential surface that can be obtained from heat capacity, expansion, and compressibility information about the liquid phase

$$\mu^{L} = \mu^{L}(T^{L}, P^{L}) \tag{7.7}$$

Figure 7.3*b* is a schematic representation of the chemical potential of the liquid phase as a function of T and P.

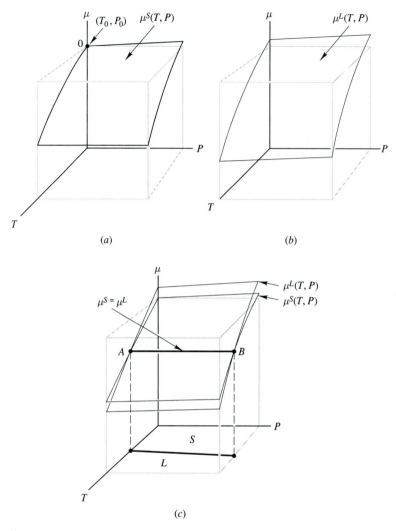

FIGURE 7.3
Sketches of the surface representing the chemical potential as a function of temperature and pressure for the solid phase (*a*) and the liquid phase (*b*) are superimposed in (*c*).

The chemical potentials of these two phases may be compared at any given (T, P) combination *if and only if* the reference state used in their computation is the *same*. Thus, some condition, defined by a choice for temperature (T_0), pressure (P_0), and *phase form* (α or L), must be chosen for the system and the chemical potentials of the α and L phases computed relative to that reference state. In Fig. 7 the reference state is chosen to be the solid phase at the condition (T_0, P_0). With this constraint, it is possible to construct surfaces for both phases on one graph, Fig. 7.3*c*.

The two surfaces intersect along a space curve *AB*. At any point on that space curve, the temperatures, pressures, and chemical potentials of the two phases are identical. These three conditions,

$$T^s = T^L \qquad P^s = P^L \qquad \mu^s = \mu^L$$

are precisely the conditions that must be met in order for the solid and liquid phases to coexist in equilibrium, Eqs. (5.26)–(5.28). Thus, the curve of intersection of the two chemical potential surfaces is the locus of points for which the solid and liquid phases are in heterogeneous equilibrium. The projection of this line onto the (P, T) plane is the phase boundary delineating the (s+L) two phase equilibrium and the limits of stability of the solid and liquid phases.

Figure 7.4 illustrates the construction when the chemical potential surface for the gas phase is added to these considerations. The curve of intersection of the solid and gas surfaces, COD, represents the (s+G) two phase equilibrium [the sublimation curve $C'O'D'$ on the (P, T) plane]. The curve produced by the intersection of the liquid and gas surfaces, labelled EOF, describes the (L+G) equilibrium [the vaporization curve $E'O'F'$ on the (P, T) plane].

All three surfaces intersect at a single point O, which is also common to the three curves that represent the two phase equilibria. At this unique point, temperatures, pressures, and chemical potentials of all three phases are the same and all three phases coexist in equilibrium. The projection of this point onto the (P, T) plane, O', is the triple point for the three phases (s+L+G).

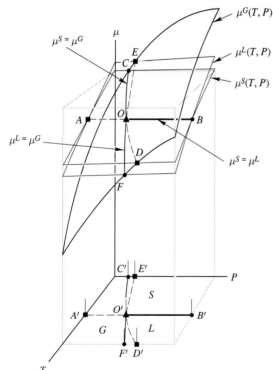

FIGURE 7.4

Superposition of chemical potential surfaces for solid, liquid and gas phases demonstrates the relationship between the structure of the phase diagram and the underlying thermodynamics.

7.1.3 Calculation of Chemical Potential Surfaces

The surface represented by the function $\mu^\alpha(T, P)$ can be computed by integrating Eq. (7.5). This integration can be carried out by fixing the value of P and integrating the term $-S^\alpha dT$ over temperature, producing an isobaric section through the surface. Repetition of this process at a sequence of values of P generates the surface. The information required to produce such an isobaric section for the α phase is the temperature dependence of the entropy of the α phase. The absolute value of the entropy of the α phase can be computed from

$$dS_P^\alpha = \frac{C_P^\alpha}{T} dT$$

or

$$S^\alpha(T) = S_{298}^\alpha + \int_{298}^T \frac{C_P^\alpha(T)}{T} dT \qquad (7.8)$$

Substitute this entropy expression into

$$d\mu^\alpha = dG^\alpha = -S^\alpha(T)dT = -\left[S_{298}^\alpha + \int_{298}^T \frac{C_P^\alpha(T)}{T} dT\right] dT$$

Integrate again from the reference temperature, 298 K, to a variable temperature, T

$$\int_{298}^T d\mu^\alpha = \int_{298}^T -\left[S_{298}^\alpha + \int_{298}^T \frac{C_P^\alpha(T)}{T} dT\right] dT$$

which gives

$$\mu^\alpha(T) - \mu^\alpha(298) = G^\alpha(T) - G^\alpha(298) = \int_{298}^T -\left[S_{298}^\alpha + \int_{298}^T \frac{C_P^\alpha(T)}{T} dT\right] dT \quad (7.9)$$

Evaluation of this integral requires the absolute entropy of the α phase at 298 K and the heat capacity of the α phase form as a function of temperature. The curve labelled α in Fig. 7.5 plots the result of such a calculation.

A similar strategy yields the free energy curve for the liquid phase. Some appropriate bookkeeping must be observed in constructing this curve. The point has been made that comparison of these chemical potential values is possible only if they are all calculated from the same reference state. The reference state chosen for Eq. (7.9) is the α phase form at 298 K. This must also be the reference state for the calculation of the free energy of the liquid phase. Thus, the curve marked L plotted in Fig. 7.5 is the function $G^L(T) - G^\alpha(298)$. The connection between the liquid and solid phases can be made at the melting point T_m, where the equality of the chemical potentials implies that

$$G^L(T_m) = G^\alpha(T_m)$$

The quantity plotted in Fig. 7.5 can be written

$$G^L(T) - G^\alpha(298) = [G^L(T_m) - G^L(T_m)] + [G^\alpha(T_m) - G^\alpha(298)] \qquad (7.10)$$

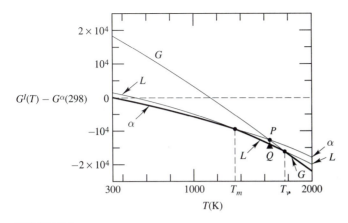

FIGURE 7.5
The temperature dependences of the molar free energy (equivalent to chemical potential) for the solid, liquid, and gas phases are computed at one atmosphere pressure. Intersections of the curves represent conditions for two phase equilibrium. All values are computed relative to that of the solid phase at 298 K.

The right-hand side merely adds and subtracts equal quantities to the expression on the left-hand side, since at the melting point $G^L T_m = G^\alpha(T_m)$. The quantity in the first bracket on the right sidehand may be evaluated by a straightforward integration of $dG = -SdT$ for the liquid phase, as was done for the α phase in deriving Eq. (7.9):

$$G^L(T) - G^L(T_m) = \int_{T_m}^T -\left[S^L(T_m) + \int_{T_m}^T \frac{C_P^L(T)}{T} dT \right] dT \qquad (7.11)$$

The absolute entropy of the liquid at the melting point can be computed from the absolute entropy of the α phase at the melting point and the entropy of fusion, $\Delta S^{\alpha \to L}(T_m)$:

$$S^L(T_m) = S_{298}^\alpha + \int_{298}^{T_m} \frac{C_P^\alpha(T)}{T} dT + \Delta S^{\alpha \to L}(T_m) \qquad (7.12)$$

The quantity in the second brackets in Eq. (7.10) can be read from the curve for the α phase at T_m. Equation (7.10) can then be plotted as a function of temperature as the liquid curve in Fig. 7.5.

A similar construction, based upon the connection between the liquid and gas phases at the vaporization temperature T_v, computes the free energy curve for the gas phase, relative to the pure α phase at 298 K. This curve is labelled G in Fig. 7.5. If the solid state exhibited allotropic forms other than α, curves for those phases would be computed in an analogous manner.

7.1.4 Competing Equilibria: Metastability

The general criterion for equilibrium derived in Chap. 5 is based upon the principle that in an isolated system entropy can only increase and thus, at equilibrium in an isolated system, the entropy is a maximum. A corollary derived in Sec. 5.5 demonstrates that in a system constrained to constant temperature and pressure, the Gibbs free energy can only decrease. Hence, at equilibrium, the Gibbs free energy is a minimum. The conditions for equilibrium derived from this extremum principle are identical to those derived from the principle that states that the entropy is a maximum at equilibrium in an isolated system, where the chemical potential, temperature, and pressure are uniform at equilibrium. The constructions in Fig. 7.4 and on the isobaric section in Fig. 7.5 show how these conditions for equilibrium determine the structure of one, two, and three phase fields in unary systems.

The free energy curves on the isobaric section of Fig. 7.5 illustrate these relationships. At each temperature in the range from 298 K to T_m, the α phase has the lowest free energy and thus is the stable phase form. Between T_m and T_v the liquid curve has the lowest free energy and liquid is stable. Above T_v the gas phase is stable. At any temperature T_1, the vertical distance between the curves, which reports the difference in free energy between the phase forms at that temperature, gives a quantitative measure of the relative stabilities of the phases.

It will also be noted that the α and gas curves intersect at the point marked P. At that point the molar free energy (or chemical potential) of the solid and gas phases are equal and the conditions for equilibrium between those two phases are satisfied. However, there exists a separate condition of the system, labelled Q on the liquid curve, which has a lower free energy and is therefore the state of stable equilibrium. The two phase equilibrium represented by the point P is said to be *metastable*; the stable equilibrium structure at that temperature is the liquid phase.

A broader perspective may be obtained by examining Fig. 7.4. At each point along the curve *COD*, the chemical potentials, temperatures, and pressures of the solid and gas phases are the same and the conditions for equilibrium are satisfied with respect to these two phases. However, points on this curve in the segment between O and D lie above the liquid surface at the same temperature and pressure. According to the extremum principle, in a comparison between states that have the same temperature and pressure, the equilibrium state is that for which the Gibbs free energy is lower. The liquid phase has a lower free energy than does an equilibrated mixture of solid and gas in this domain. Thus, for pressure-temperature combinations lying along the segment *OD,* the liquid phase form is the equilibrium state for the system. Mixtures of ($s+G$) that satisfy the conditions for equilibrium but do not possess the minimum free energy possible for the system are said to be in a state of *metastable equilibrium.*

Metastable equilibrium states play a key role in many phase transformations that are important in materials science. The sequence of thermodynamic states through which a system passes as it approaches the ultimate equilibrium condition frequently encounters states of metastable equilibrium. Indeed, if the kinetics of the transformation are slow, a system may encounter a metastable equilibrium condition and remain in that state indefinitely so that the material can be used in its metastable state.

7.2 THE CLAUSIUS-CLAPEYRON EQUATION

The curves that are the domains of two phase equilibria in (P, T) space in unary systems may each be described mathematically by some function $P = P(T)$. The Clausius-Clapeyron equation, derived in this section, is a differential form of this equation. For any pair of phases that may coexist in the unary system, integration of the Clausius-Clapeyron equation yields a mathematical expression for the corresponding phase boundary on the phase diagram. Repeated application to all the pairs of phases that may exist in the system yields all possible two phase domains. Intersections of the two phase curves produce triple points where three phases coexist. Thus, the Clausius-Clapeyron equation is the only relation required for calculating unary phase diagrams.

Consider any pair of phases α and β known to exist in the system. If the α phase is taken through any arbitrary change in state, its change in chemical potential is given by Eq. (7.5):

$$d\mu^\alpha = -S^\alpha dT^\alpha + V^\alpha dP^\alpha \qquad (7.5)$$

If the β phase is taken through an arbitrary change in state, its chemical potential obeys the analogous relation

$$d\mu^\beta = -S^\beta dT^\beta + V^\beta dP^\beta \qquad (7.13)$$

S^α and V^α are molar properties of the α phase and S^β and V^β are those for the β phase. If during this infinitesimal process, α and β are maintained in two phase equilibrium, then the changes in P, T, and μ of each phase are constrained by the conditions for equilibrium derived in Chap. 5

$$T^\alpha = T^\beta \rightarrow dT^\alpha = dT^\beta = dT \qquad (7.14)$$

$$P^\alpha = P^\beta \rightarrow dP^\alpha = dP^\beta = dP \qquad (7.15)$$

$$\mu^\alpha = \mu^\beta \rightarrow d\mu^\alpha = d\mu^\beta = d\mu \qquad (7.16)$$

If α and β are maintained in equilibrium during the process, then the relationship $T^\alpha = T^\beta$ requires that both phases experience the same temperature change. Similar statements apply to the pressures and chemical potentials.

Equations (7.14)–(7.16) imply that Eqs. (7.5) and (7.13) may be set equal to each other and that the superscripts may be dropped from the dT and dP factors.

$$d\mu^\alpha = -S^\alpha dT + V^\alpha dP = d\mu^\beta = -S^\beta dT + V^\beta dP$$

Combining like terms

$$(S^\beta - S^\alpha)dT = (V^\beta - V^\alpha)dP \qquad (7.17)$$

The coefficient of dT in this equation is the entropy per mole of the β phase form minus the entropy per mole of the α form. This quantity may be replaced by

$$\Delta S^{\alpha \rightarrow \beta} \equiv S^\beta - S^\alpha \qquad (7.18)$$

which is the *change in entropy accompanying the transformation of one mole of the α phase form to the β form* at the temperature and pressure being considered. Similarly,

$$\Delta V^{\alpha \rightarrow \beta} \equiv V^\beta - V^\alpha \qquad (7.19)$$

is the *change in molar volume* accompanying the transformation from α to β. Equation (7.17) becomes

$$\Delta S^{\alpha \rightarrow \beta} dT = \Delta V^{\alpha \rightarrow \beta} dP \qquad (7.20)$$

Rearrangement of this equation gives

$$\frac{dP}{dT} = \frac{\Delta S^{\alpha \to \beta}}{\Delta V^{\alpha \to \beta}} \tag{7.21}$$

which is one form of the Clausius-Clapeyron equation. This result implies that at any point on a phase boundary that represents a two phase equilibrium in a unary P-T diagram, the slope is equal to the ratio of the entropy change to the volume change for the associated phase transformation. Since the nature of the two phases α and β was not specified, this result is *general*, applying to all two phase boundaries. Integration of this equation, which requires information about how ΔS and ΔV vary with temperature and pressure, yields the phase boundary $P = P(T)$ for the $(\alpha + \beta)$ equilibrium.

The entropy change associated with a phase transformation is not measured directly in experiments. Calorimetric measurements yield values for the heat of transformation, such as the heat of fusion or the heat of vaporization. Since the transformation occurs isobarically under reversible conditions, the heat of transformation is equal to the enthalpy change for the process

$$Q^{\alpha \to \beta} = \Delta H^{\alpha \to \beta} = H^\beta - H^\alpha \tag{7.22}$$

This measurement can be connected with the entropy of the transformation through the following development. Recall the defined relationship for the Gibbs free energy and apply it to each phase

$$G^\alpha = H^\alpha - T^\alpha S^\alpha$$

$$G^\beta = H^\beta - T^\beta S^\beta$$

Recall that, if α and β are in equilibrium, then $\mu^\alpha = \mu^\beta$. Since the chemical potential is equal to the molar Gibbs free energy for a unary phase, Eq. (7.3), this condition implies that $G^\alpha = G^\beta$. Thus

$$G^\alpha = H^\alpha - T^\alpha S^\alpha = G^\beta = H^\beta - T^\beta S^\beta$$

Thermal equilibrium requires that $T^\alpha = T^\beta = T$. Rearranging

$$H^\beta - H^\alpha = T(S^\beta - S^\alpha)$$

or for any phase transformation in a unary system at any point (P, T) along the curve,

$$\Delta S^{\alpha \to \beta} = \frac{\Delta H^{\alpha \to \beta}}{T} \tag{7.23}$$

By Eq. (7.22) the enthalpy change is identical to the heat of transformation. The entropy change for a phase transformation in a unary system is thus the heat of transformation divided by the transformation temperature.

Insert Eq. (7.23) into the Clausius-Clapeyron equation, Eq. (7.21)

$$\frac{dP}{dT} = \frac{\Delta H^{\alpha \to \beta}}{T \Delta V^{\alpha \to \beta}} \tag{7.24}$$

This is the form of the Clausius-Clapeyron equation used most frequently in computing unary phase diagrams.

7.3 INTEGRATION OF THE CLAUSIUS-CLAPEYRON EQUATION

To integrate Eq. (7.24) to compute a phase boundary, it is necessary to develop expressions for ΔH and ΔV for the phase transformation as functions of temperature and pressure. In Chap. 4 general expressions are derived for the change in all the state functions accompanying changes in temperature and pressure, Table 4.3. The temperature and pressure dependence of ΔH and ΔV can be related to these results by applying the simple principle that the differential of a sum is the sum of the differentials. Thus, to obtain

$$\Delta H^{\alpha \to \beta} = \Delta H^{\alpha \to \beta}(T, P)$$

use the relation

$$d(\Delta H^{\alpha \to \beta}) = d(H^\beta - H^\alpha) = dH^\beta - dH^\alpha \tag{7.25}$$

Recall Eq. (4.42) in Table 4.3

$$dH = C_P dT + V(1 - T\alpha)dP \tag{4.42}$$

In this equation α is the coefficient of thermal expansion for the system, not to be confused with the *superscript* α designating a particular phase form in the present application. Evaluate Eq. (4.42) separately for the α and β phases, substitute the results into equation (7.25) and collect like terms.

$$d\Delta H^{\alpha \to \beta} = \Delta C_P dT + \Delta[V(1 - T\alpha)]dP \tag{7.26}$$

where

$$\Delta C_P \equiv C_P^\beta - C_P^\alpha \tag{7.27}$$

and

$$\Delta[V(1 - T\alpha)] = [V^\beta(1 - T\alpha^\beta)] - [V^\alpha(1 - T\alpha^\alpha)] \tag{7.28}$$

Assessment of the coefficient of dP in Eq. (7.26) for typical values of the properties involved shows that this term is negligible for pressure changes up to 100,000 atmospheres. Thus, for all practical purposes the enthalpy of transformation can be considered a function of temperature only, given by

$$d\Delta H^{\alpha \to \beta} = \Delta C_P dT \tag{7.29}$$

Heat capacity data for any substance is generally tabulated with the help of the empirical equation,

$$C_P = a + bT + \frac{c}{T^2} \tag{7.30}$$

(See Appendix D.) Each phase has its own values of the constants a, b, and c in this equation. The difference in heat capacities for two phases defined in Eq. (7.21) can be written

$$\Delta C_P = \Delta a + \Delta b T + \frac{\Delta c}{T^2} \tag{7.31}$$

where $\Delta a = a^\beta - a^\alpha$, etc. This result can be used to integrate Eq. (7.29) from some reference temperature T_0 to a variable upper limit T

$$\int_{T_0}^{T} d\Delta H = \int_{T_0}^{T} \Delta C_p(T) dT = \int_{T_0}^{T} \left[\Delta a + \Delta b T + \frac{\Delta c}{T^2} \right] dT$$

$$\Delta H(T) - \Delta H(T_0) = \left[\Delta a T + \Delta b \frac{T^2}{2} - \Delta \frac{c}{T} \right]\Bigg|_{T_0}^{T}$$

or

$$\Delta H(T) = \Delta a T + \frac{\Delta b}{2} T^2 - \Delta c \frac{1}{T} + \Delta D \tag{7.32}$$

where

$$\Delta D = \Delta H(T_0) - \left[\Delta a T_0 + \frac{\Delta b}{2} T_0^2 - \Delta c \frac{1}{T_0} \right] \tag{7.33}$$

This result may be used to evaluate the numerator in Eq. (7.24).

Evaluation of the denominator requires computation of the volume change for the transformation as a function of temperature and pressure. To obtain

$$\Delta V^{\alpha \to \beta} \Delta V^{\alpha \to \beta}(T, \ P)$$

it is convenient to use an approach that is different from that applied to the enthalpy. By definition,

$$\Delta V^{\alpha \to \beta}(T, \ P) \equiv V^\beta(T, \ P) - V^\alpha(T, \ P) \tag{7.34}$$

If β is the vapor phase and α is either a solid or liquid phase, then $V^\beta \gg V^\alpha$. (At standard temperature and pressure the molar volume of a gas occupies 22,400 cc; the molar volume of a condensed phase is typically about 10 cc.) Further, at temperatures and pressures at which the vapor phase is ordinarily encountered for most substances, deviations from ideal gas behavior are negligible. Thus, for the sublimation or vaporization curves on the phase diagram

$$\Delta V^{\alpha \to G} = V^G - V^\alpha \cong V^G = \frac{RT}{P} \tag{7.35}$$

Insertion of this result into the Clausius-Clapeyron equation together with the expression for the enthalpy of the transformation permits calculation of sublimation and vaporization curves.

If α and β are both condensed phases (solids or liquids) then information about the coefficients of expansion and compressibility for the two phases is required to compute ΔV as a function of temperature and pressure. If the pressure range under consideration is of the order of a few tens of atmospheres, then the assumption that ΔV is a constant is adequate for most calculations. This assumption is not valid if the phase diagram is to be calculated over a range of many thousands of atmospheres.

Adopt the convention that the β phase in this development is always that which is stable at the higher temperature. Then the numerator in the Clausius-Clapeyron equation (ΔS or ΔH) is positive. The sign of the slope of the $\alpha\beta$ phase boundary is thus determined by the sign of ΔV in the denominator. For sublimation and vaporization, ΔV is a large positive number and slopes are correspondingly small and positive. When both phases are condensed, $\Delta V^{\alpha-\beta}$ is *usually* positive but *may* be negative; the slope of the corresponding phase boundary has the same sign. The most familiar case in which the volume decreases in passing from the solid to liquid phase form is water. Ice cubes float in water because the density of ice is less than that of water; the molar volume of ice is larger than that of water and $\Delta V = V_{\text{water}} - V_{\text{ice}}$ is negative. The slope of the (S+L) boundary for water is negative, Fig. 7.2.

7.3.1 Vaporization and Sublimation Curves

If one of the phases is the vapor or gas phase, integration of the Clausius-Clapeyron Eq. (7.24) is accomplished by substituting Eq. (7.32) for ΔH and (7.35) for ΔV

$$\frac{dP}{dT} = \frac{\Delta H}{T \Delta V} = \frac{\left[\Delta a T + (\Delta b/2)T^2 - \Delta c(1/T) + \Delta D\right]}{T\,(RT/P)}$$

Separating the variables

$$\frac{dP}{P} = \frac{1}{R}\left[\frac{\Delta a}{T} + \frac{\Delta b}{2} - \frac{\Delta c}{T^3} + \frac{\Delta D}{T^2}\right]$$

Integrating

$$\ln\frac{P}{P_0} = \frac{1}{R}\left[\Delta a \ln T + \frac{\Delta b}{2}T + \frac{\Delta c}{T^2} - \frac{\Delta D}{T} - C_0\right] \tag{7.36}$$

where Δa, Δb, and Δc are obtained from empirical heat capacity information, as defined in Eq. (7.31) and

$$C_0 = \left[\Delta a \ln T_0 + \frac{\Delta b}{2}T_0 + \frac{\Delta c}{T_0^2} - \frac{\Delta D}{T_0}\right] \tag{7.37}$$

where P_0 and T_0 are coordinates of a known point on the phase boundary. This equation describes all such phase boundaries with precision over the range for which the empirical description of heat capacity and the ideal gas approximation are valid.

The vaporization curve, Eq. (7.36), takes on a simpler form if the range of temperatures of interest in a given application is restricted to a few hundred degrees Kelvin. Consider again Eq. (7.29) for the variation in the heat of transformation as a function of temperature. The integrated form of this equation can be written

$$\Delta H(T) = \Delta H(T_0) + \int_{T_0}^{T} \Delta C_P(T)dT \tag{7.38}$$

Heats of vaporization and sublimation are typically in the range of 100 KJ. The energy changes associated with heating the condensed and vapor phases are typically a few KJ and the differences in energy associated with their heating is even smaller. Thus, unless the temperature range is very large, in practical calculations the second term in Eq. (7.38) can be neglected and, at any temperature, $\Delta H(T)$ may be taken to be $\Delta H(T_0)$. If ΔH is thus assumed to be independent of temperature, integration of the Clausius-Clapeyron equation is simplified to

$$\frac{dP}{P} = \frac{\Delta H}{RT^2}dT$$

Intergration gives:

$$\ln\left(\frac{P}{P_0}\right) = \frac{\Delta H}{R}\left(\frac{1}{T} - \frac{1}{T_0}\right) \tag{7.39}$$

This expression predicts that if the vapor pressure in equilibrium with a condensed phase is measured at a series of temperatures, a plot of the logarithm of the vapor pressure versus the reciprocal of the temperature should be a straight line with a slope equal to $-\Delta H/R$. Figure 7.6 shows such a plot for many of the elements. All these curves obey this relationship at pressures below one atmosphere. The filled circle on most of these curves represents the triple point; the slight change in slope at that point accompanies the transition from the vaporization curve to the sublimation curve.

Figure 7.6 provides a first example of the use of thermodynamic understanding of phase equilibria to compute a thermodynamic property of the system from the form of a phase boundary. At pressures below one atmosphere, it is possible to compute the heat of vaporization (or sublimation) from the slope of the vapor pressure curve.

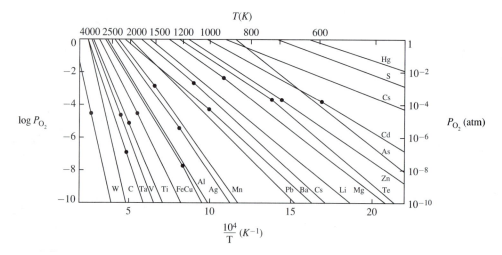

FIGURE 7.6

Compilation of vapor pressure curves for common elements, plotting the logarithm of the vapor pressure versus $1/T$. A change in slope corresponds to the melting point.

To emphasize the significance of this result, realize that no thermal (calorimetric) measurements are required to evaluate the heat of vaporization. A set of measurements of vapor pressure at a series of temperatures yields this result.

Note also that the slopes of these vaporization curves increase with increasing boiling point of the element. (Recall that the *boiling point* is defined as the temperature at which liquid and vapor are in equilibrium when the vapor pressure is equal to one atmosphere.) This trend may be traced to *Trouton's Rule*, an empirical observation that the entropy of vaporization of most elements has about the same value, about 92 J/gm-atom. This rule implies that the *enthalpy of vaporization*, which is related to the entropy by Eq. (7.23), increases in proportion to the boiling point. Since the slopes in Fig. 7.6 are proportional to the enthalpy of vaporization, they increase with the boiling point of the element.

7.3.2 Phase Boundaries Between Condensed Phases

The variation of ΔV, ΔS and ΔH with temperature and pressure is related to differences in heat capacities and coefficients of expansion and compressibility of the two phases involved. An approximate calculation of the phase boundary can be obtained by neglecting the temperature and pressure dependence of these quantities. Integration of Eq. (7.21) is then straightforward

$$P - P_0 = \frac{\Delta S}{\Delta V}(T - T_0) \tag{7.40}$$

where (P_0, T_0) is a known point on the phase boundary, frequently the point of the equilibrium temperature at $P =$ one atmosphere. With these assumptions the phase boundary is a straight line through (P_0, T_0) with slope $\Delta S/\Delta V$.

Alternatively, integration of Eq. (7.24) yields the result

$$dP = \frac{\Delta H}{\Delta V} \cdot \frac{dT}{T}$$

$$P - P_0 = \frac{\Delta H}{\Delta V} \cdot \ln\left(\frac{T}{T_0}\right) \tag{7.41}$$

This result can be shown to be equivalent to Eq. (7.40) within the bounds of the approximations applied. Expansion of the logarithm gives

$$\ln\left(\frac{T}{T_0}\right) = \left(\frac{T}{T_0} - 1\right) - \frac{1}{2}\left(\frac{T}{T_0} - 1\right)^2 + \frac{1}{3}\left(\frac{T}{T_0} - 1\right)^3 - \cdots$$

Since the range of temperature considered is small compared to T_0, higher order terms may be neglected. Then

$$P - P_0 = \frac{\Delta H}{\Delta V}\left(\frac{T - T_0}{T_0}\right)$$

which is identical with Eq. (7.40) since $\Delta S = \Delta H/T_0$.

For precise computations of phase boundaries between condensed phases, experimental values for heat capacities and coefficients of expansion and compressibility are required as functions of temperature and pressure for both phases. Then the temperature dependence of S or H, and V can be computed by applying equations such as Eq. (7.32) and the Clausius-Clapeyron equation subsequently integrated.

7.4   TRIPLE POINTS

On a P-T unary phase diagram three phases can coexist in equilibrium only at isolated points represented by the single point of intersection of the three chemical potential surfaces, as shown for the solid, liquid, and gas phases in Fig. 7.4. The triple point is also a point of intersection of the three two phase curves (s+L), (s+G), and (L+G) that form between these three phases. If the system has a fourth phase form with a chemical potential surface lying below this triple point, then the (s+L+G) triple point is *metastable* and does not appear on the stable phase diagram. If this is not the case, then the triple point is stable and does appear. In that case, the configuration around the triple point, deduced from Fig. 7.4, takes the form shown in Fig. 7.7. Each intersecting two phase field has one leg that is stable; the two phase equilibrium becomes metastable as it extends beyond the triple point. The metastable extensions always lie between the stable legs of the other pair of two phase equilibria. Thus, in a circuit around a triple point, segments of two phase equilibrium lines encountered must always alternate between stable and metastable. This structure, which can be seen graphically in Fig. 7.7, is characteristic of *all* triple points that can exist in the system.

Since a triple point (P_t, T_t) is the intersection of three, two phase equilibrium curves, it is a point that lies on all three lines. Thus, the point (P_t, T_t) simultaneously satisfies the Clausius-Clapeyron equation for all three of the two phase equilibria existing among the triplet of phases. The triple point can be computed algebraically by the simultaneous solution of any pair of these equations.

It is also characteristic of a triple point that the properties of the three pairs of phase changes, ΔS, ΔH and ΔV, are necessarily related since, for example

$$\Delta V^{\alpha \to G} = V^G - V^\alpha = V^G - V^L + V^L - V^\alpha$$

$$= (V^G - V^L) + (V^L - V^\alpha) = \Delta V^{L \to G} + \Delta V^{\alpha \to L} \qquad (7.42)$$

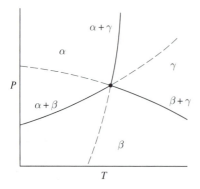

FIGURE 7.7
Sequence of stable amd metastable equilibria characteristic of all triple points on P-T phase diagrams.

Thus, the property change for the solid to gas transformation is the sum of the change from solid to liquid and liquid to gas. This result is simply an application of the principle that requires changes in state functions to be independent of the path. The change in volume (or entropy or enthalpy) for the formation of the vapor directly from the solid is identical to that obtained for a process in which the solid first melts and then the resulting liquid vaporizes.

To illustrate the computation of a triple point, consider again the $(\alpha+L+G)$ three phase equilibrium. This point is the intersection of the sublimation, melting, and vaporization curves. Equation (7.39) gives the mathematical form for the equilibria involving the vapor phase. Let the superscript (^{s}) denote properties of the sublimation curve and (^{v}) those of the vaporization curve. For the solid-vapor equilibrium

$$P^s = A^s e^{-(\Delta H^s / RT)} \tag{7.43}$$

For the liquid-vapor equilibrium,

$$P^v = A^v e^{-(\Delta H^s / RT)} \tag{7.44}$$

The triple point, (P_t, T_t), lies on both these curves. Thus

$$P_t = A^s e^{-(\Delta H^s / RT_t)} \qquad \text{and} \qquad P_t = A^v e^{-(\Delta H^v / RT_t)}$$

Solve these two equations for P_t and T_t, the coordinates of the triple point:

$$T_t = \frac{\Delta H^s - \Delta H^v}{R \ln (A^s / A^v)} \tag{7.45}$$

$$P_t = A^v e^{\Delta H^s / (\Delta H^v - \Delta H^s)} \tag{7.46}$$

If values for the constants for the sublimation and vaporization curves are known, these equations permit calculation of the solid-liquid-gas triple point.

A more practical basis for calculating the solid-liquid-gas triple point makes use of the observation that the vaporization and sublimation curves are linear on a plot of logarithm of the pressure versus $(1/T)$ in the low pressure range where this triple point appears. Construct the appropriate axes, as shown in Fig. 7.8. Note that the $(1/T)$ axis is reversed so that temperature increases from left to right. Caution; note that this change of direction on the $(1/T)$ axis changes the sign of slopes computed from the equations describing curves on this graph.

Plot the melting point and boiling point (defined to be the temperature at which the equilibrium vapor pressure is one atmosphere) on the one atmosphere line. Since the total pressure difference between the triple point and one atmosphere is a fraction of an atmosphere, the melting temperature changes only by a small fraction of one degree. Thus, on the scale of the diagram, the melting curve is a vertical line and the temperature of the triple point, T_t, is negligibly different from the melting point at one atmosphere. Use Eq. (7.39) to compute the vapor pressure at this temperature, P_t. Draw a straight line from the boiling point to (P_t, T_t); this is the (L+G) two phase line.

The sublimation curve is also a straight line on this plot. Compute the slope of the sublimation line (ΔH^s) from the heat of melting and vaporization by analogy

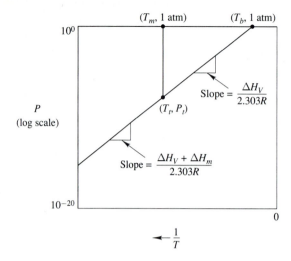

FIGURE 7.8

Illustration of the construction of a P-T phase diagram at pressures below one atmosphere. The temperature of the triple point is negligibly different from the melting point at one atmosphere.

with Eq. (7.36)

$$\Delta H^{\mathrm{s}} = \Delta H^{\mathrm{m}} + \Delta H^{\mathrm{v}} \tag{7.47}$$

and construct the line with the slope $\Delta H^{\mathrm{s}}/R$ through the triple point. If there are no allotropic forms for the solid phase, the unary phase diagram in the range below one atmosphere is complete.

Example 7.1   Phase diagram for silicon. The pertinent properties of pure silicon are given in Appendices D and E. The melting and boiling points are labelled M and B in Fig. 7.9. Substitution of $P = 1$ at $T_{\mathrm{b}} = 2750$ K into Eq. (7.44) together with $\Delta H^{\mathrm{v}} = 297$ KJ permits computation of the coefficient A^{v}

$$A^{\mathrm{v}} = P^{\mathrm{v}} e^{\Delta H^{\mathrm{v}}/RT} = (1 \text{ atm}) \cdot e^{297000/8.314 \cdot 2750} = 4.38 \times 10^5 \text{ atm}$$

The vapor pressure curve is given by

$$P^{\mathrm{v}} = 4.38 \times 10^5 (\text{atm}) \cdot e^{-(297000/8.314 \cdot T)}$$

Set the triple point temperature equal to the melting point, $T_{\mathrm{t}} = T_{\mathrm{m}} = 1683$ K. Substitute this value into the vapor pressure equation to obtain the pressure at the triple point

$$P_{\mathrm{t}} = 4.38 \times 10^5 \cdot e^{-(297000/8.314 \cdot 1683)} = 2.65 \times 10^{-4} \text{ atm}$$

Plot the triple point $(P_{\mathrm{t}}, T_{\mathrm{t}})$ as the point O and construct the lines MO and BO.
Apply Eq. (7.47) to compute the heat of sublimation.

$$\Delta H^{\mathrm{s}} = \Delta H^{\mathrm{m}} + \Delta H^{\mathrm{v}} = 46.4 + 297 = 343.4 \left(\frac{\mathrm{J}}{\mathrm{gm\text{-}atom}} \right)$$

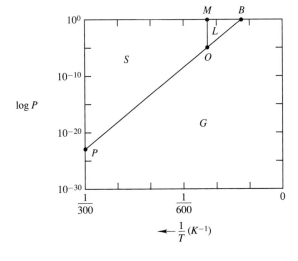

FIGURE 7.9

Illustration of the calculation of the phase diagram at low pressures for silicon.

The triple point (P_t, T_t) also lies on the sublimation curve. Substitute this information, together with the heat of sublimation, into Eq. (7.43) to evaluate the coefficient A^s

$$A^s = P_t e^{343400/8.314 \cdot 1683} = 1.21 \times 10^7 \text{ atm}$$

The sublimation curve is given by

$$P^s = 1.21 \times 10^7 e^{-(343400/8.314 \cdot T)}$$

To plot the sublimation curve, choose a temperature value of 300 K and use the above equation to compute the vapor pressure over the solid at that temperature. This point is labelled P in Fig. 7.9. A straight line from O to P is the sublimation curve. Label the solid, liquid, and vapor fields. The phase diagram for silicon below one atmosphere, computed from melting and boiling points and the heats of fusion and vaporization, is complete. The calculated triple point occurs at $(2.65 \times 10^{-4} \text{ atm}, 1683 \text{ K})$.

7.5 COMPUTER CALCULATIONS OF (P,T) UNARY PHASE DIAGRAMS

Figure 7.10 shows the experimentally determined P-T phase diagram for pure bismuth. On the pressure scale the vapor phase region is flattened against the temperature axis. A systematic computation of this complicated diagram requires:

1. A list of all the possible phase forms that may appear: I, II, III, . . . , L, G.

2. Heat capacity and coefficients of expansion and compressibility for each of the phases, so that its chemical potential surface $\mu(T, P)$ can be computed.

3. Transition temperatures, heats, and volume changes for transitions at one atmosphere or at other specific conditions, so that the relative positions of the chemical potential surfaces can be fixed.

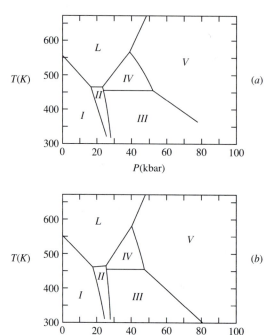

FIGURE 7.10

Experimentally determined P-T phase diagram for pure bismuth (*a*) is compared with a computed diagram in which temperature dependence on the parameters is neglected (*b*).

Each phase has its own chemical potential surface as sketched in Fig. 7.3*a*. If there are p phases in the system [$(p - 2)$ solid phase allotropic forms, liquid and vapor], then there are p chemical potential surfaces.

Pairwise intersections of these surfaces describe two phase equilibria. The number of possible two phase equilibria is equal to the number of ways in which p things can be arranged two at a time, namely $p!/[2!(p - 2)!]$. Every set of three phases produces a configuration like that shown in Fig. 7.4, yielding a triple point qualitatively like that shown in Fig. 7.7; there are $p!/[3!(p - 3)!]$ possible triple points in the system. Many, if not most, of these calculated triple points prove to be *metastable*, that is, lie above one or more chemical potential surfaces of phases other than the three defining the triple point. Alternatively, such calculated points may be outside the range of physically meaningful thermodynamic states, for example, in the range where either T or P is negative.

For example, in Fig. 7.10 bismuth exhibits five solid phase forms, liquid and vapor and $p = 7$. There are $7!/(3!4!) = 35$ triple points possible in this system. On the diagram shown [including the (I-L-G) triple point and an evident extrapolated intersection of the I-II and II-III lines to form a I-II-III triple point below room temperature], there are only six stable triple points. The remaining 29 do not appear. Stable triple points can be identified by comparing the value of the chemical potential characteristic of the three phases involved at that (P_t, T_t) with the chemical potentials of the remaining phases in the system at the same (P_t, T_t). If and only if the triple point chemical potential lies below that of all the remaining phases, then the calculated triple point is stable and appears in the final diagram.

Once the stable triple points are identified and plotted, the two phase curves that connect them can be constructed, completing the diagram.

7.6 ALTERNATE REPRESENTATIONS OF UNARY PHASE DIAGRAMS

The topology of a unary phase diagram in (P, T) space may be described as a "simple cell structure." The areas of the cells are one phase fields, cell boundaries are two phase domains, and cell vertices are triple points representing three phase equilibria. This simple construction arises because the variables used to describe the state of the phases in the system, P and T, are both "thermodynamic potentials." These properties are precisely those that appear explicitly in the conditions for equilibrium. A two phase field $(\alpha + \beta)$ is a line (a region of zero width) precisely because the pressure and temperature of the α phase in the two phase system is required to be the *same* as that of the β phase if the two phases are in equilibrium. The corresponding states of the two equilibrated phases are represented by the same point in (P, T) space.

This is not true if some property other than P or T is used in the description of the state of the two phases. Suppose the properties of the participating phases are described in terms of their pressure P and molar volume V, or in other words, suppose the diagram is plotted in (P, V) space. While the conditions for equilibrium require the pressure to be the same in both phases, the molar volumes V^α and V^β in general are not the same. The resulting plot of the phase relationships is very different in appearance from the simple (P, T) plot, Fig. 7.11.

In this representation a two phase equilibrium is portrayed by a pair of points, one giving the pressure and molar volume of the α phase and the other that of the β phase. The conditions for equilibrium require that the pressure be the same in both phases but the volumes are different. Thus, a line connecting the pair of points, called

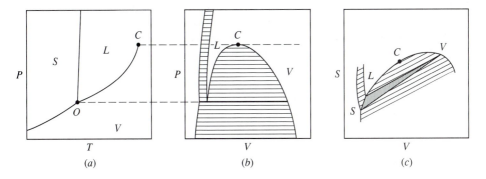

FIGURE 7.11
Alternate constructions for unary phase diagrams compare the plot in (*a*) (T, P) space with the same diagram in (V, P) space (*b*) and in (V, S) space (*c*).

a *tie line*, is horizontal (P is constant) on such a plot. The collection of states that represents all of the possible conditions for equilibrium between the α and β phases consists of two lines, one describing the variation of V^α with P and the other the variation of V^β with P. The space between the lines is filled with horizontal tie lines connecting pairs of equilibrated states.

The liquid plus vapor two phase field terminates at a point C on the (P, T) diagram called the *critical point*. The molar volume of the gas phase decreases with increasing pressure and temperature while that of the liquid increases until, at the critical point, the molar volumes coincide and the properties of the two phases become indistinguishable. This behavior is clearly represented on a (P, V) phase diagram, Fig. 7.11b, but is not particularly evident on the (P, T) diagram where it appears as a single point terminating the vapor pressure curve.

A three phase field, represented in (P, T) space by a triple point at which three, two phase lines meet, becomes a horizontal tie line at which three, two phase areas intersect in a (P, V) diagram. Each of the three phases has its particular value of molar volume. The line is horizontal because P is a potential, required to be the same in the three phase at equilibrium.

Figure 7.11c shows the same phase diagram plotted in (S, V) space. Neither S nor V is a thermodynamic potential. As a consequence, in an equilibrated two phase structure, both the entropy and the volume are different in the two phases. Thus, two phase fields consist of pairs of points connected by tie lines but the tie lines are in general not horizontal. At a triple point on the (P, T) diagram, where three phases are in equilibrium, the values of S and V may be different for all three phases. Thus this condition is represented by a *tie triangle* with corners at the points corresponding to the (S, V) values for each of the three phases.

The reader may have experience with phase diagrams in introductory courses constructed for binary (two component) and ternary (three component) systems. The common binary phase diagram is usually presented for a fixed value of pressure, constructed in (T, X_2) space, where X_2 is the mole fraction of component 2. These diagrams are topologically similar in construction to the (P, V) diagram presented for a unary system in Figure 7.11b *precisely because in both cases one variable (T) is a thermodynamic potential while the other (X_2) is not.*

Similarly, the usual presentation of a ternary phase diagram, plotted on the Gibbs composition triangle (see Sec. 9.4.4) at constant T and P, has the appearance of the diagram in Figure 7.11c. In this case the variables are the atom fractions of two components, neither of which is a thermodynamic potential. Thus, three phase fields are triangles and two phase fields are areas filled with tie lines.

Each representation of a unary phase diagram presents specific information about the relationships between the phases of matter that is not contained in other representations. The (P, T) diagram is the simplest representation. The (P, V) diagram helps visualize volume changes associated with processes represented by paths on the diagram as they pass through the fields in the structure but does not present information about the corresponding *temperatures*. The area under such a path is related to the reversible work done by the system and thus may be useful in understanding cycles in heat engines that involve phase changes like vaporization and condensation. The

(S,V) diagram presents information about yet another property but loses the information about both the pressure *and* temperature of two and three phase equilibria.

7.7 SUMMARY OF CHAPTER 7

In a unary system the chemical potential of a phase is identical with its molar free energy. Chemical potential surfaces may in principle be computed for each phase form that exists in the system by integrating

$$d\mu = dG = -SdT + VdP \qquad (7.4)$$

for that phase using heat capacities and coefficients of expansion and compressibilities.

Intersections of the chemical potential surfaces of the phases produce two phase curves and three phase triple points. Projections of these lines and points onto the (P, T) plane produces the unary phase diagram in (P, T) space.

Combination of the conditions for equilibrium between two unary phases α and β

$$T^\alpha = T^\beta \qquad P^\alpha = P^\beta \qquad \mu^\alpha = \mu^\beta$$

yields the Clausius-Clapeyron equation

$$\frac{dP}{dT} = \frac{\Delta S^{S\alpha \to \beta}}{\Delta V^{\alpha \to \beta}} = \frac{\Delta H^{\alpha \to \beta}}{T \Delta V^{\alpha \to \beta}} \qquad (7.21)$$

Integration of this equation gives the $(\alpha + \beta)$ phase boundary in (P, T) space.

The intersection of a pair of two phase boundaries that share a common phase [for example, $(\alpha + \beta)$ and $(\alpha + L)$] identifies the triple point at which the three phases [for example $(\alpha + \beta + L)$] coexist at equilibrium.

Knowledge of the relationships that exist between thermodynamic properties and phase diagrams permits calculation of some thermodynamic properties from phase boundaries. For example, the heat of vaporization can be computed from the vapor pressure curve.

Unary phase diagrams can be estimated from very limited data by assuming the vapor phase behaves ideally and ΔH is independent of temperature. More precise calculations of phase diagrams are available through applications software developed for this purpose.

If the phase diagram is plotted in (P, T) space, both of which are *thermodynamic potentials*, meaning that they are explicitly required to be equal in the conditions for equilibrium, then a simple cell structure results. If one of the variables chosen is not a potential, two phase fields broaden into areas filled with horizontal (constant potential) tie lines. If neither of the variables representing the state of the phases in the system is a potential, then three phase fields broaden into tie triangles and the tie lines in the two phase fields are not constrained to any fixed direction.

PROBLEMS

7.1. In developing the conditions for equilibrium in a unary heterogeneous system, why is it necessary to first devise the apparatus for handling *open* systems?

7.2. Consider a system consisting of *three phases*, S, L, and G. Follow the sequence of steps below to derive the conditions for equilibrium in this three phase system.

(*a*) Write expressions for the change in entropy for each phase separately.

(*b*) Write the expression for the change in entropy for the three phase system without constraints.

(*c*) Write the isolation constraints for this three phase system including the conservation of energy, volume, and total number of moles.

(*d*) Use the isolation constraints to write dU'^G, dV'^G, and dn^G in terms of the other variables.

(*e*) Use these relations to eliminate dU'^G, dV'^G, and dn^G from the expression for the entropy of the three phase system.

(*f*) Collect terms.

(*g*) Set the coefficients of the six independent differentials equal to zero.

(*h*) Write the conditions for equilibrium.

7.3. Compute and plot the chemical potential surface $\mu(T, P)$ for a monatomic ideal gas in the range 5 K $< T <$ 1000 K and 10^{-5} atm $< P <$ 10 atm.

(*a*) On the same graph sketch plausible surfaces for the chemical potential of the liquid and solid phases as functions of temperature and pressure.

(*b*) Use these sketches to illustrate the construction of the phase diagram for a system that exhibits these three phases.

7.4. Prior to Eq. (7.29) it is asserted that the pressure dependent term in $\Delta H(T, P)$ is negligible for ordinary pressures. Look up some typical values for the thermodynamic properties involved in this pressure dependent term and identify the conditions under which it is, in fact, negligible.

7.5. Sketch curves representing the variation of the molar Gibbs free energy with temperature at the pressure corresponding to a triple point for an element. Repeat this sketch for a pressure slightly above and slightly below the triple point.

7.6. A relatively useful and simple integrated form of the Clausius-Clapeyron equation is obtained [Eq. (7.39)] if the temperature dependence of the heat of vaporization is neglected. Estimate the temperature interval over which this assumption is valid for a typical element like nickel.

7.7. At one atmosphere pressure pure water ice melts at 0°C. At 10 atm, the melting point is found to be -0.08°C. The density of water at 0°C is 1.000 gm/cc, while that of ice is 0.917 gm/cc. From this information, estimate the entropy of fusion of ice.

7.8. Estimate the melting point of the HCP (ϵ) phase form of pure titanium at one atmosphere pressure. Note that ϵ is metastable above 1155 K at one atmosphere. Take $\Delta S^{\epsilon \to \beta} = 3.43 \left(\frac{J}{mol\ K} \right)$, $\Delta S_m = 9.6 \left(\frac{J}{mol\ K} \right)$ $T_m = 2000$ K.

7.9. At one atmosphere pressure pure germanium melts at 1232 K and boils at 2980 K. The pressure at the triple point (S, L, G) is 8.4×10^{-8} atm. Estimate the heat of vaporization of germanium.

7.10. Assuming that the pressure dependence of the transformation temperatures can be neglected below one atmosphere, calculate the pressures at the triple points (α, γ, G), (γ, δ, G) and (δ, L, G), for pure iron. Assume also that all heats of transformation are temperature independent. For iron $T_v = 3008$ K, $T_m = 1808$ K, $T^{\gamma\delta} = 1673$, $T^{\alpha\gamma} = 1180$ K. Corresponding heats of transformation are respectively 354.1, 16.15, 0.63 and 0.91 (KJ/gm atom).

7.11. Thallium exists in the following forms: vapor (V), liquid (L), face centered cubic (α), body centered cubic (β), and hexagonal (ϵ). Estimate and plot a phase diagram for thallium from the following information:

Transformation	ΔH (J/mol)	ΔS (J/mole-K)	ΔV (cc/mol)
$L \rightarrow V$	152,000	...	...
$\beta \rightarrow L$	4,300	.4	0.39
$\epsilon \rightarrow \beta$	380	0.75	−0.02
$\alpha \rightarrow \epsilon$	−320	0.17	0.10
$\alpha \rightarrow \beta$	60	0.92	0.08

At one atmosphere the stable phase forms are:

ϵ(0 K to 500 K); β(500 K to 576 K); L(576 K to 1730 K);   above 1730 K, V

(The FCC α form is stable only at low temperatures and high pressures.)

(a) List all the possible two phase equilibria that *could exist* in this system.

(b) Compute and plot the unary phase diagram for thallium in the range from 10^{-15} atm to 100 atm and from 300 K to 2000 K. For the part of the diagram above 1 atm plot (T, P) axes. Below 1 atm plot the axes as $(-1/T, \log P)$. Assume all ΔC_P terms are negligible.

Consider only the following equilibria: $(\alpha+\beta)$, $(\alpha+\epsilon)$, $(\beta+\epsilon)$, $(\beta+L)$, $(\beta+V)$, $(\epsilon+V)$, and $(L+V)$; the others are metastable.

7.12. Suppose that the formation of the β phase in thallium is sluggish and, during times that are practical for experimental measurements, it does not nucleate and form. Recalculate the *metastable phase* diagram for thallium below one atmosphere pressure assuming the β phase is *absent*.

7.13. Sketch a plausible phase diagram for bismuth, Fig. 7.10, on (P, V) axes. Make a similar sketch in the (S, V) plane.

CHAPTER

8

MULTICOMPONENT, HOMOGENEOUS NONREACTING SYSTEMS: SOLUTIONS

The chemical content of a system is most directly described by specifying the number of moles of each chemical component that it contains; n_k is the number of moles of component k, which is an *extensive* property of the system. Recall that one mole is simply a fixed number of units of the component specified by Avogadro's number, N_o, equal to 6.023×10^{23} atoms or molecular units per mole. The corresponding intensive property that defines not the content but the *composition* of the system is the *mole fraction* of component k, written X_k and defined to be n_k/n_T, where n_T is the total number of moles all of the components in the system.

Systems alter the number of moles of each component they contain in two ways.

1. Atoms or molecules are transferred across the boundary of the system.
2. Chemical reactions occur within the boundary of the system.

Reacting systems are treated in Chap. 11. *Open multicomponent nonreacting systems*, which have a boundary that permits the flow of matter, are the subject of this chapter.

The central thermodynamic concept required for the description of multicomponent systems is the chemical potential μ_k, introduced for unary systems in Chap. 5.

160

In illustrating the strategy for finding conditions for equilibrium, it was deduced that the condition for chemical equilibrium in a unary two phase system is the equality of the chemical potentials of the component in the two phases. Chemical potential can also be defined and evaluated for each component in a multicomponent mixture *or solution*. Analogous definitions for properties of components in mixtures can also be devised for other thermodynamic properties such as volume, entropy, etc.

To define and evaluate these properties it is necessary to devise a way of assigning an appropriate part of the total value of a thermodynamic property of a multicomponent system to each of the components it contains. A strategy for making such a distribution among the components is developed first in this chapter and leads to the general notion of *partial molal properties*. Each extensive property, U', S', V', H', F', and G', has its corresponding set of partial molal properties for a solution. Procedures for evaluating these partial molal properties from experimental information are derived, together with relationships that exist among these properties.

The concept of the *activity* of a component in a solution and a closely related concept, the *activity coefficient*, are also developed in this chapter. Because these quantities are defined in terms of the chemical potential, they also occupy a central role in the description of the thermodynamic behavior of solutions. Indeed, it is demonstrated that if experimental values for the chemical potential, the activity, or the activity coefficient of any one of the components in the system is known as a function of temperature, pressure, and composition, then *all* the thermodynamic properties of the solution can be computed as functions of temperature, pressure, and composition.

The behavior of the partial molal properties when the system is nearly pure, or in *dilute solutions*, can be shown to be general and simple. These laws of dilute solutions are presented and discussed.

The remainder of the chapter develops some models for solution behavior, illustrates strategies for computing thermodynamic properties from models, and reviews useful applications for such models.

8.1 PARTIAL MOLAL PROPERTIES

This development of a strategy for appropriately assigning part of the total value of an extensive thermodynamic property to each component in a multicomponent system is general and can be applied to all defined extensive properties. It is perhaps easiest to visualize this idea in its application to the volume of the system.

Consider a solution that contains c components or independently variable chemical species. Some of these components may be elements, some molecules. The system has some total volume V'. The problem at hand is, "What is a useful way of assigning a part of the total volume to each of the components present?" How many cubic centimeters does it make sense to associate with component 1, 2, etc.? There are a number of ways of making such an assignment of volumes. For example, the total volume could simply be multiplied by the fraction of the total number of molecules in the system that are component 1, 2, etc. Thus, if half of the atoms in the system are component A, one quarter B and one quarter C, then half of the volume would be attributed to A, one-fourth to B and one-fourth to C. The problem with this assignment is evident: it assumes that the volumes per atom of A, B, and C are all equal. Differences in the contribution of the atoms of A, B, and C cannot be deduced.

A variety of other schemes could be devised. However, the strategy based upon the definition of *partial molal properties* given below has proven to be most useful for reasons that will become clear as the concept is developed.

8.1.1   Definition of Partial Molal Properties

The volume of a system is a state function. If the system is multicomponent and open, then in addition to volume changes that may result from changing the temperature and pressure of the system, alterations in volume can occur independently (even if the temperature and pressure are held constant) if material is added to or removed from the system. Thus the state function, V', is a function not only of T and P but also of the number of moles of each component in the system: $n_1, n_2, \ldots, n_c$. Mathematically,

$$V' = V'(T, P, n_1, n_2, \ldots, n_c) \tag{8.1}$$

If the system is taken through an arbitrary infinitesimal change in state, which explicitly includes the possibility of changing the number of moles of each component, the change in volume may be written

$$dV' = \left(\frac{\partial V'}{\partial T}\right)_{P,n_k} dT + \left(\frac{\partial V'}{\partial P}\right)_{T,n_k} dP + \left(\frac{\partial V'}{\partial n_1}\right)_{T,P,n_2,\ldots,n_c} dn_1$$

$$+ \left(\frac{\partial V'}{\partial n_2}\right)_{T,P,n_1,n_3,\ldots,n_c} dn_2 + \cdots + \left(\frac{\partial V'}{\partial n_c}\right)_{T,P,n_1,n_2,\ldots,n_{c-1}} dn_c \tag{8.2}$$

Write the string of similar terms as a sum:[1]

$$dV' = \left(\frac{\partial V'}{\partial T}\right)_{P,n_k} dT + \left(\frac{\partial V'}{\partial P}\right)_{T,n_k} dP + \sum_{k=1}^{c} \left(\frac{\partial V'}{\partial n_k}\right)_{T,P,n_j \neq n_k} dn_k \tag{8.3}$$

The coefficients of dT and dP remain simply related to the coefficients of thermal expansion and compressibility, defined in equations (4.11) and (4.12) and are now applied to expansions and contractions of the solution under consideration. The coefficient of each of the changes in number of moles can be written:

$$\overline{V}_k \equiv \left(\frac{\partial V'}{\partial n_k}\right)_{T,P,n_j \neq n_k} \qquad (k = 1, 2, \ldots, c) \tag{8.4}$$

---

[1]The notation in the subscript of the composition term, $n_j \neq n_k$, is shorthand to describe the condition that, in taking each derivative, the numbers of moles of all of the components except n_k are held constant. Also, when n_k is used as a subscript in a derivative it signifies that *all* the components are held constant.

There is a coefficient for each component in the system. These quantities are defined to be the *partial molal volumes* for each component in the system and are expressed in units of (volume/mole).

An analogous definition can be devised for any of the extensive properties of the system. Use the symbol B' for any of the properties U', S', V', H', F', G'. Then, for an arbitrary change in temperature, pressure, and chemical content the change in the extensive property B' is

$$dB' = M dT + N dP + \sum_{k=1}^{c} \overline{B}_k dn_k \tag{8.5}$$

The "partial molal B for component k" is the corresponding coefficient of dn_k

$$\overline{B}_k \equiv \left(\frac{\partial B'}{\partial n_k} \right)_{T,P,n_j \neq n_k} \qquad (k = 1, 2, \ldots, c) \tag{8.6}$$

Thus, the additional thermodynamic apparatus required to treat multicomponent systems consists of a set of terms, one for each component in the system, with appropriately defined coefficients, which are the partial molal properties.

8.1.2 Consequences of the Definition of Partial Molal Properties

Consider a process in which the temperature and pressure are held constant and the system is formed by adding n_1 moles of component 1, n_2 moles of 2, n_3 of 3 and so on until a final state consisting of a homogeneous mixture of all of the components at the initial temperature and pressure is achieved. At any step during the process, Eq. (8.3) applies with $dT = 0$ and $dP = 0$:

$$dV'_{T,P} = \sum_{k=1}^{c} \overline{V}_k dn_k \tag{8.7}$$

This equation is the first consequence of the definition of partial molal properties.

Computation of the change in volume for the whole finite process requires integration of this equation. The integration procedure depends upon a knowledge of the process sequence, the order in which the components are added to the mixture. In this case, integration of Eq. (8.7) is complicated by the fact that, for example as component 2 is added to a system initially containing n_1 moles of component 1, the composition, specified by the mole fraction of component X_2, changes continuously. Hence, the integrands for the two terms $\overline{V}_1$ and $\overline{V}_2$ change throughout this step in the process. To carry out the integration it would seem that complete knowledge of how $\overline{V}_1$ and $\overline{V}_2$ with composition is required. No simple, general equation can be expected to result from this integration.

An alternate strategy for integrating Eq. (8.7) makes use of two principles:

1. Since $\overline{V}_k$ is an intensive property, it can only depend upon other intensive properties.

2. Changes in state functions can be computed by finding the simplest reversible path between two end states and computing the change for that path.

For the process under consideration, visualize the addition of all c components *simultaneously* in the proportions found in the final mixture. Thus, during the process the intensive properties (T, P and the set of X_k values) remain fixed and each of the $\overline{V}_k$ terms is constant. In this case, integration is straightforward

$$V = \sum_{k=1}^{c} \int_{0}^{n_k} \overline{V}_k dn_k = \sum_{k=1}^{c} \overline{V}_k \int_{0}^{n_k} dn_k$$

$$V' = \sum_{k=1}^{c} \overline{V}_k n_k \tag{8.8}$$

This conclusion, that the total volume for the system is the weighted sum of the partial molal volumes, can be extended without complication to any extensive property:

$$B' = \sum_{k=1}^{c} \overline{B}_k n_k \tag{8.9}$$

Accordingly, the second consequence of the definition of partial molal properties is the most rudimentary requirement of any strategy for assigning a part of a total property to each of the components and that is that the sum of the contributions must add up to the whole.

A third consequence of the definition of partial molal properties is referred to as the *Gibbs-Duhem equation*. Beginning with Eq. (8.9), compute the differential of B', still holding temperature and pressure constant in order to focus on the role of change in chemical content,

$$dB' = \sum_{k=1}^{c} d(\overline{B}_k n_k)$$

since the differential of a sum is the sum of the differentials. Differentiating the product $(\overline{B}_k n_k)$ yields

$$dB' = \sum_{k=1}^{c} [\overline{B}_k dn_k + n_k d\overline{B}_k]$$

which can be written

$$dB' = \sum_{k=1}^{c} \overline{B}_n dn_k + \sum_{k=1}^{c} n_k d\overline{B}_k$$

Compare this result with Eq. (8.5) with dT and dP set equal to zero. The first summation is equal to the left side of the equation. Accordingly, the second summation must be zero:

$$\sum_{k=1}^{c} n_k d\overline{B}_k = 0 \tag{8.10}$$

This result is called the *Gibbs-Duhem equation*. It demonstrates that the partial molal properties are not all independent. In particular, in a binary system in which this equation has only two terms, this equation provides the basis for computing values of partial molal properties for one component when values for the others have been determined. This procedure, known as a *Gibbs-Duhem integration*, is developed in a later section.

8.1.3   The Mixing Process

Temperature, pressure, volume and, according to the third law, entropy, all have absolute values in thermodynamics. In contrast, there is no universally valid state of the system for which any of the energy functions U', H', F' and G' have a zero value. Energies of a system are always evaluated *with respect to some reference state*. Problems that involve these functions in general deal only with *changes* in their values for processes. In the treatment of multicomponent open systems, the most common process considered in defining the energy functions for a solution is called the *mixing process*.

The initial state for the mixing process can be viewed as a collection of containers, each holding an arbitrary quantity of a pure component in some specified phase form (gas, liquid, solid in some crystal form) and at the temperature and pressure of the solution to be formed. This initial state of any particular component is called its *reference* state for the formation of the solution. It is this state to which values for the energy and other thermodynamic functions of the solution are referred. Reference states are particularly important in comparing energies of solutions of different phase forms, such as solid and liquid solutions. Such comparisons are at the base of the construction of phase diagrams (see Chap. 10). Evidently such comparisons are valid only if the reference states for each component is chosen to be the *same* for both solutions being compared. The mixing process is the change in state experienced by the system when appropriate amounts of the pure components in their reference states are mixed together forming a homogeneous solution brought to the same temperature and pressure as the initial state. Thus, the mixing process is the formation of a solution from its pure components at constant temperature and pressure.

Let the superscript (^{o}) denote values of properties in the reference state. Thus, B_k^o is the value per mole of the property B for pure component k. The total value of the property B' for the solution formed is given by Eq. (8.9). The value of the B' for the initial unmixed state is the sum of the values for each of the pure components

$$B'^o = \sum_{k=1}^{c} B_k^o n_k \tag{8.11}$$

The change in B when (n_1, n_2, ..., n_c) moles of pure 1, 2, ..., c are mixed at constant temperature and pressure is the value for the final state minus that for the initial state

$$\Delta B'_{\text{mix}} = B'_{\text{soln}} - B'^o$$

$$\Delta B'_{\text{mix}} = \sum_{k=1}^{c} \overline{B}_k n_k - \sum_{k=1}^{c} B_k^o n_k = \sum_{k=1}^{c} (\overline{B}_k - B_k^o) n_k \tag{8.12}$$

Introduce the notation

$$\Delta \overline{B}_k \equiv \overline{B}_k - B_k^o \tag{8.13}$$

which measures the change experienced by one mole of component k when it is transferred from its reference state to the surroundings it experiences in the solution with the composition under consideration. Equation (8.12) may be written

$$\Delta B'_{\text{mix}} = \sum_{k=1}^{c} \Delta \overline{B}_k n_k \tag{8.14}$$

Thus, $\Delta B'_{\text{mix}}$ is the weighted sum of the changes experienced in the mixing process by the individual components.

The variation of $\Delta B'_{\text{mix}}$ with the composition of the solution can be computed by differentiating Eq. (8.12)

$$d\Delta B'_{\text{mix}} = \sum_{k=1}^{c} \left[\overline{B}_k dn_k + n_k d\overline{B}_k - B_k^o dn_k - n_k dB_k^o \right]$$

The summation of the second term on the right side is zero by the Gibbs-Duhem equation, Eq. (8.10). The fourth term is zero because the B_k^o values are properties of the reference state and are not altered by changing the composition of the solution, which is the process under consideration. Thus,

$$d\Delta B'_{\text{mix}} = \sum_{k=1}^{c} (\overline{B}_k - B_k^o) dn_k = \sum_{k=1}^{c} \Delta \overline{B}_k dn_k \tag{8.15}$$

Differentiate Eq. (8.14) completely.

$$d\Delta B'_{\text{mix}} = \sum_{k=1}^{c} (\Delta \overline{B}_k dn_k + n_k d\Delta \overline{B}_k)$$

The first term on the right side is identical with the right-hand side of Eq. (8.15). Accordingly, the second term must be equal to zero

$$\sum_{k=1}^{c} n_k d\Delta \overline{B}_k = 0 \tag{8.16}$$

This result is a form of the Gibbs-Duhem equation, Eq. (8.10), applied to the mixing process. Thus, the three consequences of the definition of partial molal properties embodied in Eqs. (8.5), (8.9), and (8.10) have analogues for the mixing process in equations (8.14)–(8.16).

8.1.4 Molar Values of the Properties of Mixtures

Because the definition of partial molal properties involves consideration of an open system, the set of equations derived above is formulated for a system containing an arbitrary number of moles of the components. It is frequently useful to normalize the description of properties of mixtures and express them on the basis of one mole of solution formed. This is achieved in a straightforward way by dividing the derived equations by n_T, the total number of moles of all components in the solution. The *value per mole* of each property is designated as before by dropping the prime superscript (′) to give U, S, V, H, F, and G. Dividing both sides of Eqs. (8.5), (8.9) and (8.10) by n_T yields:

$$dB = \sum_{k=1}^{c} \overline{B}_k dX_k \tag{8.17}$$

$$B = \sum_{k=1}^{c} X_k \overline{B}_k \tag{8.18}$$

$$\sum_{k=1}^{c} X_k d\overline{B}_k = 0 \tag{8.19}$$

Similar operations performed on equations (8.14)–(8.16) yield expressions for the molar values for the mixing process

$$d\Delta B_{\text{mix}} = \sum_{k=1}^{c} \Delta \overline{B}_k dX_k \tag{8.20}$$

$$\Delta B_{\text{mix}} = \sum_{k=1}^{c} \Delta \overline{B}_k X_k \tag{8.21}$$

$$\sum_{k=1}^{c} X_k d\Delta \overline{B}_k = 0 \tag{8.22}$$

These results are summarized in Table 8.1.

TABLE 8.1
Summary of consequences of the definition of partial molal properties

Arbitrary quantity of system	Per mole of system
$\Delta B'_{\text{mix}} = \sum_{k=1}^{c} \Delta \overline{B}_k n_k$	$\Delta B_{\text{mix}} = \sum_{k=1}^{c} \Delta \overline{B}_k X_k$
$d\Delta B'_{\text{mix}} = \sum_{k=1}^{c} \Delta \overline{B}_k dn_k$	$d\Delta B_{\text{mix}} = \sum_{k=1}^{c} \Delta \overline{B}_k dX_k$
$\sum_{k=1}^{c} n_k d\Delta \overline{B}_k = 0$	$\sum_{k=1}^{c} X_k d\Delta \overline{B}_k = 0$

8.2 EVALUATION OF PARTIAL MOLAL PROPERTIES

Partial molal properties can be evaluated from experimental data of two broad types:

1. Measurements of the corresponding total property of the solution, B or ΔB_{mix}, as a function of composition
2. Measurements of the partial molal property for one of the components, $\overline{B}_k$, or $\Delta \overline{B}_k$ as a function of composition.

In the developments in this section the analysis is limited to systems with two components (binary systems). Methods exist for extending these analyses to multicomponent systems [1], pp. 56–57.

8.2.1 Partial Molal Properties from Total Properties

If the total value of a property B or ΔB_{mix} is known as a function of composition for a solution at some temperature and pressure, then it is possible to compute from this information the values of the corresponding partial molal properties $\overline{B}_k$ and $\Delta \overline{B}_k$ of each of the components in the mixture as a function of composition. For a binary system Eqs. (8.20) and (8.21) each have only two terms:

$$d\Delta B_{\text{mix}} = \Delta \overline{B}_1 dX_1 + \Delta \overline{B}_2 dX_2 \tag{8.23}$$

$$\Delta B_{\text{mix}} = \Delta \overline{B}_1 X_1 + \Delta \overline{B}_2 X_2 \tag{8.24}$$

The mole fractions sum to 1.

$$X_1 + X_2 = 1 \tag{8.25}$$

Accordingly,

$$dX_1 + dX_2 = 0$$

and

$$dX_1 = dX_2 \tag{8.26}$$

Equation (8.23) may be written

$$d\Delta B_{\text{mix}} = \Delta \overline{B}_1 (-dX_2) + \Delta \overline{B}_2 dX_2 = (\Delta \overline{B}_2 - \Delta \overline{B}_2)dX_2$$

The coefficient of dX_2 in this equation is the derivative[2]

$$\frac{d\Delta B_{\text{mix}}}{dX_2} = \Delta \overline{B}_2 - \Delta \overline{B}_1$$

---

[2]Strictly speaking, this derivative is the partial derivative taken at constant temperature and pressure but, since T and P are considered as constants throughout this discussion, it can be treated in this context as a total derivative.

Equations (8.24) and (8.27) can be considered as two simultaneous linear equations in the unknowns $\Delta\overline{B}_1$ and $\Delta\overline{B}_2$. Solve Eq. (8.27) for $\Delta\overline{B}_1$

$$\Delta\overline{B}_1 = \Delta\overline{B}_2 - \frac{d\Delta B_{mix}}{dX_2}$$

Substitute the result into Eq. (8.24)

$$\Delta B_{mix} = X_1\left[\Delta\overline{B}_2 - \frac{d\Delta B_{mix}}{dX_2}\right] + X_2\Delta\overline{B}_2$$

$$= (X_1 + X_2)\Delta\overline{B}_2 - X_1\frac{d\Delta B_{mix}}{dX_2}$$

Solve for $\Delta\overline{B}_2$, noting that $X_1 + X_2 = 1$,

$$\Delta\overline{B}_2 = \Delta B_{mix} + (1 - X_2)\frac{d\Delta B_{mix}}{dX_2} \tag{8.28}$$

The analogous result holds for component 1

$$\Delta\overline{B}_1 = \Delta B_{mix} + (1 - X_1)\frac{d\Delta B_{mix}}{dX_1} \tag{8.29}$$

since interchange of the subscripts in Eq. (8.23) and (8.27) does not alter them.

To apply this result to a system in which ΔB_{mix} has been determined experimentally as a function of composition, it is necessary to perform a statistical analysis to fit the data with a mathematical function and take its derivative. Note that

$$\frac{d\Delta B_{mix}}{dX_2} = \frac{d\Delta B_{mix}}{dX_1} \cdot \frac{dX_1}{dX_2} = -\frac{d\Delta B_{mix}}{dX_1} \tag{8.30}$$

Substitution of the function and its derivative into Eqs. (8.28) and (8.29) then yields both partial molal properties as a function of composition.

Example 8.1. Compute the partial molal enthalpies of the two components in a binary solution with an enthalpy of mixing given by

$$\Delta H_{mix} = aX_1X_2 \tag{8.31}$$

Solution. This is the simplest possible mathematical form for any function ΔB_{mix}, since such a function is required to pass through zero at the pure components ($X_1 = 1$ and $X_2 = 1$). Recognize that there is only one independent mole fraction because the fractions are related through Eq. (8.25). Thus this equation could be written as a function of only X_2:

$$\Delta H_{mix} = a(1 - X_2)X_2 = a(X_2 - X_2^2)$$

To evaluate the partial molal enthalpies, it is necessary to compute the terms in Eqs. (8.28) and (8.29). Compute the derivative

$$\frac{d\Delta H_{mix}}{dX_2} = a(1 - 2X_2) \tag{8.32}$$

Note that this total derivative is *not the same* as the partial derivative

$$\left(\frac{\partial \Delta H_{\text{mix}}}{\partial X_2}\right)_{X_1} = aX_1 = a(1 - X_2)$$

This partial derivative can be computed mathematically but has no thermodynamic meaning since it is impossible to vary X_2 while holding X_1 constant in a binary system, $X_2 = (1 - X_1)$.

To find the partial molal enthalpy for component 2, substitute expressions in Eqs. (8.31) and (8.32) into equation (8.28).

$$\Delta \overline{H}_2 = [aX_1 X_2] + X_1 \cdot [a(1 - 2X_2)]$$

The factor aX_1 is common to both terms.

$$\Delta \overline{H}_2 = aX_1 [X_2 + 1 - 2X_2]$$

$$= aX_1(1 - X_2) = aX_1 X_1$$

$$\Delta \overline{H}_2 = aX_1^2 \tag{8.33}$$

Because in this case, Eq. (8.31) is symmetrical in the variables X_1 and X_2, the analogous expression for component 1 has the same form.

$$\Delta \overline{H}_1 = aX_2^2 \tag{8.34}$$

In this introductory example, $\Delta \overline{H}_1$ and $\Delta \overline{H}_2$ have the same mathematical form because ΔH_{mix} is a symmetrical function of X_1 and X_2, that is interchanging the subscripts 1 and 2 in a $X_1 X_2$ yields the same function. This is not generally true and as a consequence, the functional form of $\Delta \overline{H}_1$ is in general different from that found for $\Delta \overline{H}_2$.

A check of the procedure can be obtained by substituting the two partial molal enthalpies into the expression for the total change on mixing.

$$\Delta H_{\text{mix}} = X_1(aX_2^2) + X_2(aX_1^2) = aX_1 X_2(X_2 + X_1)$$

Since $(X_1 + X_2) = 1$

$$\Delta H_{\text{mix}} = aX_1 X_2$$

which recovers Eq. (8.31).

8.2.2 Graphical Determination of Partial Molal Properties

Figure 8.1 plots ΔB_{mix} as a function of the mole fraction of component 2. As usual, B can be any extensive property. The value of the partial molal B for component 2 at any composition X_2^o can be obtained from this information by applying Eq. (8.28). Each of the factors in this equation is represented graphically in Fig. 8.1. The slope of the curve at P is the total derivative $(d\Delta B_{\text{mix}}/dX_2)$. Graphically, this slope is the ratio of the length BC to PB.

$$\frac{d\Delta B_{\text{mix}}}{dX_2} = \frac{BC}{PB}$$

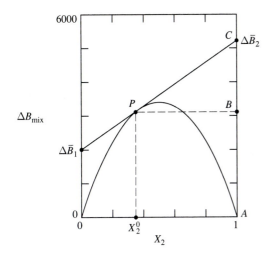

FIGURE 8.1
Graphical representation of the relation between total mixing properties ($\Delta \overline{B}_{mix}$) and the partial molal properties ($\Delta \overline{B}_1$ and $\Delta \overline{B}_2$). Reading the intercepts as the tangent lines rotates along the curve provides an easy visualization of how the partial molal properties vary with composition.

The coefficient of the derivative in Eq. (8.28), $(1 - X_2)$, is the line segment PB on the graph.

$$(1 - X_2) = PB$$

The first term in the Eq. (8.28), ΔB_{mix} located at P in Fig. 8.1, is equal to the length of the segment AB on the graph.

$$\Delta B_{mix} = AB$$

Combining these factors in Eq. (8.28) gives the graphical representation of $\Delta \overline{B}_2$.

$$\Delta \overline{B}_2 = \Delta B_{mix} + (1 - X_2)\frac{d\Delta B_{mix}}{dX_2} = AB + PB \cdot \frac{BC}{PB} = AB + BC$$

$$\Delta \overline{B}_2 = AC$$

The sum of segments AB and BC is the segment AC, which is the intercept of the tangent line drawn at P on the 2-side of the graph. By the same argument, using different subscripts and applying Eq. (8.29), $\Delta \overline{B}_1$ is the intercept of the same tangent line on the ordinate on the 1-side of the graph. Thus, to determine both partial molal properties for a solution of a given composition, construct a tangent to the ΔB_{mix} curve at that composition and read the intercepts on the two sides of the graph.

This graphical procedure is not recommended for use in place of the numerical calculation of partial molal properties. However, it does provide an extremely useful tool for visualizing how the partial molal properties vary with composition and for comparing properties of two solutions of the same components that have different phase forms, such as solid versus liquid solutions. Such comparisons are particularly useful in developing the thermodynamic understanding of binary phase diagrams, presented in Chap. 10.

8.2.3 Evaluation of the PMPs of One Component from Measured Values of PMPs of the Other

For some solution properties, particularly the Gibbs free energy, it is more convenient to measure the value of one of the *partial molal properties* than to measure the total change on mixing. In this section it is demonstrated that, given such information about component 2, the corresponding partial molal property for component 1 can be computed. Once both partial molal properties are known, the total value for the mixture can be computed.

The strategy is based upon integration of the Gibbs-Duhem equation, Eq. (8.22). For a binary system

$$X_1 d\Delta\overline{B}_1 + X_2 d\Delta\overline{B}_2 = 0 \tag{8.35}$$

Rearrange to solve for $d\Delta\overline{B}_1$:

$$d\Delta\overline{B}_1 = -\frac{X_2}{X_1} d\Delta\overline{B}_2$$

Integration of this equation is simplified if the lower limit is chosen to be $X_2 = 0$ (or $X_1 = 1$). The left hand side integrates to

$$\int_{X_2=0}^{X_2} d\Delta\overline{B}_1 = \Delta\overline{B}_1 \Big|_{X_2=0}^{X_2} = \Delta\overline{B}_1(X_2) - \Delta\overline{B}_1(X_2 = 0)$$

At $X_2 = 0$ (pure component 1) $\Delta\overline{B}_1 = \overline{B}_1(X_2 = 0) - B_1^0 = B_1^0 - B_1^0 = 0$. Thus the left hand side of the equation integrates to $\Delta\overline{B}_1(X_2)$. The right side of the equation must be integrated between the same limits:

$$\Delta\overline{B}_1 = \int_{X_2=0}^{X_2} -\frac{X_2}{X_1} d\Delta\overline{B}_2$$

It is assumed that the input for this calculation is the functional relationship between $\Delta\overline{B}_2$ and composition, X_2 usually obtained by statistical fit of a functional form to a set of experimental data points. Thus

$$d\Delta\overline{B}_2 = \frac{d\Delta\overline{B}_2}{dX_2} dX_2$$

and the integrated form of the Gibbs-Duhem equation can be written

$$\Delta\overline{B}_1 = -\int_{X_2=0}^{X_2} \frac{X_2}{X_1} \cdot \frac{d\Delta\overline{B}_2}{dX_2} dX_2 \tag{8.36}$$

It is crucial in this calculation to recognize that there is only one independent compositional variable in this binary system. Thus, the derivative in the integrand is the total derivative with respect to X_2 as argued previously. This is not the same as the partial derivative with respect to X_2 at constant X_1.

Example 8.2. Work Example 8.1 in the opposite direction by assuming the input information is the result obtained for the partial molal enthalpy of component 2, Eq. (8.33).

$$\Delta\overline{H}_2 = aX_1^2 \tag{8.33}$$

Compute the partial molal enthalpy for component 1 and the enthalpy of mixing for this solution.

Solution. In order to apply Eq. (8.36) it is first necessary to evaluate the total derivative from the input function.

$$\frac{d\Delta H_2}{dX_2} = \frac{d\Delta \overline{H}_2}{dX_1} \cdot \frac{dX_1}{dX_2} = \frac{d\Delta \overline{H}_2}{dX_1} \cdot (-1) = -2aX_1$$

Substitute this result into Eq. (8.36) and integrate.

$$\Delta \overline{H}_1 = -\int_0^{X_2} \frac{X_2}{X_1}(-2aX_1)dX_2$$

$$\Delta \overline{H}_1 = 2a\int_0^{X_2} X_2 dX_2 = 2a\left(\frac{X_2^2}{2}\right)\Bigg|_0^{X_2} = aX_2^2$$

which is the result that was obtained in Eq. (8.34). Insert both partial molal properties into Eq. (8.21) to compute the total heat of mixing

$$\Delta H_{\text{mix}} = X_1 \Delta \overline{H}_1 + X_2 \Delta \overline{H}_2 = X_1(aX_2^2) + X_2(aX_1^2)$$

Simplify

$$\Delta H_{\text{mix}} = aX_1 X_2(X_2 + X_1) = aX_1 X_2$$

These results agree with those derived from an alternate input assumption in Example 8.1.

Thus, it is clear that if any one of the three properties ΔB_{mix}, $\Delta \overline{B}_1$, or $\Delta \overline{B}_2$ is known as a function of composition, the other two can be computed. If $\Delta B_{l\text{mix}}$ is given as a function of composition, Eqs. (8.28) and (8.29) are used to compute $\Delta \overline{B}_1$ and $\Delta \overline{B}_2$. Alternatively, if either of the partial molal properties, say $\Delta \overline{B}_1$, is given as a function of composition, the other two properties can be computed by the Gibbs-Duhem integration, Eq. (8.36), and the condition that ΔB_{mix} is the weighted sum of the partial molal properties, Eq. (8.21).

8.3 RELATIONSHIPS AMONG PARTIAL MOLAL PROPERTIES

Each of the four categories in the classification of relationships developed for the total properties of systems in Chap. 4, the laws, definitions, coefficient relations, and Maxwell relations, has counterparts in relations among partial molal properties of the components in a system. Most of these relationships can be derived by applying what might be called the "partial molal operator" to the relations among the total properties. This operator, which derives partial molal properties from their corresponding total property, is contained implicitly in the definition of partial molal properties, Eq. (8.6). Define the operator as

$$\left(\frac{\partial}{\partial n_k}\right)_{T,P,n_j}$$

The notation n_j is understood to be shorthand for all variables from the set of compositional variables except n_k. Application of this operator to any total property B' yields the corresponding partial molal property.

As an example of the application of this strategy to a *definitional relationship*, recall the definition of enthalpy

$$H' = U' + PV' \tag{8.37}$$

Apply the partial molal operator to both sides of this equation

$$\left(\frac{\partial H'}{\partial n_k}\right)_{T,P,n_j} = \left(\frac{\partial U'}{\partial n_k}\right)_{T,P,n_j} + P\left(\frac{\partial V'}{\partial n_k}\right)_{T,P,n_j} + V'\left(\frac{\partial P}{\partial n_k}\right)_{T,P,n_j}$$

Note that the last term is zero since the pressure is defined as constant in the partial molal operator. Apply the general definition of partial molal properties, Eq. (8.6)

$$\overline{H}_k = \overline{U}_k + P\overline{V}_k \tag{8.38}$$

which is the relationship among the partial molal properties of the components in a solution that mimics the definition of enthalpy. Application of the same strategy to the definitions of Helmholtz and Gibbs free energies gives their counterparts

$$\overline{F}_k = \overline{U}_k - T\overline{S}_k \tag{8.39}$$

$$\overline{G}_K = \overline{H}_k - T\overline{S}_k \tag{8.40}$$

To illustrate the application of this strategy to *coefficient relationships* derived for total properties, recall the combined statements of the first and second laws for the Gibbs free energy function, Eq. (4.10)

$$dG' = -S'dT + V'dP + \delta W' \tag{4.10}$$

The coefficient relations for this equation are

$$-S' = \left(\frac{\partial G'}{\partial T}\right)_{P,n_k} \qquad \text{and} \qquad V' = \left(\frac{\partial G'}{\partial P}\right)_{T,n_k} \tag{8.41}$$

where the compositional variables, contained implicitly in $\delta W'$ in this equation, are explicitly held constant in taking these derivatives. Apply the partial molal operator to both sides of these equations

$$-\left(\frac{\partial S'}{\partial n_k}\right)_{T,P,n_j} = \left[\frac{\partial}{\partial n_k}\left(\frac{\partial G'}{\partial T}\right)_{P,n_k}\right]_{T,P,n_j}$$

and

$$\left(\frac{\partial V'}{\partial n_k}\right)_{T,P,n_j} = \left[\frac{\partial}{\partial n_k}\left(\frac{\partial G'}{\partial P}\right)_{T,n_k}\right]_{T,P,n_j}$$

Interchange the order of differentiation on the right sides of both equations

$$-\left(\frac{\partial S'}{\partial n_k}\right)_{T,P,n_j} = \left[\frac{\partial}{\partial T}\left(\frac{\partial G'}{\partial n_k}\right)_{T,P,n_j}\right]_{P,n_k}$$

and
$$\left(\frac{\partial V'}{\partial n_k}\right)_{T,P,n_j} = \left[\frac{\partial}{\partial P}\left(\frac{\partial G'}{\partial n_k}\right)_{T,P,n_j}\right]_{T,n_k}$$

Examine the resulting relations and identify the partial molal properties they contain

$$\overline{S}_k = \left(\frac{\partial \overline{G}_k}{\partial T}\right)_{P,n_k} \qquad \text{and} \qquad \overline{V}_k = \left(\frac{\partial \overline{G}_k}{\partial P}\right)_{T,n_k} \qquad (8.42)$$

Thus, the coefficient relations have counterparts in relations among the partial molal properties.

The Maxwell relation corresponding to Eq. (4.10) is

$$-\left(\frac{\partial S'}{\partial P}\right)_{T,n_k} = \left(\frac{\partial V'}{\partial T}\right)_{P,n_k} \qquad (8.43)$$

Again, apply the partial molal operator to both sides of this equation

$$-\left[\frac{\partial}{\partial n_k}\left(\frac{\partial S'}{\partial P}\right)_{T,n_k}\right]_{T,P,n_j} = \left[\frac{\partial}{\partial n_k}\left(\frac{\partial V'}{\partial T}\right)_{P,n_k}\right]_{T,P,n_j}$$

Interchange the order of differentiation

$$-\left[\frac{\partial}{\partial P}\left(\frac{\partial S'}{\partial n_k}\right)_{T,P,n_j}\right]_{T,n_k} = \left[\frac{\partial}{\partial T}\left(\frac{\partial V'}{\partial n_k}\right)_{T,P,n_j}\right]_{P,n_k}$$

Recognize the partial molal properties contained

$$-\left(\frac{\partial \overline{S}_k}{\partial P}\right)_{T,n_k} = \left(\frac{\partial \overline{V}_k}{\partial T}\right)_{P,n_k} \qquad (8.44)$$

which is the analogue of the Maxwell relation, Eq. (8.43).

To obtain a relation among partial molal properties that is the analogue of a combined statement of the first and second laws, consider the variation of the partial molal Gibbs free energy with temperature and pressure, keeping the composition of the solution constant

$$\overline{G}_k = \overline{G}_k(T, P) \qquad (8.45)$$

Since $\overline{G}_k$ is a state function, its differential is

$$d\overline{G}_k = \left(\frac{\partial \overline{G}_k}{\partial T}\right)_{P,n_k} dT + \left(\frac{\partial \overline{G}_k}{\partial P}\right)_{T,n-k} dP$$

These coefficients have been evaluated in Eq. (8.42). Substitute the results

$$d\overline{G}_k = -\overline{S}_k dT + \overline{V}_k dP \qquad (8.46)$$

which is the partial molal analogue of Eq. (4.10). Other forms of the combined statement can be derived from this relation. To obtain the analogue for the enthalpy function, take the differential of the definitional relation, Eq. (8.40), and set it equal to the right side of Eq. (8.46)

$$d\overline{G}_k = d\overline{H}_k - T d\overline{S}_k - \overline{S}_k dT = -\overline{S}_k dT + \overline{V}_k dP$$

Rearranging

$$d\overline{H}_k = Td\overline{S}_k + \overline{V}_k dP \tag{8.47}$$

which is the partial molal property analogue to the combined statement of the first and second laws for the enthalpy function, Eq. (4.6).

These examples serve to illustrate the potential for generating a host of relations among the partial molal properties of solutions. Indeed, virtually every relation derived in Chap. 4 has such a counterpart. These relationships are not widely applied, however, primarily because the focus of attention in the description of solutions is upon their variation with temperature and pressure since these are the variables usually controlled in studies of multicomponent systems.

8.4 CHEMICAL POTENTIAL IN MULTICOMPONENT SYSTEMS

The idea of the chemical potential is introduced first in Chap. 5 in applying the strategy for finding conditions for equilibrium to unary systems. In this section the concept is broadened to include the description of multicomponent systems. In the process it is demonstrated that if the chemical potential is known as a function of temperature, pressure, and composition for one of the components, *then all of the partial molal and total properties of the system can be computed.*

The definition of the thermodynamic state of an open homogeneous multicomponent system requires specification of $(c + 2)$ variables because the state of the system may be altered by changing the number of moles of each of the c independent compositional variables for the system. The internal energy can be written

$$U' = U'(S', \ V', \ n_1, \ n_2, \dots, n_k, \dots, n_c) \tag{8.48}$$

The change in U' for an arbitrary change in the state of the system has $(c + 2)$ terms

$$dV' = TdS' - PdV' + \mu_1 dn_1 + \mu_2 dn_2 + \cdots + \mu_c dn_c$$

$$dU' = TdS' - PdV' + \sum_{k=1}^{c} \mu_k dn_k \tag{8.49}$$

Define the coefficient of each compositional variable to be the *chemical potential* for that component. Thus,

$$\mu_k \equiv \left(\frac{\partial U'}{\partial n_k}\right)_{S',V',n_j} \tag{8.50}$$

Compare Eq. (8.49) with the original combined statement for the first and second laws, Eq. (4.4). For open multicomponent systems the additional apparatus required is evidently contained in

$$\delta W' = \sum_{k=1}^{c} \mu_k dn_k \tag{8.51}$$

Expressions for the other three energy functions follow from their combined statements, Eqs. (4.6), (4.8), and (4.10)

$$dH' = TdS' + V'dP + \sum_{k=1}^{c} \mu_k dn_k \qquad (8.52)$$

$$dF' = -S'dT - PdV' + \sum_{k=1}^{c} \mu_k dn_k \qquad (8.53)$$

$$dG' = S'dT + V'dP + \sum_{k=1}^{c} \mu_k dn_k \qquad (8.54)$$

Application of the coefficient relation to each of these equations shows that the chemical potential can be expressed as any of the following four derivatives

$$\mu_k = \left(\frac{\partial U'}{\partial n_k}\right)_{S',V',n_j} = \left(\frac{\partial H'}{\partial n_k}\right)_{S',P,n_j} = \left(\frac{\partial F'}{\partial n_k}\right)_{T,V',n_j} = \left(\frac{\partial G'}{\partial n_k}\right)_{T,P,n_j} \qquad (8.55)$$

All four derivatives appear to be similar to partial molal quantities. However only one is in fact a partial molal property, namely the Gibbs free energy derivative.

$$\mu_k = \left(\frac{\partial G'}{\partial n_k}\right)_{T,P,n_j} = \overline{G}_k \qquad (8.56)$$

None of the other expressions for μ_k are partial molal properties because, in their evaluations, temperature and pressure are not held constant. This constraint, (T, P) constant, is explicitly contained in the definition of partial molal properties, Eq. (8.6). Thus

$$\mu_k = \left(\frac{\partial H'}{\partial n_k}\right)_{S',P,n_j} \neq \overline{H}_k$$

and

$$\overline{H}_k = \left(\frac{\partial H'}{\partial n_k}\right)_{T,P,n_j} \neq \mu_k$$

The equality of chemical potential and the partial molal Gibbs free energy provides the basis for expressing all partial molal properties of a component k in terms of its chemical potential. From the coefficient relations, Eq. (8.42)

$$\overline{S}_k = -\left(\frac{\partial \overline{G}_k}{\partial T}\right)_{P,n_k} = -\left(\frac{\partial \mu_k}{\partial T}\right)_{P,n_k} \qquad (8.57)$$

$$\overline{V}_k = \left(\frac{\partial \overline{G}_k}{\partial P}\right)_{T,n_k} = \left(\frac{\partial \mu_k}{\partial P}\right)_{T,n_k} \qquad (8.58)$$

The enthalpy can be evaluated by rearranging the definitional relation, Eq. (8.40)

$$\overline{H}_k = \overline{G}_k + T\overline{S}_k$$

$$\overline{H}_k = \mu_k - T\left(\frac{\partial \mu_k}{\partial T}\right)_{P,n_k} \tag{8.59}$$

The partial molal internal energy is also obtained from the definitional relation

$$\overline{U}_k = \overline{H}_k - P\overline{V}_k = \mu_k - T\left(\frac{\partial \mu_k}{\partial T}\right)_{P,n_k} - P\left(\frac{\partial \mu_k}{\partial P}\right)_{T,n_k} \tag{8.60}$$

and the Helmholtz free energy function from Eq. (8.39)

$$\overline{F}_k = \overline{U}_k - T\overline{S}_k = \mu_k - P\left(\frac{\partial \mu_k}{\partial P}\right)_{T,n_k} \tag{8.61}$$

Thus, if the chemical potential of component k is measured as a function of temperature and pressure so that its temperature and pressure derivatives can be computed, all the partial molal properties of component k can be evaluated. These relationships are summarized in Table 8.2. Relationships analogous to each of these expressions can be written for the set of $\Delta \overline{B}_k$ values simply by substituting $\Delta \mu_k = \mu_k - \mu_k^0$, where μ_k^0 is the chemical potential of k in its reference state, wherever μ_k appears.

Again limit consideration to a binary system. Equality of chemical potential with the partial molal Gibbs free energy is the basis for computing the chemical potential of one component from the experimental measurements of the other through a Gibbs-Duhem integration. The Gibbs-Duhem equation for the Gibbs free energy is

$$X_1 d\overline{G}_1 + X_2 d\overline{G}_2 = 0$$

Since $\Delta \overline{G}_k = \Delta \mu_k, d\Delta \overline{G}_k = d\Delta \mu_k$:

$$X_1 d\Delta \mu_1 + X_2 d\Delta \mu_2 = 0 \tag{8.62}$$

TABLE 8.2
Relations of partial molal properties to the chemical potential

$$\overline{G}_k = \mu_k$$

$$\overline{S}_k = -\left(\frac{\partial \mu_k}{\partial T}\right)_{P,n_k}$$

$$\overline{V}_k = \left(\frac{\partial \mu_k}{\partial P}\right)_{T,n_k}$$

$$\overline{H}_k = \mu_k - T\left(\frac{\partial \mu_k}{\partial T}\right)_{P,n_k}$$

$$\overline{U}_k = \mu_k - T\left(\frac{\partial \mu_k}{\partial T}\right)_{P,n_k} - P\left(\frac{\partial \mu_k}{\partial P}\right)_{T,n_k}$$

$$\overline{F}_k = \mu_k - P\left(\frac{\partial \mu_k}{\partial P}\right)_{T,n_k}$$

The integrated form of this Gibbs-Duhem equation for the chemical potential is adapted from Eq. (8.36)

$$\Delta\mu_1 = -\int_{X_2=0}^{X_2} \frac{X_2}{X_1} \cdot \frac{d\Delta\mu_2}{dX_2} dX_2 \tag{8.63}$$

This result gives $\Delta\mu_1$ when $\Delta\mu_2$ is a known function of composition at any temperature and pressure. Thus, for a binary system, given the chemical potential of one of the components as a function of temperature, pressure, and composition, all the partial molal properties of both components can be computed. With this information, values for the total properties of the solution can be calculated by applying Eq. (8.21).

Although it is beyond the scope of this introductory text, extensions of the Gibbs-Duhem integration strategy exist for multicomponent systems. It has been demonstrated that, given values of a partial molal property for *one* component in a multicomponent system as a function of composition at any temperature and pressure, it is possible to compute the value of that property for *all* the remaining components; no additional information is required. A review of these procedures is presented in Lupis [8.1, pp. 282–285]. Thus if $\Delta\mu_k$ is determined for one component, the chemical potential may be computed for all the others. Combination of this information with the relationships in Table 8.2 demonstrates that, given the chemical potential of component k as a function of temperature, pressure, and composition, all the partial molal properties of all components and all total properties can be evaluated. The thermodynamic behavior of the system is completely determined.

Since the chemical potential is equal to the partial molal Gibbs free energy, the variation of chemical potential with temperature and pressure is identical to that for $\overline{G}_k$. This relationship is derived in Sec. 8.3, Eq. (8.46).

$$d\overline{G}_k = d\mu_k = -\overline{S}_k dT + \overline{V}_k dP \tag{8.64}$$

The corresponding relation holds for $\Delta\overline{G}_k$.

$$d\Delta\overline{G}_k = d\Delta\mu_k = -\Delta\overline{S}_k dT + \Delta\overline{V}_k dP \tag{8.65}$$

8.5 FUGACITIES, ACTIVITIES AND ACTIVITY COEFFICIENTS

Experimental measurements of the thermodynamic behavior of solutions are not aimed at the direct determination of chemical potentials, although this quantity lies at the core of the description of such systems. Common practice prefers another property, called the *activity of component k*, which is defined in terms of the chemical potential by the equation

$$\mu_k - \mu_k^0 = \Delta\mu_k \equiv RT \ln a_k \tag{8.66}$$

The argument of the logarithm, a_k, is the *activity of k* in a solution at a given temperature, pressure, and composition; R and T have their usual meanings. Activity is a unitless quantity, as is the mole fraction of component k. A closely related quantity, called the *fugacity*, is defined for mixtures of gases.

Another convenient measure of solution behavior, called the *activity coefficient of component k*, written γ_k, is defined by the equation

$$a_k = \gamma_k X_k \tag{8.67}$$

From Eq. (8.65)

$$\mu_k - \mu_k^0 = RT \ln \gamma_k X_k \tag{8.68}$$

The activity coefficient is also a unitless quantity.[3]

The origin of the term "activity" for a_k may be made clear through Eq. (8.67). If $\gamma_k = 1$, the activity of component k is equal to its mole fraction and the behavior of k, from the point of view of its chemical potential, is completely determined by its composition. If $\gamma_k > 1$, then $a_k > X_k$ and in the evaluation of its chemical potential, component k "acts as if" the solution contains more of k than the mole fraction suggests. Similarly, if $\gamma_k < 1$ so that $a_k < X_k$, the component acts as if there is less of it present than the composition suggests.

The origin of the logarithmic form of the relation of activity to mole fraction is developed in this section. As the activity concept is applied, first to heterogeneous systems in Chap. 9 and 10, then to reacting systems in Chap. 11 and finally to complex systems in Chaps. 12 and 13, the utility and convenience of this choice for the form of the relationship is made abundantly clear.

8.5.1   Properties of Ideal Gas Mixtures

It is not surprising that the earliest attempts to understand the thermodynamics of mixtures focused on the behavior of mixtures of ideal gases. Calculation of the changes in thermodynamic properties for the mixing process in this case is straightforward.

For a collection of ideal gases the mixing process is the change in state accompanying the process in which n_k moles of each of the pure gases, at a pressure P and temperature T, are mixed to form a homogeneous solution at the same temperature and total pressure. Figure 8.2 may be used to visualize this process. The initial condition of the system is represented in Figure 8.2a. The box that ultimately contains the mixture of gases is partitioned into segments, each containing the number of moles of one of the pure gases that forms the mixture. Since by definition the mixing process occurs at constant temperature and pressure, the gas in each of these compartments is at the temperature T and pressure P of the final mixture. In Fig. 8.2b the partitions

---

[3]Alternate definitions of the activity coefficient are in wide use in the chemical literature depending upon the quantity used to report composition of the solution. For example, if composition is reported in molar concentration c_k (moles of k/cc of solution) an "activity coefficient" ϕ_k can be defined by an equation analogous to Eq. (8.67):

$$a_k = \phi_k c_k$$

In this case, since a_k is unitless, ϕ_k must have units of (cc/mole). The unitless coefficient defined in Eq. (8.67) will be used throughout this text.

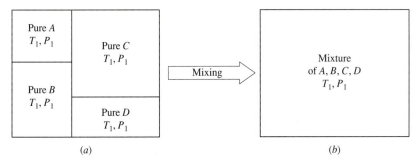

FIGURE 8.2
Illustration of the mixing process. (*a*) The initial state is a collection of systems each containing an appropriate number of moles of one of the pure components at the temperature T_1 and pressure P_1. (*b*) The partitions are removed and the components mix to form a single homogeneous solution at the same temperature T_1 and total pressure P_1.

are removed, the components mix and the homogeneous gas mixture forms. Focus on the process experienced by one of the components k. Initially, this component is in its pure state defined by (T, P). In Chap. 6 it is demonstrated that the particles in an ideal gas mixture do not interact. All their energy is associated with kinetic energy of motion. Accordingly, in the mixed state each component in the mixture behaves as if it occupies the entire volume of the system with no other components present. If the total pressure is P then, according to Dalton's law of partial pressures, each component contributes to that pressure in proportion to the relative number of molecules that it represents. Each component may be viewed as exerting a *partial pressure* P_k given by

$$P_k = X_k P \tag{8.69}$$

The total pressure is the sum of these partial pressures

$$P = \sum_{k=1}^{c} P_k = \sum_{k=1}^{c} X_k P = P \sum_{k-1}^{c} X_k = P$$

since the sum of the mole fractions is one.

Focus on the change experienced by an individual component k during the mixing process. Initially, as pure k it exists at the temperature T and pressure P. In the mixture component k remains at the temperature T and exerts a pressure equal to $P_k = X_k P$. Thus, for molecules of component k, the mixing process is equivalent to an isothermal expansion from an initial pressure P to a final pressure P_k. The change in chemical potential experienced by component k can be obtained by integrating Eq. (8.64) at constant temperature

$$d\mu_k = -\overline{S}_k dT + \overline{V}_k dP = \overline{V}_k dP \tag{8.70}$$

Terms involving dT are absent because the mixing process is isothermal and $dT = 0$. Integration requires evaluation of the partial molal volume of component k in an ideal gas mixture. The volume of an ideal gas mixture is

$$V' = n_T \frac{RT}{P} = (n_1 + n_2 + \cdots + n_k + \cdots + n_c) \frac{RT}{P}$$

From the definition of partial molal properties

$$\overline{V}_k = \left(\frac{\partial V'}{\partial n_k} \right)_{T,P,n_j} = (1) \frac{RT}{P} \tag{8.71}$$

Insert this result into Eq. (8.70) and integrate:

$$\mu_k - \mu_k^0 = \int_P^{P_k} \overline{V}_k dP = \int_P^{P_k} \frac{RT}{P} dP = RT \ln \frac{P_k}{P}$$

Recall that the partial pressure $P_k = X_k P$

$$\Delta \mu_k = RT \ln \left(\frac{X_k P}{P} \right) = RT \ln X_k = \Delta \overline{G}_k \tag{8.72}$$

Compare this result with the definitions of activity and activity coefficient, Eqs. (8.66) and (8.67). Evidently, for any component in an ideal gas mixture, the activity is equal to the mole fraction and the activity coefficient is equal to one.

The relations summarized in Table 8.2 can be applied to evaluate all the other partial molal properties of an ideal gas mixture. The derivative with respect to temperature is:

$$\left(\frac{\partial \Delta \mu_k}{\partial T} \right)_{P,n_k} = R \ln X_k \tag{8.73}$$

The pressure derivative is zero.

$$\left(\frac{\partial \Delta \mu_k}{\partial P} \right)_{T,n_k} = 0 \tag{8.74}$$

Substitution of these results into the equations summarized in Table 8.2 gives

$$\Delta \overline{S}_k = - \left(\frac{\partial \Delta \mu_k}{\partial T} \right)_{P,n_k} = -R \ln X_k \tag{8.75}$$

$$\Delta \overline{V}_k = \left(\frac{\partial \Delta \mu_k}{\partial P} \right)_{T,n_k} = 0 \tag{8.76}$$

$$\Delta \overline{H}_k = \Delta \mu_k + T \left(\frac{\partial \Delta \mu_k}{\partial T} \right)_{P,n_k} = RT \ln X_k + T(-R \ln X_k) = 0 \tag{8.77}$$

$$\Delta \overline{U}_k = \Delta \overline{H}_k - P \Delta \overline{V}_k = 0 - 0 = 0 \tag{8.78}$$

$$\Delta \overline{F}_k = \Delta \overline{U}_k - T \Delta \overline{S}_k = 0 - T(-R \ln X_k) = RT \ln X_k \tag{8.79}$$

Although these results are derived for mixtures of gases, they can be adapted to the description of liquid and solid solutions. Mixtures obeying these relations, whether solid, liquid, or gas, are in general called *ideal solutions*. These equations and the corresponding summed relations for the total properties of ideal solutions are summarized in Table 8.3.

This collection of relationships can be used to evaluate all the properties of an ideal solution if its temperature and composition are given. In the formation of an ideal solution there is no heat of mixing ($\Delta H_{\text{mix}} = 0$), no volume change ($\Delta V_{\text{mix}} = 0$), and no change in internal energy ($\Delta U_{\text{mix}} = 0$). The effects that differ from zero (ΔG_{mix}, ΔF_{mix}, and ΔS_{mix}) all derive from the change in entropy experienced by the components in going from the unmixed state to the homogeneous solution. Since no heat is transferred into the system during the mixing process, there is no entropy exchange with the surroundings for an ideal gas. The entropy change computed from Eq. (8.75) is the entropy produced by this irreversible process.

This mixing behavior is summarized for a binary system in Fig. 8.3, which plots all the properties of an ideal solution as a function of composition and temperature. Three characteristics of this behavior may be noted:

1. All the plots are symmetrical with respect to composition; substitution of component 1 for component 2 and vice versa produces the same equations.
2. Slopes of the plots of ΔS_{mix}, ΔG_{mix}, and ΔF_{mix} versus composition are vertical at the sides of the diagram because they contain the logarithm function. The derivative of $\ln x$ is $1/x$, which approaches infinity as x approaches zero.
3. The entropy of mixing is independent of temperature; at any given composition ΔG_{mix} and ΔF_{mix} vary linearly with the absolute temperature.

This ideal solution model has proven to be most useful as a basis for comparison of the properties of a real solution with those that an ideal solution would exhibit at the same temperature, pressure, and composition. Indeed, as developed in Sec. 8.5.4, it is very useful to characterize the behavior of real solutions by their "departure from ideal solution behavior."

TABLE 8.3
Properties of an ideal solution

Partial molal property	Total property	
$\Delta \overline{G}_k = RT \ln X_k$	$\Delta G_{\text{mix}} = RT \sum_{k=1}^{c} X_k \ln X_k$	(8.72)
$\Delta \overline{S}_k = -R \ln X_k$	$\Delta S_{\text{mix}} = -R \sum_{k=1}^{c} X_k \ln X_k$	(8.75)
$\Delta \overline{V}_k = 0$	$\Delta V_{\text{mix}} = 0$	(8.76)
$\Delta \overline{H}_k = 0$	$\Delta H_{\text{mix}} = 0$	(8.77)
$\Delta \overline{U}_k = 0$	$\Delta U_{\text{mix}} = 0$	(8.78)
$\Delta \overline{F}_k = RT \ln X_k$	$\Delta F_{\text{mix}} = RT \sum_{k=1}^{c} X_k \ln X_k$	(8.79)

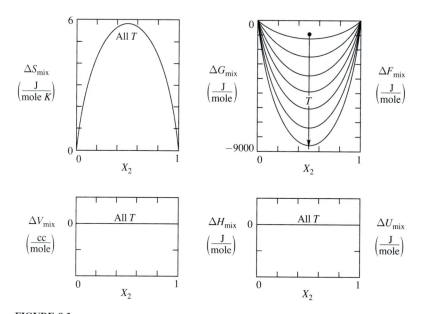

FIGURE 8.3
Properties of an ideal solution including ΔG_{mix}, ΔS_{mix}, ΔV_{mix}, ΔH_{mix}, ΔU_{mix} and ΔF_{mix}. Maximum temperature shown is 1400 K.

8.5.2 Mixtures of Real Gases: Fugacity

Attempts to apply the ideal gas model to mixtures of real gases reveal deviations from the predicted behavior. For many practical applications these deviations can be neglected. For precise analyses or in property ranges for which the ideal gas model is not a viable approximation, it is necessary to devise a more general formalism for describing the mixing behavior of real gases. One strategy for handling this extension to real gas mixtures is based upon the concept of the *fugacity*.

The change in chemical potential for component k upon mixing can be obtained in general by integrating Eq. (8.64). For real gases the partial molal volume is not given by Eq. (8.71) but must be determined experimentally. Define a function α_k, which reports the deviation of the measured partial molal volume at a given temperature and pressure from that which would be computed for an ideal gas of the same temperature, pressure, and composition:

$$\alpha_k = \overline{V}_k - \frac{RT}{P} \tag{8.80}$$

The change in chemical potential during mixing for a component in a mixture of real gases is

$$\Delta \mu_k = \int_P^{P_k} \left(\alpha_k + \frac{RT}{P} \right) dP$$

$$\Delta\mu_k = \int_P^{P_k} \alpha_k dP + RT \ln \frac{P_k}{P} \tag{8.81}$$

Thus, if α_k is determined, the chemical potential can be evaluated.

The fugacity, written f_k, is a property of a component in a gas mixture. It has units of pressure and is defined so that the form of the equation for the chemical potential change on mixing, Eq. (8.72), is retained.

$$\mu_k - \mu_k^0 = \Delta\mu_k \equiv RT \ln \frac{f_k}{P} \tag{8.82}$$

Thus the fugacity occupies the role that the partial pressure plays in ideal gas mixtures. The fugacity of a component in a real gas can be evaluated by setting Eq. (8.81) and (8.82) equal to each other

$$RT \ln \frac{f_k}{P} = \int_P^{P_k} \alpha_k dP + RT \ln \frac{P_k}{P}$$

and solving for f_k. Algebraic manipulation yields

$$f_k = P_k e^{\left[\int_P^{P_k} \alpha_k dP\right]} \tag{8.83}$$

As the deviation from ideal behavior, reported in α_k, approaches zero, the fugacity of component k approaches its partial pressure.

Determination of the fugacity of one of the components in a gas mixture as a function of composition, temperature and pressure permits calculation of the chemical potential for that component through Eq. (8.82). It is demonstrated in Sec. 8.4 that if the chemical potential is known as a function of temperature, pressure and composition for one of the components, then all the properties of the solution can be evaluated. Thus measurement of the fugacity of one component over a range of temperature, pressure, and composition is sufficient to describe the behavior of real gas mixtures completely.

8.5.3 Activity and the Behavior of Real Solutions

Comparison of the definition of activity, Eq. (8.66), with that for fugacity, Eq. (8.82), demonstrates that these quantities have the simple relationship

$$a_k = \frac{f_k}{P} \tag{8.84}$$

where P is the pressure in the reference state for component k. Either quantity provides a viable description of the behavior of gas mixtures. However, the fugacity concept is limited in its application to gas mixtures because its definition gives it the character of pressure. The activity concept is not so constrained and therefore can be applied to describe the characteristics of gas, liquid, and solid solutions.

Recall the definition of activity, Eq. (8.66):

$$\Delta\mu_k = RT \ln a_k = \Delta\overline{G}_k \tag{8.85}$$

Table 8.2 may be used to relate all the partial molal properties of a component in a mixture to its activity. The temperature derivative is

$$\left(\frac{\partial \Delta \mu_k}{\partial T}\right) = R \ln a_k + RT \left(\frac{\partial \ln a_k}{\partial T}\right)_{P,n_k} \tag{8.86}$$

The derivative with respect to pressure is

$$\left(\frac{\partial \Delta \mu_k}{\partial P}\right)_{T,n_k} = RT \left(\frac{\partial \ln a_k}{\partial P}\right)_{T,n_k} \tag{8.87}$$

Substitution of these results into the equations presented in Table 8.2 produces the set of equations summarized in Table 8.4.

Focus again on a binary system. If the activity of component 2 is measured as a function of composition, then the activity of component 1 can be computed through a Gibbs-Duhem integration. Combine the definition of activity, Eq. (8.66) with the Gibbs-Duhem integration for the chemical potential, Eq. (8.63):

$$\ln a_1 = - \int_{X_2=0}^{X_2} \frac{X_2}{X_1} \cdot \frac{d \ln a_2}{d X_2} d X_2 \tag{8.93}$$

Some difficulties may arise in carrying out this integration because as a_2 approaches zero, $\ln a_2$ approaches infinity. Analytical techniques are available to circumvent this problem (see references 8.2 and 8.3).

Extensions of the Gibbs-Duhem strategy to multicomponent systems permit calculation of the activities of all the components from experimental measurements for one component. Thus, like chemical potential and fugacity, complete knowledge of the activity for one component is sufficient to characterize the solution thermodynamics of a multicomponent system.

TABLE 8.4
Relationships between the partial molal properties of component k and its activity

$$\Delta \overline{G}_k = RT \ln a_k \tag{8.85}$$

$$\Delta \overline{S}_k = -R \ln a_k - RT \left(\frac{\partial \ln a_k}{\partial T}\right)_{P,n_k} \tag{8.88}$$

$$\Delta \overline{V}_k = RT \left(\frac{\partial \ln a_k}{\partial P}\right)_{T,n_k} \tag{8.89}$$

$$\Delta \overline{H}_k = -RT^2 \left(\frac{\partial \ln a_k}{\partial T}\right)_{P,n_k} \tag{8.90}$$

$$\Delta \overline{U}_k = -RT^2 \left(\frac{\partial \ln a_k}{\partial T}\right)_{P,n_k} - PRT \left(\frac{\partial \ln a_k}{\partial P}\right)_{T,n_k} \tag{8.91}$$

$$\Delta \overline{F}_k = RT \ln a_k - PRT \left(\frac{\partial \ln a_k}{\partial P}\right)_{T,n_k} \tag{8.92}$$

8.5.4 Use of the Activity Coefficient to Describe the Behavior of Real Solutions

For reasons that become evident in this section, the description of solution thermodynamics based on the activity coefficient is perhaps the most convenient of those reviewed in this chapter. The definition of the activity coefficient γ_k of component k and its relation to chemical potential are presented in Eqs. (8.67) and (8.68).

$$a_k \equiv \gamma_k X_k \tag{8.67}$$

$$\Delta\mu_k = RT \ln \gamma_k X_k = \Delta\overline{G}_k \tag{8.68}$$

In general, the activity coefficient is a function of temperature, pressure, and composition, and must be determined experimentally for each solution. Using the properties of the logarithm function, Eq. (8.68) can be written

$$\Delta\mu_k = \Delta\overline{G}_k = RT \ln \gamma_k + RT \ln X_k \tag{8.94}$$

The second term on the right side is the partial molal Gibbs free energy of mixing for an ideal solution. The first term can then be characterized as reporting the "departure from ideal behavior" of component k in the mixture. The first term is defined as the "excess" contribution to the partial molal Gibbs free energy of k, and is designated by the superscript (^{xs}). The second term is an "ideal" contribution, designated by the superscript (^{id}). The total Gibbs free energy of mixing may be written

$$\Delta\overline{G}_k = \Delta\overline{G}_k^{xs} + \Delta\overline{G}_k^{id} \tag{8.95}$$

where

$$\Delta\overline{G}_k^{xs} = RT \ln \gamma_k \tag{8.96}$$

and

$$\Delta\overline{G}_k^{id} = RT \ln X_k \tag{8.97}$$

If the activity coefficient is larger than 1, component k "acts" as if it has more k than the composition suggests. In this case $\ln \gamma_k$ and consequently the excess free energy is positive and the system is said to exhibit a "positive departure from ideal behavior." For $\gamma_k < 1$, $\Delta\overline{G}_k^{xs}$ is negative and a "negative departure" describes the behavior of the component.

In order to express the remaining partial molal properties in terms of the activity coefficient, it is necessary to apply the definition, Eq. (8.94), to the set of relationships presented in Table 8.2. The temperature derivative of Eq. (8.94) gives

$$\left(\frac{\partial \Delta\mu_k}{\partial T}\right)_{P,n_k} = R \ln \gamma_k + RT \left(\frac{\partial \ln \gamma_k}{\partial T}\right)_{P,n_k} + R \ln X_k \tag{8.98}$$

The derivative with respect to pressure is

$$\left(\frac{\partial \Delta\mu_k}{\partial P}\right)_{T,n_k} = RT \left(\frac{\partial \ln \gamma_k}{\partial P}\right)_{T,n_k} \tag{8.99}$$

Substitution of these results into the equations in Table 8.2 yields the expressions for the partial molal properties summarized in Table 8.5.

The expression for any one of these partial molal properties can be decomposed into a set of terms that involve only the activity coefficient and may be called the "excess" part of the property, and an "ideal" part identical with the value for an ideal solution. In the case of the volume, enthalpy, and internal energy, recall that the corresponding changes for an ideal solution are zero. Thus, for these functions the total property is all "excess."

In a binary system a Gibbs-Duhem equation can be devised for the activity coefficients. Recall again the Gibbs-Duhem equation for chemical potentials

$$X_1 d\Delta\mu_1 + X_2 d\Delta\mu_2 = 0 \tag{8.62}$$

and the relation between chemical potential and activity coefficient, Eq. (8.94). Write the differential of the latter equation

$$d\Delta\mu_k = RT(d\ln\gamma_k + d\ln X_k)$$

Substitute the result into Eq. (8.62)

$$X_1 \cdot RT(d\ln\gamma_1 + d\ln X_1) + X_2 \cdot RT(d\ln\gamma_2 + d\ln X_2) = 0$$

Note that

$$X_1 d\ln X_1 + X_2 d\ln X_2 = X_1\frac{dX_1}{X_1} + X_2\frac{dX_2}{X_2} = dX_1 + dX_2 = 0$$

Simplify, to yield

$$X_2 d\ln\gamma_1 + X_2 d\ln\gamma_2 = 0 \tag{8.105}$$

TABLE 8.5
Relationships between the partial molal properties of component k and its activity coefficient

Total excess ideal	
$\Delta \overline{G}_k = RT\ln\gamma_k + RT\ln X_k$	(8.94)
$\Delta \overline{S}_k = -R\ln\gamma_k - RT\left(\dfrac{\partial\ln\gamma_k}{\partial T}\right)_{P,n_k} - R\ln X_k$	(8.100)
$\Delta \overline{V}_k = RT\left(\dfrac{\partial\ln\gamma_k}{\partial P}\right)_{T,n_k} + 0$	(8.101)
$\Delta \overline{H}_k = -RT^2\left(\dfrac{\partial\ln\gamma_k}{\partial T}\right)_{P_k} + 0$	(8.102)
$\Delta \overline{U}_k = -RT^2\left(\dfrac{\partial\ln\gamma_k}{\partial T}\right)_{P,n_k} - PRT\left(\dfrac{\partial\ln\gamma_k}{\partial P}\right)_{T,n_k} + 0$	(8.103)
$\Delta \overline{F}_k = RT\ln\gamma_k - PRT\left(\dfrac{\partial\ln\gamma_k}{\partial P}\right)_{T,n_k} + RT\ln X_k$	(8.104)

which is the Gibbs-Duhem equation for the activity coefficients in a binary system. Its integrated form is similar to that for the activity, Eq. (8.93).

$$\ln \gamma_1 = - \int_{X_2=0}^{X_2} \frac{X_2}{X_1} \cdot \frac{d \ln \gamma_2}{d X_2} d X_2 \tag{8.106}$$

Thus, to compute the activity coefficient of component 1 from a set of measurements for component 2, use statistical methods to fit a function to the data for component 2 and substitute the result into Eq. (8.106).

Extensions of the Gibbs-Duhem strategy, similar to those that apply to the activity function, exist for computing activity coefficients of all the components in a multicomponent system based on measurements for one of the components. Thus, a model or direct experimental determination of the activity coefficient for one component provides the information necessary to compute all the solution properties of the system. Of the variety of formal descriptions that satisfy this purpose presented in this chapter, that based upon the activity coefficient provides the best access to an understanding of thermodynamic behavior of solutions.

8.6   THE BEHAVIOR OF DILUTE SOLUTIONS

Begin with pure component 1 and visualize the addition of a few atoms of component 2 forming a dilute solution of the *solute*, component 2, in the *solvent*, component 1. In this range of compositions the average solvent atom experiences the same surroundings that it has in the pure state; only a very small fraction of solvent atoms neighbor solute atoms. Thus, the only significant influence that the addition of solute atoms has upon the properties of solvent atoms is to slightly reduce their numbers. Accordingly, the solvent atoms act as if they were in an ideal solution. This behavior gives rise to an experimentally observed limiting law that applies to all solutions, called *Raoult's law for the solvent*. Here assuming the solvent to be component 1

$$\lim_{X_1 \to 1} a_1 = X_1 \tag{8.107}$$

In the same composition range, every solute atom is completely surrounded by solvent atoms. That is, until a sufficient number of solute atoms is added so that their spheres of influence begin to interact, each solute atom added to the solution makes the same contribution to the properties of the system. In this range the average properties of the solute atoms are proportional to their concentration. However, this relationship is different for different solutes added to the same solvent; this proportionality is thus characterized by a constant that is specific to the solute-solvent combination in the system. This behavior gives rise to an experimentally observed limiting law that also applies to all solutions, called *Henry's law for the solute*, here assumed to be component 2.

$$\lim_{X_2 \to 0} a_2 = \gamma_2^o X_2 \tag{8.108}$$

The coefficient γ_2^o, called the *Henry's law constant*, is the activity coefficient for the solute and is seen to be independent of composition in the dilute range. The value of this constant depends on both the solute and the solvent in the system and, for a given solute-solvent combination, varies with temperature and pressure. If the value of the Henry's law constant is determined as a function of temperature and pressure, then all the thermodynamic properties of a dilute solution can be computed.

These limiting laws are illustrated graphically in Fig. 8.4, which show plots of the activities of both components in a binary system as a function of composition at a fixed temperature and pressure. The dilute solution limits correspond to the behavior shown on these plots at the sides of these diagrams. The activity plot for the solvent approaches 1 in the limit of the pure solvent along the line that has a slope of 1, in accordance with Eq. (8.91). For the solute, the activity approaches zero in the dilute limit along a line that has a slope equal to the Henry's law constant for that solution, in accordance with Eq. (8.98). The range of composition over which the solution obeys these limiting laws, is a "dilute" solution, varies widely among systems that have been studied, ranging from a few parts per million to several percent.

The two limiting laws are not independent of each other. It can be shown that if either is assumed, then the other can be derived through a Gibbs-Duhem integration of the activity coefficient.

These laws have practical utility because they are general, meaning that they are not based upon a model and are valid for all systems in the dilute range. Many practical applications involve dilute solutions. These laws permit predictions of some aspects of the behavior of the system because the form of the dependence of properties upon composition can be derived without experimental evaluation of the Henry's law coefficient for the system. Examples illustrating their application to the calculations of phase diagrams and to chemical equilibria in dilute reacting systems are presented in Chaps. 10 and 11.

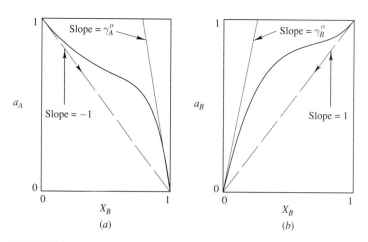

FIGURE 8.4
Variation of activity with composition illustrates the limiting laws for dilute solution behavior.

8.7 SOLUTION MODELS

The analysis of experimental measurements of the thermodynamic behavior of solutions can be carried out empirically by fitting convenient mathematical functions to each data set. A more fundamental understanding of such information is obtained if the analysis is based upon a model for the behavior of solutions. The mathematical analysis may be similar in these two approaches; in the latter case the data set is used to evaluate parameters in the model that have theoretical significance.

The simplest model for solution behavior is the ideal solution described in Sec. 8.5.1. This model contains no adjustable parameters and given the composition and temperature of the solution, all of its properties can be computed. As a consequence, the ideal solution model is incapable of describing differences that may exist between two solutions of the same composition at the same temperature. This section presents solution models that expand upon this simple model, introducing an increasing number of parameters as the level of sophistication and flexibility is increased.

8.7.1 The Regular Solution Model

One class of solution models is based upon the concept of the *regular solution*. In its simplest form the definition of a regular solution contains two components.

1. The entropy of mixing is the same as that for an ideal solution:

$$\Delta \overline{S}_k^{\text{rs}} = \Delta \overline{S}_k^{\text{id}} = -R \ln X_k \qquad (8.109)$$

for all components at all temperatures and pressures.

2. The enthalpy of solution is not zero, as in an ideal solution, but is some function of composition.

$$\Delta \overline{H}_k^{\text{rs}} = \Delta \overline{H}_k(X_1, X_2, \ldots) \qquad (8.110)$$

A statement equivalent to 1. above is: the excess entropy of mixing is zero for a regular solution.[4] As a consequence of this definition, since the excess partial molal Gibbs free energy is

$$\Delta \overline{G}_k^{\text{xs}} = \Delta \overline{H}_k^{\text{xs}} - T \Delta \overline{S}_k^{\text{xs}}$$

then

$$(\Delta \overline{G}_k^{\text{xs}})^{\text{rs}} = (\Delta \overline{H}_k^{\text{xs}})^{\text{rs}} - T(0) = \Delta \overline{H}_k(X_1, X_2, \ldots) \qquad (8.111)$$

---

[4]In more sophisticated versions of the regular solution formalism, this assumption is relaxed to allow incorporation of a configurational contribution to the excess entropy [8.4].

Thus, in a regular solution, because the excess entropy of mixing is defined to be zero, the excess Gibbs free energy is equal to the enthalpy of mixing and is a function only of composition. Accordingly, the sign of the departure from ideal behavior (positive or negative) that characterizes the behavior of the solution is determined by the sign of the heat of mixing.

It follows from the definition of the activity coefficient and Eq. (8.94) that

$$\Delta G_k^{\text{XS}} = RT \ln \gamma_k = \Delta \overline{H}_k \tag{8.112}$$

so that the activity coefficient can be evaluated from a model for the heat of mixing.

$$\gamma_k = e^{\Delta \overline{H}_k / RT} \tag{8.113}$$

It is shown in Sec. 8.5.4 that if the activity coefficient is known, then all the properties of a solution can be calculated. Thus, application of the regular solution model focuses on the evaluation of the heat of mixing as a function of composition.

Equations analogous to Eqs (8.99)–(8.101) also hold for the total mixing properties of the solution.

The definition of the regular solution also requires that the heat of mixing for such a solution cannot be a function of temperature. If the heat of mixing were a function of temperature then, by Eq. (8.111), the excess free energy of mixing would also be a function of temperature and the derivative of this quantity with respect to temperature would not be zero. Since, by the coefficient relationship, Eq. (8.42), for example, the derivative of the free energy with respect to temperature is the negative of the entropy, it can be concluded that if the heat of mixing is a function temperature, then the excess entropy of mixing is not zero. This violates the definition of a regular solution. Models have been devised for which the heat of mixing depends upon temperature, as may be required to describe some experimental observations, but such models are not regular solutions.

The simplest regular solution model that can be devised contains a single adjustable parameter in its description of the heat of mixing

$$\Delta H_{\text{mix}} = a_0 X_1 X_2 \tag{8.114}$$

where a_0 is a constant (not to be confused with a_k, the activity of component k). Mathematical forms simpler than this expression are inadmissible because mixing properties must pass through zero at $X_1 = 0$ and $X_2 = 0$. The Gibbs free energy of mixing obtained from this model is

$$\Delta G_{\text{mix}} = a_0 X_1 X_2 + RT (X_1 \ln X_1 + X_2 \ln X_2) \tag{8.115}$$

The sign of the single adjustable parameter a_0 determines the sign of the excess free energy of mixing and thus the nature of the departure from ideal behavior.

Figure 8.5 illustrates the variation of ΔS, ΔH, and ΔG versus composition and temperature for this model for a positive departure from ideal behavior. This figure demonstrates the essential symmetry of the properties computed from this model with respect to composition. Note that ΔS and ΔH are not functions of temperature in this model. The temperature dependence of ΔG_{mix} is completely contained in the coefficient of the $T \Delta S_{\text{mix}}$ term. At a given composition, the free energy of mixing

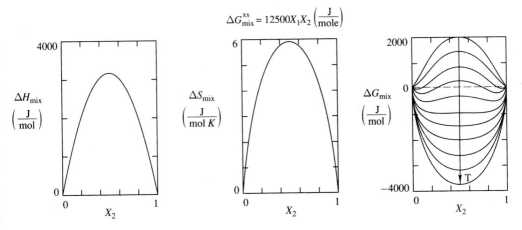

FIGURE 8.5
Variation of thermodynamic mixing properties with composition and temperature for the simplest regular solution model for a positive departure from ideal behavior. $a_0 = 12500$ J/mole. Maximum temperature is 1200 K.

varies linearly with temperature. For positive departures from ideal behavior at low temperatures, it is observed that the free energy of mixing curve develops an additional maximum and minimum. In Chap. 10 it is demonstrated that this behavior leads to the development of a *miscibility gap* in the solution, which appears as a particular construction of a two phase field on a phase diagram containing such a solution.

The expression for ΔH_{mix} contained in Eq. (8.114) is used in Example 8.1 as an application of the procedure for computing partial molal properties from the total properties of a solution. It was demonstrated that

$$\Delta \overline{H}_1 = a_0 X_2^2 \qquad \text{and} \qquad \Delta \overline{H}_2 = a_2 X_1^2 \qquad (8.116)$$

Thus, according to Eq. (8.113), the activity coefficients for such a solution are

$$\gamma_1 = e^{a_0 X_2^2 / RT} \qquad \text{and} \qquad \gamma_2 = 3^{a_0 X_1^2 / RT} \qquad (8.117)$$

Again, it can be seen that the sign of the parameter a_0 determines whether the activity coefficients are greater or less than 1 and hence determines the sign of the departure from ideal behavior that characterizes the model.

Consider the values of the activity coefficients in the limit of dilute solutions. For example, for a dilute solution of solute component 2 in the solvent component 1 this limit is characterized by X_2 approaching zero or X_1 approaching 1. From Eq. (8.116), in this limit the activity coefficient for the solute component 2 becomes the Henry's law coefficient

$$\gamma_2^0 = e^{a_0 / RT} \qquad (8.118)$$

A similar evaluation for the dilute solution at the other end of the composition range yields the Henry's law coefficient for component 1 when it is dilute in component 2.

$$\gamma_1^o = e^{a_0/RT} \tag{8.119}$$

Thus, this simple solution model gives the same value of the Henry's law coefficient at both ends of the composition range. This result exposes a clear limitation in the application of this one parameter solution model that derives from the symmetrical form it assumes for the heat of mixing, Eq. (8.114). It implies that the properties associated with an atom of component 2 surrounded by atoms of component 1 in a dilute solution are identical with the properties of an atom of component 1 surrounded by atoms of component 2. Since components 1 and 2 are composed of different atoms, this situation is in general unrealistic, although it may provide a useful approximation when these components are similar.

In spite of its evident limitations, this simplest of the regular solution models has proven useful as a tool for visualizing the behavior of solutions. For example, all the various possible classes of phase diagrams can be obtained by applying this model to each phase form that may exist in the system, although quantitative fits to real phase diagrams require more sophisticated models. Initial attempts at computer calculations of phase diagrams were formulated on the basis of this model [8.5]. These applications are demonstrated in detail in Chap. 10.

The flexibility of the regular solution model may be increased by adding terms to the expression for the heat of mixing and introducing an additional model parameter with each term. Equation (8.114) becomes

$$\Delta H_{\text{mix}} = X_1 X_2 (a_0 + a_1 X_2 + a_2 X_2^2 + \cdots) \tag{8.120}$$

With a sufficient number of terms and a corresponding number of adjustable parameters this equation can provide a statistically valid fit to any collection of empirical data for the heat of mixing. The corresponding partial molal enthalpies can be written

$$\Delta \overline{H}_1 = X_2^2 (b_0 + b_1 X_2 + b_2 X_2^2 + \cdots) \tag{8.121}$$

and

$$\Delta \overline{H}_2 = X_1^2 (c_0 + c_1 X_2 + c_2 X_2^2 + \cdots) \tag{8.122}$$

where the coefficients b_i and c_i can be computed from the set of a_i coefficients in Eq. (8.120). Figure 8.6 illustrates this model flexibility by plotting heats and free energies of mixing for selected values of a_0 and a_1 in a model where terms beyond the second linear term are neglected. These calculations assume a regular solution that requires that the coefficients in these equations be independent of temperature.

8.7.2 Nonregular Solution Models

The simplest of the models that extend beyond the regular solution model can be obtained by introducing a temperature dependent factor into the expression for the

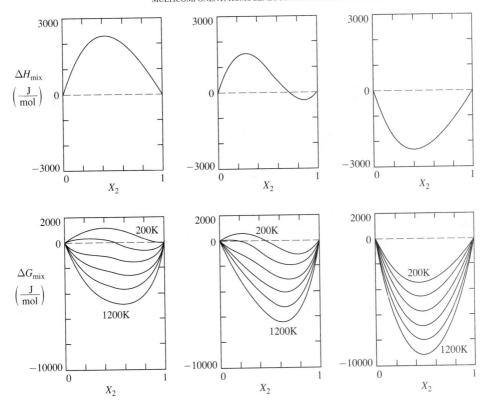

FIGURE 8.6
Mixing properties of a two parameter regular solution model with $\Delta H_{\text{mix}} = X_1 X_2 (a_1 X_1 + a_2 X_2)$. ($a$):
$a_1 = 12500$, $a_2 = 5500$; (b): $a_1 = 12500$, $a_2 = -5500$; (c) $a_2 = -12500$, $a_2 = -5500$.

excess free energy of mixing. The temperature dependence introduces a second parameter into the model, which may take the form

$$\Delta G^{\text{xs}}_{\text{mix}} = a_0 X_1 X_2 \left(1 + \frac{b}{T} \right) \tag{8.123}$$

The excess entropy of mixing is given by

$$\Delta S^{\text{xs}}_{\text{mix}} = -\left(\frac{\partial \Delta G^{\text{xs}}_{\text{mix}}}{\partial T} \right)_{P,n_k} = \frac{a_0 b}{T^2} X_1 X_2 \tag{8.124}$$

The heat of mixing computes as

$$\Delta H^{\text{xs}}_{\text{mix}} = \Delta G^{\text{xs}}_{\text{mix}} + T \Delta S^{\text{xs}}_{\text{mix}}$$

$$\Delta H^{\text{xs}}_{\text{mix}} = \Delta H_{\text{mix}} = a_0 X_1 X_2 \left(1 + \frac{2b}{T} \right) \tag{8.125}$$

The next step in the hierarchy of increasingly sophisticated models simply adds a term to the composition dependence.

$$\Delta G_{\text{mix}}^{\text{xs}} = X_1 X_2 (a_0 + a_1 X_2) \left(1 + \frac{b}{T} \right) \qquad (8.126)$$

Models like those presented in Eqs. (8.125) and (8.126) prove to be useful in developing descriptions of the thermodynamics of solutions for computer calculations of phase diagrams [8.1, pp. 438–475, and 8.5].

8.7.3 Atomistic Models for Solution Behavior

The parameters contained in these models can be given physical significance on the basis of atomistic models for the behavior of solutions. The most direct of these atomistic models has been called the *quasichemical theory of solutions*, which is developed in some detail in this section. Other approaches, based upon the statistical thermodynamics of the distribution of atoms over the sites in a crystal are more sophisticated [8.1, pp. 435–503, and 8.3–8.6].

The quasichemical theory of solutions bears its name because it views a solution as a large molecule, with each pair of adjacent atoms treated as if connected by a chemical bond. In a binary system consisting of two kinds of atoms A and B, there are possible only three classes of nearest neighbor pairs of atoms that form such bonds:

$$A - A \qquad B - B \qquad A - B$$

Each type of bond is assumed to be endowed with its characteristic value of energy:

$$e_{\text{AA}} \qquad e_{\text{BB}} \qquad e_{\text{AB}}$$

In each case, the value of the energy corresponds to the formation of the bond from atoms that are originally so far apart that they are not interacting, as if they are in the vapor state.

Figure 8.7 is a plot of the variation of the energy of a system of two atoms as a function of their separation distance x. The slope of this plot is the negative of the force acting between the pair. Large values of x correspond to two atoms in the vapor state; since the vapor is the reference state for the plot, the energy of the pair is defined to be zero for large x. As the atoms are brought together, their electron clouds begin to interact and they are attracted toward each other. At very small x the ion cores begin to interact and the atoms are repelled. The distance d, which corresponds to the minimum energy of the pair, is the equilibrium separation distance between the atoms in the condensed state. The corresponding energy e_{ij}, a large negative number on this plot, is the energy associated with each bond of type ij in the solution.

In the quasichemical viewpoint all the internal energy of the solution is assumed to be contained in these interactions between neighboring pairs. This is a relatively naive view in comparison with more rigorous treatments of the formation of crystals from the vapor, which holds that the electron cloud develops a distribution of energy values associated with a density of allowable states. One consequence of this simple view is that the energy of a particular bond is independent of all the surrounding

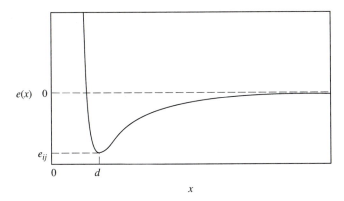

FIGURE 8.7
Variation of the energy of a pair of atoms as a function their distance of separation. The equilibrium distance, d, between the pair is characterized by the minimum in their energy.

atoms except the pair that forms the bond. Thus, the energy of each bond type is independent of composition. Let the numbers of each type of bond in the system be

$$P_{AA} \qquad P_{BB} \qquad P_{AB}$$

Then the internal energy of the solution, the energy change associated with condensing the solution from a vapor of the same composition, is

$$U_{\text{soln}} = P_{AA}e_{AA} + P_{BB}e_{BB} + P_{AB}e_{AB} \tag{8.127}$$

The number of bonds of each type is related to the composition and to the coordination number z, of the solution. The coordination number is the number of nearest neighbors that surround an atom in a crystal. (In a liquid solution there is a distribution of coordination numbers; z can be taken to be the average number of nearest neighbors in this case.) In a simple cubic crystal lattice, $z = 6$; for body centered cubic crystals, $z = 8$; in face centered cubic and hexagonal close packed structures, $z = 12$. In a system containing N_o atoms the total number of bonds P_T is

$$P_T = \frac{1}{2}N_o z \tag{8.128}$$

The factor of 1/2 is necessary because each bond is shared by two atoms; if one simply multiplied the number of atoms by the average number of near neighbors, each bond would be counted twice.

For a system of a given composition the numbers of bonds of each of the three types are not independently variable. Each atom of A is at one end of a near neighbor pair or bond. Each AA bond contains two ends incident upon A atoms and each AB bond contains one A-end. The total number of bond ends that terminate upon A atoms is equal to the total number of A atoms multiplied by z, the number of bond ends per atom. Thus

$$2P_{AA} + P_{AB} = zN_A = zN_oX_A \tag{8.129}$$

The analogous argument focused on the B atoms in the system gives

$$2P_{BB} + P_{AB} = zN_B = zN_oX_B \qquad (8.130)$$

Solve these two equations for P_{AA} and P_{BB}:

$$P_{AA} = \frac{1}{2}[X_AN_oz - P_{AB}] \qquad (8.131)$$

$$P_{BB} = \frac{1}{2}[X_BN_oz - P_{AB}] \qquad (8.132)$$

Thus in the characterization of the number of bonds of each type in a binary system it is only necessary to evaluate the number of unlike bonds, P_{AB}. There are no simplifying assumptions about the nature of the arrangement of atoms in the system in this description. Equations (8.131) and (8.132) can be used to eliminate P_{AA} and P_{BB} in the expression for the energy of the solution, Eq. (8.127).

$$U_{soln} = P_{AB}\left[e_{AB} - \frac{1}{2}(e_{AA} + e_{BB})\right] + \frac{1}{2}N_oz[X_Ae_{AA} + X_Be_{BB}] \qquad (8.133)$$

The change in internal energy for the mixing process for the solution is defined to be

$$\Delta U_{mix} = U_{soln} - \left[X_1U_A^0 + X_BU_B^0\right] \qquad (8.134)$$

where U_A^0 and U_B^0 are the internal energies per mole of pure A and pure B, each assumed to have the same crystal structure as the solution. Consider N_o atoms of pure A. The system contains only AA bonds. The number of bonds is $(1/2)N_oz$; the energy to form each from a vapor of pure A is e_{AA}. Thus

$$U_A^0 = \frac{1}{2}N_oze_{AA} \qquad (8.135)$$

For pure B,

$$U_B^0 = \frac{1}{2}N_oze_{BB} \qquad (8.136)$$

Substitute these results into Eq. (8.134)

$$\Delta U_{mix} = P_{AB}\left[e_{AB} - \frac{1}{2}(e_{AA} + e_{BB})\right] + \frac{1}{2}N_oz\left[X_Ae_{AA} + X_Be_{BB}\right]$$

$$- \left[X_A\frac{1}{2}N_oze_{AA} + X_B\frac{1}{2}N_oze_{BB}\right]$$

and simplify

$$\Delta U_{mix} = P_{AB}\left[e_{AB} - \frac{1}{2}(e_{AA} + e_{BB})\right] \qquad (8.137)$$

For condensed phases, the internal energy of mixing is negligibly different from the enthalpy of mixing since

$$\Delta H_{mix} = \Delta U_{mix} + P\Delta V_{mix} \cong \Delta U_{mix} \qquad (8.138)$$

so that Eq. (8.137) can be taken as the heat of mixing for the solution. Thus the quasichemical theory predicts that the heat of mixing of a solution is proportional to the number of unlike bonds it contains and to a parameter that reports the difference in energy between unlike (AB) bonds and the average energy of like bonds in the structure. This result is general in the sense that it assumes no restrictions on how the atoms are arranged in the solution. The primary limitation in the theory is the assignment of all energy effects to near neighbor pairs so that e_{AA}, e_{BB}, and e_{AB} are independent of composition.

The number of unlike bonds in the system, P_{AB}, takes on central significance in this approach to modeling thermodynamic behavior of solutions. In a mixture of a given composition the number of unlike near neighbor pairs reflects the arrangement of atoms in the system. For example, if like atoms tend to cluster together, then P_{AB} is small, Fig. 8.8a. If, on the other hand, atoms of the two types find themselves in an ordered structure, then most near neighbor pairs are unlike atoms, Fig. 8.8c. In a "random" mixture, Fig. 8.8b, P_{AB} has some intermediate value. This concept of a random mixture is used as a point of reference in describing atomic arrangements in mixtures since ideal solutions, which are also used as a reference in describing behavior, can be shown to be random mixtures. Further, the computation of P_{AB} for a random mixture is straightforward. Systems with P_{AB} values that are less than the random value for their composition are said to exhibit a tendency toward *clustering*. Those for which P_{AB} is larger than the random value evince a tendency toward *ordering*.

The development that follows is limited to the description of random solutions. In the end it is shown that this leads to an inconsistent result, except for an ideal solution. More specifically, it is shown that if the heat of mixing is not zero, then the arrangement of atoms cannot be random. This implies that the regular solution model, as described in Sec. 8.7.1, contains a conceptual flaw; more sophisticated versions of the regular solution model are required to describe the behavior of real solutions. Nonetheless, the regular solution model remains useful for two reasons:

1. It provides a valid approximate description for some real solutions, particularly at high temperatures where the entropy term ($T \Delta S_{mix}$) dominates in the Gibbs free energy of mixing.

2. It provides a mathematically tractable model for introducing concepts that are fundamental to the understanding of its more sophisticated versions.

The arrangement of atoms in a solution is defined to be random if the probability that any site chosen at random is occupied by an A atom is simply equal to the fraction of all of the sites in the structure that are occupied by A atoms, X_A. There is no preference for A atoms to occupy a particular site in the system. Similarly, the probability that a site is occupied by a B atom is X_B in a random mixture. Consider a pair of neighboring sites, labelled I and II. The atoms on these sites produce an AA bond if two simultaneous independent events are satisfied: site I is occupied by an A atom and site II is occupied by an A atom. Thus, the probability that the I-II site combination is an AA bond is the product of the probability that site I is occupied by

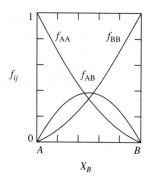

(a) Clustered

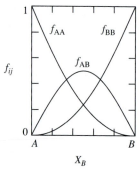

(b) Random

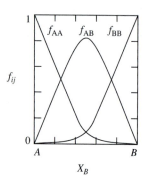

(c) Ordered

FIGURE 8.8
Variation of the numbers of each of the three bond types in a binary solution for (a) a clustered structure, (b) a random structure and (c) an ordered structure.

an A atom and the probability that site II is similarly occupied.

$$f_{AA} = X_A X_A = X_A^2 \tag{8.139}$$

A BB bond results if both sites are occupied by B atoms.

$$f_{BB} = X_B X_B = X_B^2 \tag{8.140}$$

An AB bond may result from either of two distinguishable arrangements: site I has an A atom and site II has a B, or site I has a B atom and site II has an A. Thus

$$f_{AB} = X_A X_B + X_A X_B = 2 X_A X_B \tag{8.141}$$

These bond probabilities can be interpreted as the fraction of bonds in the system that belong to each type. As a check of these formulations, the sum of these fractions should equal 1.

$$f_{AA} + f_{BB} + f_{AB} = X_A^2 + X_B^2 + 2X_A X_B = (X_A + X_B)^2 = 1$$

The number of bonds of each type in a random mixture is simply the fraction of bonds of that type multiplied by the total number of bonds, $(1/2)N_o z$. Of particular interest is the number of unlike bonds

$$P_{AB} = \tfrac{1}{2} N_o z f_{AB} = N_o z X_A X_B \tag{8.142}$$

Insertion of this computation of the number of unlike bonds into Eq. (8.138) gives the heat of mixing for a random mixture.

$$\Delta H_{\text{mix}} = N_o z X_A X_B \left[e_{AB} - \frac{1}{2}(e_{AA} + e_{BB}) \right] \tag{8.143}$$

In order to focus on the composition dependence in this expression, it can be written

$$\Delta H_{\text{mix}} = a_0 X_A X_B \tag{8.144}$$

in which

$$a_0 = N_o z \left[e_{AB} - \tfrac{1}{2}(e_{AA} + e_{BB}) \right] \tag{8.145}$$

The entropy of mixing of a random solution can be shown to be identical with the ideal entropy of mixing. In the solution, N_A atoms of A and N_B atoms of B are distributed independently on the N_0 available sites. The number of ways the atoms can be arranged is

$$\Omega = \frac{N_o!}{N_A! N_B!} \tag{8.146}$$

Apply Boltzmann's hypothesis, Eq. (6.3) to compute the entropy of this distribution

$$S = k \ln \Omega = k \ln \frac{N_o!}{N_A! N_B!} = k[\ln N_o! - (\ln N_A + \ln N_B!)]$$

Use Stirling's approximation:

$$S = k[(N_o \ln N_o - N_o) - (N_A \ln N_A - N_A) - (N_B \ln N_B - N_B)]$$

Simplify the result, recognizing that $N_o = N_A + N_B$:

$$S = k\left[(N_A + N_B) \ln N_o - N_A \ln N_A - N_B \ln N_B \right]$$

$$= k\left[-N_A(\ln N_A - \ln N_o) - N_B(\ln N_B - \ln N_o) \right]$$

$$= -k\left[N_A \ln \frac{N_A}{N_o} + N_B \ln \frac{N_B}{N_o} \right]$$

$$= -kN_o(X_A \ln X_A + X_B \ln X_B)$$

$$S = -R(X_A \ln X_A + X_B \ln X_B) \tag{8.147}$$

Since the value of Ω for the unmixed components is 1, this result can be interpreted as the entropy of mixing of a random solution, ΔS_{mix}. This expression is identical with the entropy of mixing of an ideal solution. From the definition of a regular solution it is seen that this is also the entropy of mixing of a regular solution.

The free energy of mixing of a random solution can be computed by inserting Eq. (8.144) and (8.147) into the definitional relationship

$$\Delta G_{\text{mix}} = \Delta H_{\text{mix}} - T \Delta S_{\text{mix}}$$

$$\Delta G_{\text{mix}} = a_0 X_A X_B + RT (X_A \ln X_A + X_B \ln X_B) \tag{8.148}$$

Compare this result with Eq. (8.115), the simplest of the regular solution models. It is seen that this model implies that the atoms are arranged randomly in the solution and that the model parameter a_0 is determined by the energies of the three types of bonds in the system.

The first term on the right side of Eq. (8.148) is also the excess free energy of mixing and thus determines the sign of the departure of the behavior of the system from an ideal solution. Examination of the form deduced for a_0, Eq. (8.145), shows that all factors are positive except the quantity contained in brackets. Figure 8.9 illustrates possible combinations of bond energies that determine the sign of a_0. In mathematical terms

Positive Departure:

$$a_0 > 0 \rightarrow e_{AB} > \tfrac{1}{2}(e_{AA} + e_{BB}) \tag{8.149}$$

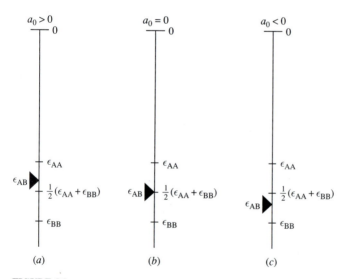

FIGURE 8.9
Illustration of the relative values of bond energies that lead to (a) positive, (b) zero, and (c) negative departures from ideal solution behavior.

Negative Departure:

$$a_0 < 0 \rightarrow e_{AB} < \tfrac{1}{2}(e_{AA} + e_{BB}) \tag{8.150}$$

Ideal Solution:

$$a_0 = 0 \rightarrow e_{AB} = \tfrac{1}{2}(e_{AA} + e_{BB}) \tag{8.151}$$

These considerations point to a conceptual difficulty in the strict application of the regular solution model with its implicit assumption of random mixing embodied in the definition. Consider, for example, a random solution exhibiting a positive departure from ideal behavior. According to the quasichemical theory that produced the inequality (8.149), like bonds have a lower energy than unlike bonds in such a system. Thus the system could lower its heat of mixing, and hence the free energy of mixing, by increasing the number of like bonds above the random value given by Eqs. (8.145) and (8.146). In a system constrained to constant temperature and pressure, a process that lowers the Gibbs free energy is a spontaneous process. Thus the atoms can be expected to arrange themselves in configurations that have more like bonds than a random solution, meaning that the system tends to exhibit clustering. If this process occurs, then the arrangement of atoms is no longer random and the entropy of mixing is not simply the random value that corresponds to ideal and regular solutions.

The analogous argument can be made for a system that exhibits a negative departure from ideal behavior. Inequality (8.150) requires that like bonds have a lower energy than unlike bonds in such a system. The enthalpy and thus Gibbs free energy can be lowered by increasing the number of unlike bonds or, in other words, by tending to produce an ordered arrangement of atoms.

More sophisticated models have been devised to incorporate deviations from randomness into a computation of the entropy of mixing. Swalin [8.4] provides an introductory discussion of short and long range order parameters and their relation to the entropy of mixing of ordered solutions. Models based upon statistical thermodynamics and the description of clusters of atoms, rather than just near neighbor pairs, include the central atoms model discussed by Lupis [8.1] and the cluster variation model due to Fontaine [8.6]. Statistically based models that predict the behavior of interstitial solutions in crystals are also discussed by Lupis [8.1]. Devereux [8.3] and Lupis [8.1] also present extensions of these models to ternary and higher order systems. However, the regular solution model remains a useful tool for introducing the subject of solution models and the strategies involved in their development.

8.8   SUMMARY OF CHAPTER 8

Partial molal properties of a component in a mixture are defined by

$$\overline{B}_k \equiv \left(\frac{\partial B'}{\partial n_k} \right)_{T,P,n_j \neq n_k} \tag{8.6}$$

Consequences of this definition, summarized in Table 8.1, include:

$$d\Delta B_{\text{mix}} = \sum_{k=1}^{c} \Delta \overline{B}_k dX_k \tag{8.20}$$

$$\Delta B_{\text{mix}} = \sum_{k=1}^{c} \Delta \overline{B}_k X_k \tag{8.21}$$

$$\sum_{k=1}^{c} X_k d\Delta \overline{B}_k = 0 \tag{8.22}$$

Partial molal properties of components in a binary system can be computed from the corresponding total property of the solution

$$\Delta \overline{B}_k = \Delta B_{\text{mix}} + (1 - X_k)\frac{d\Delta B_{\text{mix}}}{dX_k} \tag{8.28}$$

A Gibbs-Duhem integration

$$\Delta \overline{B}_1 = -\int_{X_2=0}^{X_2} \frac{X_2}{X_1} \cdot \frac{d\Delta \overline{B}_2}{dX_2} dX_2 \tag{8.36}$$

yields the partial molal property of component 1 when values are known for component 2.

Relationships can be derived among partial molal properties that are analogues to the laws, the definitions, coefficient relations, and Maxwell relations.

All the properties of a solution can be computed if either the chemical potential, the activity, defined by

$$\mu_k - \mu_k^0 \equiv RT \ln a_k \tag{8.66}$$

or the activity coefficient, defined by

$$a_k \equiv \gamma_k X_k \tag{8.67}$$

is known as a function of temperature, pressure and composition. These relationships are summarized in Tables 8.2, 8.3, 8.4, and 8.5. The definition of the activity coefficient permits decomposition of each partial molal property, as well as each total property of a solution, into an ideal and an excess component.

A hierarchy of solution models, beginning with the ideal and regular solutions and progressing through a series of increasingly complex phenomenological solution models, provides a useful basis for the analysis of the behavior of real solutions.

The quasichemical model for solutions yields an expression for the heat of mixing in terms of the bond energies in the system

$$\Delta H_{\text{mix}} = P_{\text{AB}} \left[e_{\text{AB}} - \tfrac{1}{2}(e_{\text{AA}} + e_{\text{BB}}) \right] \tag{8.143}$$

Application of this model to random solutions demonstrates an essential flaw in the simple regular solution model but nonetheless, does yield insight into the connection between atomic arrangements and departures from ideal behavior.

PROBLEMS

8.1. Titanium metal is capable of dissolving up to 30 atomic percent oxygen. Consider a solid solution in the system TiO containing an atom fraction, $X_O = 0.12$. The molar volume of this alloy is 10.64 cc/mol. Calculate:
(a) The weight percent of O in the solution.
(b) The molar concentration (gm-atoms/cc) of O in the solution.
(c) The mass concentration (gm/cc) of O in the solution.

Use these calculations to deduce *general* expressions for weight percent, and molar and mass concentrations of a component in a binary solution in terms of the atom fraction X_2, the molar volume, V, and the gram atomic weights, AW_1 and AW_2, of the elements involved.

8.2. Review the consequences of the definition of partial molal properties that make it a convenient measure of the contribution of each component to the total value of the thermodynamic properties of a solution.

8.3. Given that the volume change on mixing of a solution obeys the relation

$$\Delta V_{\text{mix}} = 2.7 X_1 X_2^2 \left(\frac{\text{cc}}{\text{mol}} \right)$$

(a) Derive expressions for the partial molal volumes of each of the components as functions of composition.
(b) Demonstrate that your result is correct by using it to compute ΔV_{mix}, demonstrating that the equation above is recovered.

8.4. Use the partial molal volumes computed in Problem 8.3 to demonstrate that the Gibbs-Duhem equation holds for these properties in this system.

8.5. In the system Pandemonium (Pn)–Condominium (Cn), the partial molal heat of mixing of Pandemonium can be fitted by the expression

$$\Delta \overline{H}_{\text{Pn}} = 12,500 X_{\text{Pn}}^2 X_{\text{Cn}} \left(\frac{\text{J}}{\text{mole}} \right)$$

Calculate and plot the function that describes the variation of the heat of mixing with composition for this system.

8.6. For an ideal solution it is known that, for component 2

$$\Delta \overline{G}_2 = RT \ln X_2$$

Use the Gibbs-Duhem integration to the derive corresponding relation for component 1.

8.7. Recall the definitional relationship for the enthalpy function:

$$H' = U' + PV'$$

Use the partial molal operator and the definition of the properties of the pure components to derive the analogous relationship

$$\Delta \overline{H}_k = \Delta \overline{U}_k + P \Delta \overline{V}_k$$

8.8. The excess free energy of mixing in face centered cubic solid solutions of aluminum and zinc is well described by the relation

$$\Delta G_{\text{mix}}^{\text{xs}} = X_{\text{Al}} X_{\text{Zn}} (9600 X_{\text{Zn}} + 13200 X_{\text{Al}}) \cdot \left(1 - \frac{T}{4000} \right)$$

Compute and plot curves for ΔG_{mix} as a function of composition for a sequence of temperatures ranging from 300 K to 700 K.

8.9. Using the relation given in Problem 8.8, calculate and plot the activity of Zn in an FCC solid solution of these elements at 500 K.

8.10. The system A–B forms a regular solution at high temperatures with a heat of mixing given by the relation

$$\Delta G_{mix}^{xs} = (1 - bT)(1 - cP)X_A X_B$$

Suppose $b = 2 \times 10^{-4}$ K^{-1} and $c = 2 \times 10^{-5}$ (atm)$^{-1}$.

(a) Derive expressions for *all* the mixing properties for this system.

(b) Use these relations to evaluate all the properties of the solution at 100 atm pressure, 550 K and a composition $X_B = 0.35$.

8.11. Criticize the following reported finding: The system A–B forms a regular solution at high temperatures with the heat of mixing found to obey the relation

$$\Delta H_{mix} = -14,500 X_A X_B \left(1 - \frac{350}{T}\right)$$

8.12. Given that Henry's law holds for the solute of a dilute real solution, derive Raoult's law for the solvent.

8.13. The system A–B forms a regular solution with the heat of mixing given by:

$$\Delta H_{mix} = -13,500 X_A X_B \left(\frac{J}{mole}\right)$$

(a) Derive expressions for the Henry's law constant for A as a solute in B and B as a solute in A.

(b) Plot both Henry's law constants as a function of temperature.

8.14. The system A–B can be described by the quasichemical model. The heat of vaporization of pure A is 98,700 J/mol; that of pure B is 127,000 J/mol. At a solution composition $X_B = 0.40$ at 750 K the activity of component A is found to be 0.53. Estimate the energy of an AB bond in this system.

8.15. The system A–B exhibits a tendency toward ordering, with the number of unlike near neighbor pairs larger than the random value by 30 percent at all compositions. Compute and plot the number of AA, AB, and BB bonds in the system.

8.16. The A–B system exhibits a measured heat of mixing given by the relationship

$$\Delta H_{mix} = X_A X_B (7,500 X_A + 18,200 X_B)$$

The bond energies (J/bond) for this system are estimated to be

$$e_{AA} = -6.5 \times 10^{-20} \qquad e_B = -5.3 \times 10^{-20} \qquad e_{AB} = 5.4 \times 10^{-20}$$

Compute and plot the fractions of AA, AB, and BB bonds in the system as a function of composition.

REFERENCES

1. Lupis, C. H. P.: *Chemical Thermodynamics of Materials*, Elsevier Science Publishing Co., Inc., New York, N.Y., p. 56–57, 1983.
2. Darken, L. S. and R. W. Gurry: *Physical Chemistry of Metals*, McGraw-Hill, Inc., New York, N.Y., 1951.

2. Darken, L. S. and R. W. Gurry: *Physical Chemistry of Metals*, McGraw-Hill, Inc., New York, N.Y., 1951.
3. Devereux, Owan F.: *Topics in Metallurgical Thermodynamics*, John Wiley & Sons, New York, N.Y., p. 282–286, 1983.
4. Swalin, R. A.: *Thermodynamics of Solids*, John Wiley & Sons, New York, N.Y., p. 148–164, 1972.
5. Kaufman L. and H. Bernstein: *Computer Calculations of Phase Diagrams*, Academic Press, New York, N.Y., 1970.
6. DeFontaine, D.: "Configurational Thermodynamics of Solid Solutions," in *Solid State Physics*, Vol. 34, H. Ehrenreich, F. Seitz and D. Turnbull, eds., Academic Press, New York, N.Y., 1979.

CHAPTER
9

MULTICOMPONENT HETEROGENEOUS SYSTEMS

The logical progression through the hierarchy of classes of thermodynamic systems builds upon the treatment of unary heterogeneous systems in Chap. 7 and multicomponent homogeneous systems in Chap. 8 to the class of systems that are the subject of this chapter: multicomponent, heterogeneous systems. This class of systems is important in materials science because most commercial materials contain a number of components in a microstructure that is an aggregate of two or more phases. Control of composition and the arrangement of the phases in the structure is tantamount to controlling properties. Also, many reacting systems involve the interaction of components in more than one phase. For example, oxidation of a metal involves at least three phases: the metal, the gas containing oxygen, and the oxide that forms. Microelectronic devices consist of multicomponent connectors, sometimes layers of two or three phases, that connect electronically active components of another phase, all laid down on a substrate that may be yet another phase. It is evident that the treatment of multicomponent, multiphase systems has broad application in technology and science.

The apparatus necessary for describing multicomponent, multiphase systems is developed in this chapter. This apparatus is then applied in the general strategy for finding conditions for equilibrium. This set of relations that must exist between the thermodynamic properties when such a system is in equilibrium is then used as a basis for deriving the classic *Gibbs phase rule*. The Gibbs phase rule is a very general relationship that is the basis for the construction of phase diagrams, the primary thinking tool used to understand the behavior of multiphase, multicomponent systems.

The construction of phase diagrams is illustrated for a variety of representations of unary, binary, and ternary systems.

The conditions for equilibrium derived in this chapter are the set of equations that relate the limit of stability of phases, graphically presented in a phase diagram, to the thermodynamics of the system. Thus, as is shown for *unary* two phase systems in Chap. 7, calculation of phase diagrams from thermodynamic information starts with these equations. The same relationships provide a basis for estimating thermodynamic properties of some of the phases involved from an experimental determination of the phase diagram. Connections of this kind are also demonstrated in Chap. 7 for phase boundaries in a unary system, where it is shown that the heat of vaporization can be determined from the phase boundary between liquid and vapor. Similar applications are illustrated in calculations of phase diagrams for binary and ternary systems in Chap. 10.

9.1   THE DESCRIPTION OF MULTIPHASE, MULTICOMPONENT, NONREACTING SYSTEMS

Chapter 8 shows that the thermodynamic apparatus necessary to describe a homogeneous multicomponent system consists of the addition of a string of terms that incorporates variation of the number of moles of each of the components in the system. For a system with c components, the combined statement of the first and second laws becomes

$$dU' = T dS' - P dV' + \sum_{k=1}^{c} \mu_k dn_k \qquad (8.49)$$

Consider a system made up of p distinguishable phase forms, each of which contains the same set of c components. Figure 9.1 illustrates a microstructure with three phases, each containing two components. Each of these phases, viewed as a system, exchanges heat, work, and matter with its surroundings, or more specifically with the other phases in the system and the exterior of the whole system. Focus on the α phase, Fig. 9.1b. Equation (8.49) describes the change in internal energy experienced by the α phase when the state of the whole multiphase system is altered in an arbitrary way. Use the superscript ($^\alpha$) to denote properties of the α phase:

$$dU'^\alpha = T^\alpha dS'^\alpha - P^\alpha dV'^\alpha + \sum_{k=1}^{c} \mu_k^\alpha dn_k^\alpha \qquad (9.1)$$

A relationship like this holds in each phase in the system.[1]

---

[1] In order to write these equations it is necessary to assume that the intensive properties of the α phase, that is, T^α, P^α and μ_k^α, are uniform within the phase. This is equivalent to assuming that each phase in the system is in *internal equilibrium*; exchanges between the phases derive from differences that may exist in their intensive properties. The treatment of phases with nonuniform intensive properties is the subject of Chap. 14, Continuous Systems.

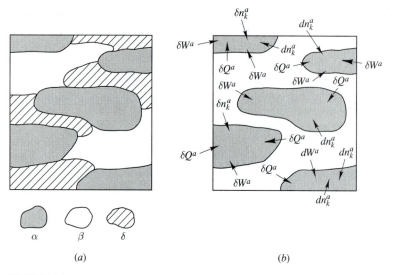

FIGURE 9.1

A microstructure with three phases, (*a*). The α phase exchanges heat, work and material with the other phases and with the surroundings (*b*).

The treatment of the behavior of the whole multiphase system focuses on the *extensive properties* of the system. The strategy is simple and makes use of the definition of extensive properties. For the extensive properties V', S', U', H', F', and G', the value of the property for the system is the sum of the values for the separate parts

$$B'_{\text{sys}} = B'^{\text{I}} + B'^{\text{II}} + \cdots + B'^{\alpha} + \cdots + B'^{P}$$

$$B'_{\text{sys}} = \sum_{\alpha=1}^{P} B'^{\alpha} \tag{9.2}$$

where B' can be any of the extensive properties. If the system is taken through an arbitrary change in state, then the change in B'_{sys} is simply the sum of the changes that each phase experiences because the differential of a sum is the sum of the differentials

$$dB'_{\text{sys}} = \sum_{\alpha=1}^{P} dB'^{\alpha} \tag{9.3}$$

This simple principle is the basis for handling the description of multiphase systems.

As an application of this strategy consider the change in internal energy. The change in internal energy of the multiphase system when it is taken through an arbitrary infinitesimal process is

$$dU'_{\text{sys}} = \sum_{\alpha=1}^{P} dU'^{\alpha} \tag{9.4}$$

The change in internal energy experienced by each phase is determined by its own properties and is given by Eq. (9.1). Combining Eqs. (9.1) and (9.4) yields

$$dU'_{sys} = \sum_{\alpha=1}^{P} \left[T^\alpha dS'^\alpha - P^\alpha dV^\alpha + \sum_{k=1}^{c} \mu_k^\alpha dn_k^\alpha \right] \tag{9.5}$$

The entropy change for the system plays a key role in the strategy for deriving the conditions for equilibrium developed in the next section. For a single multicomponent phase, the change in entropy for an arbitrary change in state is obtained by solving Eq. (9.1) for dS'^α.

$$dS'^\alpha = \frac{1}{T^\alpha} dU'^\alpha + \frac{P^\alpha}{T^\alpha} dV'^\alpha - \frac{1}{T^\alpha} \sum_{k=1}^{c} \mu_k^\alpha dn_k^\alpha \tag{9.6}$$

When the whole multiphase system is taken through an arbitrary process, its change in entropy is given by Eq. (9.3).

$$dS'_{sys} = \sum_{\alpha=1}^{P} dS'^\alpha \tag{9.7}$$

Substitute Eq. (9.6):

$$dS'_{sys} = \sum_{\alpha=1}^{P} \left[\frac{1}{T^\alpha} dU'^\alpha + \frac{P^\alpha}{T^\alpha} dV'^\alpha - \frac{1}{T^\alpha} \sum_{k=1}^{c} \mu_k^\alpha dn_k^\alpha \right] \tag{9.8}$$

Thus, the thermodynamic apparatus needed to describe multiphase systems is based simply on the definition of extensive properties and the mathematical relationship tthat the differential of a sum is the sum of the differentials.

9.2 CONDITIONS FOR EQUILIBRIUM

Recall again the general criterion for equilibrium and the associated strategy for finding conditions for equilibrium presented in Chap. 5. The criterion states that, in an isolated system, the entropy of the system is a maximum at equilibrium. Implementing this principle requires derivation of the condition of the system that yields a maximum in the entropy function when its variations are constrained by the conditions that it remain isolated from its surroundings.

To derive the conditions for equilibrium in a system containing p phases, it is only necessary to consider the equilibrium between any *two* of these phases. The results obtained for a two phase system can then be extended to a system with an arbitrary number of phases by some simple inductive reasoning. Thus the system under consideration consists of two phases, α and β, each containing c components. The change in entropy of each phase is given by Eq. (9.3) with appropriate superscripts. If the two phase system is taken through an arbitrary change in state, the change in entropy is given by

$$dS'_{sys} = dS'^\alpha + dS'^\beta \tag{9.9}$$

or, more explicitly,

$$dS'_{sys} = \frac{1}{T^\alpha}dU'^\alpha + \frac{P^\alpha}{T^\alpha}dV'^\alpha - \frac{1}{T^\alpha}\sum_{k=1}^{c}\mu_k^\alpha dn_k^\alpha$$

$$+ \frac{1}{T^\beta}dU'^\beta + \frac{P^\beta}{T^\beta}dV'^\beta - \frac{1}{T^\beta}\sum_{k=1}^{c}\mu_k^\beta dn_n^\beta \qquad (9.10)$$

If the system is isolated from its surroundings, then not all of the $2(2+c)$ variables in this equation are independent; some are related through the isolation constraints.

If, during this process, the two phase system is isolated from its surroundings, then, whatever changes occur within the system, its internal energy U'_{sys}, volume V'_{sys}, and the number moles of each of its components, $n_{k,sys}$, cannot change. These isolation constraints can be stated in mathematical form as

$$dU'_{sys} = 0 = dU'^\alpha + dU'^\beta \rightarrow dU'^\beta = -dU'^\alpha \qquad (9.11)$$

$$dV'_{sys} = 0 = dV'^\alpha + dV'^\beta \rightarrow dV'^\beta = -dV'^\alpha \qquad (9.12)$$

$$dn_{k,sys} = 0 = dn_k^\alpha + dn_k^\alpha \rightarrow dn_k^\beta = -dn_k^\alpha \qquad (k = 1, 2, \ldots, c) \qquad (9.13)$$

Evidently, whatever change occurs in one of these variables must be compensated by an opposite change in the other since the total values of the properties for the system are constrained not to change because the system is isolated.

Equations (9.11) through (9.13) can be used to eliminate dependent variables in the expression for the change in entropy, Eq. (9.10), applied to an isolated system. Substitute for dU'^β, dV'^β, and dn_k^β:

$$dS'_{sys,iso} = \frac{1}{T^\alpha}dU'^\alpha + \frac{P^\alpha}{T^\alpha}dV'^\alpha - \frac{1}{T^\alpha}\sum_{k=1}^{c}\mu_k^\alpha dn_k^\alpha$$

$$+ \frac{1}{T^\beta}(-dU'^\alpha) + \frac{P^\beta}{T^\beta}(-dV'^\alpha) - \frac{1}{T^\beta}\sum_{k=1}^{c}\mu_k^\beta(-dn_k^\alpha)$$

Collecting terms

$$dS'_{sys,iso} = \left(\frac{1}{T^\alpha} - \frac{1}{T^\beta}\right)dU'^\alpha + \left(\frac{P^\alpha}{T^\alpha} - \frac{P^\beta}{T^\beta}\right)dV'^\alpha - \sum_{k=1}^{c}\left(\frac{\mu_k^\alpha}{T^\alpha} - \frac{\mu_k^\beta}{T^\beta}\right)dn_k^\alpha$$

$$(9.14)$$

These $(2+c)$ variables remaining can be changed independently in an isolated system. The condition for an extremum is obtained by setting each of the coefficients in this expression equal to zero

$$\frac{1}{T^\alpha} - \frac{1}{T^\beta} = 0 \rightarrow T^\alpha = T^\beta \qquad \text{(Thermal Equilibrium)} \qquad (9.15)$$

$$\frac{P^\alpha}{T^\alpha} - \frac{P^\beta}{T^\beta} = 0 \rightarrow P^\alpha = P^\beta \qquad \text{(Mechanical Equilibrium)} \qquad (9.16)$$

$$\frac{\mu_k^\alpha}{T^\alpha} - \frac{\mu_k^\beta}{T^\beta} = 0 \rightarrow \mu_k^\alpha = \mu_k^\beta \quad (k = 1, 2, \ldots, c) \quad \text{(Chemical Equilibrium)} \quad (9.17)$$

Equation (9.17) holds for each of the c components in the system. These equations represent the conditions that must be satisfied in order for any two phases to coexist in thermodynamic equilibrium. They are the *conditions for equilibrium* in a multicomponent two phase system.

In order to extend this result to a system containing an arbitrary number of phases, it is only necessary to consider the relationship between the phases two at a time. For example, consider a system consisting of three phases, α, β, and ϵ. When this system attains equilibrium, α is equilibrated with β, β with ϵ, and α with ϵ. Consider a system composed of the two phases β and ϵ. Equilibrium in this two phase system is achieved when equations like Eqs. (9.16) through (9.18) are satisfied for a $(\beta + \epsilon)$ system

$$T^\beta = T^\epsilon \qquad P^\beta = P^\epsilon \qquad \mu_k^\beta = \mu_k^\epsilon \quad (k = 1, 2, \ldots, c) \quad (9.18)$$

Combining these equations with those obtained for the $(\alpha + \beta)$ equilibrium gives the conditions for equilibrium in a three phase system

$$T^\alpha = T^\beta = T^\epsilon \tag{9.19}$$

$$P^\alpha = P^\beta = P^\epsilon \tag{9.20}$$

$$\mu_k^\alpha = \mu_k^\beta = \mu_k^\epsilon \quad (k = 1, 2, \ldots, c) \tag{9.21}$$

In a system composed of p phases at equilibrium, each pair of phases is in equilibrium and relationships corresponding to Eqs. (9.15) through (9.17) exist for that pair. Thus, the set of conditions that must be satisfied when a system composed of p phases and c components comes to equilibrium is

$$T^\mathrm{I} = T^\mathrm{II} = \cdots = T^\alpha = \cdots = T^P \tag{9.22}$$

$$P^\mathrm{I} = P^\mathrm{II} = \cdots = P^\alpha = \cdots = P^P \tag{9.23}$$

$$\mu_1^\mathrm{I} = \mu_1^\mathrm{II} = \cdots = \mu_1^\alpha = \cdots = \mu_1^P \tag{9.24a}$$

$$\mu_2^\mathrm{I} = \mu_2^\mathrm{II} = \cdots = \mu_2^\alpha = \cdots = \mu_2^P \tag{9.24b}$$

$$\vdots$$

$$\mu_c^\mathrm{I} = \mu_c^\mathrm{II} = \cdots = \mu_c^\alpha = \cdots = \mu_c^P \tag{9.24c}$$

Stated in words; for a multicomponent, multiphase system to come to equilibrium, the temperature, the pressure, and the chemical potential of each component must be the same in all the phases. These equations form the basis for the construction and calculation of phase diagrams.

9.3 THE GIBBS PHASE RULE

Perhaps the most rudimentary piece of information about a complicated thermodynamic system is the number of independent variables required to describe its state. In a unary, homogeneous, closed, nonreacting, otherwise simple system this number is two: if, for example, the temperature and pressure are specified, the state of the system is determined. To describe the state of a solution it is necessary to specify its composition through a set of mole fractions in addition to its temperature and pressure. However, as the complexity of a system increases, the number of variables that must be assigned values in order to fix its state does not necessarily increase in proportion. For example, in Chap. 7 it is shown that, in a unary two phase system at equilibrium, relationships between the variables dictate that, if the pressure in one of the phases is specified, the pressure in the other phase must be the same and, furthermore, the temperature of both phases is determined. In a three phase unary system, there is *no* freedom of choice in assigning values to state variables; the pressure and temperature of all three phases is determined.

Evidently the answer to the question "How many independent variables characterize this system?" is not trivial. In his classic paper published in 1875, titled "On the Equilibrium of Heterogeneous Substances," J. Willard Gibbs presents the solution to this basic problem as part of his development of the foundations of chemical thermodynamics. The number of independent variables that a system has is the number of variables to which values must be assigned in order to define its thermodynamic state. Gibbs called this number the *number of degrees of freedom, f,* for the system.

The term "degrees of freedom" is borrowed from mathematical usage where it is applied in the consideration of *systems of equations relating many variables.* As a simple example, consider the following system of equations:

$$4x + 3y - z = 9$$
$$5x - 9y + 2z = -3$$
$$x + y + z = 1$$

There are three relationships among three variables. A unique set of values for the variables x, y and z can be found so that this system of equations is satisfied. There is no freedom of choice for these variables; their values are *determined* for this system of equations; its number of degrees of freedom is *zero*.

Consider next the set of equations

$$2u - 3v + x - y + 3z = 10$$
$$-3u + 5v - 2x + 4y + z = -8$$

These two equations in five unknowns cannot be solved to yield unique values for the five variables. Indeed, it is possible to choose arbitrary values for, say, u, v, and z and *then* find values of x and y for which the system of equations is satisfied. This system of two equations in five variables thus has *three independent variables* and mathematically, it is said to exhibit *three degrees of freedom.*

These considerations are not limited to systems of linear equations. Every system of well-behaved algebraic equations exhibits this behavior. In general, a system of n such equations in m variables has at least one solution. The solution may not be unique. For example, a quadratic equation in a single unknown has two solutions, both of which satisfy the equation. Computer software based upon numerical iteration techniques that solve these problems with efficiency is widely available. However, a system consisting of n equations relating m variables, where $m > n$, does *not* have at least one solution; it has $(m - n)$ too many variables. If arbitrary numerical values are independently assigned to these $(m - n)$ variables, then the system is left with n equations in n unknowns and can be solved. For a system of n equations among m variables one may define the *number of degrees of freedom* as

$$f = m - n \qquad (9.25)$$

This is the number of variables to which values can be freely assigned without jeopardizing the validity of the equations describing the behavior in the system. Thus f may be thought of as the number of independent variables in a system of equations and the remaining n variables, which may be computed once the f independent values have been assigned, are then *dependent* variables.

With this mathematical concept in mind, the number of degrees of freedom of a multicomponent, heterogeneous, nonreacting, otherwise simple system can be computed. It is necessary to enumerate the variables that the system has, list the number of equations that relate these variables when the system has attained equilibrium, and then subtract. Since the conditions for equilibrium are formulated in terms of the intensive properties of the system (T, P, μ_k), it is necessary to formulate these counts of variables and equations in terms of intensive properties.

Consider first one phase in the p phase system. The state of this phase can be fixed by assigning values to its temperature, pressure, and composition. It is important to recognize that, for the description in terms of intensive properties, there are $(c - 1)$ compositional variables X_k because in any given phase the mole fractions of the components sum to one. Thus the variables that describe its state are $(T, P, X_2, X_3, \ldots, X_c)$; here it is assumed that X_1 has been computed from the remaining mole fractions.[2] The number of variables in this list is $[2 + (c - 1)] = [1 + c]$. Since all the phases are assumed to contain the same components, each phase is described by this number of variables. For the system of p phases, each with c components, the variables are

$$T^{\mathrm{I}}, \ P^{\mathrm{I}}, \ X_2^{\mathrm{I}}, \ X_3^{\mathrm{I}}, \ldots, \ X_c^{\mathrm{I}} \qquad \text{For Phase I}$$

---

[2]Some texts list the intensive properties appropriate to the description of a single phase as $(T, P, \mu_2, \mu_3, \ldots, \mu_c)$, since these are the variables explicitly contained in the conditions for equilibrium. Note that μ_1 is also *not independent* of the other chemical potentials, since they are related in a given phase by the Gibbs-Duhem equation. In either case, the number of intensive variables required to specify the state of a single phase is the same: $c + 1$.

$$T^{\mathrm{II}}, \ P^{\mathrm{II}}, \ X_2^{\mathrm{II}}, \ X_3^{\mathrm{II}}, \dots, X_c^{\mathrm{II}} \qquad \text{For Phase II}$$

$$\vdots$$

$$T^{\alpha}, \ P^{\alpha}, \ X_2^{\alpha}, \ X_3^{\alpha}, \dots, X_c^{\alpha} \qquad \text{For Phase } \alpha$$

$$\vdots$$

$$T^{P}, \ P^{P}, \ X_2^{P}, \ X_3^{P}, \dots, X_c^{P} \qquad \text{For Phase } p$$

Each of the rows has $(1 + c)$ variables listed; there are p rows, one for each phase. The total number of variables in the system is thus

$$m = p(1+c) \tag{9.26}$$

When the system comes to equilibrium, the conditions for equilibrium, Eqs. (9.22)–(9.24), must be satisfied. The number of independent equations contained in this system of equations can be obtained by simply counting equals signs. Each row contains $(p-1)$ *independent* equations; there are $(2+c)$ rows of equations representing conditions for thermal, mechanical, and chemical equilibrium. Thus the number of equations relating the variables in the system is

$$n = (p-1)(2+c) \tag{9.27}$$

The number of degrees of freedom in this system of equations is obtained by substituting Eqs. (9.26) and (9.27) into Eq. (9.25)

$$f = m - n = [p(1+c)] - [(p-1)(2+c)]$$

$$= p + pc - 2p - pc + 2 + c$$

$$f = c - p + 2 \tag{9.28}$$

which is the *Gibbs phase rule*. Thus the calculation of the number of independent variables in a system consisting of p phases and c components is straightforward. The implications of this result for the construction of phase diagrams are developed in the next section.

9.4   THE STRUCTURE OF PHASE DIAGRAMS

Phase diagrams are graphical representations of the domains of stability of the various classes of structures (one phase, two phase, three phase, etc.) that may exist in a system at equilibrium. Phase diagrams are most commonly constructed in temperature-pressure-composition space. Other coordinate systems, though not yet as widely used, may find increasing practical application.

For single phase regions in a unary system the number of degrees of freedom $f = (1 - 1 + 2) = 2$ (for example, T and P); for binary systems, $f = (2 - 1 + 2) = 3$. In a c component system, f for a single phase region is $f = (c - 1 + 2) = (c + 1)$. Further, for any given system, the phase rule shows that f decreases as the number of phases p increases. It is thus clear from the phase rule that the regions of stability of single phases have the highest number of degrees of freedom in any system with a given number of components. That is, single phase regions require the largest number of variables for their specification. Accordingly, the graphical space in which the phase diagram is constructed must have $(c + 1)$ independent coordinates so that the full range of behavior of the single phase regions can be represented. In a unary system the diagram can be plotted in (T, P) space; binary diagrams require a three dimensional space, most commonly with (T, P, X_2) coordinates; complete representation of a ternary system requires a four dimensional space (for example, T, P, X_2, X_3); and so on.

The printed page is two dimensional and thus most quantitative phase diagrams are represented as *sections* taken through the multidimensional space required for their full representation. (Projected views of three dimensional diagrams are useful for teaching purposes or for visualizing the relationships among phases described by three variables. However useful such views are, they cannot be used to read quantitative information about the domains of the phases.) A section is obtained by fixing a value for one or more of the independent variables in the system. Phase diagrams for unary, binary, and ternary systems plotted with the most commonly chosen variables are shown in Fig. 9.2. Since $(c + 1) = 2$ for a unary system, the complete representation of all possible states can be plotted on the printed page, Fig. 9.2a. Most binary diagrams are plotted for a constant value of the pressure on the system, usually chosen to be one atmosphere, Fig. 9.2b. Ternary diagrams are represented for a fixed value of both temperature and pressure, Fig. 9.2c.

These sections are well behaved (easy to interpret) if they are taken at constant values of the *thermodynamic potentials T, P*, and μ_k, those variables explicitly contained in the conditions for equilibrium, Eqs. (9.22)–(9.24). In two phase, or three phase, or higher order regions on the diagram, it is precisely these vari-

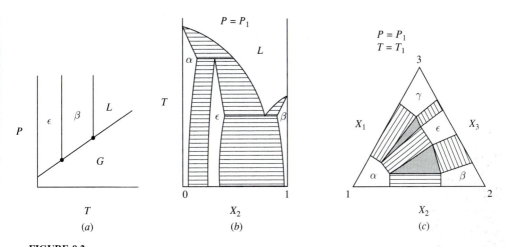

FIGURE 9.2
Common representations for (a) unary $(c = 1)$, (b) binary $(c = 2)$ and (c) ternary $(c = 3)$ phase diagrams.

ables that are required to be constant by the conditions for equilibrium. The states of the phases participating in these multiphase equilibria all lie in the plane of the diagram if and only if the diagram is plotted by holding constant one or more of these potentials since equilibrated states all have the same values of the variables chosen to be constant. Thus the compositions, temperatures, and pressures of the three phases that coexist in equilibrium in a ternary system, Fig. 9.2c, are all represented in the plane of the diagram. Sections taken with other variables constant (e.g., X_2, X_3, S, or V) have multiphase regions in which the participating states are *out of the plane representing the diagram*; values of these variables for phases in equilibrium are in general not the same. The relationships among equilibrated states are therefore not represented on a section obtained by holding a variable such as V or X_2 constant.

9.4.1   Phase Diagrams Plotted in Thermodynamic Potential Space.

The simplest form of phase diagram for any number of components is obtained when it is plotted in a space with thermodynamic potentials $(T, P, \mu_2, \mu_3, \ldots, \mu_c)$ as coordinates. It is convenient to replace each chemical potential axis with the corresponding value of activity.[3] Thus representation of a phase diagram in $(T, P, a_2, \ldots, a_c)$ space yields the simplest though not necessarily most useful representation. A phase diagram for a c component system plotted in potential space is a simple cell structure. Further, sections through such diagrams, obtained by assigning a constant value to one of the potentials, (e.g., P, T, or μ_k), are also simple cell structures. The notion embodied in the phrase "cell structure" is demonstrated in Fig. 9.3a for a binary system. A three dimensional (P, T, a_2) space is required for full representation of a binary system. The five phases presumed to exist in this system are α, β, and ϵ solid allotropic forms and gas (G) and liquid (L). Each of the five single phase domains that exist in this system are represented by the volumes of the cells in this space. Two phase fields are the surfaces that separate these cells, the cell boundaries. Three phase fields are triple lines where three cells meet and four phase fields are the discrete quadruple points at which four cells meet.

   Phase diagrams plotted in potential space have this simple configuration precisely because the variables describing the state of each phase are the thermodynamic potentials, which, according to the conditions for equilibrium, are *equal* for two or more phases in equilibrium. Figure 9.3b is a section through the phase diagram shown in Fig. 9.3a at constant pressure. The point Q lying on the line separating the α and ϵ domains in Fig. 9.3b represents one combinationof the three variables $(T, P,$

---

[3]For dilute solutions the chemical potential μ_k of a solute approaches $-\infty$ as X_k approaches zero; a chemical potential axis thus spans the range from $-\infty$ to 0. In the same situation the activity of a solute, a_k, simply approaches zero as X_k approaches zero, and the activity axis varies between 0 and 1.

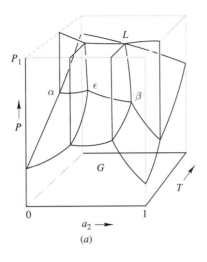

(a)

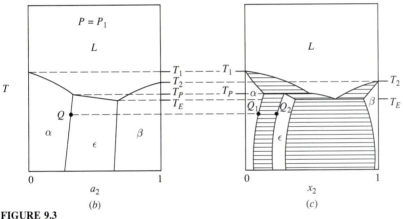

(b)   (c)

FIGURE 9.3
(a) A phase diagram plotted on coordinates that are thermodynamic potentials (P, T and a_2 in this case) is a simple cell structure. (b) A section taken at constant potential, pressure in this case, is also a cell structure. (c) The constant pressure section shown in part (b) represented in the familiar (X_2, T) space.

and a_2) at which the two phases α and ϵ coexist in equilibrium. The combination of the three variables that describes the state of the α phase (T^α, P^α, and a_2^α) in this two phase system and the combination that represents the state of the ϵ phase (T^ϵ, P^ϵ, and a_2^ϵ) are the *same point* because at equilibrium $T^\alpha = T^\alpha$, $P^\alpha = P^\epsilon$ and, from $\mu_2^\alpha = \mu_2^\alpha$, it follows that $a_2^\alpha = a_2^\epsilon$ (assuming the reference state for component 2 is the same for both phases). In contrast, if the variables chosen to plot the diagram are (T, P, and X_2), Fig. 9.3c, then the same equilibrium state between the α and ϵ phases is represented by a *pair of separated points*, Q_1 and Q_2, since X_2^α and X_2^ϵ, the *compositions*, are not the same for the two equilibrated phases.

9.4.2 Unary Systems

The structure of phase diagrams in unary systems is explored in Sec. 7.6. Fig. 9.4 is analogous to Fig. 7.12 and shows three representations of the same system in coordinates for which

1. Both are thermodynamic potentials (T, P), Fig. 9.4*a*
2. One is a potential and the other is not (V, P), Fig. 9.4*b*
3. Neither coordinate is a potential (V, S), Fig. 9.4*c*.

The diagram in (T, P) space is a simple cell structure with cells for single phase regions, linear cell boundaries representing states of two phase equilibrium, and triple points depicting the domains of coexistence of three phases. In (V, P) space, Fig. 9.4*b*, two phase regions broaden into areas with individual coexisting equilibrium states represented as a pair of points on the each of the boundaries connected by a horizontal (constant potential P) line, called a *tie line*. Three phase equilibria consist of three separated points in (V, P) space (the molar volumes of the three phases are different),

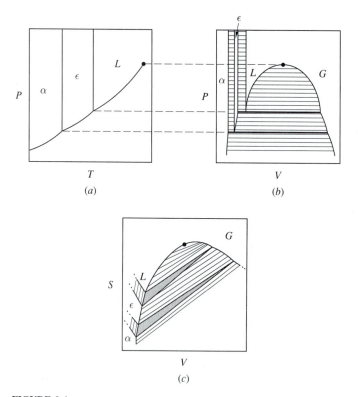

FIGURE 9.4

Three representations of the same unary phase diagram: (*a*) both variables are potential (T, P); (*b*) one variable is a potential, P in (V, P), and (*c*) neither variable (S nor V) is a thermodynamic potential.

all lying on a tie line that is horizontal because three phases in equilibrium have the same pressure.

In (V, S) space, Fig. 9.4c, domains of stability of two phases also consist of areas on the diagram. Any pair of phases in equilibrium is represented by two points on opposite boundaries of a two phase region [e.g., (V^α, S^α) and (V^ϵ, S^ϵ)]. However, the tie line that connects these states is in general not horizontal because S^α and V^α are not equal to S^α and V^ϵ. Three phase fields are represented by *tie triangles* because none of the three phases are expected to have the same values of S and V.

It is now possible to place these constructions in context with the Gibbs phase rule. In a unary system, $c = 1$ and the Gibbs phase rule becomes

$$f_1 = 1 - p + 2 = 3 - p \tag{9.29}$$

An exhaustive list of the possibilities is:

$$p = 1 \qquad f_1 = 3 - 1 = 2$$
$$p = 2 \qquad f_1 = 3 - 2 = 1$$
$$p = 3 \qquad f_1 = 3 - 3 = 0$$

It is not possible for more than three phases to coexist at equilibrium in a unary system.

In (T, P) space the number of degrees of freedom for a given number of phases is equal to the dimensionality of the domain that can contain that number of phases. A single phase, for which $f_1 = 2$, is represented by a two dimensional domain or an area on the diagram. A two phase region for which $f_1 = 1$ is represented by a one dimensional domain, that is, by a curve on the graph. In Chap. 7 it is demonstrated that this line is given by the Clausius-Clapeyron equation. Three phase regions, for which $f_1 = 0$, appear as triple points where three, two phase lines intersect. These constructions for two and three phase fields are now analyzed.

Specification of the states of a pair of phases requires four variables: $(T^\alpha, P^\alpha, T^\epsilon, P^\epsilon)$. If these states are equilibrated, these variables are related by three equations $(T^\alpha = T^\epsilon, P^\alpha = P^\epsilon, \mu^\alpha = \mu^\epsilon)$. The system has $(4 - 3) = 1$ degree of freedom. If a value is assigned to any of the four variables, say T^α, the other three can be computed from the three equations. By incrementing the value assigned to the independent variable T^α, the complete set of states in which α and ϵ can exist in equilibrium is obtained. The representation of the values of T^α and P^α that correspond to this set of states for the α phase is a curve $P^\alpha = P^\alpha(T^\alpha)$ in (T, P) space. The set of states corresponding to the ϵ phase is also represented by a curve $P^\epsilon = P^\epsilon(T^\epsilon)$. In (T, P) space *these two curves coincide* because the conditions for two phase equilibrium require that $T^\alpha = T^\epsilon$ and $P^\alpha = P^\epsilon$. Thus the locus of points corresponding to the $(\alpha + \epsilon)$ equilibrium states appears as a *single curve* in (T, P) space.

Specification of a state in which three phases coexist requires evaluation of the states of all three phases. This requires the values for six variables, such as $(T^\alpha, P^\alpha, T^\epsilon, P^\epsilon, T^G, P^G)$ for an $(\alpha + \epsilon + G)$ field. If these three phases are in equilibrium, there are six relations among these variables, $(T^\alpha = T^\epsilon = T^G; P^\alpha = P^\epsilon = P^G; \mu^\alpha = \mu^\epsilon = \mu^G)$. This system of six variables and six equations has *no degrees* of freedom and no independent variables. The states of the three phases are uniquely determined. If the states are represented in terms of T and P, then the values of these variables describing each of the three phases coincide and the three different states plot as the *same point* in (T, P) space. Since the conditions for equilibrium for these three phases include as a subset the conditions for equilibrium between the α and ϵ phase, those for ϵ and G, *and* those for α and G, these three coincident points must also lie on the corresponding two phase curves. Thus the curves that represent these three two phase equilibria must all pass through the point that describes the three phase equilibrium.

A unary phase diagram in (T, P) space is seen to consist of one phase areas bounded by two phase lines that intersect three at a time in triple points. The overall construction, Fig. 9.4a, has the configuration of a simple cell structure.

The same system has a different appearance and provides a different set of information if plotted in (V, P) space, Fig. 9.4b. In each of the cases considered, that is, one, two, and three phase equilibria, the number of variables and the number of relations remain the same. Thus single phase regions in a unary system have two degrees of freedom. However, in this representation the variables V and P are chosen to describe the state rather than T and P. Single phase regions plot as areas.

The specification of the state of two phases requires four variables, now taken to be $(V^\alpha, P^\alpha, V^\epsilon, P^\epsilon)$. If the two phases are in equilibrium, three relations hold among these variables: $(T^\alpha = T^\epsilon; P^\alpha = P^\epsilon; \mu^\alpha = \mu^\epsilon)$. There is $(4 - 3) = 1$ independent variable and one degree of freedom. If any of the four variables is specified, say V^α, the other three can in principle, be computed. This requires that relationships for T and μ as functions of V must be known for each phase. Chapter 4 provides a general procedure for deriving such relations. By incrementing V^α over the full range of interest, the collection of all the states in which α and ϵ are in equilibrium is obtained. The representation of the set of values that correspond to states that the α phase can exhibit in equilibrium with ϵ plots as a curve $P^\alpha = P^\alpha(V^\alpha)$ in (V, P) space. Corresponding states for the ϵ phase plot as $P^\epsilon = P^\epsilon(V^\epsilon)$. Unlike the plot in (T, P) space, *these two curves do not coincide*. A given equilibrated $(\alpha + \epsilon)$ state plots as two discrete points, (V^α, P^α) for the α phase and (V^ϵ, P^ϵ) for the ϵ phase, connected by a tie line, which shows that they are related. The conditions for equilibrium do not require that V^α and V^ϵ be equal. However, these conditions *do* require that $P^\alpha = P^\epsilon$. Thus the pair of points that represent states of α and ϵ that coexist in equilibrium has the same value of P and, if P is plotted on the y-axis, the tie line is *horizontal*. Accordingly, in this representation, and in any plot in which one variable is a potential and the other is not, two phase fields plot as areas. The curves bounding the areas are the possible states for the two phases that can exist in equilibrium. Pairs of states of the two phases that are in equilibrium are connected by a tie line, which is horizontal in this construction.

Since three phase equilibria possess zero degrees of freedom in a unary system, the six variables required to specify the states of the three phases, $(V^\alpha, P^\alpha, V^\epsilon, P^\epsilon, V^G, P^G)$, are determined by the six conditions for equilibrium. The state of each phase is represented by a point in (V, P) space. The conditions for mechanical equilibrium, $P^\alpha = P^\epsilon = P^G$, require that these three points all have the same pressure. If P is the y-axis, the three equilibrium states all lie on the same horizontal line. Subsets of the conditions for three phase equilibrium correspond to the conditions for two phase equilibrium; the three, two phase equilibria, $(\alpha + \epsilon)$, $(\alpha + G)$, and $(\epsilon + G)$, must intersect at this three phase tie line. Thus three phase equilibria in (V, P) space, or any space in which one of the variables is a potential and the other is not, appear as three points connected by a horizontal tie line at which the three corresponding two phase fields intersect.

A unary system represented in (V, P) space (more generally, a space in which one variable is a potential and the other is not) has one phase fields that are areas. The two phase fields consist of areas bounded by curves that represent all the possible states of the two phases that can exist in equilibrium; specific equilibrated states are connected by horizontal (more generally, constant potential) tie lines that fill the two phase field. Three phase equilibria are represented by three points on a horizontal tie line.

The case in which *neither* of the variables used to describe states of the systems is a potential is illustrated by choosing V and S as variables, Fig. 9.4c. The constructions are analogous to the (V, P) representation outlined above with one important exception; the tie lines that connect corresponding states are not constrained to be horizontal. One phase fields are areas, as in all of the cases examined. Two phase fields consist of two curves, one of the form $S^\alpha = S^\alpha(V^\alpha)$, the other $S^\epsilon = S^\epsilon(V^\epsilon)$. Specific two phase systems are represented by a pair of points, one (S^α, V^α) on the α side of the region, the other (S^ϵ, V^ϵ) on the ϵ side. These points define a tie line that can be arbitrarily oriented in the (S, V) plane.

A three phase equilibrium is represented by three points that not only do not lie on a horizontal line but are not collinear at all. These three points define a *tie triangle* in this representation, Fig. 9.4c. The lines that form the sides of the triangle connecting α with ϵ, α with G, and ϵ with G are each terminal tie lines of the corresponding two phase field.

Plotted in a space in which neither of the state variables are potentials, a unary phase diagram has areas for one phase fields, pairs of curves connected by continuously rotating tie lines for two phase regions and tie triangles formed by the intersection of three, two phase fields for three phase regions.

The constructions described in detail for a unary system apply to two dimensional sections through binary, ternary, and higher order systems, provided only that the sections are formed by holding one or more of the thermodynamic potentials constant. The topology of each plot depends upon the nature of the variables used to represent the state of the system. If both variables are potentials, then a cell structure results with the rules of construction represented in Fig. 9.4a. If only one is a potential, the structure has the appearance of Fig. 9.4b. If neither is a potential, then the rules governing the construction of Fig. 9.4c apply. These rules of construction have their roots in the *mathematics* of graphical representation of systems of equations, and not in the thermodynamics of the systems involved.

9.4.3 Binary Phase Diagrams

The complete representation of phase relationships in a binary system requires a space with $(c + 1) = (2 + 1) = 3$ dimensions. The simplest structure results when the three variables chosen are potentials, (P, T, a_2). Figure 9.3 shows an example of the kind of cell structure obtained for a binary system with three solid phases, α β, and ϵ, in addition to the liquid and gas phases. Figure 9.5a shows another example of such a cell structure. In this simpler case the system exhibits only two solid phase forms, α and β. Figure 9.5b is a sketch of the same system plotted in (T, P, X_2) space, that is, with the activity of the independent component replaced by its mole fraction. The (T, P, X_2) diagram is clearly more complicated than the cell structure. It is not possible to read quantitative information from either of these diagrams, which are simply projections of a three dimensional representation on the two dimensional printed page.

Quantitative information about phase relationships in binary systems is generally represented by taking a section through the three dimensional diagram at constant potential and usually at constant pressure. Constructing an isobaric section is equivalent to assigning a value to one of the independent variables P for the system. Figures 9.6a and b show sections through the phase diagrams sketched in Fig. 9.5 at fixed values of P. Figure 9.6c is the same diagram sketched as it would appear in (V, X_2) space. Since these graphs are two dimensional, the information they present can be read directly. Note that the topology of the three diagrams in Fig. 9.6 is comparable to the three diagrams sketched in Fig. 9.4.

The rules that govern the construction of one, two, and three phase fields in both Fig. 9.4 and 9.6 have identical origins as presented in detail in Sec. 9.4.2 for unary systems. This is because the number of degrees of freedom for one, two, and three phase regions is the same in both plots. For a binary system $c = 2$ and the Gibbs

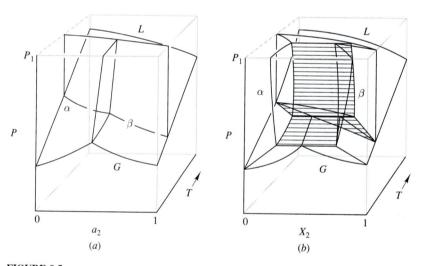

FIGURE 9.5

A complete phase diagram for a binary system that has two solid phase forms, α and β, represented as a simple cell structure in potential space (T, P, a_2) (a) and (b) as a more complicated structure in (T, P, X_2) space.

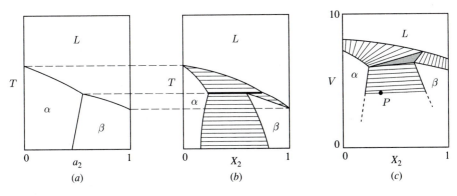

FIGURE 9.6
Constant pressure sections through the diagrams sketched in Figure 9.5 represented in (a) (T, a_2) potential space, and (b) (T, X_2) space, and (c) in (V, X_2) space. The most common form of representation of binary phase diagrams is given in (b).

phase rule, Eq. (9.28) becomes

$$f_2 = 2 - p + 2 = 4 - p \tag{9.30}$$

Choose a value for one of the variables (P); this reduces the number of degrees of freedom by one:

$$f_2' = 3 - p \tag{9.31}$$

This result is identical to Eq. (9.29) for a unary system.

9.4.4 Ternary Phase Diagrams

Since for a ternary system $c = 3$, the number of independent variables required to specify the state of a single phase is $(c + 1) = 4$. Complete representation of a ternary system requires a four dimensional space. While such spaces can be manipulated mathematically, they cannot be visualized in a world limited to three dimensions. Fixing one variable, say P, yields a diagram that can be constructed in three dimensions. Textbooks and compilations of diagrams can display these three dimensional diagrams as projections into the two dimensional printed page. It has been pointed out that such diagrams may be pedagogically useful but it is not possible to read quantitative information on such projections.

To represent quantitative information on the printed page it is necessary to reduce the number of degrees of freedom for a one phase system to two. For a ternary system this is accomplished by assigning values to *two* of the four variables that describe its state. It has been pointed out that diagrams retain simple rules of interpretation if and only if the variables that are fixed are thermodynamic potentials. Accordingly, in the representation of phase relationships in ternary systems on the two dimensional printed page, temperature and pressure are usually constant.

Figure 9.7 shows three representations of the same phase diagram for a ternary system plotted on isobaric, isothermal sections. Depiction of the variation of the be-

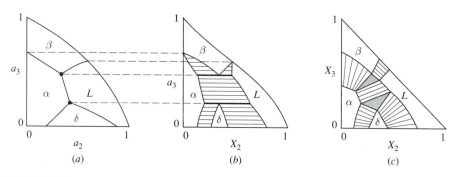

FIGURE 9.7

An isobaric, isothermal section through a four dimensional ternary phase diagram yields two dimensional diagrams which can be plotted (a) in (a_2, a_3) space, (b) in (X_2, a_3) space, or (c) in (X_2, X_3) space.

havior of the system with temperature requires a series of such diagrams constructed at a sequence of temperatures. The topology and rules of construction of each of these three representations of ternary phase equilibria are identical with those presented for unary systems in Fig. 9.4. For a ternary system $c = 3$ and the number of degrees of freedom is

$$f_3 = 3 - p + 2 = 5 - p \qquad (9.32)$$

Assigning values to two of the variables in the system reduces the number of degrees of freedom on a section to

$$f_3' = 5 - p - 2 = 3 - p \qquad (9.33)$$

a result identical to Eq. (9.29) for a unary system.

The most common format for presentation of ternary phase diagrams fixes T and P and uses X_2 and X_3, the compositions, as variables to describe the state of the system, Fig. 9.7c. Two different versions of the presentation of the composition variables are in common use, Fig. 9.8. In Fig. 9.8a the two independent composition variables, here taken as X_2 and X_3, are plotted on orthogonal axes in the standard Cartesian system. Since the dependent composition $X_1 = 1 - (X_2 + X_3)$, lines of constant X_1 plot at a slope of -1. The origin $(0, 0)$ corresponds to pure component 1. The boundaries of the diagram, defined by the fact that physically meaningful ranges of the three composition variables are restricted between 0 and 1, are the two Cartesian axes and the line with a slope of -1 passing through $(1, 0)$ and $(0, 1)$. The composition corresponding to any point in this triangle is easily read from the two axes. This representation of ternary compositions has found application in describing the behavior of dilute solutions or near one corner of the system where one of the components is readily viewed as the dependent compositional variable.

The second representation of composition in ternary systems, illustrated in Fig. 9.8b, is more widely used in displaying ternary phase diagrams. This compositional coordinate system is called the *Gibbs triangle*. It is constructed as an equilateral triangle with sides of unit length. The interior of the triangle is constructed with lines at 0°, 60°, and 120°, that is with lines parallel to the sides of the triangle. Each

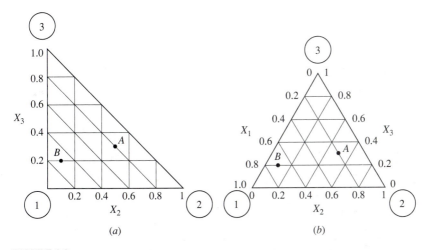

FIGURE 9.8
Alternate representations of the composition plane for isobaric isothermal sections through ternary systems:
(*a*) orthogonal composition axes, and (*b*) the Gibbs triangle. The compositions *A* and *B* are the same on
both diagrams.

corner represents a pure component; each side represents a binary system in which
the concentration of the component at the opposite corner is zero. Lines parallel to
the side opposite to a corner for any given component are the locus of points within
the composition space for which that component has a constant value. In Fig. 9.8*b*
the lines at $0°$ represent constant values of component 3, with X_3 labelled on the 2-3
side of the diagram. The lines at $60°$ represent constant values of the mole fraction of
component 2 with numerical values for X_2 labelled on the 12 side; those at $120°$ are
constant X_1 values, as labelled on the 1-3 side of the diagram. The composition of a
particular point in the space can be read by constructing lines parallel to these three
directions and reading the scales on the sides of the diagram. For the point labelled
A in this diagram, the composition is ($X_1 = 0.2$, $X_2 = 0.5$, $X_3 = 0.3$). For the point
B, the values are ($X_1 = 0.7$, $X_2 = 0.1$, $X_3 = 0.2$). These compositions are plotted for
comparison on Fig. 9.8*a*.

The Gibbs triangle is widely used to describe ternary compositions because it
gives equal weight to all three components. The Cartesian coordinate system tends
to assign particular importance to the dependent composition variable, here shown
as component 1. Some analyses of behavior require evaluation of slopes of phase
boundaries; derivatives taken in the Cartesian system are more easily interpreted than
in the Gibbs triangle system.

The most common representation of phase relationships in ternary systems plots
the composition axes as mole fractions (or in atomic percent, which is equivalent) on
a Gibbs triangle at a fixed temperature and pressure. Since the mole fraction axes are
not potentials, these diagrams share the principles of construction and interpretation
with those illustrated for a unary system in Fig. 9.4*c* and for a binary system in
Fig. 9.6*c*. Single phase fields are areas with two degrees of freedom. Two phase

regions consist of a pair of curves with corresponding equilibrated states connected by tie lines, consistent with this single degree of freedom. Three phase regions are invariant with no degrees of freedom; the fixed compositions of the three phases form the corners of a tie triangle.

9.5   THE INTERPRETATION OF PHASE DIAGRAMS

In plotting the phase diagrams in Figs. 9.4, 9.6, and 9.7 the maximum number of degrees of freedom is reduced to two, so that the diagrams could be presented on the printed page. Any state of the system can be represented by a point in the diagram obtained by assigning a value to each of the two variables on the axes. For every such point the phase diagram tells

1. How many phases exist in the system at equilibrium,
2. Which phases they are,
3. Their thermodynamic states expressed in terms of the variables on the axes on the diagram,
4. For diagrams represented in (b) and (c) of these figures, the relative amounts of these phases.

Information about the relative amounts of the phases at equilibrium is not available for diagrams plotted in potential coordinates.

In the simple cell structures shown in (a) of each of these figures, a point can lie in one of the areas, on the boundary between areas, or at a triple point. Within any area, the equilibrium state consists of a single phase corresponding to the label for that area with properties corresponding to the coordinates of the point chosen. If the point lies on a boundary, the equilibrium structure consists of the two phases that the boundary separates, each with the properties corresponding to the point. If the point lies on a triple point in the phase diagram, the equilibrium state consists of the three phases that meet at that point, each with the values of the potentials corresponding to the triple point.

Phase diagrams that are constructed in a two variable space with one axis a potential, and the other not, obey the rules of construction shown in (b) of Fig. 9.4, 9.6, and 9.7. If a point representing the state of the system lies in a single phase region, then the equilibrium state consists of that single phase with the properties corresponding to the plotted point. If a point representing a state of interest lies within the area of a two phase field, then it must lie on a particular tie line, Fig. 9.9, since the envelope of tie lines fills the two phase area. A system with the properties designated by the point P in Fig. 9.9 consists of the two phases ϵ and L when it comes to equilibrium. These phases have the properties designated by the points A and B. Thus their temperatures are $T^\epsilon = T^L$ (the potentials are equal at equilibrium) and their compositions are X_2^ϵ and X_2^L. The relative amounts of the two phases that exist at equilibrium can also be determined in this case by applying the *lever rule*.

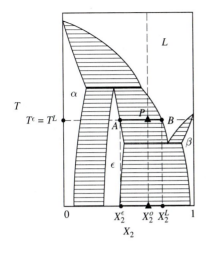

FIGURE 9.9
The line AB is a tie line connecting equilibrium compositions for the ϵ and L phases at the temperature of the line. Relative quantities of the two phases can be determined by the lever rule.

9.5.1 The Lever Rule for Tie Lines

The point P in Fig. 9.9 illustrates the lever rule. This point represents an equilibrium state of the system specified in terms of its average composition X_2^o and its temperature T. This state consists of a mixture of two phases, ϵ with properties $(X_2^\epsilon, T^\epsilon)$ and the liquid with properties (X_2^L, T^L). In this particular case T is a potential and the conditions for equilibrium require that $T^\epsilon = T^L$. The relative amounts of the ϵ and liquid phases are given by the *lever rule*, which is now derived.

Suppose that this system has n_T total moles of the components. When it achieves equilibrium it is a mixture of the ϵ and L phases. Let n^ϵ be the number of moles of the components contained in the ϵ phase at equilibrium and n^L the number of moles of the L phase. The average composition of the two phase mixture is given by X_2^o; the total number of moles of component 2 in the system is thus $n_T X_2^o$. The number of moles of component 2 in the ϵ phase in the system is $n^\epsilon X_2^o$; that of the L phase is $n^L X_2^L$. Conservation of component 2 requires that

$$n_T X_2^o = n^\epsilon X_2^\epsilon + n^L X_2^L \tag{9.34}$$

Dividing both sides of the equation by n_T

$$X_2^o = \frac{n^\epsilon}{n_T} X_2^\epsilon + \frac{n^L}{n_T} X_2^L = f^\epsilon X_2^\epsilon + f^L X_2^L$$

where f^ϵ and f^L are the fractions of all of the moles of components in the system that are in the ϵ and L phases when the system comes to equilibrium. Since these are fractions, $f^\epsilon = 1 - f^L$ and

$$X_2^o = f^\epsilon X_2^\epsilon + (1 - f^\epsilon) X_2^L = X_2^L + f^\epsilon (X_2^\epsilon - X_2^L)$$

Solve this equation for the fraction of the total number of moles in the system that are in the ϵ phase at equilibrium.

$$f^\epsilon = \frac{X_2^L - X_2^o}{X_2^L - X_2^\epsilon} \tag{9.35}$$

The fraction of moles in the L phase is

$$f^L = 1 - f^\epsilon = \frac{X_2^o - X_2^\epsilon}{X_2^L - X_2^\epsilon} \tag{9.36}$$

This result can be interpreted graphically in Fig. 9.9 as the ratio of two lengths on the tie line

$$f^\epsilon = \frac{PB}{AB} \quad \text{and} \quad f^L = \frac{AP}{AB} \tag{9.37}$$

For the point P in Fig. 9.9 the relative number of moles contained in the ϵ phase, f^ϵ, is given by (PB/AB) and f^L is given by (AP/AB). These arguments yield a simple way of evaluating the quantities of the phases that correspond to the state of a system represented by a point P in a two phase field. The lever rule can be applied to any tie line in any two phase field, including those illustrated in (c) of Fig. 9.4, 9.6, and 9.7.

The lever rule gives the relative amounts of the two phases that the tie line connects, expressed in units of moles of phase per mole of system if the properties used to designate the states of the phases are expressed in values per mole. This is the case in all the examples treated in this chapter. For some purposes it is useful to plot phase diagrams using other units, such as weight percent rather than mole percent or atom fraction, or specific volume (volume per *gram*) rather than molar volume. In these cases the lever rule still applies but the fractions f^α and f^β obtained are *weight* fractions of the α and β phases.

With this in mind it is possible to illustrate why this construction is called the *lever rule*. If the quantities are reported in terms of weight or mass, then the lever rule can be visualized through a simple mechanical analog. Imagine the tie line as a rigid lever and the point P to be its fulcrum, Fig. 9.10. Pans for the α and β phases hang from the ends of the lever at the points A and B, respectively. For any location of the fulcrum, P, the expressions $f^\alpha = (PB/AB)$ and $f^\beta = (AP/AB)$ give the relative weights of the phases that must be placed in the pans to balance the system. If the fulcrum is near the α side of the lever, a small quantity of the β phase balances a large amount of α. If P is near the β end, a small amount of α balances a large quantity of β. For a sequence of systems for which the state of the two phase mixture represented

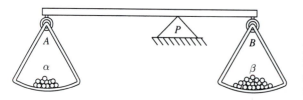

FIGURE 9.10
Mechanical illustration of the lever rule applied to tie lines in two phase fields.

by P moves from A to B, the fraction of the number of moles in the system contained in the β phase at equilibrium changes from 0 to 1. This analog does not strictly apply when quantities are reported in terms of the number of moles of the components because in the mechanical analog, the balance is achieved in a gravitational field that acts on the *mass* of the phases rather than their number of moles. The geometrical construction is valid for either mass or molar units. The mechanical analog that aids in visualizing the construction is valid only for mass units.

The lever rule construction applies to every tie line that exists in a two phase field, no matter how many variables are used to specify the states of the two phases involved. For example, in a ternary system the complete specification of the state of a phase requires four variables, such as $(T^\alpha, P^\alpha, X_2^\alpha, X_3^\alpha)$, needing a four dimensional space for its representation. Two phase fields in this four dimensional phase diagram are four dimensional volumes filled with a collection of tie lines connecting pairs of equilibrated states. A state of the system represented by the point P would be described by values of the four variables (T, P, X_2, X_3). The point P must lie on some tie line in the two phase field connecting states for the α and β phases, each represented by points A and B in this four dimensional space. The relative amounts of the two phases existing at equilibrium is still given by the ratios expressed in Eq. (9.37).

9.5.2   The Lever Rule for Tie Triangles

In the representation of phase diagrams in (a) of Figs. 9.4, 9.6, and 9.7 where both axes are potentials, three phase fields plot as a single point. In the diagrams as plotted in (b) of these figures, the three phase fields are horizontal lines. The identities of the three phases that coexist and their thermodynamic states can be read from these representations. However, it is not possible to deduce the *relative quantities* of the three phases that coexist at equilibrium in these representations.

In contrast, if the diagram were plotted on axes in which *neither* of the variables are potentials, the three phase regions would appear as tie triangles, (as in (c) of Fig. 9.4, 9.6, and 9.7). A representative three phase region, a tie triangle, is shown in Fig. 9.11. Any state of the system P that lies within the triangle is composed of the three phases α, β, and ϵ at equilibrium. The states of the three phases are given by the corners of the triangle, labelled A, B and C. The relative amounts of the α, β, and ϵ, phases corresponding to P are given by a generalization of the lever rule, based on the same principle of conservation of moles of each of the components. This construction is illustrated without proof in Fig. 9.11.

To determine the quantity of a particular phase present in an equilibrated system represented by the point P, draw lines from each of the corners through P as in Fig. 9.11. It can be shown that the fraction of the number of moles in the structure contained in each of the three phases is

1. f^α, given by the ratio of the segment lengths (PD/AD)
2. f^β, given by the ratio (PE/BE)
3. f^ϵ, given by the ratio (PF/CF).

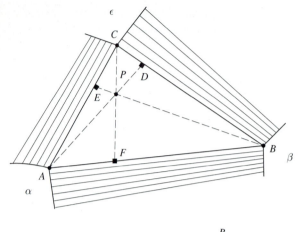

FIGURE 9.11
Use of the tie triangle to determine relative amounts of the three phases in equilibrium for a selected average composition at P.

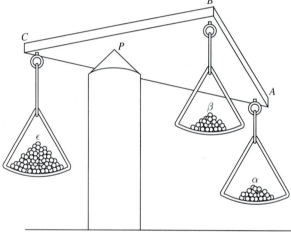

FIGURE 9.12
Mechanical illustration of the generalization of the lever rule applied to a tie triangle.

This result essentially derives from the requirement that the number of moles in the whole system is the sum of the number in each phase.

A mechanical analog can be visualized for the tie triangle if the quantities involved are reported in terms of the weight of the phases. Imagine the triangle as a rigid frame with a fulcrum located at P and pans hanging from each of the points A, B and C, Figure 9.12. The fractions computed from this construction are the relative weights of each of the three phases that would keep the system in balance.

9.6 APPLICATIONS OF PHASE DIAGRAMS IN MATERIALS SCIENCE

Each of the kinds of phase diagrams reviewed in this chapter provides its own view of the information contained in a multicomponent, multiphase system. The information displayed about the phases involved is determined by the choice of axes on which the diagram is plotted. It is demonstrated that whether both, one, or none of the state variables chosen to represent the system are thermodynamic potentials determines the rules governing the construction and interpretation of the resulting diagram. For any given state of the system, any point on the diagram, the phase diagram tells how many phases are present, which ones they are and what their properties are, expressed in terms of the variables chosen to describe their states. For some representations lever rules also provide information about the relative amounts of the phases in two or three phase regions.

Perhaps the most common application of phase diagrams as a tool in materials science is in attempts to understand the changes in internal structure, that is *microstructure*, that may accompany some process to which a material is subject. If the process is carried out slowly so that the system is essentially equilibrated at each step along the way, then each successive state along the path can be represented as a point on the corresponding phase diagram. As the state of the system moves on the diagram through the various one, two and three phase fields, the phase diagram predicts the state of the system at each point. This information can be used to provide insight into the changes that must occur in the system in order to produce each new condition predicted by the phase diagram.

A classic example is provided by the slow solidification of a liquid material system with composition X_2^o, Fig. 9.13. Initially the material is melted and equilibrated in the liquid state at some temperature, T_0. As the casting cools, its temperature drops. The state of the system crosses into the $(\alpha + L)$ field. At the temperature T_1 the phase diagram reports the equilibrium state as a small quantity of the solid α phase with composition X_2^α in a liquid of a composition slightly richer in component 2 than

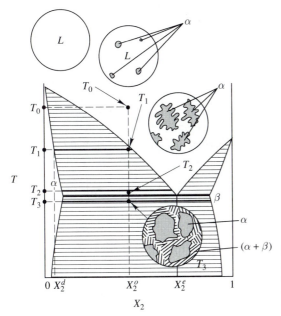

FIGURE 9.13

Sequence of states and microstructural evolution predicted for the solidification of a material of hypoeuctic composition in a eutectic system.

X_2^o. This requires that particles of the α phase must nucleate and begin to grow. As the temperature reduces toward T_2, the balance shifts on the lever rule, changing the composition of the α phase and increasing its amount at the expense of the liquid. The phase diagram predicts that at T_2 the equilibrated system consists of about 40% α and 60% liquid. The liquid has the composition, X_2^e. At T_3, just below the three phase field, the system consists of a mixture of two solid phases, $(\alpha + \beta)$, with relative amounts about 70% α and 30% β determined from the lever rule. It is readily inferred that the liquid that existed at T_2 has solidified, *isothermally* (the temperature change from T_2 to T_3 is arbitrarily small during this process) producing a mixture of α and β phases. This final solidification process usually occurs by nucleation of the two solid phases and their simultaneous cooperative growth into the liquid phase, consuming it. The final structure is composed of the 40% that existed as α at T_2 with the remaining 60% of the system a mixture of α and β phases, frequently as alternating platelets. Thus the phase diagram provides a framework for understanding the evolution of the microstructure during this solidification process. The pattern of behavior that would result for materials of different compositions (X_2^o values) can be readily deduced. Materials that exhibit this kind of phase diagram, called a *eutectic* system, are in fact observed to develop these microstructures when slowly solidified.

The strategy involved in predicting the sequence of events outlined in the last paragraph is straightforward. Plot each successive state that occurs in the process on the phase diagram for the system. Use the diagram to answer the questions, "Which phases are present?", and "How much of each?" Then ask, "What changes in structure must the system experience to proceed from its previous condition to its present one?" Finally ask, "What processes are required to achieve this change in structure?" This strategy can be applied to systems with phase diagrams of arbitrary complexity.

This simple strategy applies if the system is changing state slowly so that each successive state is not importantly different from equilibrium. Usually the rates of change are such that the material is never in its equilibrium state during processing. Composition, temperature, and pressure (more generally, stress) gradients exist in the system. Its state cannot be represented by a point in a state space because its intensive properties are not uniform. In this case, the phase diagram can provide a guide to the state toward which the system is changing, and may still predict, although qualitatively, an overall sequence of changes that the system can be expected to traverse.

Even where the system is far from equilibrium the phase diagram can provide useful quantitative information in the analysis of the microstructural processes the material experiences. As an example, consider the phase diagram shown in Fig. 9.14a. The material with composition X_2^o is first heated and held until it equilibrates at the temperature T_a. It is then quenched to the temperature T_p and held for a length of time. If the quench is rapid so that no transformation occurs during the temperature change, the starting structure at the reaction temperature T_p is 100% α phase with composition X_2^o. This structure is *supersaturated*; that is, it contains more solute than X_2^o, the maximum amount that the α phase can hold when equilibrated at T_p. The equilibrium state at T_p is a mixture of α and β with the compositions respectively X_2^α and X_2^β and the lever rule predicts the final structure contains about 10% β and

90% α. Thus the state toward which the system is striving is dispersion of small β particles in a matrix of the α phase.

Figure 9.14b is a sketch of the composition profile that would be expected to exist *during* the growth of a β particle. Far from the particle ($\rho \gg r$) the composition has not yet changed and remains at X_2^o. One of the essential uses of phase diagrams is in the prediction of the composition values at a two phase ($\alpha + \beta$) interface during growth. The principle that is applied is the *assumption of local equilibrium* at the interface. More explicitly, this means that the conditions for thermodynamic equilibrium apply for a volume element that straddles the interface, containing a volume element of α on one side and β on the other. This principle asserts that there are no discontinuities in the thermodynamic potentials, T, P, and μ_k, across the interface, meaning that, the conditions for thermodynamic equilibrium are satisfied *locally* at the interface. Since the phase diagram is a graphical representation of these conditions, at one atmosphere pressure and $T = T_p$, these conditions are satisfied uniquely by the compositions X_2^α and X_2^β obtained from the phase diagram in Fig. 9.14a. These composition values provide the boundary conditions for the solution of the diffusion problem, which ultimately predicts the rate at which the particle grows. The resulting composition distribution sketched in Fig. 9.14b shows a concentration gradient in the α phase. Solute (component 2 in this case) flows by diffusion *down the gradient*

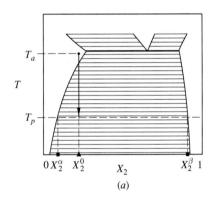

(a)

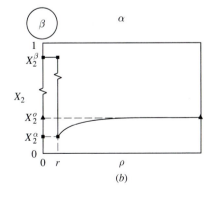

(b)

FIGURE 9.14
The two phase field ($\alpha + \beta$) in this phase diagram, (a) is characteristic of systems that may exhibit precipitation processes. (b) Is the concentration profile that may be expected exist in the matrix near a growing precipitate particle.

toward the growing β particle. Since the β particle is rich in component 2, this flow is precisely what is needed to support the growth of the β particle. While other factors may complicate the growth behavior of particles in real systems (interfaces tend to have a structure that must be maintained during growth, slow interface reaction rates may violate the assumption of local equilibrium, capillarity or stress induced effects may operate) this growth scenario is in fact found in simple precipitation reactions.

One representation of an isothermal, isobaric section through a ternary phase diagram can be constructed by choosing one compositional variable as a chemical potential, μ_3. Figure 9.15a shows an example of such a diagram for the Ni-Co-O system [9.1]. The rules of construction of such a diagram correspond to Fig. 9.7b. The other compositional axis is most conveniently chosen to be the ratio $X_{Ni}/(X_{Ni}+X_{Co})$. This variable is chosen to represent composition because for a fixed value of oxygen composition it varies from 0 to 1 as the possible range of compositions is traversed. The potential utility of this diagram becomes apparent when it is recognized that tie

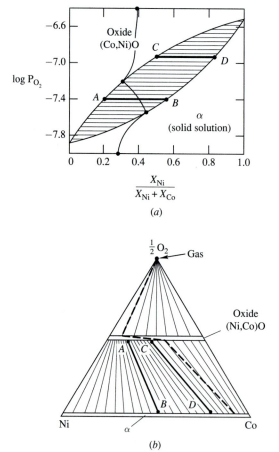

FIGURE 9.15

Isobaric, isothermal ternary diagram for the Ni-Co-O system plotted in a composition–chemical potential (oxygen potential) space (a) and in a composition space (b) [9.1].

lines are described by the condition $\mu_O^\alpha = \mu_O^\beta$, for example. Thus on such a plot with μ_O as the ordinate, all of the tie lines are *horizontal*, corresponding to a fixed value of μ_O. Further, because one of the conditions for three phase equilibrium is $\mu_O^\alpha = \mu_O^\beta = \mu_O^\epsilon$, three phase regions also appear as horizontal lines. The consequence of these observations is that a ternary isotherm plotted with chemical potential as one axis exhibits the same constructions as does an ordinary and familiar binary (T, X_2) diagram. Figure 9.15*b* contrasts the standard Gibbs triangle construction with mole fraction axes with the same diagram plotted on the axes defined in this paragraph. The version of a ternary phase diagram shown in Fig. 9.15*a* finds particular application when one component is a reactive gas like oxygen, chlorine, or nitrogen. In such a system controlling partial pressures of the gaseous component is equivalent to controlling its chemical potential. Indeed, the ordinate on the diagram can be replaced by $\log P_{O_2}$, which for a fixed temperature is proportional to μ_O. The analysis of behavior under controlled atmospheres is thus simplified with this version of ternary diagram. Figure 9.16 illustrates a similar plot for a slightly more complex ternary system, Fe-Cr-O [9.1]. The heavy broken line is used in Problem 9.10.

9.7   SUMMARY OF CHAPTER 9

The thermodynamic apparatus for describing multiphase systems is based upon the principle that, for extensive properties, the value for the system is the sum of the values contributed by each phase. The change in any extensive property for a multiphase system is then simply the sum of the changes in that property occurring in each phase.

Application of the general strategy for finding conditions for equilibrium in a multiphase, multicomponent system yields the results:

$$T^I = T^{II} = \cdots = T^\alpha = \cdots = T^P \tag{9.22}$$

$$P^I = P^{II} = \cdots = P^\alpha = \cdots = P^P \tag{9.23}$$

$$\mu_1^I = \mu_1^{II} = \cdots = \mu_1^\alpha = \cdots = \mu_1^P \tag{9.24a}$$

$$\mu_1^I = \mu_2^{II} = \cdots = \mu_2^\alpha = \cdots = \mu_2^P \tag{9.24b}$$

$$\vdots$$

$$\mu_c^I = \mu_c^{II} = \cdots = \mu_c^\alpha = \cdots = \mu_c^P \tag{9.24c}$$

The number of degrees of freedom in a multiphase, multicomponent system is equal to the number of independent variables it exhibits and is defined as the difference between the number of variables and the number of relations corresponding to the conditions for equilibrium in the system. Gibbs showed that for a system with c components and p phases,

$$f = c - p + 2 \tag{9.28}$$

The domains of stability of phases in unary and multicomponent systems can be displayed in a variety of ways. Quantitative representation of such information is

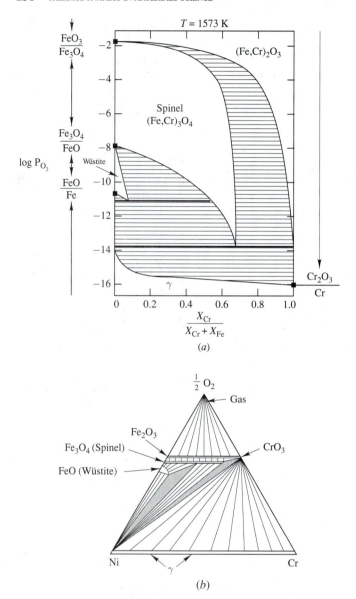

FIGURE 9.16
Potential-composition diagram for the ternary system, Fe-Cr-O [9.1].

usually limited to two dimensional plots achieved by fixing an appropriate number of the potentials in the system.

1. If both axes are potentials, then one phase regions are areas, two phase domains are lines and three phase regions are points and the diagram is a simple cell structure.

2. If only one axis is a potential (e.g., T in binary isobaric diagrams or μ_3 in ternary isothermal isobaric isotherms) and the other axis is not a thermodynamic potential, then single phase regions remain areas but two phase regions consist of a pair of phase boundaries connected by horizontal tie lines and three phase regions appear as a single horizontal tie line.

3. If neither axis of the diagram is a potential, then two phase regions have a structure similar to that found in the case with one axis plotted as a potential but the tie lines are not constrained to be horizontal; three phase regions appear as tie triangles.

In two and three phase fields the relative amounts of the phases that exist in equilibrium can obtained by applying the *lever rule* to a tie line or a tie triangle.

In general, two phase regions bound single phase domains and intersect to produce three phase equilibria. Thus the calculation of phase diagrams need only focus on two phase equilibria.

Phase diagrams are an important tool in interpreting microstructural processes that occur in materials. If processing is slow enough the phase diagram exhibits the sequence of states the system traverses. For processes further from equilibrium, the assumption of local equilibrium applied at interfaces between the phases invokes phase diagram information to visualize boundary conditions for the flows that occur.

PROBLEMS

9.1. Use Eq. (9.3) to write out a general expression for the change in *volume* of a three phase $(\alpha + \beta + \gamma)$, two component (A and B) system. Include all twelve terms.

9.2. Follow the general strategy for finding conditions for equilibrium for the system described in Problem 9.1.
   (*a*) Write out an explicit expression for the change in entropy.
   (*b*) Write the isolation constraints for this three phase system.
   (*c*) Use the isolation constraints to eliminate dependent variables in the expression for the entropy.
   (*d*) Collect like terms.
   (*e*) Count the number of independent variables.
   (*f*) Set the coefficients equal to zero.
   (*g*) Write the conditions for equilibrium.

9.3. Consider a system with components Cu, Ni, and Zn that contains 4 phases: α, β, γ, and L.
   (*a*) List the variables required to specify the state of each phase: count them.
   (*b*) List the conditions for equilibrium for this system and count them.
   (*c*) From these counts, compute the number of degrees of freedom.
   (*d*) Compare this result with the number of degrees of freedom computed from the Gibbs phase rule.

9.4. Sketch the phase diagram for pure water in (P, V) space. Be careful to incorporate the observation that solid water *shrinks* upon conversion to the liquid state. Discuss complications in the structure of the diagram that derive from this fact.

9.5. Sketch a (a_2, P) section through the phase diagram for the binary system shown in Fig. 9.5*a* *at a constant temperature*. Choose a temperature that lies between the triple points of the pure components. Use the resulting cell structure to sketch a plausible (X_2, T) diagram for this system.

9.6. Consider the phase diagram drawn in Fig. 10.20, plotted in (T, X_2) space. Sketch a plausible phase diagram for this system in (T, a_2) coordinates.

9.7. Sketch an isothermal isobaric phase diagram for the A - B - C system in (a_B, a_C) space; assume this system exhibits four different phases at the temperature of interest. Sketch equivalent phase diagrams for this system
(a) in (X_B, a_C) space;
(b) on the Gibbs triangle in (X_B, X_C) space.

9.8. Sketch or otherwise reproduce the phase diagram shown in Figure 9.7c. In the two and three fields construct contours that represent a constant molar fraction of the phases involved. Use the lever rule to construct contours at $f^I = 0.2, 0.4, 0.6$ and 0.8.

9.9. Prove the lever rule construction for tie triangles.

9.10. A Co-Ni alloy with $X_{Ni} = 0.20$ is heated in air at 1600 K. The phase diagram for this system is shown in Fig. 9.15. The oxygen potential must vary from nearly 0 in the gas phase to a large negative number in the alloy. This variation can contain no discontinuities if the oxidation process is diffusion controlled. The curve on Figure 9.15 represents the sequence of states the system has at some point in the oxidation process.
(a) Sketch a *microstructure* that could correspond to this sequence.
(b) Label the interfaces in your microstructure and evaluate the compositions at the interfaces in the system.

9.11. Figure 9.14 illustrates a phase diagram and composition profile that could characterize the early stages of precipitation of β from α for an alloy of composition X_2^o that was solution treated then quenched to the temperature T_2.
(a) Sketch a β precipitate particle in its α matrix.
(b) Sketch the composition profile that would result if this sample were taken to equilibrium at T_p.
(c) Use the structure in part (b) as the *starting structure*. Imagine that the sample is heated rapidly (upquenched) to T_a and held. Use the principle of local equilibrium and the phase diagram to determine the interface compositions at T_1. Sketch the composition profile that must develop from the starting structure obtained in part (b).
(d) Argue that this composition is precisely what is needed to dissolve the β particle.

REFERENCES

1. Pelton, A. D. and H. Schmalzreid; *Metallurgical Transactions*, vol. 4, p. 1395 (1973).

CHAPTER

10

THERMODYNAMICS OF PHASE DIAGRAMS

A phase diagram is a map that presents the domains of stability of phases and their combinations. A point in this space, which represents a state of the system that is of interest in a particular application, lies within a specific domain on the map. Reading the map then tells you that, at that state when it comes to equilibrium,

1. What phases are present;
2. The states of those phases;
3. The relative quantities of each phase.

Phase diagrams are a primary thinking tool in materials science because they provide the basis for predicting or interpreting the changes in internal structure of a material that accompany its processing or subsequent service. A system that is equilibrated in some initial domain on the map and then placed in another domain with a different equilibrium structure, such as by simply changing its temperature, undergoes a series of microstructural transformations that take it toward its new equilibrium state given by the phase diagram. Microstructures with optimum properties can be selected by interrupting this process and quenching in the structure, which is then the state in which the material is used. In an inverse application of this strategy, a material that has failed or malfunctioned can be examined in the context of its phase diagram to infer what processes it may have experienced and thus unravel the sequence of events

241

that led to its failure. The phase diagram is often the first place to look in any attempt to understand how a material behaves.

In recent years there has been a rapid expansion in the development of strategies for calculating phase diagrams. While a significant fraction of binary metallic and ceramic phase diagrams have been explored experimentally, the information is not always complete or consistent. Coupling such studies with thermodynamic measurements and model calculations provides a firm basis for establishing measured diagrams and for extrapolating into domains for which measurements are not available. Only a small fraction of ternary systems have been explored since the number of possible ternary systems is much larger than binary systems. Thus strategies that permit extrapolation from binary information to estimate ternary and higher order phase diagrams are potentially very useful. Also, a knowledge of the underlying thermodynamics provides the basis for understanding the factors that determine the kinetics of processes like nucleation and growth during microstructural transformations. The connections between phase diagrams and the underlying thermodynamics are developed in detail in this chapter.

The most useful thinking tool for visualizing these connections is the *free energy-composition* or *(G-X) diagram*. The interplay between the free energy of mixing of the phases in a structure and the form of the phase diagram is developed to illustrate how two and three phase fields are generated in binary phase diagrams. The role played by intermediate phases in significantly affecting the structure of phase diagrams is explored. The peculiar *(G-X)* curve that leads to the formation of a miscibility gap in a phase diagram is presented. Calculations of phase diagrams based upon simple solution models illustrate these developments and lay the foundation for presenting computer calculations of phase diagrams from thermodynamic data bases and sophisticated solution models. The connections thus established permit inverting the strategy to estimate thermodynamic properties from experimentally determined phase diagrams. Generalization of these concepts to describe ternary phase diagrams is undertaken at the end of the chapter.

10.1 FREE ENERGY-COMPOSITION (*G-X*) DIAGRAMS

Visualization of the connection between domains of stability of the phases that may exist in a system and the underlying thermodynamics is most conveniently accomplished with the free energy-composition (*G-X*) diagram. For a given phase this diagram is a plot of the Gibbs free energy of mixing versus the mole fraction of component 2 at a fixed pressure and temperature. Figure 8.3 illustrates such a plot for an ideal solution at a series of temperatures. Each of these curves has the familiar mathematical form

$$\Delta G_{\text{mix}} = RT(X_1 \ln X_1 + X_2 \ln X_2) \tag{10.1}$$

(see Table 8.3). It is symmetrical about $X_2 = 0.5$ and has a vertical slope at $X_2 = 0$ and $X_2 = 1$. The value at the minimum (at $X_2 = 0.5$) is $-RT \cdot \ln 2$. Thus, its magnitude at the minimum increases linearly with absolute temperature.

The form of the $(G\text{-}X)$ curve for real solutions is given by

$$\Delta G_{\text{mix}} = \Delta G_{\text{mix}}^{\text{xs}} + RT(X_1 \ln X_1 + X_2 \ln X_2) \tag{10.2}$$

where the excess free energy of mixing can be positive or negative and in general depends upon composition, temperature, and pressure. Figure 10.1 illustrates how the ideal free energy of mixing curve can be altered by the excess free energy contribution. A variety of models for the excess free energy are examined in Chap. 8.

Each phase that may exist in a system has its own free energy-composition diagram. The competition for domains of stability of the phases and interactions that produce two and three phase fields that separate them can be visualized by comparing the $(G\text{-}X)$ curves for all the phases in the system. For such a comparison of free energies of mixing to be valid, it is absolutely essential that the energetics of each component in all the phases be referred to the *same* reference state. This point is self evident, however, its implementation is not trivial.

10.1.1   Reference States for $(G\text{-}X)$ Curves

The $(G\text{-}X)$ curves that can be constructed for each phase are derived for the mixing process in which the solution is formed from the pure components in some initial unmixed condition. This initial condition for each component is called the *reference state* for that component. Specification of the reference state requires assignment of four attributes:

1. Pressure.
2. Temperature.
3. Composition.

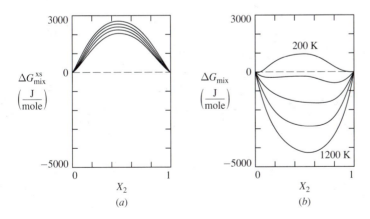

FIGURE 10.1
Free energy-composition diagrams for a flexible three parameter solution model plotted as a function of temperature

4. Phase Form.

A reference state must be defined for each *component* in the system. The choices made for these four factors in the solution models developed in Chap. 8 were implicit; it is now necessary to be explicit about these assumptions. The expressions derived for the free energy of mixing in those solution models assumes that the condition of each component at the start of the mixing process is:

1. Pressure = same pressure as the solution
2. Temperature = same temperature as the solution
3. Composition = pure component
4. Phase Form = same phase form as the solution.

Thus in describing *a liquid* solution of components 1 and 2 at one atmosphere and 750 K, the ideal solution model assumes that the reference state for component 1 is pure *liquid* 1 at 750 K and one atmosphere; that for component 2 is assumed to be pure *liquid* 2 at 750 K and one atmosphere. If the ideal solution model were used to describe a *solid* solution of componants 1 and 2 at 750 K and one atmosphere, the implicitly assumed reference state for 1 is pure *solid* 1 at 750 K and one atmosphere; that for 2 is pure *solid* 2 at 750 K and one atmosphere. A direct comparison of the energetics of mixing of components 1 and 2 in the solid solution with that in the liquid solution on the basis of these models is *not valid*.

Such a comparison requires that the choice of reference state for component 1 be the same for both the liquid and solid solutions and that the choice made for 2 must be the same for both solutions. To achieve this comparison, it is necessary to change the reference states in the calculation of mixing behavior of one of the solutions. The choice of which reference states are chosen and which are changed is a matter of convenience. It is only essential that they be made the same for component 1 in both phases and the state chosen for component 2 must also be the same in both phases. The procedure that must be applied to change a reference state is straightforward. It requires a knowledge of the difference in Gibbs free energies between the two reference states.

Expressions for the Gibbs free energy of mixing can be written in a way that makes the choice of reference state explicit:

$$\Delta G^\alpha_{\text{mix}}\{\alpha; \alpha\} = X^\alpha_1 \left(\overline{G}^\alpha_1 - G^{o\alpha}_1\right) + X^\alpha_2 \left(\overline{G}^\alpha_2 - G^{o\alpha}_2\right) \tag{10.3}$$

$$\Delta G^L_{\text{mix}}\{L; L\} = X^L_1 \left(\overline{G}^L_1 - G^{oL}_1\right) + X^L_2 \left(\overline{G}^L_2 - G^{oL}_2\right) \tag{10.4}$$

The notation in brackets indicates the choice of reference states for components 1 and 2, respectively, and this is made more explicit by the superscripts on the G^o values on the right sides of these equations. To compare the mixing behavior of these two solutions, it is necessary to make the reference states the same. This can be accomplished in any of four ways:

	I	II	III	IV
Ref. state for α solution:	$\{\alpha; \alpha\}$	$\{\alpha; L\}$	$\{L; \alpha\}$	$\{L; L\}$
		or	or	or
Ref. state for L solution:	$\{\alpha; \alpha\}$	$\{\alpha; L\}$	$\{L; \alpha\}$	$\{L; L\}$

Any of these four choices satisfies the requirements that the reference state for component 1 is the same in both phases and the reference state for component 2 is the same in both phases. Each also requires that two reference states be changed in Eqs. (10.3) and (10.4).

To illustrate the procedure for changing reference states, choose the reference state labelled II. It is therefore required to evaluate

$$\Delta G^{\alpha}_{\text{mix}}\{\alpha; L\} = X_1^{\alpha}\left(\overline{G}_1^{\alpha} - G_1^{o\alpha}\right) + X_2^{\alpha}\left(\overline{G}_2^{\alpha} - G_2^{oL}\right) \qquad (10.5)$$

$$\Delta G^{L}_{\text{mix}}\{\alpha; L\} = X_1^{L}\left(\overline{G}_1^{L} - G_1^{o\alpha}\right) + X_2^{L}\left(\overline{G}_2^{L} - G_2^{oL}\right) \qquad (10.6)$$

for this choice of reference states. Careful comparison with Eq. (10.3) shows that the $G_2^{o\alpha}$ term has been replaced by G_2^{oL}; in Eq. (10.4), the G_1^{oL} term has been replaced by $G_1^{o\alpha}$, as required by the specified reference states on the left side of these equations. In this pair of equations, the G^o terms are the same for component 1 (both are α phase) and the same for component 2 (both are L). Mixing behavior of these two phases, when formulated in this way, can now be compared.

It remains to connect Eq. (10.5) with the corresponding solution model Eq. (10.3) with its assumed reference states. In order to achieve this, write Eq. (10.5) and simply add and subtract $G_2^{o\alpha}$ inside the brackets in the second term:

$$\Delta G^{\alpha}_{\text{mix}}\{\alpha; L\} = X_1^{\alpha}\left(\overline{G}_1^{\alpha} - G_1^{o\alpha}\right) + X_2^{\alpha}\left(\overline{G}_2^{\alpha} - G_2^{o\alpha} + G_2^{o\alpha} - G_2^{oL}\right)$$

$$= X_1^{\alpha}\left(\overline{G}_1^{\alpha} - G_1^{o\alpha}\right) + X_2\left(\overline{G}_2^{\alpha} - G_2^{o\alpha}\right) + X_2^{\alpha}\left(G_2^{o\alpha} - G_2^{oL}\right)$$

The first two terms on the right side are identical with the right side of Eq. (10.3); that is, the first two terms represent the value for ΔG_{mix} that would be computed from the solution model with its choice of reference states. The last term in parentheses involves

$$G_2^{o\alpha} - G_2^{oL} = \Delta G_2^{oL \to \alpha}$$

which is the change in free energy per mole when pure component 2 is transformed from the liquid to the α phase form at the temperature and pressure of interest. Equation (10.5) may be written

$$\Delta G^{\alpha}_{\text{mix}}\{\alpha; L\} = \Delta G^{\alpha}_{\text{mix}}\{\alpha; \alpha\} + X_2^{\alpha}\Delta G_2^{oL \to \alpha} \qquad (10.7)$$

For the liquid solution, a similar strategy yields

$$\Delta G^{L}_{\text{mix}}\{\alpha; L\} = \Delta G^{L}_{\text{mix}}\{L; L\} + X_1^{L}\Delta G_1^{o\alpha \to L} \qquad (10.8)$$

First terms on the right sides of these equations can be computed from a solution model, such as Eq. (10.1) or more generally Eq. (10.2). The second terms require information about the difference in Gibbs free energy between the liquid and α phases for the pure components.

Figure 10.2a illustrates the effect that changing the reference state for component 2 has upon the shape of the (G-X) curve for the α phase. It is evident from Eq. (10.7) that a change in the reference state in this case adds a term that varies linearly with mole fraction. In Fig. 10.2a the dashed line is a plot of that linear term. The endpoints of the line occur at

$$X_2 = 0, \qquad \Delta G_{\text{mix}} = 0$$

$$X_2 = 1, \qquad \Delta G_{\text{mix}} = \Delta G_2^{oL \to \alpha}$$

The resulting $\Delta G_{\text{mix}}^{\alpha}$ curve hangs from the points $(0, 0)$ and $(1, \Delta G_2^{oL \to \alpha})$ on the sides of the diagram. The construction derived from the change in reference state for the liquid phase is shown in Fig. 10.2b. Here the hanging points correspond to

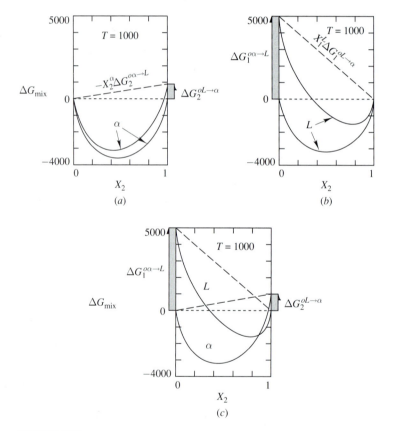

FIGURE 10.2
Effect of change in reference states of pure 1 and pure 2 upon the shape of the G-X diagram for (a) the solid solution, α, and (b) the liquid solution L. The comparison of mixing behavior in (c) is now valid because the reference states are consistent.

the evaluation of the second term in Eq. (10.8) at $X_2 = 0$ and 1, and are the points $(0, \Delta G_1^{o\alpha \to L})$ and $(1, 0)$. Both curves now report the free energy of mixing for their respective phases starting from pure 1 in the α phase form and pure 2 in the liquid phase form. These mixing curves can now be legitimately superimposed on the same diagram, Fig. 10.2c, and the mixing behavior of the α and liquid solutions may be compared.

These discussions have centered on one of the four possible choices for consistent reference states as an example. An alternate choice adds different linear terms to the expressions for $\Delta G_{\text{mix}}\{i; j\}$ and different dashed lines to the $(G\text{-}X)$ graphs. As a general principle, the choice of reference states for the components in a particular phase determines the hanging points for the corresponding $(G\text{-}X)$ curve at $X_2 = 0$ and $X_2 = 1$. If, for a given component, the reference state has the same phase form as the solution being considered, then the $(G\text{-}X)$ curve hangs from the origin. If the phase form of the reference state is different, then the curve hangs from an ordinate value corresponding to the free energy difference between the phase form of the solution and that of the reference state.

A change in reference state for a solution alters the calculated values of the activities and activity coefficients for the components in that solution. This is implicit in the definition of the activity since it contains the reference state value for the chemical potential. Recall the definition of activity

$$\mu_k - \mu_k^o = RT \ln a_k \qquad (8.66)$$

Apply it to component 2 in the α phase

$$\mu_2^\alpha - \mu_2^{o\alpha} = \overline{G_2^\alpha} - G_2^{o\alpha} = RT \ln a_2^\alpha \qquad (10.9)$$

Now suppose the activity of component 2 in the α phase is required in a context in which it is desirable to define the reference state for component 2 to be some other phase, say the β phase:

$$\overline{G_2^\alpha} - G_2^{o\beta} = RT \ln a_2^{'\alpha} \qquad (10.10)$$

where $a_2^{'\alpha}$ is evidently not the same as a_2^α. Add and subtract $G_2^{o\alpha}$ to the left side of this equation:

$$\overline{G_2^\alpha} - G_2^{o\alpha} + G_2^{o\alpha} - G_2^{o\beta} = RT \ln a_2^{'\alpha}$$

$$RT \ln a_2^\alpha + \Delta G_2^{o\beta \to \alpha} = RT \ln a_2^{'\alpha}$$

Solving for the new activity value

$$a_2^{'\alpha} = a_2^\alpha e^{\Delta G_2^{\beta \to \alpha}/RT} \qquad (10.11)$$

Substitution of the definition of activity coefficient demonstrates that this correction for change in reference state is essentially contained in the activity coefficient.

$$a_2^{'\alpha} = \gamma_2^{'\alpha} X_2^\alpha = \gamma_2^\alpha X_2^\alpha e^{\Delta G_2^{\beta \to \alpha}/RT}$$

$$\gamma_2^{'\alpha} = \gamma_2^\alpha e^{\Delta G_2^{\beta \to \alpha}/RT} \qquad (10.12)$$

Thus, changing the reference state for a component has the effect of multiplying the activity (or activity coefficient) at any composition by a constant factor for a composition series at a fixed temperature. Graphically, this stretches the ordinate by a constant factor, Fig. 10.3. The value of the factor may vary with temperature in a complicated way, as determined by the temperature dependence of the exponential factor in Eq. (10.12).

10.1.2   The Common Tangent Construction and Two Phase Equilibrium

In Chap. 9 it is demonstrated that the structure of a two phase field in a binary system is determined by the conditions for equilibrium:

$$T^\alpha = T^\beta \qquad P^\alpha = P^\beta \qquad \mu_1^\alpha = \mu_1^\beta \qquad \mu_2^\alpha = \mu_2^\beta$$

A graphical representation of this set of equations on a $(G\text{-}X)$ diagram is illustrated by the *common tangent construction* in Fig. 10.4. The diagram is drawn at a fixed value of

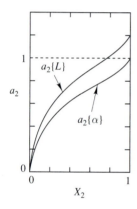

FIGURE 10.3
Variation of activity of component 2 in the α phase with composition computed from a solution model for the reference choice $\{\alpha\}$ is compared with the same information computed with the reference choice $\{L\}$. Assume $\Delta G_2^{o\alpha \to L}$ is negative.

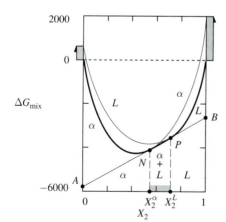

FIGURE 10.4
The $G\text{-}X$ curves for two different phases are plotted with consistent reference states. If these curves cross, then a unique line can be drawn that is tangent to both curves. The compositions at these tangent points satisfy the conditions for equilibrium between the phases.

pressure and temperature. Thus the conditions for mechanical and thermal equilibrium between the pair of phases is valid for all points on the graph. To make the illustration more concrete suppose β is the liquid phase. The conditions for chemical equilibrium between the α and liquid phases are satisfied uniquely at the pair of compositions corresponding to the points N and P in Fig. 10.4.

To demonstrate this assertion, recall the construction in Fig. 8.1. Given ΔB_{mix} for any extensive property as a function of composition, the partial molal properties, $\Delta \overline{B}_1$ and $\Delta \overline{B}_2$, corresponding to any composition X_2 can be read as the intercepts on the sides of the diagram of the tangent line drawn to the curve at X_2. On a $(G\text{-}X)$ plot, these intercepts are the partial molal Gibbs free energies of the components, which are equal to their chemical potentials, reported relative to the reference states for which the diagram is plotted. Because the $(G\text{-}X)$ curves for the two phases cross, it is possible to find a unique pair of compositions, labelled X_2^α and X_2^L in Fig. 10.4, through which a common tangent line, a single line tangent to both curves, may be constructed. The intercepts for that line are labelled A and B in Fig. 10.4. By the construction presented in Fig. 8.1 the point A represents the chemical potential for component 1 in an α solution of composition X_2^α and at the same time gives the chemical potential of component 1 in a liquid solution of composition X_2^L. More explicitly,

$$\Delta \mu_1^\alpha = \mu_1^\alpha(X_2^\alpha) - \mu_1^{o\alpha} = \Delta \mu_1^L = \mu_1^L(X_2^L) - \mu_1^{o\alpha}$$

Note that on this graph the reference states for component 1 were carefully chosen to be the same (the α phase form in this case). Thus,

$$\mu_1^\alpha(X_2^\alpha) = \mu_1^L(X_2^L) \tag{10.13}$$

for the compositions corresponding to N and P. The common tangent line also gives the same intercept for the α and liquid phases on the 2-side of the diagram at the point B. Read the intercepts

$$\Delta \mu_2^\alpha = \mu_2^\alpha(X_2^\alpha) - \mu_2^{oL} = \Delta \mu_2^L = \mu_2^L(X_2^L) - \mu_2^{oL}$$

Since the reference state for component 2 is chosen to be the liquid phase form for both solutions

$$\mu_2^\alpha(X_2^\alpha) = \mu_2^L(X_2^L) \tag{10.14}$$

Equations (10.13) and (10.14) are precisely the conditions for chemical equilibrium between the two phases. Thus the construction of a tangent line common to two $(G\text{-}X)$ curves for a pair of phases identifies the compositions of those two phases that coexist in equilibrium at the temperature and pressure for which the $(G\text{-}X)$ curves are drawn. These compositions represent the states of the two phases at opposite ends of the tie line that marks the two phase field in the phase diagram for the pressure and temperature chosen.

It is possible to view the relation between the common tangent construction and two phase equilibria in a different light. It is shown in Chap. 5 that the criterion for equilibrium in a system that is constrained to constant temperature and pressure is a minimum in the Gibbs free energy function. Stated another way, in a comparison all the possible states that a system may exhibit that have the same temperature and pressure, the equilibrium state is that which has the lowest value of the Gibbs free energy. The $(G\text{-}X)$ curves represent the thermodynamic behavior of solutions at constant temperature and pressure; thus this criterion applies to this representation of solution behavior.

As a simple illustration of the application of this criterion, consider the $(G\text{-}X)$ curve for a single solid phase α, Fig. 10.5. Focus on a system with the average composition X_2^o. It is possible to prepare a system with this average composition as a mechanical mixture of two solid solutions with compositions and free energies per mole corresponding to A and B in Fig. 10.5. The lever rule can be applied to give the relative amounts of these solutions that would be contained in mechanical mixture of two solid solutions in order to have an average composition X_2^o. By applying the lever rule it is seen that a mixture of (EF/DF) parts of solution A and (DE/DF) parts of solution B have the required average composition. The free energy of this mechanical mixture lies along the line AB at the point C. The point M represents the free energy of a single homogeneous solid solution of the same composition. M lies below C, meaning that it has a lower Gibbs free energy than this mechanical mixture of two solutions. Because the $(G\text{-}X)$ curve is everywhere concave upward, this conclusion holds for *every pair* of solution compositions A and B that straddle the composition X_2^o so that a mixture of some proportion can yield the average composition X_2^o. Thus, the point M, which corresponds to the free energy of a single homogeneous solution of composition X_2^o, has the lowest value of Gibbs free energy of all of the possible configurations that the system could assume.

Now apply the same principle to a system that can exhibit regions of stability for two different phases, Fig. 10.6. According to the common tangent construction illustrated in Fig. 10.4, the compositions at N and P can exist in two phase equilibrium. Consider a system that has the composition X_2^o lying between X_2^α and X_2^β. A system that is a mechanical mixture of α and β of these two compositions has a free energy of mixing from the reference states that lies along the line joining N and P. For a mixture

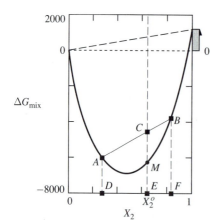

FIGURE 10.5
If the $G\text{-}X$ curve for a solution is concave upward at every point then, at any composition, X_2^o, a single homogeneous solution has a lower free energy than any mechanical mixture of two solutions with the same average composition.

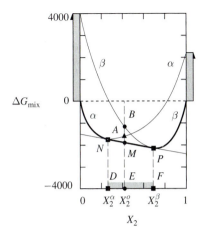

FIGURE 10.6

Consistent G-X curves are plotted for two phases, α and β. For any composition X_2^o between the points of common tangency a two phase mixture of α and β with compositions X_2^α and X_2^β has a lower free energy than either homogeneous phase at X_2^o, or any mixture of solution compositions other than X_2^α and X_2^β.

with proportions (EF/DF) of α and (DE/DF) of β, which has an average composition given by the point E, the free energy is given by the point M. Consider first a system that is all α phase at the composition X_2^o; its free energy of mixing is designated by the point A in Fig. 10.6. Similarly, a system that is all β phase of this composition has a free energy given by point B. Systems having the average composition X_2^o prepared from any combination of mixtures of solutions of the α and β phases, all having free energies of mixing that lie along the lines joining their composition points on the α and β curves, have Gibbs free energy values lying above the point M. It is concluded that the two phase mixture represented by point M, consisting of (EF/DF) parts of α of composition X_2^α and (DF/EF) parts of β of composition X_2^β, has the lowest value of the Gibbs free energy of all possible configurations that have the average composition X_2^o. Thus, this two phase mixture is the equilibrium state for a system of this composition, a result that is in agreement with the conclusion derived from the equality of the chemical potentials in Fig. 10.4.

As a general principle, if a system exhibits an arbitrary number of phases, Fig. 10.7, then the sequence of equilibrium configurations for the system is a combination of single and two phase regions given by the curve segments and common tangent lines that trace the minimum possible value of ΔG_{mix} across the composition range. In Fig. 10.7 this trace is given by the segments OA-AB-BC-CD-DE-EF-FP. This representation has been called the *taut string construction*. The free energy curves of the phases are viewed as rigid and a string tied at O and through P is pulled taut around these curves. The alternating curved and straight segments of the curve formed by the string give the state at each composition that has the minimum Gibbs free energy.

10.1.3   Two Phase Fields on Binary Phase Diagrams

The common tangent construction gives the compositions at the phase boundaries of a two phase field at a fixed temperature. To generate the complete two phase field on an isobaric binary phase diagram, it is only necessary to repeat this construction at a

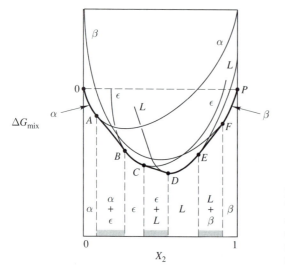

ΔG_{mix}

X_2

FIGURE 10.7
Taut string construction shows the
sequence of equilibrium conditions
across the composition range in a system
that may exhibit four different phase
forms, α, ϵ, L and β.

set of temperatures that span the range for which the diagram is plotted. In developing
this construction it is found that two different aspects of the thermodynamic behavior
determine the configuration:

1. The *mixing behavior* of the two competing phases reported in the excess and ideal
   free energies of mixing, Eq. (10.2).
2. The temperature variation of the relative stabilities of the two phase forms for the
   pure components as reported in their difference in Gibbs free energy.

   The relative stabilities of the phase forms for the pure components can be obtained from information about the variation of the Gibbs free energy of the phases with
temperature at constant pressure. It is demonstrated in Sec. 7.1.3 that this information
can be computed for any component given heat capacities of the phases, entropies of
the transformations, and the absolute entropy of one of the phases at 298K. Figure
10.8 is a sketch of this information as a function of temperature for the solid and
liquid phases for (a) component 1 and (b) component 2. In each case the curves cross
at the melting point where the free energies are equal. Below the melting point, α has
the lower free energy and is stable; above the melting point the liquid phase is stable.
At any temperature T_1, a quantitative measure of the difference in Gibbs free energy
between the phases is given by the vertical distance between the two curves.
   Examples of the variation of mixing behavior with composition and temperature
for a given phase are presented for some solution models in Fig. 8.3, 8.4, and 8.5.
Since $\Delta G_{\text{mix}} = \Delta H_{\text{mix}} - T \Delta S_{\text{mix}}$, as the temperature increases, the contribution from
the second term increases and the free energy of mixing becomes more negative.
A qualitatively different mixing pattern develops at sufficiently low temperatures for
systems with a positive departure from ideal behavior, Fig. 8.4. Consequences of this
unusual behavior are explored at the end of this section.

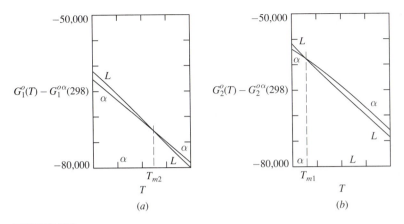

FIGURE 10.8
Molar free energies are plotted as function of temperature for the liquid and solid phases for (*a*) pure component 1 and (*b*) pure component 2.

The pattern of thermodynamic behavior that generates a two phase ($\alpha + L$) field can be displayed by computing and plotting (G-X) curves at a sequence of temperatures that lie between the melting points, Fig. 10.9. At each temperature for each component, choose as the reference state the phase form that is stable for that component. As a convention for plotting such diagrams, choose component 1 to be the one with the higher melting point. With this convention, in the temperature range between the melting points, the reference state of pure 1 is the solid phase α and that for component 2 is the liquid phase L. Thus, on each (G-X) diagram, the mixing curve for α passes through 0 on the 1-side of the diagram and that for the liquid passes through 0 on the 2-side. At a particular temperature T the hanging point for the α curve on the 2-side of the diagram is obtained from Fig. 10.8b by reading the free energy difference between liquid and solid at T. Similarly, the hanging point for the liquid curve on the 1-side lies a distance above the origin corresponding to the vertical distance between solid and liquid curves plotted for component 1 in Fig. 10.8a, evaluated at T. The mixing curves adapted from figures like Figs. 8.3, 8.4, or 8.5, with the straight line additions corresponding to the reference state corrections, hang from these points.

At the melting point of component 1 the α and L curves intersect at the origin because the hanging points for component 1 coincide, and $G_1^{o\alpha} = G_1^{oL}$ in Fig. 10.8a. As the temperature decreases the hanging points on the 1-side move apart while those on the 2-side move toward each other. As a consequence, the point at which the two mixing curves cross moves across the composition scale on the diagram. Since it is a geometric necessity that the common tangent construction straddles the crossing point for the two curves, the composition interval between the tangents, which defines the two phase field, moves across the diagram from left to right as the temperature decreases. Finally, at the melting point of component 2, the crossing point for the α and L curves coincides with the origin on the 2-side of the diagram.

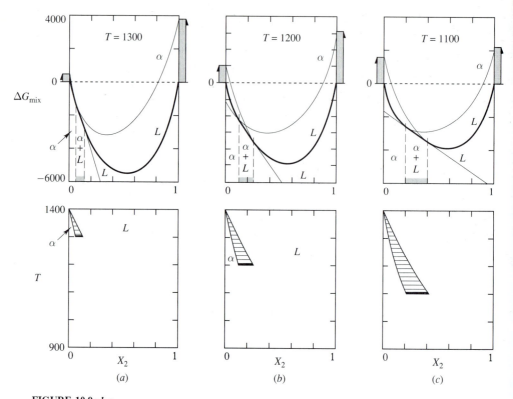

FIGURE 10.9a,b,c

Temperature sequence of G-X curves demonstrates the pattern of thermodynamic behavior that generates a simple two phase field on a binary phase diagram in (X_2, T) space. $a_0^{\alpha} = 1000$ (J/mol): $a_0^{L} = 6000$ (J/mol).

This pattern of behavior is characteristic of every monotonic two phase field, no matter what the phases may be. The quantitative details depend upon the components and phases involved but the qualitative pattern is pervasive. However, not all two phase fields are monotonic; some exhibit an extremum (maximum or minimum) in the phase boundaries. Figure 10.10 shows a two phase field with a minimum together with a set of plausible $(G$-$X)$ diagrams that would produce this configuration. In this case the behavior of the pure components is similar to that which characterizes monotonic two phase fields. The minimum derives primarily from a difference in mixing behavior for the α and L phases. The liquid phase has a much more negative departure from ideal behavior than does the solid. As a result, for temperatures below the melting point of component 2, the liquid curve crosses the solid curve *twice*, Fig. 10.10*b*. Associated with each crossing is a common tangent construction and thus, a two phase field. With continued decreasing temperature the liquid phase becomes increasingly more unstable with respect to the solid for the pure components, causing the liquid curve to move upward relative to the curve for the solid phase. The crossing points and their associated two phase fields move toward each other. At T_3 the two mixing curves touch each other at a single point; the pair of two phase fields have merged

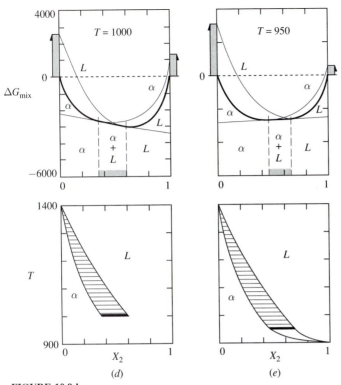

FIGURE 10.9d,e

Temperature sequence of G-X curves demonstrates the pattern of thermodynamic behavior that generates a simple two phase field on a binary phase diagram in (X_2, T) space. $a_0^\alpha = 1000$ (J/mol): $a_0^L = 6000$ (J/mol).

to a single point. At lower temperatures the solid curve lies below the liquid at all compositions and the liquid phase does not exist at equilibrium. These constructions demonstrate the characteristics of such two phase fields. Both phase boundaries, the liquidus and the solidus, must have minima that coincide in temperature and composition.

Two phase fields that exhibit a maximum (or, in fact a maximum *and* a minimum) are observed and arise from a similar construction.

Another configuration of the two phase field frequently encountered in phase diagrams is called the *miscibility gap*. This class of two phase field is unique in that it develops in a particular range of compositions from a single phase, rather than from the competition between two phase forms, as is characteristic of the fields in Figs. 10.9 and 10.10. Miscibility gaps develop at sufficiently low temperatures within phases that exhibit a positive departure from ideal behavior. Figure 10.11 illustrates this behavior.

Assume that the excess free energy is positive and not very sensitive to temperature. Then, as the temperature is decreased, the contribution from the ideal free energy of mixing, $\Delta G_{mix}^{id} = -T \Delta S_{mix}^{id}$, decreases and at sufficiently low temperature the total free energy of mixing curve develops an undulation, Fig. 10.11b. The tem-

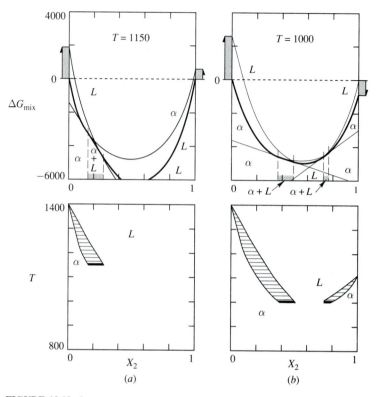

FIGURE 10.10a,b

Pattern of thermodynamic behavior that generates a two phase field with a minimum. $a_0^\alpha = 6000$ (J/mol); $a_0^L = -2000$ (J/mol).

perature at which this undulation first appears is called the *critical temperature* for the miscibility gap, usually labelled T_c. A common tangent line can be constructed and in this case the tangents occur at two points on the *same curve*. The minimum free energy arguments apply and a mixture of two solutions of the same phase form but different compositions, given by the tangent points, has a lower free energy in the composition interval between the tangent points than any other mixture of solutions. Thus the equilibrium condition has the unusual characteristic that it is a two phase mixture (the compositions are discretely different) but both phases have the same structure (e.g., liquid, or FCC, or BCC, etc.). Further decrease in temperature further reduces the ideal mixing contribution, expands the undulation and with it the width of the two phase field, Figures 10.11c, d.

An interesting property of miscibility gaps is illustrated in Fig. 10.12. The undulation in the (G-X) curve contains a region, bounded by the inflection points M and N, in which the free energy of mixing curve is *concave downward*. As the temperature is increased, the inflection points move toward each other and merge at the critical point. The domain in the miscibility gap thus generated is called the *spinodal region;* it is a characteristic of all miscibility gap structures. The boundary of the spinodal

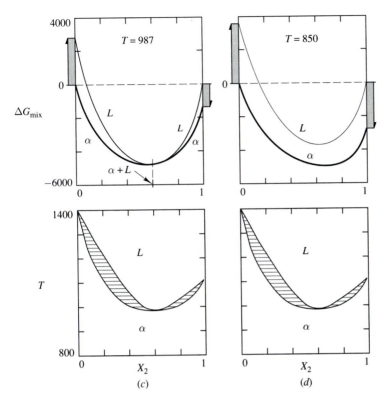

FIGURE 10.10c,d
Pattern of thermodynamic behavior that generates a two phase field with a minimum. $a_0^\alpha = 6000$ (J/mol); $a_0^L = -2000$ (J/mol).

region is identified with the inflection points on the $(G\text{-}X)$ curve; recall from analytical geometry that the inflection points are given by the condition that the second derivative of the function that the curve represents is zero.

Solutions existing in this range are capable of unusual and useful behavior. Consider, for example, the variation of chemical potential with composition in this region. In normal solutions with $(G\text{-}X)$ curves that are concave upward, the chemical potential of component 2 increases with concentration of 2, Fig. 10.13a. This can be visualized by recalling the graphical construction that permits determination of partial molal properties from the intercept of a line tangent to the mixing curve, Fig. 8.1. If the curve is concave upward, as is usually the case (see, for example, any of the curves in Figs. 10.4 to 10.9), the tangent to a point on the curve rotates counterclockwise as the point moves toward the 2-side of the diagram. Thus the intercept, which gives $\Delta\mu_2$, moves upward and $\Delta\mu_2$ increases with X_2 as in Fig. 10.13a. Inside the spinodal region, where the $(G\text{-}X)$ curve is locally concave *downward*, the tangent rotates clockwise as the composition moves toward the 2-side of the diagram, Fig. 10.12b. Thus within this region an increase in concentration of component 2 leads to a *decrease* in its chemical potential, Fig. 10.13b.

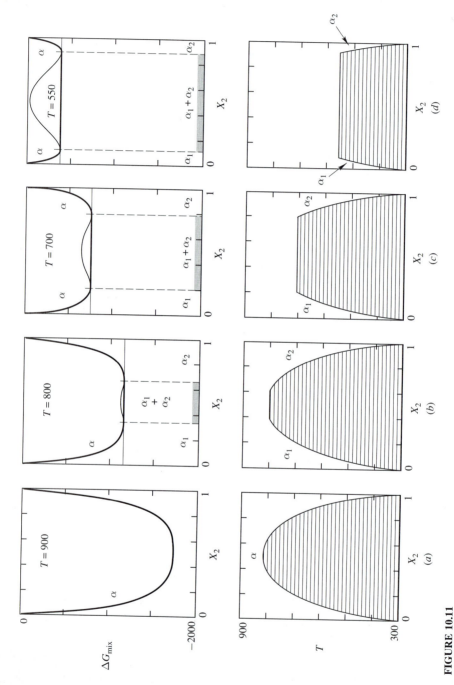

FIGURE 10.11
Pattern of thermodynamic behavior that generates a miscibility gap. $a_0^L = 13700$ (J/mol).

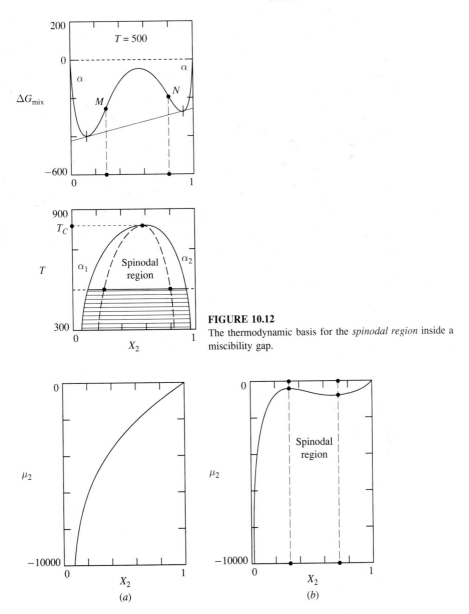

FIGURE 10.12
The thermodynamic basis for the *spinodal region* inside a miscibility gap.

FIGURE 10.13
Variation of chemical potential with composition (*a*) in a normal system, and (*b*) inside the spinodal region in a miscibility gap. Both curves are computed from regular solution models at 650 K. For (*a*), $a_0 = 4000$; for (*b*) $a_0 = 12500$ (J/mol).

One consequence of this unusual behavior is *uphill diffusion* inside the spinodal region. Normally, if a system does not have a uniform composition, atoms move in a direction to eliminate the nonuniformity and flow of a component occurs from regions rich in that component to regions that are less rich. This flow of atoms of one component through the system is called *diffusion*. In ordinary systems, in which the chemical potential and concentration increase or decrease together, this direction of flow also corresponds to atom motion from regions of high chemical potential to regions of low chemical potential (i.e., down the chemical potential gradient). Since the ultimate equilibrium condition of the system is a uniform chemical potential (as is demonstrated rigorously in Chap. 14), this flow is spontaneous, taking the system toward equilibrium because it reduces differences in chemical potential. Inside the spinodal region, where the chemical potential and composition vary in opposite directions, a flow from a zone with high chemical potential toward one with a lower value is still spontaneous; the system acts to eliminate differences in chemical potential. However, this case corresponds to motion of atoms from a region with less concentration of the component toward a region that is more concentrated, and flow is *up* the concentration gradient. Thus the drift of the system toward equilibrium is accompanied by the development of an increasingly nonuniform composition distribution. The system *spontaneously unmixes*.

Another way of visualizing this phenomenon is shown in Fig. 10.14. Suppose that a system with composition X_2^o is initially equilibrated at a temperature T_a above the critical temperature of the miscibility gap so that it has a uniform composition. The sample is then quenched to the temperature T_1, which places it inside the spinodal region. A sample that has a uniform composition X_2^o existing at the temperature T_1 has a free energy of mixing given by the point P in Fig. 10.14 Now consider a mixture of two solutions, indicated by the points M and N in this figure. The free energy of this nonuniform structure lies along the line MN. If the average composition of the system is X_2^o, its free energy is given by the point Q. Note that Q lies below P because the curve is concave downward. Thus the mixture of two solutions ($M + N$) is more stable than the uniform system at P. Since this is true even when M and N are arbitrarily close to P, even infinitesimal fluctuations in composition are stable relative to a uniform system. Fluctuations in composition that form amplify with time by uphill diffusion, ultimately forming the stable two phase mixture with compositions given by the common tangent construction at R and S. This process can occur only inside a

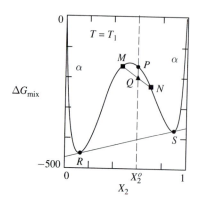

FIGURE 10.14
Inside the spinodal region mixtures of two solutions are more stable than an initially uniform system. This spontaneous unmixing of an initially uniform phase in the initial state of a *spinodal decomposition*.

spinodal region and is called *spinodal decomposition*. The resulting structure may be technologically useful because the decomposition typically occurs at low temperatures and produces a very fine, nanometer-scale microstructure.

10.1.4   Three Phase Equilibria

Consider next a binary system that is capable of forming three different phases, say α, L, and β, where the crystal structure of β is different from that of α. The description of the thermodynamic behavior of such a system requires information about the mixing behavior and relative stabilities of the pure components for all three phases. It is then possible to plot, on the same $(G\text{-}X)$ diagram, ΔG_{mix} curves for each of the three phases and compare their free energies. In this comparison it is essential that the behavior of component 1 in all three phases be referred to the *same reference state* and the same statement must also hold for component 2.

Interactions between the three mixing curves taken two at a time produce the three, two phase fields that are possible in this system: $(\alpha + L)$, $(\beta + L)$, and $(\alpha + \beta)$. If, at a given temperature two $(G\text{-}X)$ curves cross, then the common tangent construction exists for that pair of curves and the phase boundaries can be obtained by the construction developed in the last section. The full two phase field can be generated by scanning over the temperature range of interest in the phase diagram. As described in Chap. 9, intersections of these two phase fields produce the three phase equilibria that can exist in an isobaric binary phase diagram.

The two phase field interactions that lead to a *eutectic diagram* are shown in Fig. 10.15a. The $(\alpha + L)$ and $(\beta + L)$ fields slope in opposite directions. The point E at which their liquidus curves intersect defines the eutectic composition and temperature. A constant temperature line through E represents the equilibrium $(\alpha + \beta + L)$; it intersects the solidus curve for the $(\alpha + L)$ field at A and the solidus curve for $(\beta +$

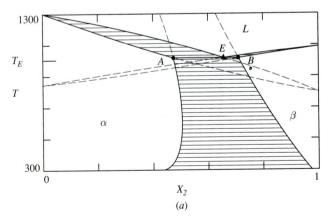

FIGURE 10.15a
Intersection of three two phase fields that share the same three phases (α, β, L) may produce (a) a eutectic phase diagram, or (b) a *peritectic* diagram (see next page).

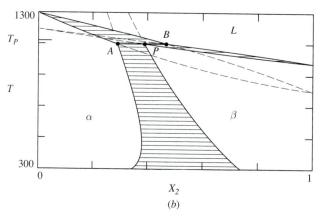

FIGURE 10.15*b*

L) at *B*. The points *A* and *B* are compositions of α and β that are in equilibrium with the liquid and therefore with each other. Accordingly, the boundaries of the ($\alpha + \beta$) field must also pass through the points *A* and *B*. The parts of the ($\alpha + L$) and ($\beta + L$) fields that extend below the eutectic line, shown as dashed lines in Fig. 10.15*a*, are *metastable* two phase equilibria, as is the part of the ($\alpha + \beta$) field that extends above the line.

The sequence of (*G-X*) diagrams that underlies this construction is shown in Fig. 10.16. As temperature decreases and the liquid phase becomes progressively less stable with respect to the solid phases, the liquid phase hanging points and curve move upward in (*G-X*) space. The crossing points between α and *L*, and β and *L*, move toward each other and the associated common tangent lines rotate toward each other. At the eutectic temperature the common tangent lines for ($\alpha + L$) and ($\beta + L$) come into coincidence to satisfy the conditions for the three phase ($\alpha + \beta + L$) equilibrium. The common tangent line for the ($\alpha + \beta$) equilibrium is also coincident with these lines at the eutectic temperature. Below the eutectic temperature the ($\alpha + L$) and ($\beta + L$) common tangent constructions still exist but now the ($\alpha + \beta$) line has a lower free energy, Fig. 10.16*d*. Thus below T_E the ($\alpha + L$) and ($\beta + L$) equilibria are *metastable*. These metastable equilibria are technologically important because they play a key role in the process by which a eutectic liquid solidifies, determining the rate of the process and the scale of the microstructure that results.

If the ($\alpha + L$) and ($\beta + L$) fields slope in the same direction, then a *peritectic* phase diagram results, Fig. 10.15*b*. Here the point of intersection of the liquidus curves for ($\alpha + L$) and ($\beta + L$), labelled *B* in the diagram, determines the temperature of the three phase equilibrium. The solid phases in equilibrium with this liquid composition are labelled *A* and *P*. these solid phases must be in equilibrium with each other and therefore identify the compositions of the ($\alpha + \beta$) equilibrium at the peritectic temperature T_P. The underlying (*G-X*) relationships are illustrated in Fig. 10.17. In this case, as the liquid (*G-X*) curve moves upward the ($\alpha + L$) tangent line touches the

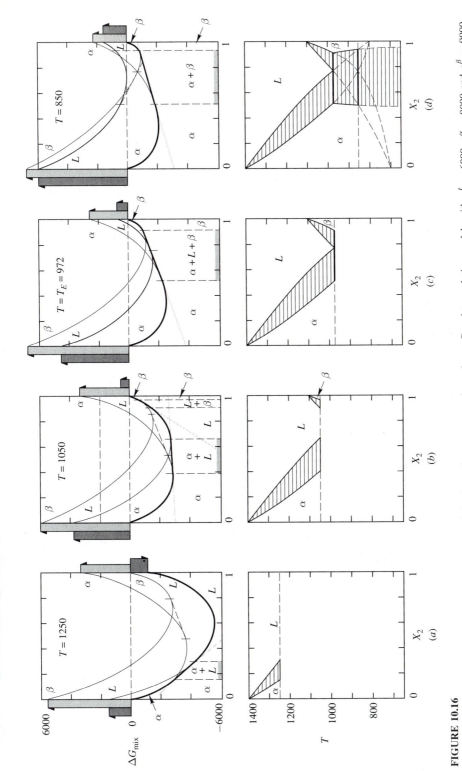

FIGURE 10.16
Pattern of thermodynamic behavior that generates a eutectic phase diagram among three phases. Regulator solution models with $a_0^L = 6000$, $a_0^\alpha = 8000$ and $a_0^\beta = 9000$ (J/mol).

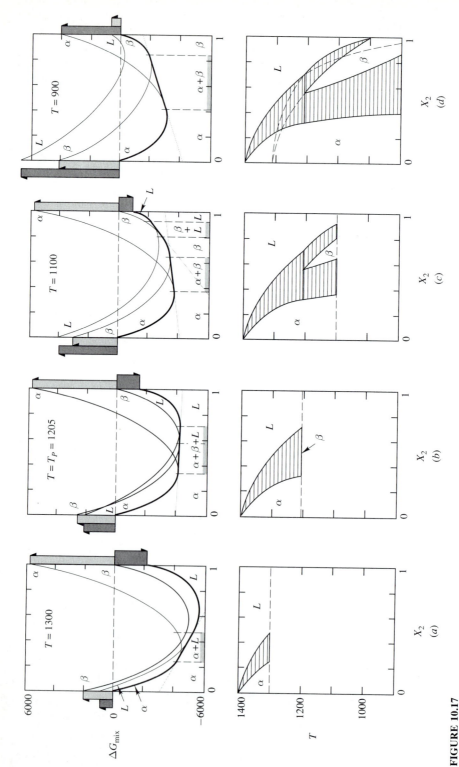

FIGURE 10.17
Pattern of thermodynamic behavior that generates a peritectic phase diagram among three phases. Regular solution models with $a_0^L = 10000$, $a_0^\alpha = 1500$ and $a_0^\beta = 6000$ (J/mol).

264

β curve first at a single point, which identifies the three phase peritectic equilibrium. The now stable $(\beta + L)$ equilibrium rapidly moves to the side of the diagram and vanishes at the melting point of the β phase and the $(\alpha + \beta)$ field descends to room temperature.

It is possible to produce eutectic and peritectic diagrams from only *two* phase forms (α and L) if the α phase contains a miscibility gap. In each case the critical temperature of the miscibility gap must lie at a temperature that is sufficiently high so that the undulating $(G-X)$ curve for the α phase interacts with the liquid curve; the miscibility gap is then metastable above the three phase line and its maximum is not observed experimentally. The sequence of $(G-X)$ diagrams corresponding to such a eutectic system is shown in Fig. 10.18. Development of a similar sequence for a peritectic diagram is left as an exercise for the reader. To determine whether a given diagram of these forms is made from three phase forms α, β, and L, or from only two phase forms, one of which contributes a miscibility gap, it is only necessary to check the crystal structures of the solid phases of the pure components from which the system is constructed. If they are the same, then the two phase region connecting the solid phases is a miscibility gap and the underlying thermodynamics is that presented in Fig. 10.18. Copper and silver form a simple eutectic phase diagram; however copper and silver are both face centered cubic crystals. Thus the low temperature two phase region in this system is in actuality a miscibility gap.

10.1.5 Intermediate Phases

Three phases are required to produce an ordinary eutectic or peritectic diagram such as those shown in Fig. 10.15–10.17. The majority of binary systems exhibit more than just three phases. Those solid solutions that are formed by adding solute to the pure components, such as the phases at the sides of the diagram, are called *terminal solid solutions*. If one of the components has stable allotropic forms, then additional terminal solid solutions exist in the temperature regions where these phase forms exist. In addition there may exist in the system phase forms that are different from any of those exhibited by the pure components. These are called *intermediate phases*. Each additional phase form has its own $(G-X)$ diagram that potentially interacts with all of the others to generate the phase diagram.

At any given temperature, the sequence of stable one and two phase fields traversed as X_2 passes from 0 to 1, is determined by the collection of $(G-X)$ diagrams for all of the phases that the system may exhibit, all carefully referred to the same reference states for the pure components. The taut string construction illustrated in Fig. 10.7 gives the sequence of alternating one and two phase fields that have the minimum free energy and are thus stable. The $(G-X)$ curves for the terminal phases always have a vertical slope at the sides of the diagram. As the temperature changes the curves move with respect to each other in $(G-X)$ space. Crossing points move and common tangents rotate. Occasionally common tangent lines come into coincidence passing through the unique temperature corresponding to a three phase field, either eliminating a phase or introducing a new one to the list of stable phases as the temperature changes.

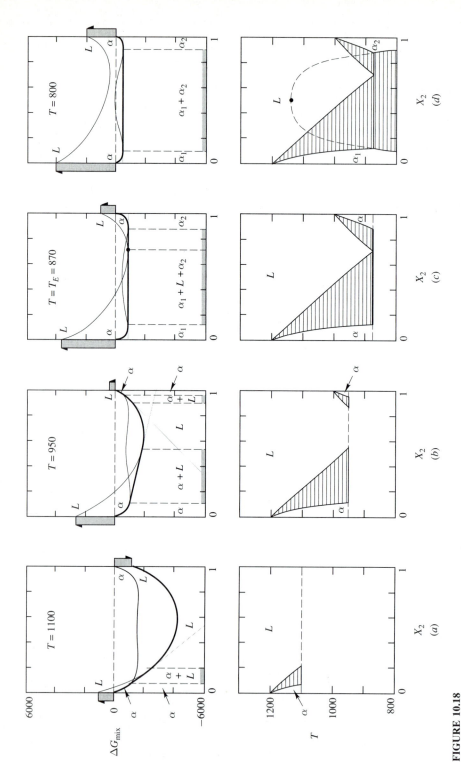

FIGURE 10.18

Pattern of thermodynamic behavior that generates a eutectic phase diagram in which the solid-solid two phase field is a miscibility gap in the α phase. Regular solution models with $a_0^L = 9000$, $a_0^L = 19000$ (J/mol); $T_c = 1143$ K.

Figure 10.19 shows a collection of plausible $(G\text{-}X)$ diagrams that correspond to a phase diagram with a single intermediate phase ϵ. Between T_1 and T_2 the $(\alpha + L)$ and $(L + \epsilon)$ tangents rotate into coincidence, forming the $(\alpha + L + \epsilon)$ eutectic. The liquid curve lifts off this line leaving the two solid phases. Between T_2 and T_3 the β curve touches the $(\epsilon + L)$ equilibrium line forming the peritectic $(\epsilon + \beta + L)$ equilibrium. Then, the β curve penetrates past the $(\epsilon + L)$ line to form two phase fields $(\epsilon + \beta)$ and $(\beta + L)$. Between T_3 and T_4 the $(\alpha + \epsilon)$ and $(\epsilon + \beta)$ tangent lines rotate into coincidence to form the $(\alpha + \epsilon + \beta)$ eutectoid line. The ϵ curve lifts off the line, leaving the $(\alpha + \beta)$ field at the lowest temperatures.

Intermediate phases frequently have the character of a *chemical compound*, particularly in ceramic systems. That is, the composition of the phase does not exhibit a significant deviation from some fixed ratio like A_2B or AB_3. On the phase diagram such an intermediate phase appears as a *line compound*, Fig. 10.20; that is, although the single phase region labelled ϵ has a structure and a width that is qualitatively similar to that shown for the ϵ phase in Fig. 10.19, this width is too small to be resolved on the scale at which the diagram is plotted. The $(G\text{-}X)$ curve for this line compound can be plotted as a single point for practical purposes; this point is labelled P in Fig. 10.20. The value of ΔG_{mix} at P can be thought of as the free energy of formation of the line compound *from the reference states*. Some complicated diagrams with many intermediate phases can be constructed from information about free energies of formation of each of the compounds as a function of temperature.

An important characteristic of the thermodynamics of line compounds is illustrated in Fig. 10.20. Small changes in composition across the very limited range of variation available to a line compound are accompanied by large changes in chemical potential of the components. When ϵ is equilibrated with the α phase its value of X_2 is be slightly less than its stoichiometric value. At that composition, the chemical potential of component 2 in ϵ is given by point M, the intercept on the 2-side of the diagram of the common tangent line connecting α and ϵ. For ϵ equilibrated with the liquid, the value of X_2 in the ϵ is only slightly larger than stoichiometric; the chemical potential of component 2 is given by the point N. Thus, within line compounds, changes in composition that are too small to resolve on the scale of the phase diagram are accompanied by significant changes in chemical potentials.

This behavior reflects the fact that in most solid state line compounds the two components occupy specific sites in the crystal lattice of the phase. Each component has its own *sublattice*. The compound is stoichiometric because the ratios of the two types of sites are fixed by the geometry of the crystal structure. Deviations from this fixed ratio can occur only by introducing *defects* into the crystal structure. For example, a composition slightly enriched in component 2 can be achieved by leaving some of the sites in the sublattice for component 1 vacant. Conversely, atoms of component 2 can be squeezed into positions between the normal lattice sites, called *interstitial positions*. Analogous structural defects can accommodate deviations in stoichiometry toward the 1-side of the diagram. The energy to form such lattice defects is large on the scale of thermodynamic energies. Hence, large changes in chemical potential are associated with their formation. The thermodynamics of defects in crystals is discussed in detail in Chap. 13.

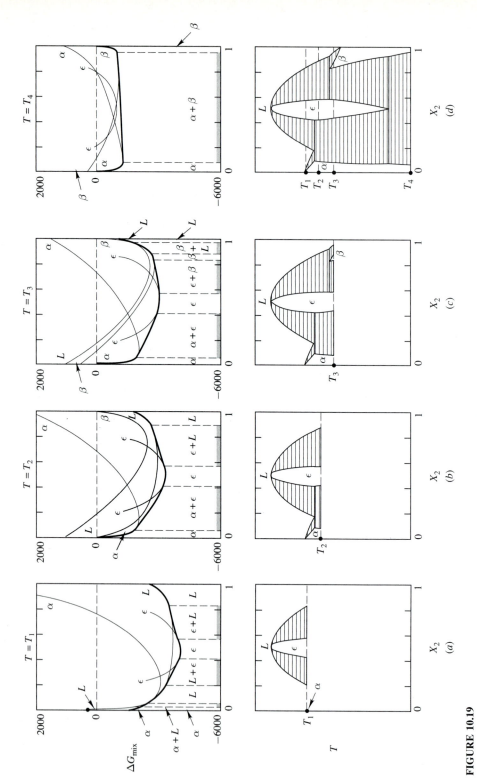

268

FIGURE 10.19
Free energy-composition diagrams that correspond to a phase diagram with a single intermediate phase.

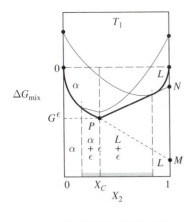

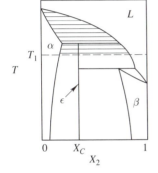

FIGURE 10.20
Free energy-composition diagram characteristic of a phase diagram with an intermediate phase that forms as a *line compound*.

10.1.6 Metastable Phase Diagrams

Consider a system that can contain p phases, G, L, α, β, ϵ, ..., p. Pairwise intersections between the $(G\text{-}X)$ curves for those phases produce two phase fields. The number of possible two phase fields that can result is at least $p!/[2! \cdot (p-2)!]$ (the number of ways that p items can be arranged in two categories) plus the number of miscibility gaps the system can have. Parts of at least some of these two phase fields are stable and appear on the equilibrium phase diagram, defining it, as in the solid line portions of the two phase fields in Figs. 10.15a and 10.15b. Other parts are *metastable*, indicated for example by the dashed line portions of the two phase fields in Fig. 10.15, implying that an alternate structure has a lower free energy in that part of the diagram and is the stable equilibrium construction. Some two phase fields can be metastable over their entire range of existence and thus not appear anywhere on the stable diagram.

Metastable structures frequently play an important role in understanding how microstructures develop in materials science. It has already been pointed out that the metastable extensions of the $(\alpha + L)$ and $(\beta + L)$ equilibria in a eutectic system play a key role in the theory of eutectic solidification and in the control of the resulting microstructure. Formation of solute rich zones during processing to develop and control precipitation hardening in some systems is frequently associated with the absence of a metastable miscibility gap in the phase diagram. The development of microcrystalline or metallic glass structures in rapid solidification technology is sometimes associated with the absence of a stable phase that does not form because it is very slow to nucleate or grow.

If during the time of an experiment a particular phase that is known to be stable in the system does not form, then the behavior of the system can be interpreted with the use of a metastable phase diagram. A metastable phase diagram can be developed from information about the stable diagram simply by leaving one or more of the stable phase (G-X) curves out of the construction of the diagram. Figure 10.21 presents an example of the application of this strategy. The stable diagram is shown in Figure 10.21a. In constructing Fig. 10.21b, it is assumed that the ϵ phase does not form and the α, β and L curves are unchanged in Fig. 10.21b. In this example, a simple eutectic phase diagram results when the ϵ phase is deleted from the (G-X) constructions with the result that the liquid phase exists at significantly lower temperatures than in the stable diagram, favoring the formation of a low temperature glass phase in this system.

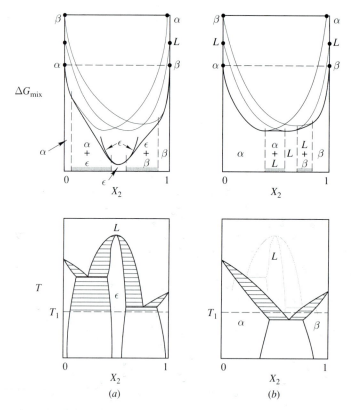

FIGURE 10.21
The metastable phase diagram (b) results from the stable diagram in (a) if the intermediate phase ϵ is slow to form.

10.2 THERMODYNAMIC MODELS FOR BINARY PHASE DIAGRAMS

Two kinds of information are needed in order to compute a set of $(G\text{-}X)$ curves from which to deduce a phase diagram: the thermodynamics of mixing behavior and the relative free energies of the pure components. The ideal and regular solution models and their descendants provide the basis for describing mixing behavior. Relative stabilities of the pure component phase forms require absolute entropies, heat capacities, and entropies of the transformations that can occur for these components. This information can be used to derive expressions that describe the variation of the chemical potentials of the components in each phase at any temperature. Substitution of these derived model expressions for the chemical potentials into the two conditions for chemical equilibrium yields two equations in the unknown phase boundary compositions. These equations can be solved, in principle, to compute X_2^α and X_2^β at that temperature. Repetition of this strategy at a sequence of temperatures generates the $(\alpha + \beta)$ field. Repetition for the other possible combinations of two phases in the system generates the complete set of two phase fields and their intersections. The stable and metastable parts of the diagram can then be inferred from the known stable parts of the unary phase diagrams, (i.e., from their connections to the sides of the diagram). This strategy is first demonstrated for the simplest of models, which assumes that the mixing behavior of each phase is ideal and that the stabilities of the pure components behave in a simple way. The procedure is then illustrated for the simplest regular solution model. Modifications to diagrams brought into play by the existence of intermediate phases that may be treated as line compounds complete this discussion.

10.2.1 Ideal Solution Models for Phase Diagrams

To compute the phase boundaries in a two phase field from a solution model, it is necessary to derive expressions for the chemical potentials of the components in both phases. Consider a system consisting of the α and β phases. Recall the expression for the chemical potential of a component in an ideal solution

$$\Delta \mu_k^\alpha = \mu_k^\alpha - \mu_k^{o\alpha} = RT \ln X_k^\alpha \tag{8.72}$$

Use this expression to write explicit expressions for the chemical potentials of the components in the α phase in a binary system

$$\mu_1^\alpha = \mu_1^{o\alpha} + RT \ln X_1^\alpha = G_1^{o\alpha} + RT \ln X_1^\alpha \tag{10.15}$$

$$\mu_2^\alpha = \mu_2^{o\alpha} + RT \ln X_2^\alpha = G_2^{o\alpha} + RT \ln X_2^\alpha \tag{10.16}$$

Write the analogous expressions for the components in a β phase solution that is ideal

$$\mu_1^\beta = \mu_1^{o\beta} + RT \ln X_1^\beta = G_1^{o\beta} + RT \ln X_1^\beta \tag{10.17}$$

$$\mu_2^\beta = \mu_2^{o\beta} + RT \ln X_2^\beta = G_2^{o\beta} + RT \ln X_2^\beta \tag{10.18}$$

Choose a value of temperature, T. The conditions for equilibrium of the α and β phases are

$$\mu_1^\alpha = \mu_1^\beta \quad \text{and} \quad \mu_2^\alpha = \mu_2^\beta$$

Set Eq. (10.15) equal to Eq. (10.17)

$$\mu_1^\alpha = G_1^{o\alpha} + RT \ln X_1^\alpha = G_1^{o\beta} + RT \ln X_1^\beta = \mu_1^\beta$$

Choose X_2 as the compositional variable in each phase

$$G_1^{o\alpha} + RT \ln(1 - X_2^\alpha) = G_1^{o\beta} + RT \ln(1 - X_2^\beta)$$

Rearranging

$$\frac{1 - X_2^\beta}{1 - X_2^\alpha} = e^{-(\Delta G_1^{o\alpha \to \beta}/RT)} \equiv K_1(T) \tag{10.19}$$

Set Eq. (10.16) equal to Eq. (10.18)

$$\mu_2^\alpha = G_2^{o\alpha} + RT \ln X_2^\alpha = G_2^{o\beta} + RT \ln X_2^\beta = \mu_2^\beta$$

Rearranging

$$\frac{X_2^\beta}{X_2^\alpha} = e^{-(\Delta G_2^{o\alpha \to \beta}/RT)} \equiv K_2(T) \tag{10.20}$$

The functions $K_1(T)$ and $K_2(T)$ are implicitly defined in these equations and they contain only information about the relative stabilities of the pure components embodied in $\Delta G_1^o(T)$ and $\Delta G_2^o(T)$.

Equations (10.19) and (10.20) are simply restatements of the conditions for equilibrium that define the $(\alpha + \beta)$ system when both solutions are ideal. They are simultaneous equations that are linear in X_2^α and X_2^β. Their solution is straightforward.

$$X_2^\alpha = \frac{K_1 - 1}{K_1 - K_2} \tag{10.21}$$

$$X_2^\beta = K_2 X_2^\alpha = K_2 \frac{K_1 - 1}{K_1 - K_2} \tag{10.22}$$

Equation (10.21) is the functional form $X_2^\alpha = X_2^\alpha(T^\alpha)$ relating composition to temperature for the phase boundary on the α side of the $(\alpha + \beta)$ field and Eq. (10.22) is the analogous function $X_2^\beta = X_2^\beta(T^\beta)$ for the β side. To compute the ideal solution phase diagram, it only remains to evaluate ΔG_k^o for pure 1 and 2 for the α to β transformation as a function of temperature so that $K_1(T)$ and $K_2(T)$ can be computed.

Estimates of these temperature functions are easily obtained. For any pure component the definitional relationship is

$$\Delta G^o(T) = \Delta H^o(T) - T \Delta S^o(T) \tag{10.23}$$

where ΔH^o and ΔS^o represent, respectively, the enthalpy and entropy for the transformation at any temperature. At constant pressure, these temperature functions are given by

$$\Delta H^o(T) = \Delta H^o(T_0) + \int_{T_o}^{T} \Delta C_P^o(T) dT \tag{10.24}$$

and

$$\Delta S^o(T) = \Delta S^o(T_0) + \int_{T_0}^{T} \frac{\Delta C_P^o(T)}{T} dT \tag{10.25}$$

where ΔC_P^o is the difference in heat capacity between the α and β phase forms for the pure component. If T_0 is the equilibrium temperature for α and β for that component, then $\Delta H^o(T_0)$ and $\Delta S^o(T_0)$ are, respectively, the enthalpy (heat) and entropy of the transformation for the pure component at that equilibrium temperature. Further, recall that these two quantities are related through Eq. (7.17)

$$\Delta T^o(T_0) = T_0 \Delta S^o(T_0) \tag{7.17}$$

Now assume that the difference in heat capacity between the two phase forms is sufficiently small so that the integrals in Eqs. (10.24) and (10.25) can be neglected with respect to the magnitudes of the other terms. This is equivalent to the assumption that the heat and entropy of the transformation are independent of temperature. Equation (10.23) can be written

$$\Delta G^o(T) = \Delta H^o(T_0) - T \Delta S^o(T_0) = T_0 \Delta S^o(T_0) - T \Delta S^o(T_0)$$

$$\Delta G^o(T) = \Delta S^o(T_0)[T_0 - T] \tag{10.26}$$

Adopt the convention that α is the stable phase at low temperature and β is to be at high temperature; then ΔS^o is positive. For $T < T_0$, $\Delta G^o(T)$ is positive for the transformation $\alpha \to \beta$ and the α form is stable and for $T > T_0$, $\Delta G^o(T)$ is negative and the β phase is stable.

With the temperature dependence of the difference in free energies of the pure components thus estimated to be linear, it becomes possible to evaluate the functions $K_1(T)$ and $K_2(T)$ in Eqs. (10.21) and (10.22) as

$$K_1(T) = e^{-(\Delta S_1^o(T_{01} - T)/RT)} \tag{10.27}$$

and

$$K_2(T) = e^{-(\Delta S_2^o(T_{02} - T)/RT)} \tag{10.28}$$

Thus in the ideal solution model of a two phase field the adjustable parameters are:

$$T_{01}, \quad T_{02}, \quad \Delta S_1^o, \quad \Delta S_2^o, \quad (\text{or } \Delta H_1^o, \quad \Delta H_2^o)$$

To make the example more concrete, if α is the solid phase and β the liquid, then the data required are the melting points and entropies of fusion of the two pure components. It is relatively simple to write an application program to compute and plot an ideal solution model two phase field. Inputs to the program are the melting points and entropies of fusion of the two pure components. Increment the temperature over the range between the melting points. For each temperature write relations for each ΔG_k^o in terms of the entropy and melting temperature and express $K_k(T)$ in terms of ΔG_k^o for both components. Then put these values into Eqs. (10.21) and (10.22) to compute and plot $X_2^\alpha(T)$ and $X_2^\beta(T)$.

The ideal solution model is not very flexible because the mixing contribution has no adjustable parameters. Given the two equilibrium temperatures that pin the ends of the two phase field, the only remaining adjustable parameters are the entropies of the transformation for pure component 1 and pure component 2. Figure 10.22 shows the limited pattern of two phase field structures that can be obtained from the ideal solution model for various combinations of the entropy differences. If both transformation entropies are small, the field is narrow and straight. If both are large, the field is broad and straight. If one transformation entropy is large and the other small, the field is narrow opposite the side with the small entropy change and broad opposite the side with the large change in entropy. All fields are monotonic, such that the ideal solution model is incapable of yielding a two phase field with a maximum or minimum.

Consider an ideal solution model for a binary system with three phases α, β, and L. Three, two phase fields are possible: $(\alpha + \beta)$, $(\alpha + L)$, and $(\beta + L)$. If it

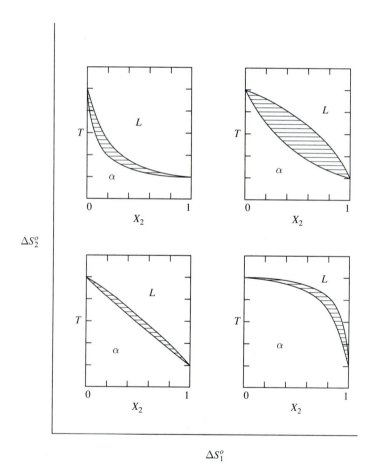

FIGURE 10.22

The ideal solution model can generate a relatively limited pattern of two phase fields. The diagrams shown correspond to different combinations of the entropies of fusion of the pure components.

is assumed that the mixing behavior of all three phases is ideal, then the procedure just developed can be applied separately to compute the three, two phase fields in the system. In this case, it is necessary to know T_0 and ΔS^o values for *all three, two phase equilibria* for each component. Values for these six pieces of information are not independent, however, since, for example,

$$\Delta S_1^{o\alpha \to L} = \Delta S_1^{o\alpha \to \beta} + \Delta S_1^{o\beta \to L} \tag{10.29}$$

A similar relation holds for component 2, as well as for the enthalpies of transformation. Thus, of the twelve parameters contained in an ideal solution model for a three phase system, only eight are independent. In constructing a model for such a system it is essential to be aware of these constraining relationships. The three, two phase fields can be computed from this information by applying Eqs. (10.21) and (10.22) with values of K_1 and K_2 appropriately evaluated. The diagrams shown in Fig. 10.15a and 10.15b are ideal solution phase diagrams computed for a eutectic system and a peritectic system. The rapid broadening evident in the $(\alpha + \beta)$ fields at low temperatures derives from the behavior of K_1 and K_2 as $T \to 0$ K.

10.2.2   Regular Solution Model for Phase Diagrams

The simplest of the regular solution models, introduced in Sec. 8.7.1, assumes that

$$\Delta G_{\mathrm{mix}}^{\mathrm{xs}} = \Delta H_{\mathrm{mix}} = a_0 X_1 X_2 \tag{10.30}$$

with the result that

$$\Delta \overline{H}_k = \Delta \overline{G}_k^{\mathrm{xs}} = a_0 (1 - X_k)^2 \qquad (k = 1, 2) \tag{10.31}$$

and provides a much more flexible description of two phase fields than does the ideal solution model. This model introduces one additional parameter for each phase, a_0^α, a_0^β, etc., which, according to the quasichemical theory discussed in Sec. 8.7.3, is determined by the relative energies of like and unlike bonds in the phase. Within a single phase this regular solution model is capable of generating a miscibility gap. In other two phase equilibria, depending upon the values of the mixing parameters, the model can develop two phase field shapes that have a maximum or minimum as well as monotonic structures.

The chemical potential of component k in phase α can in general be written as

$$\mu_k^\alpha = \mu_k^{o\alpha} + \Delta \overline{G}_k^{\mathrm{xs}\alpha} + RT \ln X_k^\alpha \tag{10.32}$$

where the excess partial molal Gibbs free energy is a function of composition and temperature. For the simple regular solution model, according to Eq. (10.31),

$$\mu_k^\alpha = \mu_k^{o\alpha} + a_0^\alpha (1 - X_k^\alpha)^2 + RT \ln X_k^\alpha \tag{10.33}$$

In deriving the equations that describe a two phase $(\alpha + \beta)$ field, it is first necessary to write out the chemical potentials of both components in both phases:

$$\mu_1^\alpha = \mu_1^{o\alpha} + a_0^\alpha (X_2^\alpha)^2 + RT \ln(1 - X_2^\alpha) \tag{10.34}$$

$$\mu_2^\alpha = \mu_2^{o\alpha} + a_0^\alpha (1 - X_2^\alpha)^2 + RT \ln X_2^\alpha \tag{10.35}$$

$$\mu_1^\beta = \mu_1^{o\beta} + a_0^\beta (X_2^\beta)^2 + RT \ln(1 - X_2^\beta) \tag{10.36}$$

$$\mu_2^\beta = \mu_2^{o\beta} + a_0^\beta (1 - X_2^\beta)^2 + RT \ln X_2^\beta \tag{10.37}$$

Compare these equations to Eqs. (10.15)–(10.18) for the ideal solution model. The conditions for equilibrium require that Eq. (10.34) be set equal to Eq. (10.36) and that Eq. (10.35) be equal to Eq. (10.37). After some simplification,

$$\frac{\Delta G_1^{o\alpha \to \beta}}{RT} + \frac{a_0^\beta}{RT}(X_2^\beta)^2 - \frac{a_0^\alpha}{RT}(X_2^\alpha)^2 + \ln \frac{(1 - X_1^\beta)}{(1 - X_2^\alpha)} = 0 \tag{10.38}$$

$$\frac{\Delta G_1^{o\alpha \to \beta}}{RT} + \frac{a_0^\beta}{RT}(1 - X_2^\beta)^2 - \frac{a_0^\alpha}{RT}(1 - X_2^\alpha)^2 + \ln \frac{(X_2^\beta)}{(X_2^\alpha)} = 0 \tag{10.39}$$

In a model calculation of a two phase field, the parameters a_0^α and a_0^β must be input together with the temperature dependence of the ΔG^o values for the pure components estimated from Eq. (10.26) or its more sophisticated versions. Thus, these quantities are constants at any given temperature. Then Eq. (10.38) and (10.39), which are explicit expressions for the conditions for chemical equilibrium, constitute a pair of simultaneous equations in the unknowns X_2^α and X_2^β, which are the phase boundary compositions at that temperature. Unlike the ideal solution model, it is not possible to obtain an analytical solution to this pair of equations. Iterative numerical techniques are required. Lupis [10.1] provides a useful discussion of some of these numerical methods. Such procedures are available in standard mathematics applications programs and can be efficiently computed. To calculate the phase boundaries for the complete two phase field, choose a temperature, evaluate the ΔG^o values, solve the equations, and plot the result. Increment the temperature and repeat the algorithm until the field is complete.

A computer program written to compute phase boundaries for a regular solution model is easily generalized to handle an arbitrary solution model. It is only necessary to input the more general expressions for the partial molal excess free energies in Eqs. (10.32).

If a_0 is positive for a given phase, then at sufficiently low temperatures that phase exhibits a miscibility gap. This construction is illustrated in Fig. 10.11. For the simple regular solution model the excess free energy of mixing is a symmetrical parabola given by Eq. (10.30). Since the ideal contribution also has a symmetrical form about $X_2 = 0.5$, ΔG_{mix} is also symmetrical. The $(G\text{-}X)$ curves computed from this model at several temperatures are shown in Fig. 8.5. The phase boundaries can be computed with the numerical algorithm already described.

The boundary of the spinodal region and the critical temperature of the miscibility gap can be derived from this model without resorting to numerical techniques. The spinodal region is defined by the condition that the ΔG_{mix} curve is concave downward. At any given temperature the limits of this condition are defined by the inflection points in the $(G\text{-}X)$ curve where its curvature changes sign. Mathematically, the inflection point on a curve representing a function is described by the condition that the second derivative of the function is zero. In this case the function is

$$\Delta G_{\text{mix}} = a_0 X_1 X_2 + RT(X_1 \ln X_1 + X_2 \ln X_2)$$

Take the second derivative, noting again that X_1 and X_2 are related so that this is the *total* derivative in the context of these two variables.

$$\frac{d^2 \Delta G_{\text{mix}}}{dX_2^2} = -2a_0 + \frac{RT}{X_1 X_2}$$

To find the inflection point in the curve, which defines the boundary of the spinodal region at any temperature, set this derivative equal to zero and rearrange the result.

$$X_1 X_2 = \frac{RT}{2a_0} \tag{10.40}$$

This equation describes how the compositions at the boundary of the spinodal region vary with temperature.

Thus, the spinodal boundary is seen to be a parabola symmetrical about $X_2 = 0.5$. As the temperature is increased these inflection points move toward each other and at the critical temperature T_c they meet at $X_2 = 0.5$. Inserting this condition into Eq. (10.40)

$$(0.5)(0.5) = \frac{RT_c}{2a_0}$$

And solving for the critical temperature

$$T_c = \frac{a_0}{2R} \tag{10.41}$$

Thus, in the simple regular solution model, a positive heat of mixing for any phase implies that it exhibits a miscibility gap below a temperature given by Eq. (10.41). Whether this miscibility gap is stable, metastable, or interacts with other phases to form additional two phase fields is determined by the usual comparisons available through the (G-X) curves for the system. For example, Fig. 10.18 illustrates a case in which the critical temperature lies in the region in which the liquid phase is thermodynamically stable. Interactions between the miscibility gap and the liquid phase produce a eutectic phase diagram.

10.2.3 The Midrib Curve

Consider the (G-X) curves for two phases at some temperature T_1. The system forms a two phase field at the temperature in question if and only if the two curves cross at some point. A common tangent can then be constructed connecting the two curves and through the conditions for equilibrium, the boundaries of the two phase field can be defined. The point at which the two (G-X) curves cross necessarily lies within the two phase field. It is defined by the condition

$$\Delta G_{\text{mix}}^{\alpha}\{\alpha; \beta\} = \Delta G_{\text{mix}}^{\beta}\{\alpha; \beta\} \tag{10.42}$$

where the selected reference states must be the same for both phases. If the temperature range is scanned to trace out the $(\alpha + \beta)$ field, this crossing point traces out a curve that lies within that field. This *midrib* curve, though not necessarily at the center of the two phase field, is guaranteed to lie within it. It provides a useful means of locating the two phase field in (X_2, T) space and has other applications of technological importance. Properties of midrib curves for the simplest regular solution model are explored in this section.

Focus on an $(\alpha + L)$ two phase field. Choose α to be the reference state for component 1 and L to be that for component 2. The free energies of mixing of these two phases are given by:

$$\Delta G^{\alpha}_{mix}\{\alpha; L\} = a_0^{\alpha} X_1^{\alpha} X_2^{\alpha} + RT(X_1^{\alpha} \ln X_1^{\alpha} + X_2^{\alpha} \ln X_2^{\alpha}) - X_2^{\alpha} \Delta G_2^{o\alpha \to L} \quad (10.43)$$

$$\Delta G^{L}_{mix}\{\alpha; L\} = a_0^{L} X_1^{L} X_2^{L} + RT(X_1^{L} \ln X_1^{L} + X_1^{L} \ln X_2^{L}) + X_1^{L} \Delta G_1^{o\alpha \to L} \quad (10.44)$$

The last term in each equation accommodates the change in reference state required. At any temperature T^* the midrib point is given by the point of intersection of these curves, Eq. (10.42). Note also that at this point of intersection the value of X_2 is the same for both phases; thus

$$X_2^{\alpha} = X_2^{L} = X^* \quad \text{and} \quad X_1^{\alpha} = X_1^{L} = (1 - X^*)$$

When Eq. (10.43) and (10.44) are substituted into Eq. (10.42) the ideal mixing terms cancel

$$a_0^{\alpha} X^*(1 - X^*) - X^* \Delta G_2^{o\alpha \to L} = a_2^{L} X^*(1 - X^*) + (1 - X^*)\Delta G_1^{o\alpha \to L}$$

Rearranging

$$(a_0^{\alpha} - a_0^{L})X^*(1 - X^*) = (1 - X^*)\Delta G_1^{o\alpha \to L} + X^* \Delta G_2^{o\alpha \to L}$$

Apply Eq. (10.26) to estimate the free energies of the pure components:

$$(a_0^{\alpha} - a_0^{L})X^*(1 - X^*) = (1 - X^*)\Delta S_1^{o\alpha \to L}(T_{01} - T^*)$$

$$+ X^* \Delta S_2^{o\alpha \to L}(T_{02} - T^*)$$

Introduce $\Delta a_0 = a_0^{L} - a_0^{\alpha}$ and simplify the notation for the entropy factors:

$$-\Delta a_0 X^*(1 - X^*) - (1 - X^*)T_{01}\Delta S_1^o - X^* T_{02}\Delta S_2^o = -T^* \left[(1 - X^*)\Delta S_1^o + X^* \Delta S_2^o\right]$$

Solve for T^* to obtaining the equation for the midrib curve

$$T^* = \frac{\Delta a_0 X^*(1 - X^*) + (1 - X^*)T_{01}\Delta S_1^o + X^* T_{02}\Delta S_2^o}{(1 - X^*)\Delta S_1^o + X^* \Delta S_2^o} \quad (10.45)$$

As a check, examine the behavior as $X^* \to 0$. Terms with X^* as a factor are zero and $(1 - X^*) \to 1$; $T^* \to T_{01}$ as $X^* \to 0$, as required by the phase diagram. Similarly, as $X^* \to 1$, $T^* \to T_{02}$. Thus, the midrib curve passes through the melting points at the sides of the diagram (or more generally for any $\alpha + \beta$ field, the transformation temperatures for the pure components).

To identify the limiting conditions for formation of a two phase field with a maximum or minimum, it is useful to examine the slope of the midrib curve at the sides of the phase diagram. The derivative of Eq. (10.45) with respect to X^* yields

$$\frac{dT^*}{dX^*} = \frac{\Delta a_0 \left[(1 - X^*)^2 \Delta S_1^o - (X^*)^2 \Delta S_2^o \right] + [T_{02} - T_{01}] \Delta S_1^o \Delta S_2^o}{\left[(1 - X^*) \Delta S_1^o + X^* \Delta S_2^8 \right]} \tag{10.46}$$

The slope of the $T^*(X^*)$ curve at the sides of the diagram can be found by examining the limiting cases. As $X^* \to 0$, $T^* \to T_{01}$, the melting point on the pure 1 side of the diagram. Equation (10.46) simplifies for this limiting case to

$$\frac{dT^*}{dX^*} = \frac{\Delta a_0 + \Delta S_2^o \Delta T_0}{\Delta S_1^o} \tag{10.47}$$

where $\Delta T_0 = T_{01} - T_{02}$ that is, the difference between the melting points of the pure components. Similarly, in the limit as $X^* \to 1$, $T^* \to T_{02}$, and the slope is given by

$$\frac{dT^*}{dX^*} = \frac{-\Delta a_0 + \Delta S_1^o \Delta T_0}{\Delta S_2^o} \tag{10.48}$$

Retain the convention that the component with the higher melting point is selected to be component 1. If the slope of the midrib curve at $X^* = 0$ is negative, then the two phase field is either monotonic or has a minimum. However, if the slope at $X^* = 0$ is *positive*, then the resulting two phase field must exhibit a maximum. Thus the condition that defines a two phase field with a maximum in this model is

$$\Delta a_0 = a_0^L - a_0^\alpha > \Delta S_2^o \Delta T_0 \tag{10.49}$$

Examine the sign of the slope of the midrib curve at $X^* = 1$. If this slope is negative, then the two phase field can either be monotonic or exhibit a maximum. However, if the slope is *positive* at $X^* = 1$, then the two phase field must exhibit a minimum. The numerator in Eq. (10.48) must be positive, or

$$\Delta a_0 = a_0^L - a_0^\alpha < -\Delta S_1^o \Delta T_0 \tag{10.50}$$

It is concluded that, at least in the simplest regular solution model, a comparison of the difference in solution parameter values with the entropies of fusion of the pure components determines whether a two phase field is monotonic or exhibits an extremum. Figure 10.10 is an example of a phase diagram computed from this model with parameters that satisfy inequality (10.50).

The midrib curve also comes into play in interpreting some kinetic phenomena. Focus on the portion of a two phase field and its $(G\text{-}X)$ curves sketched in Fig. 10.23. A sample of composition X' is initially heated to T_a and brought to equilibrium as the stable β phase. During processing this sample is quenched to T_1 and held at that temperature. When it arrives at T_1, still in the β phase form, its thermodynamic condition is given by the point P on Fig. 10.23. The Gibbs free energy of the equilibrium structure at T_1 is the two phase mixture given by the point R. However, it is possible for the metastable condition labelled Q to form spontaneously from the starting point at P since Q has a lower free energy than P. Further, since the equilibrium two phase mixture is made up of two solutions with compositions different from X', formation

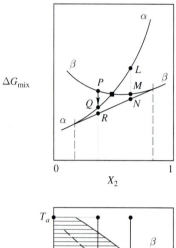

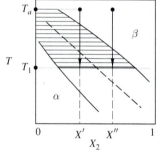

FIGURE 10.23
Role played by the crossing point of two *G-X* curves in delimiting possible transformation sequences during processing.

of this state requires diffusion of the components through the solid phases to change their compositions. Solid state diffusion is a relatively slow process. In contrast, formation of the state labelled Q requires no composition change but only a change in crystal structure from β to α. Depending upon the kinetics of this process, it may thus be possible to form α of the same composition as β and the final structure may then form from this intermediate state with a microstructure entirely different from that which may form by the direct decomposition of β.

In contrast, if the starting alloy composition were X'', *on the other side of the midrib point*, the as-quenched condition at T_1 is labelled M. The α phase of the same composition as this quenched in β phase has a *higher* free energy given by L, which lies above M. Evidently α of the same composition X'' *cannot form* directly from β for this composition and the transformation can only proceed directly to the two phase mixture labelled N. It is evident that the midrib separates a two phase field into two regions, one in which a structural change without a composition change may occur on the way to the stable structure, and a second region in which this possibility is ruled out.

The midrib curve locates the two phase field in (X_2, T) space, gives its curvature, and reveals whether it is monotonic or has a maximum or minimum. Equation (10.45), which describes this curve for the simplest regular solution model, can be derived and plotted without resorting to numerical analysis. This result is easily generalized to include more sophisticated solution models.

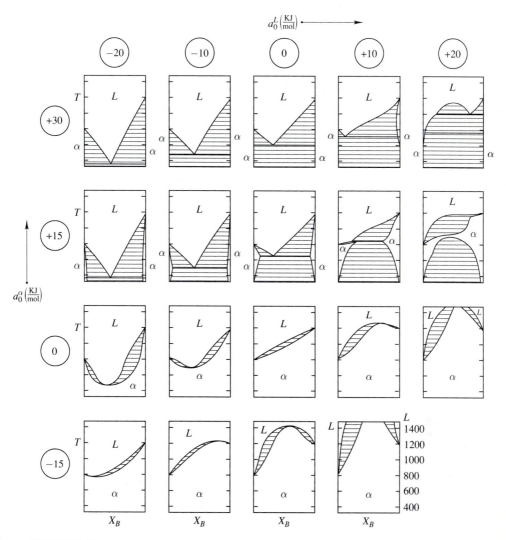

FIGURE 10.24
Pattern of phase diagrams that can be generated from only two phases, α and L, with the simplest regulation solution model.

10.2.4 Pattern of Regular Solution Phase Diagrams with Two Phases

The pattern of phase diagrams that can be generated from two phases α and L in the simplest regular solution model is laid out in Fig. 10.24. All the diagrams in this figure have the same melting points and entropies of fusion for the pure components. The pattern that develops is entirely due to variations in the mixing parameters a_0^α and a_0^L. The top row of diagrams has a fixed positive value of a_0^α; and each diagram thus has a miscibility gap in the solid solution with a known critical temperature T_c. The diagrams in the bottom row are calculated with a small value of a_0^α. The value of Δa_0 chosen for the first column of diagrams satisfies inequality (10.50) and produces an $(\alpha + L)$ field with a minimum. The middle column has Δa_0 values that produce monotonic $(\alpha + L)$ fields and the last column has a value of Δa_0 that produces a maximum in the $(\alpha + L)$ field. The requirement that $a_0^L > a_0^\alpha$ in Fig. 10.24 implies that the liquid phase also has a miscibility gap with a critical temperature above that for the solid phase. It is evident that a wide variety of phase diagrams can be produced with only two phases and their miscibility gaps by this simplest regular solution model.

10.2.5 Diagrams with Three or More Phases

Because this model is capable of developing miscibility gaps and two phase fields with extrema, the variety of phase diagrams that can be produced when a third phase is added to the model is too broad to be discussed systematically in this context. As in the case of an ideal solution model, the strategy for calculating a diagram with three or more phases is based upon the computation of all the possible two phase fields that can appear in the structure by applying Eq. (10.38) and (10.39). Intersections of these two phase fields produce the three phase invariant equilibrium lines in the diagram. The problem of sorting out which of these equilibria are stable and which are metastable is nontrivial and increases in complexity as the number of phases in the system increases. In computer calculations of phase diagrams, this problem is circumvented by applying a procedure based directly on the principle of minimization of Gibbs free energy rather than on the conditions for equilibrium that are derived from it. This strategy is discussed in Sec. 10.5.

   An important and challenging problem appears when additional phases must be incorporated into phase diagram models. Because it is essential that the reference states for each component must be the same in all phases, it is necessary to obtain estimates of the stabilities of all the phase forms in the system relative to the reference phase for each pure component. Consider a system that contains four phases: liquid L, face centered cubic α, body centered cubic β, and hexagonal ϵ. Suppose, at a given temperature, pure component 1 in the α (FCC) phase form is chosen as the reference state for component 1. To develop a (G-X) diagram at the given temperature, it is necessary to obtain estimates of the free energies of pure component 1 for L, β, and ϵ relative to the α phase. The same information must be estimated for pure component 2. This estimate is straightforward for the liquid phase, applying the approach leading to Eq. (10.26). However, it may be that pure component 1 does not exist in the β (BCC) or ϵ (HCP) forms at *any* temperature and pressure. Consider, for example, the aluminum-zinc system. Pure aluminum is FCC (α); the terminal phase on the zinc-rich side is HCP (ϵ). Calculation of the (G-X) diagram for the HCP (ϵ) phase at any temperature T requires an estimate of the free energy difference between hexagonal and face centered cubic forms of pure aluminum at temperature T. However, aluminum does not exist in the hexagonal form at any temperature and pressure. Thus an extrapolation from a region in which pure aluminum is stable in this phase form is not possible. Alternate strategies must be devised to obtain this information. The most common approach employs experimental measurements of phase boundaries and inverts the model calculation to obtain information about ΔG_k^o, ΔS_k^o, and T_k^o for transformations that cannot occur in pure component k.

Example 10.1. Use of phase boundary information to estimate $\Delta G_{Zn}^{o\alpha \to \epsilon}$ at 1200 K. Figure 10.25 shows a portion of the copper-zinc phase diagram. Assume for the purpose of this example that both the liquid and α phases form ideal solutions. The following information is known:

Cu: $\qquad T_m = 1356K \qquad \Delta S_{Cu}^{o\alpha \to L} = 9.59 \qquad$ J/mol-k

Zn: $\qquad T_m = 692K \qquad \Delta S_{Zn}^{o\epsilon \to L} = 9.64 \qquad$ J/mol-k

At 1200 K the compositions read from the phase diagram are $X_{Zn}^{\alpha} = 0.27$ and $X_{Zn}^{L} = 0.34$. In the ideal solution model the ratio of these two compositions is simply related to $K_{Zn}(1200K)$, [see Eq. (10.20)].

$$\frac{X_{Zn}^{L}}{X_{Zn}^{\alpha}} = K_{Zn} = e^{-(\Delta G_{Zn}^{o\alpha \to L}/RT)}$$

Solution. Solve for

$$\Delta G_{Zn}^{o\alpha \to L} = -8.314 \cdot 1200 \cdot \ln\left(\frac{0.34}{0.27}\right) = -2300 \text{ J/mol}$$

Apply Eq. (10.26) to estimate $\Delta G_{Zn}^{o\epsilon \to L}$ at 1200 K

$$\Delta G_{Zn}^{o\epsilon \to L} = \Delta S_{Zn}^{o\epsilon \to L}(T_{MZn} - T) = 9.64(692 - 1200) = -4900 \text{ J/mol}$$

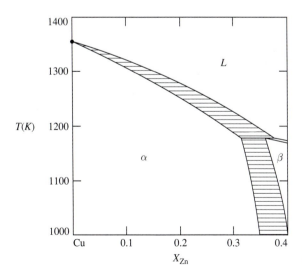

FIGURE 10.25
Portion of the copper-rich side of the copper-zinc phase diagram at high temperatures.

Note that

$$\Delta G_{Zn}^{\alpha \to \epsilon} = G_{Zn}^{o\epsilon} - G_{Zn}^{o\alpha}$$

$$= G_{Zn}^{o\epsilon} - G_{Zn}^{oL} + G_{Zn}^{oL} - G_{Zn}^{o\alpha}$$

$$= -\Delta G_{Zn}^{o\epsilon \to L} + \Delta G_{Zn}^{o\alpha \to L} =$$

$$\Delta G_{Zn}^{o\alpha \to \epsilon} = -(-4900) + (-2300) = +2600 \text{ J/mol}$$

which provides an estimate of the Gibbs free energy of FCC zinc relative to HCP zinc at 1200 K, although FCC zinc never forms in the laboratory.

This example is based on the simplest of phase diagram models in which both solutions are assumed to be ideal. For the phases involved, the strategy can be applied through more sophisticated models invoking more points on the two phase $(\alpha + L)$ field to provide a statistically based estimate of this quantity and others like it. Once a valid estimate of the energetics of FCC zinc is obtained from this diagram, and perhaps some others in which zinc also dissolves in FCC phases, then this result can be used in all phase diagrams that have zinc as a component and contain an FCC phase, including ternary and higher order systems.

10.2.6   Modelling Phase Diagrams with Line Compounds

Figures 10.19 through 10.21 illustrate the role played by intermediate phases in determining the structure of phase diagrams. The composition range of such phases is frequently too narrow to be resolved on a phase diagram and the single phase field for such a phase plots as a vertical line on the phase diagram. In this case it is usually sufficient to represent the $(G\text{-}X)$ curve for the line compound by a single point which plots the free energy of formation of the compound from the reference states for the components in the diagram at the composition of the line compound, Fig. 10.20.

It is crucial to note that in the remainder of the $G\text{-}X$ curve calculations in the system, the free energy is given in units of energy per gram atom of solution. In tabulated information, free energies of formation of compounds are generally reported as energy per mole of the compound formed. If the compound has the formula $M_u X_v$, then one mole of the compound contains $(u + v)$ gram atoms of M and X atoms. Thus the free energy value required for representation on a $(G\text{-}X)$ diagram is the free energy of formation ΔG_f^o computed from tabulated values *divided by* $(u + v)$. The value of G^ϵ corresponding to the point P in Fig. 10.20 is determined from the formation reaction for that compound.

$$2A + B = A_2 B \qquad \Delta G_f^o$$

$$G^\epsilon = \frac{\Delta G_f^o}{2 + 1} = \frac{\Delta G_f^o}{3}$$

As an illustration of the treatment of line compounds, suppose that the terminal α phase in the system shown in Fig. 10.20 is dilute and obeys the simple regular solution model. The chemical potentials of the components in a dilute solution are given by

Raoult's Law: $\qquad\qquad \Delta\mu_1^\alpha = RT \ln(1 - X_2^\alpha)$ (10.51)

Henry's Law: $\qquad\qquad \Delta\mu_2^\alpha = RT \ln \gamma_2^{o\alpha} + RT \ln X_2^\alpha$ (10.52)

where X_2 is the mole fraction of component 2 and γ^o is the Henry's law constant for the solution. In the simple regular solution model it has been shown that $RT \cdot \ln \gamma_2^o = a_0$ [see Eq. (8.119)]. Thus, in this model Henry's law becomes

Henry's Law: $\qquad\qquad \Delta\mu_2^\alpha = a_0 + RT \ln X_2^\alpha$ (10.53)

Equations (10.51) and (10.53) identify the intercepts at $X_2 = 0$ and $X_2 = 1$ of a line tangent to the $(G\text{-}X)$ curve for the α phase at any composition X_2 in its dilute range. The equation of the tangent line to the $(G\text{-}X)$ curve for the α phase can be obtained by recalling from analytical geometry the equation of a line that passes through two given points, (x_1, y_1) and (x_2, y_2)

$$(y - y_1) = \frac{y_2 - y_1}{x_2 - x_1}(x - x_1)$$

For the tangent line with intercepts given by Eq. (10.51) and (10.53), $x_1 = 0$, $x_2 = 1$ and y_1 and y_2 are given by these equations. Substitute these values into the equation for the straight line and rearrange

$$y = [a_0 + RT \ln X_2 - RT \ln(1 - X_2)]x + RT \ln(1 - X_2)$$

Since the solution is assumed to be dilute, $X_2 \ll 1$, $(1 - X_2) \cong 1$ and $\ln(1 - X_2) \cong \ln(1) = 0$. Thus,

$$y = [a_0 + RT \ln X_2]x$$ (10.54)

This is the equation of a line tangent to the $(G\text{-}X)$ curve for the α phase in its dilute range, touching the curve at the composition X_2.

The value of $X_2 = X_2^\alpha$ that describes the phase boundary on the α side of the $(\alpha + \epsilon)$ field in Fig. 10.20, lies on the particular tangent that also passes through the point P, which corresponds to $y = G^\epsilon$ and $x = X_c$, the mole fraction of component 2 in the ϵ phase. Substitute these values for y and x in Eq. (10.54)

$$G^\epsilon = [a_0 + RT \ln X_2^\alpha] X^\epsilon$$

Solve for the composition of the phase boundary

$$\ln X_2^\alpha = \frac{1}{RT}\left(\frac{G^\epsilon}{X^\epsilon} - a_0\right)$$ (10.55)

This is the equation for the phase boundary on the α side of the $(\alpha + \epsilon)$ field, known as the *solvus line* for the terminal phase α.

Apply the definitional relationship to the free energy of formation of the compound

$$\Delta G_f^\epsilon = \Delta H_f^\epsilon - T \Delta S_f^\epsilon$$

where ΔH_f^ϵ is the heat of formation of the compound and ΔS_f^ϵ is the entropy of formation of one mole of the compound. The corresponding values *per gram atom* of ϵ formed

$$G^\epsilon = H^\epsilon - TS^\epsilon$$

are obtained by dividing ΔH_f and ΔS_f by the number of gram atoms in one mole of the compound. Then

$$\ln X_2^\alpha = \frac{1}{RT}\left(\frac{H^\epsilon - TS^\epsilon}{X^\epsilon} - a_0\right)$$

$$= \left(\frac{H^\epsilon - a_0 X^\epsilon}{X^\epsilon R}\right)\frac{1}{T} - \frac{S^\epsilon}{X^\epsilon R}$$

where H^ϵ and S^ϵ are values *per gram atom* of compound formed. The solubility limit in the α phase thus varies with temperature according to the relation

$$X_2^\alpha = Ae^{(B/RT)} \tag{10.56}$$

where

$$A = e^{-(S^\epsilon / X^\epsilon R)}$$

and

$$B = \frac{H^\epsilon - a_0 X^\epsilon}{X^\epsilon}$$

Heats of formation for most compounds are negative and the associated reaction is exothermic. Thus B is negative in Eq. (10.56) and the solubility X_2^α increases with temperature. Further, the more stable the compound, or the more negative its free energy of formation, the smaller will be the solubility limit in the terminal α phase.

As the temperature is raised an intermediate compound may either *melt congruently*, as does the ϵ phase in Fig. 10.21, or *decompose* into two other phases, as in Fig. 10.20. The condition that defines a congruent melting point is depicted in Fig. 10.26. With increasing temperature the free energy of formation of the compound

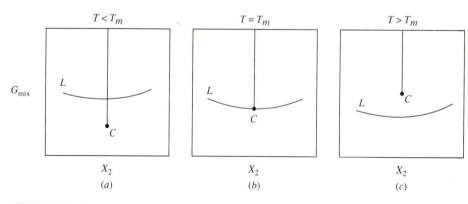

FIGURE 10.26
Condition for congruent melting of an intermediate line compound.

becomes progressively less negative. Simultaneously, the liquid $(G\text{-}X)$ curve descends in the diagram as the liquid phase increases its stability. At the congruent melting point, Fig. 10.26b, the free energy of formation of the compound is equal to the free energy of mixing of the liquid phase.

$$\Delta G^L_{\text{mix}} = G^\epsilon = H^\epsilon - T_{mc} S^\epsilon$$

$$\Delta G^{Lxs}_{\text{mix}} + RT_{mc}\left[(1 - X^\epsilon)\ln(1 - X^\epsilon) + X^\epsilon \ln X^\epsilon\right] = H^\epsilon - T_{mc} S^\epsilon$$

This assumes that the reference states for the evaluation of the heat and entropy of formation of the compound are the pure liquid phases for the components. If these quantities are referred to the pure solid states, then additional terms must be added to bring the reference states for the components into agreement. Assume that the excess free energy of mixing is not a function of temperature. Solving for the melting point

$$T_{mc} = \frac{H^\epsilon - \Delta G^{L\,\text{xs}}_{\text{mix}}}{R\left[(1 - X^\epsilon)\ln(1 - X^\epsilon) + X^\epsilon \ln X^\epsilon\right] + S^\epsilon} \tag{10.57}$$

The excess free energy in the liquid phase may be positive or negative; the remaining terms in this equation are negative. One conclusion readily obtained from this result is: the larger (more negative) the heat of formation of the compound, the higher will be its congruent melting point.

Figure 10.27 shows a phase diagram for a system in which the terminal solid solutions are dilute and all the intermediate phases appear as line compounds, along with a sketch of the $(G\text{-}X)$ diagram constructed at T_1. The entire diagram can be generated from a model for the liquid phase and enthalpies and entropies of the compounds, so that ΔG_f can be estimated as a function of temperature for each phase. This kind of phase diagram is typical of systems treated in chemistry where the participating phases are essentially stoichiometric in composition.

10.3 THERMODYNAMIC MODELS FOR THREE COMPONENT SYSTEMS

The strategies developed for modelling binary phase diagrams extend directly to the description of three component systems. Four variables (e.g., T, P, X_2, and X_3,) are required to describe the state of a single phase. Thus the $(G\text{-}X)$ curve that represents the mixing behavior of binary systems at some chosen temperature and pressure expands to a $(G\text{-}X_2\text{-}X_3)$ surface for a phase with three components, Fig. 10.28. The typical $(G\text{-}X\text{-}X)$ surface is a distorted paraboloid hanging over the Gibbs triangle from three points that represent the free energies of the pure components relative to their chosen reference states. Curves of intersection with the sides of the prism in Fig. 10.28 are $(G\text{-}X)$ curves for the corresponding binary systems.

The tangent line-intercept construction shown in Fig. 8.1 gives a graphical visualization of the partial molal properties for the components in a binary system at any composition. An analogous construction applies to surface plots of ΔB_{mix} versus composition in three component systems, Fig. 10.28. To find the partial molal properties of each of the components for any composition P construct the *tangent plane*

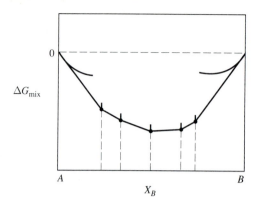

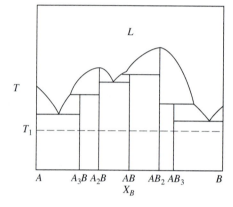

FIGURE 10.27
Typical *G-X* curve for a system with very limited terminal solubility and a number of intermediate line compounds.

to a point P' on the surface at that composition. The intercepts of that tangent plane on the three axes are the corresponding partial molal properties.

The competition between two different phase forms for stability can be represented by superimposing the two $(G\text{-}X\text{-}X)$ surfaces characteristic of these phases with the reference states for the three components carefully chosen to be the same in both phase forms, Fig. 10.29. The conditions for chemical equilibrium between two phases in a three component system require that

$$\mu_1^\alpha = \mu_1^\beta \qquad \mu_1^\alpha = \mu_2^\beta \qquad \mu_3^\alpha = \mu_3^\beta \tag{10.58}$$

If the reference states are consistently chosen, the conditions may be restated

$$\Delta\mu_1^\alpha = \Delta\mu_1^\beta \qquad \Delta\mu_2^\alpha = \Delta\mu_2^\beta \qquad \Delta\mu_3^\alpha = \Delta\mu_3^\beta \tag{10.59}$$

Thus if P^α represents a point on the α surface and P^β represents a point on the β surface, this pair of points represents a pair of compositions that can exist in equilibrium if and only if the tangent plane at P^α and that at P^β are the *same plane*. If this is true for the pair of points, then the intercepts on the three axes are the same for P^α as for P^β and the conditions for two phase equilibrium is satisfied. Thus the *common tangent construction* devised for binary systems extends to ternary two phase equilibria. However, the tangent becomes a tangent *plane* that touches both surfaces.

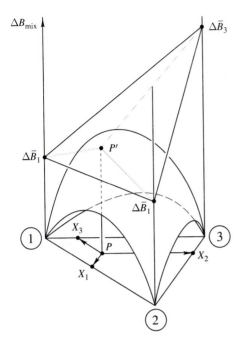

FIGURE 10.28
Mixing behavior of ternary solutions is represented by a surface over the Gibbs triangle. Tangent plane construction yields the partial molal properties of the three components corresponding to any composition in a ternary system.

The pair of points P^α and P^β project to ends of a tie line on the Gibbs triangle. As the common tangent plane rolls over the two surfaces, it identifies a continuous series of pairs of tangent points that project to form the two phase field on the composition plane.

Because the constructions involved are three dimensional, this graphical visualization of the competition between phase forms for stability in ternary systems is not as useful as is the $(G\text{-}X)$ construction for binary systems. Nonetheless, the construction is available and may provide a useful thinking tool in some applications.

Equation (10.58) forms the basis for calculating two phase fields in ternary systems from solution models for the phases involved. Since, for any component in a multicomponent system,

$$\mu_k^I = G_k^{oI} + \Delta\overline{G}_k^{xsI} + RT \ln X_k^I \qquad (I = \alpha, \beta; \quad k = 1, 2, 3) \qquad (10.60)$$

The conditions for equilibrium yield three equations of the form

$$\Delta G_k^{o\alpha \to \beta} + \left[\Delta\overline{G}_k^{xs\beta} - \Delta\overline{G}_k^{xs\alpha}\right] + RT \ln \left[\frac{X_k^\beta}{X_k^\alpha}\right] = 0 \qquad (k = 1, 2, 3) \qquad (10.61)$$

The solution model can be used to evaluate the excess free energies in terms of composition and temperature. The free energy change for the phase transformation for the pure components can be computed for each of the unary systems. This system of three equations relates four variables, X_2^α, X_3^α, X_2^β, and X_3^β; thus, there remains one degree of freedom. In order to compute the set of composition pairs that delineate the

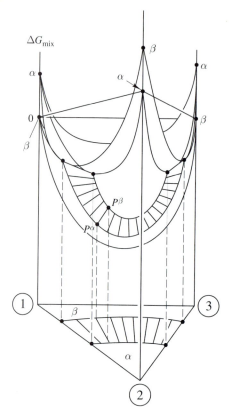

FIGURE 10.29

The common tangent plane construction describes a pair of points on two different G-X-X surfaces that satisfy the conditions for two phase equilibrium.

$(\alpha + \beta)$ equilibria, it is necessary to choose a value for one of these four variables, say X_2^α. Then using Eqs. (10.61) compute values for the other three corresponding to the chosen value of X_2^α. An iterative equation solver routine available in standard mathematics applications software can be used for this purpose. The value of X_2^α is then incremented and the calculation repeated until the full collection of tie lines that constitute the two phase field is generated.

The principle of minimization of the Gibbs free energy to determine the domains of stability of the phases also applies in ternary systems. The taut string construction is replaced by a taut elastic sheet construction. The free energy surfaces are considered rigid surfaces hanging from appropriate fixed points on the unary axes. An elastic sheet is attached at the three corners of the diagram and pulled taut against the free energy surfaces. Where it contacts the surface of a particular phase, that phase has the minimum free energy and is stable. Between the surfaces the elastic sheet traces out the common tangent plane construction that represents two phase equilibria. Where the sheet is stretched between three free energy surfaces, a three phase tie triangle represents the condition for equilibrium. This visualization is also useful in systems with line compounds so that the one phase regions are essentially limited to points on the Gibbs triangle, Fig. 10.30.

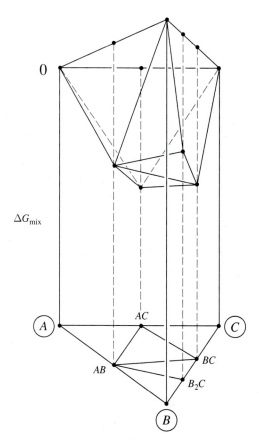

FIGURE 10.30
The taut elestic sheet construction gives the domains of stability of one, two, and three phase combinations.

10.4 CALCULATION OF PHASE DIAGRAMS IN POTENTIAL SPACE

If the coordinates used to represent the states of the phases in a system are chosen from the thermodynamic potentials, then the resulting phase diagram is a simple cell structure. The counterpart to the common binary diagram plotted in (X_2, T) space is the isobaric diagram in (a_2, T) space. The thermodynamic information necessary to compute a binary diagram in potential space is identical with that required for the more familiar (X_2, T) diagrams modelled in Sec. 10.2. Since these calculations require models for the relationship between chemical potential and composition, corresponding relations between composition and activity are implicit. No new information is required to compute an (a_2, T) diagram.

In (a_2, T) space every two phase field is represented by a single curve that sets the limits of stability for the phases it separates. The quantitative form of that curve depends on the choice of the reference state used in defining the activity of component 2. Figure 10.31 illustrates this point for a simple ideal solution phase diagram model. Figure 10.31a shows the diagram in (X_2, T) space.

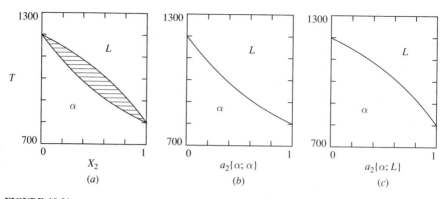

FIGURE 10.31
Ideal solution model of a phse diagram (*a*) in (X_2, T) space and (*b*) and (*c*) in (a_2, T) space. Reference states for a_2 are the α phase form in (*b*), and the L phase form in (*c*).

First, consider the case in which the α phase form is chosen as the reference state for component 2. Then, in the α phase solid solution region, according to the ideal solution model, $a_2^\alpha = X_2^\alpha$. The identical simple relation does not hold for the L phase because, in order for a_2 to be equal to X_2 in an ideal solution, the reference state for component 2 must be the same phase form as the solution, L in this case. The equation describing the α boundary of the $(\alpha + L)$ field has the form $X_2^\alpha = X_2^\alpha(T^\alpha)$ given by Eq. (10.21). Since for this choice of reference state $a_2^\alpha = X_2^\alpha$, the same equation describes the condition for $(\alpha + L)$ equilibrium in the (a_2, T) phase diagram shown in Fig. 10.31*b*. Thus, for this ideal solution model, the curve representing the $(\alpha + L)$ equilibrium for this choice of reference state for component 2 coincides with the curve on the α side of the $(\alpha + L)$ field in Fig. 10.31*a*.

The same argument demonstrates that if the L phase form is chosen as reference state for component 2 at all temperatures, the two phase field in the (a_2, T) diagram, Fig. 10.31*c*, is identical with the curve on the L side of the $(\alpha + L)$ field in Fig. 10.31*a*.

These simple identities between an appropriate phase boundary on the (X_2, T) diagram and the two phase field on the (a_2, T) diagram exist in this case because the system is an ideal solution model for which, with appropriate choice of reference state, the activity is equal to the atom fraction. More generally, a_2 and X_2 are functionally related within each phase but are not simply equal to each other. Perhaps the most straightforward strategy for obtaining a two phase field curve in an (a_2, T) diagram is to compute the composition-temperature diagram as outlined in Sec. 10.2. Then use the model relationship between activity and composition to convert one of the $X_2(T)$ curves to the corresponding $a_2(T)$.

In constructing phase diagrams in potential space, it is important to be explicit about the choice of the reference state for the component represented on the activity axis and to maintain that choice in modeling the mixing behavior of all of the phases on the diagram. Since three, phase equilibria are formed by the intersection of three, two phase equilibria, in (a_2, T) space, they appear as the point of intersection of the three curves representing the two phase equilibria involved. This is further evident from the observation that when three phases coexist, one of the conditions for equilibrium requires that the chemical potential of component 2 be the same in all three phases. Since the reference states are consistently chosen to be the same for all of the phases, this condition implies that the activity of component 2 is the same in all three phases at the triple point. Thus three phase equilibria appear as triple points where the three related two phase equilibrium curves intersect.

10.5 COMPUTER CALCULATIONS OF PHASE DIAGRAMS

The thermodynamic underpinnings of phase diagrams developed in this chapter and in Chap. 9 provide the basis for comprehensive computer calculations of binary, ternary, and higher order phase diagrams. This subject is so central to the field of materials science that a monthly journal, *Bulletin of Alloy Phase Diagrams*, focused on computer calculations of phase diagrams, has been published for a number of years. A variety of compilations of phase diagrams [10.2–10.8] exist for metallic and ceramic systems. Early versions of these compilations were based entirely on direct experimental observation of the phases that exist at equilibrium at a sequence of temperatures and compositions that blanket the $(T\text{-}X_2)$ space. More recent compilations use flexible thermodynamic models for phase diagrams to integrate both direct observations of phase equilibria and thermodynamic measurements in expanding data bases to arrive at an optimized phase diagram. Perhaps the most comprehensive of these programs is offered through the services of F*A*C*T* (Facility for the Analysis of Chemical Thermodynamics), which can be accessed on-line at the École Polytechnique, CRCT, Montreal, Quebec, Canada.

Through appropriately flexible computer models that let the user assign weighting factors to the relative importance of various forms of input information, it is possible to develop a diagram for any given system that provides a best fit with all the available data. As experience and assessments accumulate, this information is being incorporated into comprehensive data bases that strive to develop consistent descriptions for the thermodynamics of the elements, solid and liquid solutions, and intermediate compounds. For example, the thermodynamic information for pure nickel that is part of the input information in the calculation of the iron-nickel phase diagram must be the same information that provides part of the input to compute the aluminum-nickel diagram or the copper-nickel diagram, etc. Similarly, the same model that produces the binary copper-nickel phase diagram as part of the input to compute the Cu-Ni-Au ternary diagram must also form the basis for computing the Cu-Ni-Fe system.

An early comprehensive study of phase diagrams that applied computer calculations based upon thermodynamic models was carried out by Kaufman and Bernstein [10.9] in the late 1960s. His text provides a comprehensive introduction to the subject, carrying the reader systematically through unary systems, binary systems of increasing complexity, including line compounds, and ultimately to ternary systems. Equation (10.26) is assumed to describe the behavior of the pure components. The single parameter regular solution model describes mixing behavior so that Eqs. (10.38) and (10.39) define the phase boundaries for each two phase field. An algorithm based on atom sizes and electronegativities was devised to estimate values for the mixing parameters for each type of solution. Typical results of this study are displayed in Fig. 10.32. A pair of elements that are adjacent to each other in the periodic table produce a simple diagram that may be reasonably represented in this model. As the separation between elements in the periodic table increases, the diagram increases in complexity and the description with this simple model becomes unsatisfactory. The situation can be improved by adding parameters to the description of the pure components and their mixing behavior.

A complete phase diagram can be modelled by applying Eq. (10.32) to compute the phase boundaries of all possible existing two phase combinations. This strategy becomes cumbersome in systems with more than a few phases because, at any temperature, most of the possible combinations of two phases are metastable. The problem of determining which two phase fields actually appear on the diagram is nontrivial. Accordingly, most computer based calculations of phase diagrams are based on the principle of minimization of the Gibbs free energy and

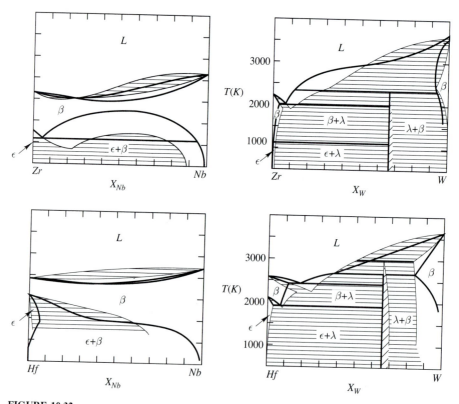

FIGURE 10.32

Representative examples of phase diagrams computed with Kaufman's strategies, comparing experimentally determined (lightly sketched, with filled two phase fields) and calculated (heavily drawn lines) diagrams.

the taut string construction rather than iterative solutions of the chemical potential equations.

Input for the taut string analysis is the collection of (G-X) curves of all the phase forms known to exist in the system at the temperature of interest. Figure 10.33 sketches such a collection of (G-X) curves for a system with four phases: α, β, ϵ, and liquid (L). In modeling these curves, care has been taken to choose consistent reference states for the components: $\{\alpha; L\}$. Functions describing the tangent line to each of these (G-X) curves are also compiled. The composition axis is divided into n equal intervals spaced a distance $\delta X = 1/n$ apart. Values of ΔG_{mix} are computed for each phase at each of these n composition values and stored in a table. Beginning at $X = 0$, determine the phase with the lowest free energy. This is the stable phase form for pure component 1; suppose it is the α phase. Increment to $X = 0 + \delta X$ and compute the tangent line to the α phase (G-X) curve at this composition. At each of the (n-1) values of X between δX and 1 compare the value of G on the tangent line with all the values of G tabulated at that composition for all the phases. If none are lower than the tangent line value, α still has the lowest free energy at this composition. Increment the composition to $2\delta X$ and repeat the procedure.

Eventually, for a tangent line constructed at some composition X', it is found that the G value for one of the other phases, ϵ in Fig. 10.33, lies below that computed

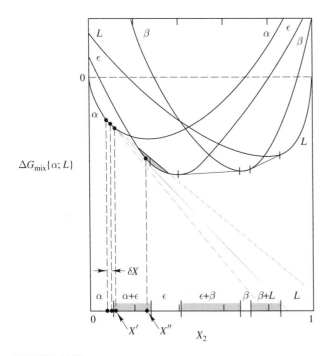

FIGURE 10.33
Illustration of one strategy for applying the taut string construction to determine the domains of stability of the phases at some temperature T_1.

for the tangent line at some composition X''. This identifies, within the accuracy of the increment δX, the limiting compositions of the $(\alpha + \epsilon)$ two phase field and guarantees that this $(\alpha + \epsilon)$ equilibrium is one of the stable two phase fields on the diagram. Thus, it is not necessary to explore compositions in the range between X' and X''. The function describing the tangent line to the ϵ phase is now called and X is incremented to $X'' + \delta X$. The value of G along the tangent line is again computed and compared with G values for all of the phases at each value of X in the interval from X'' to 1. If none are lower than the tangent line value, ϵ is the stable phase at $X'' + \delta X$ and the tangent line at the next composition, $X'' + 2\delta X$, is examined. X is incremented until another stable phase is found and the procedure is repeated until the full range of compositions is scanned. Accuracy can be improved either by increasing n or by increasing the resolution of the composition scale near phase boundaries that have been identified at low resolution in X. Once the sequence of stable phases is established at the temperature T, the temperature is incremented to $T + \delta T$ and the procedure repeated until the entire stable phase diagram is established. This kind of brute force approach, involving very large numbers of simple calculations or numerical comparisons, is very well suited to digital computers.

Extension of this algorithm to ternary and higher order systems is straightforward. In a ternary system, the $(G\text{-}X)$ curve computed from the thermodynamic solution models become $(G\text{-}X\text{-}X)$ surfaces over the Gibbs triangle. The tangent line is replaced by a tangent plane at any given composition and the composition range scanned becomes an area of the Gibbs triangle. The number of calculations required is roughly one half of the square of the number involved in a binary system calculation. The principle of minimization of the Gibbs free energy at constant temperature and pressure still determines the equilibrium configuration.

The great advantage provided by the development and application of successful thermodynamic models to computer calculation of phase diagrams is the thermodynamic information that this approach yields. With a limited amount of thermodynamic input data and a measured phase diagram, it is possible to devise a set of consistent solution models that describe the thermodynamic behavior of all the phases over the full range of composition and temperature. Although caution must be exercised in applying this information, it provides a basis for estimating thermodynamic properties of phases and two phase equilibria in regions both where they are stable and where they are metastable. This kind of information provides a firm foundation for the analysis of the sequence of processes that might occur in a transformation from an unstable state toward equilibrium as well as the rates of these processes. It provides the framework for answering the fundamental question of materials processing, "What will happen to this material if I do this to it?"

10.6 SUMMARY OF CHAPTER 10

Free energy-composition $(G\text{-}X)$ diagrams provide a graphical basis for visualizing the connection between the thermodynamics of the phases in a system and the phase diagram that it exhibits.

The common tangent construction is a graphical representation of the conditions for equilibrium in a two phase system, expressed either as an equality of the chemical potentials or as a minimization of the Gibbs free energy function.

Relative Gibbs free energies of the pure components combine with mixing behavior of the solutions to determine the location and shape of two phase fields.

Two phase fields identify the boundaries of single phase regions and intersect to form three phase fields. Determination of the two phase fields in a system is sufficient to determine the phase diagram.

At sufficiently low temperatures any phase that exhibits a positive excess free energy exhibits a miscibility gap.

The spinodal region within a miscibility gap, between the inflection points in the $(G\text{-}X)$ curve, is a region in which nonuniform systems are more stable than uniform ones. This attribute provides the basis for spinodal decomposition.

The incorporation of line compounds as intermediate phases can be adequately described by their free energy of formation (*per gram atom of compound*) as a function of temperature.

The ideal solution model can generate only monotonic two phase fields. The pattern of shapes is very limited and determined by the entropies of transformation of the pure components.

Even the simplest, one parameter regular solution model is capable of producing the full variety of phase diagram configurations. Additional terms, including a temperature dependence in the excess free energy, can be added to provide quantitative agreement with measured diagrams.

Extensive computer programs incorporating data bases, a variety of solution models, statistical bases for model parameter selection, and free energy minimization algorithms have been devised to compute binary, ternary, and even higher order phase diagrams.

The establishment of firm connections between the thermodynamics of solutions and phase diagrams, implemented by computer calculations, yields information about phases and equilibria that form the basis for the analysis, and eventually the prediction, of the processes in materials science that control microstructure and properties.

PROBLEMS

10.1. Input for calculation of Fig. 20.2: $T_{01} = 1500$ K, $T_{02} = 850$ K; $\Delta S_1 = 9$, $\Delta S_2 = 7$ (J/mol K); $a_0^\alpha = 8400$, $a_0^L = 10500$ (J/mol). Replot the $(G\text{-}X)$ curves shown in Fig. 10.2 for the following choices of reference states for components 1 and 2:

(a) $\{L; L\}$

(b) $\{L; \alpha\}$

(c) $\{\alpha; \alpha\}$

10.2. The A-B system forms regular solutions in both the β and liquid phase forms. Parameters for the heat of mixing for these phases are $a_0^\beta = -8200$ J/mole and $a_0^L = -10500$ J/mol.

(a) Compute and plot curves for ΔG_{mix} for the β and liquid phases for these models at 800 K.

Pure A melts at 1050 K with a heat of fusion of 8200 J/mol. Pure B melts at 660 K with a heat of fusion of 6800 J/mol.

(b) Change the reference states appropriately to plot and compare $\Delta G^{\beta}_{\text{mix}}\{\beta; L\}$ and $\Delta G^{L}_{\text{mix}}\{\beta; L\}$.

Neglect differences in heat capacities for the phases for the pure components.

10.3. The free energy of mixing (J/mol) of a liquid solution is modelled by

$$\Delta G_{\text{mix}}\{L; L\} = 8400 X_A X_B + RT[X_A \ln X_A + X_B \ln X_B]$$

(a) Compute and plot the activity of component B as a function of composition at 600 K for this solution.

(b) The free energy of fusion of pure B at 600 K is computed to be -1200 J/mol. Compute and plot the activity of B as a function of composition for the choice of reference states $\{L; \alpha\}$.

10.4. Use the taut string construction given for the system in Fig. 10.7 to construct a plot of the chemical potential of component 2 as a function of composition for this system.

10.5. Consider the $(G\text{-}X)$ curve plotted in Fig. 10.5. A mechanical mixture is formed by taking quantities of two solutions with compositions given by the points A and B in this figure so as to give a system with an average composition X_2^o. Prove that the Gibbs free energy of mixing for this two-solution mixture is given by the point C.

10.6. It has been argued that, if the $(G\text{-}X)$ curves for two different phases cross within the diagram, then it is *always* possible to construct a common tangent line that straddles the crossing point. What aspect of the behavior of $(G\text{-}X)$ curves guarantees that *both* tangent points in fact lie within the composition range, $0 < X_2 < 1$?

10.7. Prove that a change in the reference states of either or both of the $(G\text{-}X)$ curves producing a common tangent construction at some temperature *does not alter the compositions* at which the tangents occur.

10.8. Sketch a sequence of $(G\text{-}X)$ curves that produce a *maximum* in an $(\alpha + L)$ two phase field.

10.9. Consider an α solid solution that obeys the "simplest regular solution" model with the interaction parameter $a_0 = 12,700$ J/mol. This solution exhibits a miscibility gap below $T_c = 746$ K. At 650 K,

(a) Compute the compositions of the limits of the spinodal region.

(b) Compute and plot $\Delta \mu_2^{\alpha} = \mu_2^{\alpha} - \mu_2^{o\alpha}$ as a function of composition.

(c) Illustrate the peculiar behavior inside the spinodal region by computing and plotting

$$\frac{d\Delta \mu_2^{\alpha}}{dX_2}$$

as a function of composition.

10.10. Sketch a series of $(G\text{-}X)$ curves that produce a peritectic phase diagram in which both solid phases are the *same* phase form, α, so that the solid-solid two phase field is a miscibility gap.

10.11. Look up the phase diagram for the BaO_2-TiO_2 system. Sketch a plausible $(G\text{-}X)$ diagram for this system at $1400°C$.

10.12. In Fig. 10.27 one of the intermediate compounds forms at the composition AB_2. Sketch the phase diagram in the vicinity of this composition, expanding the composition axis by a factor of, say, 1000, so that the departures from stoichiometry can be displayed.

10.13. Refer again to the phase diagram for the BaO_2-TiO_2 system obtained for Problem 10.11. Suppose that under practical solidification conditions in this system, the $BaTi_3O_7$ phase does not form.

(a) Estimate the composition and temperature of the last liquid to solidify in this system if this phase does not form.

(b) Suppose also that the $BaTi_2O_5$ phase is sluggish to form; then what is the temperature and composition of the last liquid to solidify?

10.14. Use a mathematics application software package to compute and plot a simple ideal solution phase diagram. Write the program so that:

(a) Input is T_1^o, T_2^o, ΔS_1^o, and ΔS_2^o.

(b) Output is a plot of the two phase field.

Fix the melting points. Try several combinations of values of ΔS_1^o and ΔS_2^o and explore the kinds of diagrams that may be developed. Values of these parameters range around 1 J/mol-K for solid-solid transformations, 8 for melting and 90 for vaporization.

10.15. Use the computer program developed in Problem 10.14 as a basis to compute and plot a phase diagram for a system involving three phases, α, β, and L. Identify the stable and metastable portions of this diagram.

10.16. The system A-B obeys the simple regular solution model. A melts at 1352 K with an entropy of fusion of 6.9 J/mol-K; B melts at 1148 K with an entropy of fusion of 8.5 J/mol-K. Neglect differences in heat capacities of the pure solid and liquid phases. For the liquid phase, $a_0^L = -5,600$ J/mol; for the α phase $a_0^\alpha = -11,400$ J/mol. Find the compositions of the boundaries of the two phase ($\alpha + L$) field at 1300 K. (Note: this requires solution of two simultaneous nonlinear equations and standard mathematics applications packages have this feature.)

10.17. Compute and plot the phase boundaries and the spinodal boundaries for a solution that obeys the model

$$\Delta G_{mix}^\alpha = 10,600X_1X_2 + RT[X_1 \ln X_1 + X_2 \ln X_2]$$

10.18. Find the critical temperature for the miscibilty gap found in a regular solution with

$$\Delta H_{mix} = X_1X_2[4,850X_1 + 12,300X_2]$$

10.19. Compute and plot the midrib curves for the ($\alpha + L$) fields for two systems with the following properties:

(a)

	$T_x^o K$	ΔS_k^o (J/mol-K)
Component 1	1283	8.8
Component 2	942	6.3

$a_0^\alpha = 7,280$ J/mol $\quad a_0^L = -2,100$ J/mol

(b)

	$T_x^o K$	ΔS_k^o (J/mol-K)
Component 1	1283	8.8
Component 2	942	6.3

$a_0^\alpha = -4,800$ j/mol $\quad a_0^L = 5,200$ J/mol

10.20. At 1550 K the solubility of oxygen in zirconium is estimated to be 120 ppm. The Zr-O system forms a very stable compound, ZrO_2, an important ceramic refractory. No other oxides are stable at 1550 K. Estimate the Henry's law coefficient for oxygen in zirconium at 1550 K.

10.21. Assuming that the liquid solution of oxygen in silicon is an ideal solution, estimate the melting point of quartz, a crystal form of silica (SiO_2). The free energy of formation of quartz is

$$\Delta G_f^0 = -910,900 + 182 \ T \ \frac{J}{mol}$$

Note carefully that the reference states in this relation are crystal silicon and gaseous oxygen. O_2 boils at 90.2 K with an entropy of vaporization of 75.4 (J/mol K). Compare your estimate with the observed melting point of quartz. Comment on your result.

10.22. Consider the potential composition diagram for the Fe-Cr-O system shown in Fig. 9.16. Replot this diagram on the Gibbs triangle. Plot the tie lines in the two phase fields quantitatively.

REFERENCES

1. Lupis, C.H.P.: *Chemical Thermodynamics of Materials*, Elsevier Science Publishing Co., Inc., New York, N.Y., pp 219—229, 1983.
2. Hansen, M., *Constitution of Binary Alloys*, Second Edition, McGraw-Hill, Inc., New York, N.Y., 1958.
3. Elliott, R.P., *Constitution of Binary Alloys, First Supplement*, McGraw-Hill, Inc., New York, N.Y., 1965.
4. Shunk, F.A, *Constitution of Binary Alloys, Second Supplement*, McGraw-Hill, Inc., New York, N.Y., 1969.
5. Hultgren, R., P.D. Desai, D.T. Hawkins, M. Gleiser, and K.K. Kelley: *Selected Values of the Thermodynamic Properties of Binary Alloys*, ASM, Materials Park, Ohio, 1973.
6. Levin, E. M., C.R. Robbins, and H.F. McMurdie: *Phase Diagrams for Ceramists*, American Ceramic Society, Columbus, Ohio, 1964.
7. Levin, E.M., C.R. Robbins and H.F. McMurdie: *Phase Diagrams for Ceramists, 1969 Supplement*, American Ceramic Society, Columbus, Ohio, 1969.
8. *Bulletin of Alloy Phase Diagrams*, ASM, Materials Park, Ohio, Journal published bimonthly since 1980 in collaboration with the National Institute for Standards and Technology.
9. Kaufman, L., and H. Bernstein: *Computer Calculations of Phase Diagrams*, Academic Press, New York, N.Y., 1970.

MULTICOMPONENT, MULTIPHASE REACTING SYSTEMS

At the heart of the idea of a chemical reaction is the visualization of the chemical *molecule*. A molecule is an arrangement of atoms of some of the elements in the system into a specific geometric and energetic configuration. This configuration is so specific that it can be represented by a chemical formula: CO_2, H_2O, H_2, CH_4, Al_2O_3, and HNO_3 come quickly to mind as examples that describe succinctly the number of atoms of each element in the molecule.

In a system consisting of a mixture of molecular types, the atoms may spontaneously redistribute themselves among the various molecules that are present. This rearrangement necessarily occurs without changing the number of atoms of each element in the system. Such a rearrangement is commonly called a *chemical reaction* and a system capable of this kind of change is classified as a *reacting system*. The notion that such a process is a rearrangement of the atoms present among the molecular types in the system, requiring conservation of the number of atoms of each element, is succinctly expressed by statements that take the form

$$C + O_2 = CO_2$$

or

$$2H_2 + O_2 = 2H_2O$$

These expressions, which are usually also referred to as "chemical reactions," are in essence statements of conservation of the atoms of the elements in such systems. Equivalent statements can be obtained by dividing all the coefficients by a constant value, such that

$$H_2 + \tfrac{1}{2}O_2 = H_2O$$

is an equally valid statement of the conservation of hydrogen and oxygen atoms in a system containing these three components.

If a system consists of e elements and c components, some of which are molecules, then the number r of independent conservation statements or "reactions" that can be written for the system is

$$r = c - e \tag{11.1}$$

Thus a system that contains the elements C and O ($e = 2$) and is made up of the molecular species O_2, CO, and CO_2 ($c = 3$) exhibits a single independent chemical reaction

$$2CO + O_2 = 2CO_2$$

Such a system is termed a *univariant* reacting system.

If the system also contains elemental carbon C as a component, so that $c = 4$, representing C, O_2, CO, and CO_2, then the number of independent chemical reactions $r = 4 - 2 = 2$[1]:

$$C + O_2 = CO_2 \tag{11.1}$$

$$2C + O_2 = 2CO \tag{11.2}$$

Other reactions can be written in this system, such as,

$$C + CO_2 = 2CO \tag{11.3}$$

but this statement is not independent of the first two expressions of elemental conservation. If reactions [11.1] and [11.2] are true for the system, then reaction [11.3] must also be true because it is a linear combination of the first two conservation reactions: [11.3] = [11.2] − [11.1]. To demonstrate this relationship, reverse the statement of reaction [11.1] and add it to reaction [11.2].

$$CO_2 = C + O_2 \tag{11.1}$$

$$2C + O_2 = 2CO \tag{11.2}$$

Combining these statements

$$CO_2 + 2C + O_2 = C + O_2 + 2CO$$

---

[1] Equation numbers representing chemical reactions are enclosed in brackets, [ ], to distinguish them from numbers for mathematical equations, which are enclosed in parentheses, ( ).

Eliminate redundant terms [O_2 appears on both sides, as does C] to obtain reaction [11.3]

$$CO_2 + C = 2CO \qquad [11.3]$$

Thus this system is a *bivariant* reacting system. Systems for which $r = (c - e) > 1$ are *multivariant reacting systems*.

The thermodynamic apparatus necessary to handle multivariant reacting systems is the subject of this chapter. The treatment first focuses on reactions in the gas phase where the components unambiguously exist in molecular form. The general strategy for finding conditions for equilibrium is applied first to a univariant system to clearly identify how, in this context, a reacting system differs from a nonreacting system. The derivation leads to the familiar law of mass action and to the definition of the equilibrium constant for the reaction. Treatment of a bivariant system shows how these results generalize when more than one independent reaction may occur. Further generalization from the bivariant to the multivariant case is then straightforward.

The treatment of multiphase systems parallels that developed for describing reactions in the gas phase. In solids and liquids, even in "line compounds" that are practically stoichiometric, the chemical concept of the molecule loses the clear meaning that it has for the gas phase. Crystals composed of ordered arrays of molecular units like O_2 or CO_2 are relatively rare. Most solid "compounds" are crystals with components of ionic, covalent, and sometimes metallic bonding. Molecular formulas such as SiO_2, Al_2O_3, $NaCl$, and FeO arise from the nature of the bonding and the geometry of the crystal structure. All such compounds exhibit defects that support deviations from the strict molecular ratio that characterizes a molecule (see Chap. 13). Rigorous treatment of such systems can be best formulated through use of their phase diagrams, as developed in Chap. 10, which allow incorporation of such deviations in composition. However, the treatment of such systems can be greatly simplified if the fiction of a molecular formula is assumed. This simplification is important because it makes analysis of such complex systems more practical. These points are developed in detail in Sect. 11.2.

Treatment of multicomponent, multiphase systems as reacting systems leads to a representation of the competition between the components in the form of *predominance diagrams*. Strategies for developing such diagrams and their relation to the phase diagram for the same system are presented in Sect. 11.3.

11.1 REACTIONS IN THE GAS PHASE

Application of the general strategy for finding conditions for equilibrium to a reacting system yields new relationships among the chemical potentials, superimposed upon the set of conditions for equilibrium already derived for multicomponent, multiphase *non*reacting systems. These relationships derive directly from the conservation of elements that accompanies chemical reactions and is succinctly represented in the statement of chemical equations. The simplest illustration of this connection is embodied in the description of a univariant reacting system in which all the components are gases. This class of system is treated first.

11.1.1 Univariant Reactions in the Gas Phase

Consider a gas mixture that contains the three components O_2, CO, and CO_2. Since $r = c - e = 3 - 2 = 1$, this is a univariant system. Only one chemical reaction can be written among these components.

$$2CO + O_2 = 2CO_2$$

To find the conditions for equilibrium in such a system, first write an expression for the change in entropy that it can experience. The combined statement of the first and second laws for this one phase system is

$$dU' = TdS' - PdV' + \sum_{k=1}^{c} \mu_k dn_k \tag{11.2}$$

Rearrange this equation to express the change in entropy

$$dS' = \frac{1}{T}dU' + \frac{P}{T}dV' - \frac{1}{T}\sum_{k=1}^{c} \mu_k dn_k$$

Write the equation explicitly in terms of the three components in this application:

$$dS' = \frac{1}{T}dU' + \frac{P}{T}dV' - \frac{1}{T}\left[\mu_{CO}dn_{CO} + \mu_{O_2}dn_{O_2} + \mu_{CO_2}dn_{CO_2}\right] \tag{11.3}$$

The general criterion states that the entropy is a maximum in an isolated system. If a system is isolated from its surroundings, then

Its internal energy cannot change

$$dU' = 0 \tag{11.4}$$

Its volume cannot change

$$dV' = 0 \tag{11.5}$$

In systems considered up to now, the third isolation constraint expresses the conservation of the components in the system since, if a system is isolated, matter does not cross its boundary. This constraint has the form

$$dn_k = 0 \qquad (k = 1, 2, 3, \ldots, c) \tag{11.6}$$

However, if the system is capable of chemical reactions, then this set of conditions no longer holds in an isolated system. Even if matter cannot cross the boundary of the system, the number of moles of each of the components may still change. The atoms contained in the system may rearrange themselves over the molecular components in the process, decreasing the number of molecules of some components and increasing the number of others. Thus in an isolated system capable of chemical reactions,

$$dn_k \neq 0 \qquad (k = 1, 2, 3, \ldots, c) \tag{11.7}$$

However, it is true that *the number of gram atoms of each of the elements in the system cannot change* no matter how they rearrange themselves among the molecules. Atoms cannot be created or destroyed and, in an isolated system, are not exchanged with the surroundings. Thus, the isolation constraint that applies to a reacting system can be stated

$$dm_i = 0 \qquad (i = 1, 2, 3, \dots, e) \tag{11.8}$$

where m_i is the total number of gram atoms of element i in the system. This condition applies to each element in the system. This reformulation of the isolation constraint is the essential difference between reacting and nonreacting systems. The consequences of this difference are now developed.

The system under study contains two elements, carbon (C) and oxygen (O). It is easy to compute the number of gram atoms of carbon in the system since each molecule of carbon monoxide has one carbon atom and each molecule of carbon dioxide also has one carbon atom. Thus the total number of gram atoms of carbon atoms is

$$m_C = n_{CO_2} + n_{CO} \tag{11.9}$$

A count of the oxygen atoms in the system gives

$$m_O = n_{CO} + 2n_{CO_2} + 2n_{O_2} \tag{11.10}$$

The coefficient of each term on the right sides of these expressions corresponds to the number of atoms of the element in question contained in the corresponding molecular formula. No matter what reactions occur within the system, if the system is isolated, the number of carbon and oxygen atoms cannot change. The isolation constraints can be obtained by taking the differentials of Eqs. (11.9) and (11.10) and setting the results equal to zero.

$$dm_C = 0 = dn_{CO} + dn_{CO_2} \tag{11.11}$$

$$dm_O = 0 = dn_{CO} + 2dn_{CO_2} + 2dn_{O_2} \tag{11.12}$$

Equation (11.11) implies that in an isolated system

$$dn_{CO} = -dn_{CO_2} \tag{11.13}$$

while Eq. (11.12) gives

$$dn_{O_2} = -\frac{1}{2} \left[dn_{CO} + 2dn_{CO_2} \right]$$

Insert Eq. (11.13) and simplify:

$$dn_{O_2} = -\frac{1}{2} dn_{CO_2} \tag{11.14}$$

These relationships are embodied in the chemical equation for the reaction that states that for every mole of CO_2 formed, one mole of CO and one-half mole of O_2 must be consumed. Equations (11.13) and (11.14) are essentially restatements of the isolation constraints, Eqs. (11.11) and (11.12). These conservation equations demonstrate that, although there are three components in this system, the number of moles of only one of them can be varied independently.

The change in entropy that this system can experience when it is isolated from its surroundings is obtained by substituting the isolation constraints, Eqs. (11.4), (11.5), (11.13), and (11.14) into the expression for the entropy, Eq. (11.3)

$$dS'_{iso} = \frac{1}{T}(0) + \frac{P}{T}(0) - \frac{1}{T}\left[\mu_{CO}(-dn_{CO_2}) + \mu_{O_2}\left(-\frac{1}{2}dn_{CO_2}\right) + \mu_{CO_2}dn_{CO_2}\right]$$

which may be written

$$dS'_{iso} = -\frac{1}{T}\left[\mu_{CO_2} - \left(\mu_{CO} + \frac{1}{2}\mu_{O_2}\right)\right]dn_{CO_2} \qquad (11.15)$$

The linear combination of chemical potentials contained in brackets in this expression is defined to be the *affinity for the reaction*:

$$\mathscr{A} \equiv \left[\mu_{CO_2} - \left(\mu_{CO} + \frac{1}{2}\mu_{O_2}\right)\right] \qquad (11.16)$$

This property of the system, which through the chemical potentials depends on the temperature, pressure, and composition of the system, may be thought of as "the chemical potential of the products minus the chemical potential of the reactants" for the reaction

$$CO + \tfrac{1}{2}O_2 = CO_2 \qquad [11.4]$$

With introduction of this notation Eq. (11.15) becomes

$$dS'_{iso} = -\frac{1}{T} \cdot \mathscr{A} \cdot dn_{CO_2} \qquad (11.17)$$

If the temperature, pressure, and composition of the gas mixture are known, then the chemical potentials of the components, and hence the affinity, can be computed. Suppose for a given state this computation gives a value for $\mathscr{A}$ that is negative. This implies that, for the given state, the chemical potential of the reactants is higher than that of the products. Examine Eq. (11.17), keeping in mind that the second law requires that in an isolated system, entropy can only increase. If $\mathscr{A}$ is negative, then, in order for dS to be positive, dn_{CO_2} must be positive. Thus, if the chemical potential of the reactants is higher than that of the products, then the only process possible is an increase in the number of moles of products and *products form*. If, on the other hand, for the given composition, calculation yields $\mathscr{A} > 0$, implying the products have a higher chemical potential than the reactants, then in order for dS to be positive, dn_{CO_2} must be *negative and products decompose*.

The condition for equilibrium in this system corresponding to the maximum in its entropy is obtained by setting the coefficient of dn_{CO_2} equal to zero.

$$\mathscr{A} = \left[\mu_{CO_2} - \left(\mu_{CO} + \frac{1}{2}\mu_{O_2}\right)\right] = 0 \qquad (11.18)$$

Thus the system attains equilibrium when its composition arrives at the state for which the chemical potential of the reactants equals that of the products. For compositions on one side of this balance point, $\mu_{\text{reactants}} > \mu_{\text{products}}$ and products form; if $\mu_{\text{reactants}} < \mu_{\text{products}}$, then products decompose. This relationship constitutes the additional thermodynamic apparatus needed to treat reacting systems.

To demonstrate the generality of this result, consider a system consisting of two elements, M and X, and three components: the molecule M_aX_b, the molecule X_2, and the molecule M_rX_s. If a system composed of these three components is isolated from its surroundings, the change in entropy for this system for any internal process is

$$dS'_{\text{iso}} = -\frac{1}{T}\left[\mu_{M_aX_b}dn_{M_aX_b} + \mu_{X_2}dn_{X_2} + \mu_{M_rX_s}dn_{M_rX_s}\right] \tag{11.19}$$

The number of gram atoms of each of the elements can be expressed in terms of the number of moles of the components at any instant in time as

$$m_M = an_{M_aX_b} + (0)n_{X_2} + rn_{M_rX_s}$$

and

$$m_X = bn_{M_aX_b} + 2n_{X_2} + sn_{M_rX_s}$$

where the coefficients in each case are given by the corresponding formula for the molecular component. The isolation constraints yield

$$dm_M = 0 = adn_{M_aX_b} + rdn_{M_rX_s} \rightarrow dn_{M_aX_b} = -\frac{r}{a}dn_{M_rX_s} \tag{11.20}$$

and

$$dm_X = 0 = bdn_{M_aX_b} + 2dn_{X_2} + sdn_{M_rX_s}$$

$$dn_{X_2} = -\frac{1}{2}(bdn_{M_aX_b} + sdn_{M_rX_s}) = -\frac{1}{2}\left(-\frac{br}{a}dn_{M_rX_s} + sdn_{M_rX_s}\right)$$

$$dn_{X_2} = -\frac{(as - br)}{2a}dn_{M_rX_s} \tag{11.21}$$

Substitute these relationships into Eq. (11.19)

$$dS'_{\text{iso}} = -\frac{1}{T}\left[\mu_{M_aX_b}\left(-\frac{r}{a}dn_{M_rX_s}\right) + \mu_{X_2}\left(-\frac{as-br}{2a}dn_{M_rX_s}\right) + \mu_{M_rX_s}\right]dn_{M_rX_s}$$

$$dS'_{\text{iso}} = -\frac{1}{T}\left[\mu_{M_rX_s} - \left(\frac{r}{a}\mu_{M_aX_b} + \frac{as-br}{2a}\mu_{X_2}\right)\right]dn_{M_rX_s} \tag{11.22}$$

which may be written as Eq. (11.17) with the affinity

$$\mathscr{A} = \left[\mu_{M_rX_s} - \left(\frac{r}{a}\mu_{M_aX_b} + \frac{as-br}{2a}\mu_{X_2}\right)\right] \tag{11.23}$$

The first term represents the chemical potential of the products and the terms in parentheses represent the chemical potential for the reactants for the balanced reaction.

$$\left(\frac{r}{a}\right)M_aX_b + \left(\frac{as-br}{2a}\right)X_2 = M_rX_s \tag{11.5}$$

The reader can verify that this equation is balanced by showing that the number of M atoms is the same on both sides, as is the number of X atoms.

Equation (11.22) has the generic form

$$dS'_{\text{iso}} = -\frac{1}{T} \cdot \mathscr{A} \cdot dn_{M_r X_s} \tag{11.24}$$

The condition for equilibrium is

$$\mathscr{A} = \left[\mu_{M_r X_s} - \left(\frac{r}{a} \mu_{M_s X_b} + \frac{as - br}{2a} \mu_{X_2} \right) \right] = 0 \tag{11.25}$$

If, for a given composition $\mathscr{A} > 0$, then $dn_{M_r X_s}$ must be negative so that the entropy change is positive and products decompose. If $\mathscr{A} < 0$, products form spontaneously.

It is evident that these results hold for any chemical reaction

$$lL + mM = rR + sS \tag{[11.6]}$$

for which

$$\mathscr{A} = \mu_{\text{products}} - \mu_{\text{reactants}} \tag{11.26}$$

$$\mathscr{A} = (r\mu_R + s\mu_S) - (l\mu_L + m\mu_M) \tag{11.27}$$

The condition for equilibrium corresponding to the maximum in the entropy is $\mathscr{A} = 0$ and the direction of the reaction otherwise is determined by the sign of the affinity.

Recall that the chemical potentials of components are not usually reported directly in experimental studies. This aspect of solution behavior of a component is normally supplied in terms of the activity or, equivalently, the activity coefficient of the component. Since the affinity evidently plays a key role in the analysis of reacting systems, it is useful to rewrite the affinity, given in its general form in Eq. (11.27), in terms of the activities of the components.

Recall the definition of activity a_k of component k in a solution, Eq. (8.66):

$$\mu_k = \mu_k^o + RT \ln a_k = G_k^o + RT \ln a_k \tag{11.28}$$

where G_k^o is the Gibbs free energy per mole of component k when it is in its standard or reference state. An expression like this applies to each component in Eq. (11.27). These substitutions lead to an equivalent expression for the affinity of the reaction.

$$\mathscr{A} = \left[r(G_R^o + RT \ln a_R) + s(G_S^o + RT \ln a_S) \right] - \left[l(G_L^o + RT \ln a_L) \right.$$
$$\left. + m(G_M^o + RT \ln a_M) \right]$$

Collecting terms that are similar

$$\mathscr{A} = \left[(rG_R^o + sG_S^o) - (lG_L^o + mG_M^o) \right] + RT \left[(r \ln a_R + s \ln a_S) - (l \ln a_L + m \ln a_M) \right]$$

The first set of brackets represents the change in Gibbs free energy that would accompany the complete conversion of l moles of L and m moles of M when they are in their standard states to form r moles of R and s moles of S in their standard states. This quantity is called the *standard free energy change for the reaction* with the notation

$$\Delta G^o \equiv \left[(rG_R^o + sG_S^o) - (lG_L^o + mG_M^o) \right] \qquad (11.29)$$

The four terms in the second set of brackets may be written as a single term by employing the properties of the logarithm function, $x \ln y = \ln y^x$, and $\ln x + \ln y = \ln xy$. The expression for the affinity becomes

$$\mathscr{A} = \Delta G^o + RT \ln \left[\frac{a_R^r A_S^s}{A_L^l a_M^m} \right] \qquad (11.30)$$

The notation can be further simplified by introducing a quantity called the *proper quotient of activities for the reaction*, defined by

$$Q \equiv \frac{a_R^r a_S^s}{a_L^l a_M^m} \qquad (11.31)$$

This quantity can be thought of as the ratio of "the activity of the products" to "the activity of the reactants," always keeping in mind that these expressions refer to products of the activities of the participating components raised to powers corresponding to the coefficients in the stoichiometric equation.

With these definitions the affinity for the reaction can be written as

$$\mathscr{A} = \Delta G^o + RT \ln Q \qquad (11.32)$$

If the relations between the activities of the components and the composition of the mixture are known, then, for a given composition, a_L, a_M, a_R, and a_S can be evaluated. The proper quotient of activities Q thus has a particular value for that composition, which can be computed from this information.

Whatever its initial state, the system spontaneously evolves until it achieves the specific composition that is its equilibrium state. At that composition, Q takes on a particular value which could be written Q_{equil}. The symbol uniformly adopted to represent this quantity is K, called the *equilibrium constant* for the reaction. Thus

$$K \equiv Q_{\text{equil}} = \left[\frac{a_R^r a_S^s}{a_L^l a_m^m} \right]_{\text{equil}} \qquad (11.33)$$

K is the value that the proper quotient of activities takes on when the composition of the system achieves its equilibrium distribution. Recall that the condition for equilibrium corresponds to $\mathscr{A} = 0$, at which $Q = K$. Put these conditions into Eq. (11.32) to obtain the familiar *law of mass action*.

$$\mathscr{A} = 0 = \Delta G^o + RT \ln K$$

Thus the condition for equilibrium in a univariant reacting system can be written

$$\Delta G^o = -RT \ln K \qquad (11.34)$$

This is the *working equation* most widely used in solving practical problems that deal with chemical reactions.

The expression for the affinity at any state, Eq. (11.32), can be rewritten by inserting Eq. (11.34) for ΔG^o.

$$\mathscr{A} = -RT \ln K + RT \ln Q$$

Again applying the properties of logarithms,

$$\mathscr{A} = RT \ln \frac{Q}{K} \tag{11.35}$$

This expression for the affinity is entirely equivalent to Eqs. (11.27) or (11.32). No simplifying assumptions have been introduced, only simplifying notation. With this form of the affinity expression the conditions for spontaneous change can be reinterpreted. Recall that in the range of compositions for which the affinity is negative, products form. From Eq. (11.35), A is less than zero if $(Q/K) < 1$ and *products form*. For the range in which A is positive, products decompose; this corresponds to the range of values for which $(Q/K) > 1$ and *products decompose*. For the unique composition at which $(Q/K) = 1$, where $Q = K$, the affinity is zero and the system has the equilibrium composition. This situation is represented in Fig. 11.1.

Example 11.1. A gas mixture at one atmosphere total pressure has the following composition:

Component	H_2	O_2	H_2O
Mole fraction	0.01	0.03	0.96

At 700°C, for the reaction

$$2H_2 + O_2 = 2H_2O \qquad \Delta G^o = -393 \text{KJ}$$

Determine the direction of spontaneous change for this system.

The equilibrium constant for this reaction at 700°C (973 K) can be computed by rearranging Eq. (11.34)

$$K = e^{-(\Delta G^o/RT)} = e^{-(-393,000/8.314 \cdot 973)} = 1.25 \times 10^{21}$$

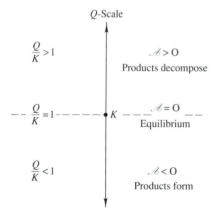

Q-Scale

$\dfrac{Q}{K} > 1$ $\qquad \mathscr{A} > 0$
Products decompose

$\dfrac{Q}{K} = 1$ —————• K — — $\mathscr{A} = 0$ — —
Equilibrium

$\dfrac{Q}{K} < 1$ $\qquad \mathscr{A} < 0$
Products form

FIGURE 11.1
Sketch of the scales for affinity and (Q/K) showing the ranges of spontaneous decomposition and formation of products.

Note that equilibrium constants are unitless quantities. A gas mixture can be considered an ideal solution; thus the activities of the components are given by their mole fractions in the mixture. The proper quotient of activities for the mixture given is

$$Q = \frac{a_{H_2O}^2}{a_{H_2}^2 a_{O_2}} = \frac{X_{H_2O}^2}{X_{H_2}^2 X_{O_2}} = \frac{(0.96)^2}{(0.01)^2(0.03)} = 3.1 \times 10^5$$

Thus $Q/K = 3.1 \times 10^5/1.3 \times 10^{21} = 2.4 \times 10^{-16} \ll 1$. According to Fig. 11.1, there is a strong tendency for products (H_2O) to form in this system.

Example 11.2. What is the equilibrium composition for the system described in Example 11.1?

  From a practical point of view, the magnitude of the equilibrium constant means that at equilibrium the numerator in the proper quotient of activities is 21 orders of magnitude larger than the denominator. This suggests that the system maximizes the H_2O content relative to the other components. However, if all the hydrogen present is converted to water, not all the oxygen is consumed. Conversion of the 0.01 moles of H_2 to water consumes only 0.005 moles of O_2. In the process 0.01 moles of additional water are formed. For each mole of the initial mixture in the system the final mixture contains negligible hydrogen, $0.96 + 0.01 = 0.97$ moles of H_2O, and $0.03 - 0.005 = 0.025$ moles of excess oxygen. The total number of moles is reduced from 1.00 to $0.97 + 0.025 = 0.995$. Expressed in mole fractions, the final composition is

Component	H_2	O_2	H_2O
Mole fraction	negligible	$\frac{0.025}{0.995} = 0.0251$	$\frac{0.970}{0.995} = 0.9749$

  The form of Eq. (11.34), which expresses the condition for equilibrium, yields an exponential relationship between the equilibrium constant and the standard free energy change for a reaction. Measured values for ΔG^o for chemical reactions of technological interest span the range from about $+100$ KJ to -1000 KJ. As a consequence, values of K range over *many orders of magnitude*. The value of Q obtained for any composition for a system of interest in a practical problem also ranges over many orders of magnitude. Thus it is likely that the value of Q computed for a given composition of practical interest differs from the value of K representing the equilibrium composition for the system by many orders of magnitude, as is the case in Example 11.1. The qualitative answer to the generic question posed in Example 11.1, namely, "Given the composition of the system, which way lies equilibrium?," is usually easily obtained. Except for systems with initial compositions that happen to lie near the equilibrium composition, Q/K is orders of magnitude greater or less than 1. In this case neither Q nor K need to be known with accuracy to answer this qualitative question.

  In contrast the generic question posed by Example 11.2, namely, "Given the initial composition of a system, what will be its composition at equilibrium?," requires accurate knowledge of the equilibrium constant and computations that incorporate conservation of the elements in the system. A general strategy for treating such problems is presented in the next section.

11.1.2 Multivariant Reactions in the Gas Phase

A system with two independent chemical reactions is the simplest example of a multivariant reacting system. Consider again a system with two elements, carbon (C) and oxygen (O). In addition to the three components considered in Sect. 11.1.1 (CO, O_2, and CO_2) this system may exhibit measurable quantities of a fourth component, carbon vapor [C(g)]. The extent to which this component can be neglected in treating gas mixtures containing carbon and oxygen can only be assessed by including this component in the analysis. In this system $e = 2, c = 4$ and $r = 4 - 2 = 2$. It is a *bivariant* reacting system.

Expressions for the number of gram atoms of carbon and oxygen take the form

$$m_C = (1)n_{C(g)} + (0)n_{O_2} + (1)n_{CO_2} + (1)n_{CO} \tag{11.36}$$

$$m_O = (0)n_{C(g)} + (2)n_{O_2} + (2)n_{CO_2} + (1)n_{CO} \tag{11.37}$$

The compositional isolation constraints become

$$dm_C = dn_{C(g)} + dn_{CO_2} + dn_{CO} = 0 \tag{11.38}$$

$$dm_O = 2dn_{O_2} + 2n_{CO_2} + dn_{CO} = 0 \tag{11.39}$$

Thus the four compositional variables are related by two linear equations. This system of equations has two degrees of freedom. Choose dn_{CO} and dn_{CO2} as independent variables. From Eq. (11.38),

$$dn_{C(g)} = -(dn_{CO_2} + dn_{CO}) \tag{11.40}$$

From Eq. (11.39), after some manipulation,

$$dn_{O_2} = -(dn_{CO_2} + \frac{1}{2}dn_{CO}) \tag{11.41}$$

The expression for the change in entropy for this isolated system contains four terms, one for each component.

$$dS'_{iso} = -\frac{1}{T} \left[\mu_{C(g)}dn_{C(g)} + \mu_{O_2}dn_{O_2} + \mu_{CO_2}dn_{CO_2} + \mu_{CO}dn_{CO} \right]$$

Use Eqs. (11.40) and (11.41) to eliminate the dependent variables:

$$dS'_{iso} = -\frac{1}{T} \left[\mu_{C(g)} \ (-dn_{CO_2} - dn_{CO}) + \mu_{O_2} \left(-dn_{CO_2} - \frac{1}{2}dn_{CO} \right) + \mu_{CO_2}dn_{CO_2} \right.$$

$$\left. + \mu_{CO}dn_{CO} \right]$$

Collecting terms in dn_{CO} and dn_{CO2}

$$dS'_{iso} = \frac{1}{T} \left[[\mu_{CO} - \left(\mu_{C(g)} + \frac{1}{2}\mu_{O_2} \right)]dn_{CO} + [\mu_{CO_2} - (\mu_{C(g)} + \mu_{O_2})]dn_{CO_2} \right]$$
$$\tag{11.42}$$

Inspection of this equation reveals that the coefficient of dn_{CO} is the affinity $\mathscr{A}_{[CO]}$ for the reaction

$$C(g) + \frac{1}{2}O_2 = CO$$

while the coefficient of dn_{CO2} is the affinity $\mathscr{A}_{[CO_2]}$ for the reaction

$$C(g) + O_2 = CO_2$$

Thus Eq. (11.42) can be written

$$dS'_{\text{iso}} = -\frac{1}{T}\left[\mathscr{A}_{[CO]}dn_{CO} + \mathscr{A}_{[CO_2]}dn_{CO_2}\right] \tag{11.43}$$

This is a general expression for the change in entropy that this system may exhibit when it is isolated from its surroundings, expressed in terms of the two independent variables in the system. The constrained maximum in S can be found by setting the coefficients of the differentials in this expression equal to zero. Thus in this bivariant system the conditions for equilibrium are

$$\mathscr{A}_{[CO]} = \mu_{CO} - (\mu_{C(g)} + \frac{1}{2}\mu_{O_2}) = 0 \tag{11.44}$$

and

$$\mathscr{A}_{[CO_2]} = \mu_{CO_2} - (\mu_{C(g)} + \mu_{O_2}) = 0 \tag{11.45}$$

The same strategy that converts the expression for the affinities in Eq. (11.27) to that in Eq. (11.32) and leads to the condition for equilibrium in a univariant reacting system given in Eq. (11.34) can be applied separately to each of these equations. The conditions for equilibrium in this bivariant system can be written as

$$\Delta G^o_{[CO]} = -RT \ln K_{[CO]} \tag{11.46}$$

and

$$\Delta G^o_{[CO_2]} = -RT \ln K_{[CO_2]} \tag{11.47}$$

The standard free energy changes and equilibrium constants are those that apply to the CO and CO_2 formation reactions [CO] and [CO_2] that describe this system.

While the derivation of the conditions for equilibrium is straightforward for this bivariant case, the conditions for spontaneous change, based upon the requirement that the entropy increase in Eq. (11.43), are not so clear cut. For example, even if both affinities have the same sign it is possible for dn_{CO} and dn_{CO2} to have opposite signs and the sum to still yield a positive change in entropy. The directions of the reactions that increase the entropy in an isolated bivariant reacting system depend explicitly upon the values of the quantities involved; no general statement may be made. This situation evidently becomes more complicated as additional independent reactions are added in a multivariant system.

A reacting system is bivariant if it is characterized by two independent chemical reactions. Other reactions among the components of the system can be written but these are not independent; they are linear combinations of the two independent reactions.

In the introduction it is shown that the reaction

$$C(g) + CO_2 = 2CO \qquad [11.7]$$

can be obtained by subtracting reaction $[CO_2]$ from reaction $[CO]$ after doubling its coefficients. Thus, these reactions are written $[11.7] = 2 \cdot [CO] - [CO_2]$. The stoichiometric statement embodied in reaction $[11.7]$ is a linear combination of those in $[CO_2]$ and $[CO]$ and is thus not an independent statement. It is also possible to write the reaction

$$2CO + O_2 = 2CO_2 \qquad [11.8]$$

For this reaction, $[11.8] = 2 \cdot [CO] - 2 \cdot [CO_2]$.

It is easy to show that the affinities for reactions $[11.7]$ and $[11.8]$ are the analogous linear combinations of the affinities for reactions $[CO]$ and $[CO_2]$. The affinity for reaction $[11.7]$ is defined as

$$\mathscr{A}_{[11.7]} = 2\mu_{CO} - (\mu_{C(g)} + \mu_{CO_2}) \qquad (11.48)$$

The linear combination of $\mathscr{A}_{[CO]}$ and $\mathscr{A}_{[CO2]}$ indicated is

$$\mathscr{A}_{[11.7]} = 2\mathscr{A}_{[CO]} - \mathscr{A}_{[CO_2]}$$

$$= 2\left[\mu_{CO} - \left(\mu_{C(g)} + \frac{1}{2}\mu_{O_2}\right)\right] - [\mu_{CO_2} - (\mu_{C(g)} + \mu_{O_2})]$$

$$= 2\mu_{CO} - 2\mu_{C(g)} - \mu_{O_2} - \mu_{CO_2} + \mu_{C(g)} + \mu_{O_2}$$

$$2\mathscr{A}_{[CO]} - \mathscr{A}_{[CO_2]} = 2\mu_{CO} - (\mu_{C(g)} + \mu_{CO_2}) = \mathscr{A}_{[11.7]} \qquad (11.49)$$

The demonstration that $\mathscr{A}_{[11.8]} = 2 \cdot \mathscr{A}_{[CO_2]} - 2 \cdot \mathscr{A}_{[CO]}$ is left as an exercise to the reader.

The relationships between the affinities also imply relationships between the standard free energy changes and the equilibrium constants for these reactions. For example, define the relationship

$$\mathscr{A}_{[k]} = m\mathscr{A}_{[i]} - n\mathscr{A}_{[j]}$$

Substitute Eq. (11.32) for each of the affinities in this Eq. (11.50),

$$\Delta G_{[k]}^o + RT \ln K_{[k]} = m(\Delta G_{[i]}^o + RT \ln K_{[i]}) - n(\Delta G_{[j]}^o + RT \ln K_{[j]})$$

$$\Delta G_{[k]}^o + RT \ln K_{[k]} = (m\Delta G_{[i]}^o - n\Delta G_{[j]}^o) + RT \ln \frac{K_{[i]}^m}{K_{[j]}^n}$$

Comparing corresponding terms

$$\Delta G_{[k]}^o = m\Delta G_{[i]}^o - n\Delta G_{[j]}^o \qquad (11.51)$$

and

$$K_{[k]} = \frac{K_{[i]}^m}{K_{[j]}^n} \qquad (11.52)$$

Thus any reaction that is a linear combination of other reactions has a standard free energy change that is the same linear combination of the ΔG^o values for the contributing reactions. The equilibrium constant for such a reaction is a product of K values for the contributing reactions raised to powers that correspond to coefficients in the linear combination relating the reactions.

Incidentally, the description of the system emerged in terms of reactions [CO] and [CO$_2$] in this derivation because the affinities for these reactions appeared in Eq. (11.42). There is nothing special about this result and it can be traced to the choice made for the independent composition variables in the conservation Eq. (11.40) and (11.41). If n_C or n_{O_2} had been chosen as independent variables, then Eqs. [11.7] and [11.8] would have appeared as the appropriate description of the system.

Since at equilibrium $\mathscr{A}_{[CO]}$ and $\mathscr{A}_{[CO2]}$ are independently zero [Eqs.] (11.44) and (11.45)] and the affinity for reaction [11.7] is a linear combination of these values, given by Eq. (11.50), then when the system reaches its equilibrium state,

$$\mathscr{A}_{[11.7]} = 0 \qquad (11.53)$$

Similarly, because $\mathscr{A}_{[11.8]}$ is a linear combination of $\mathscr{A}_{[CO]}$ and $\mathscr{A}_{[CO_2]}$, at equilibrium, the affinity for reaction [11.8] must also be zero.

$$\mathscr{A}_{[11.8]} = 0 \qquad (11.54)$$

Thus the working equations that describe the equilibrium state for the system are

$$\Delta G^o_{[j]} = -RT \ln K_{[j]} \qquad [j = 1, 2, 3, 4] \qquad (11.55)$$

Only two of these four equations are independent.

These arguments extend directly to *multivariant* reacting systems with e elements forming c components. In a system with a large number of components, there are $r = (c - e)$ independent equations of the form of Eq. (11.55) that describe the equilibrium state. However, as the preceding argument demonstrates, at equilibrium it is possible to write an equation like Eq. (11.55) *for every possible reaction that can be written among the components in the system*, but only r of these equations are independent.

The general strategy for determining the equilibrium composition in a multivariant reacting system makes use of the independent conditions for chemical equilibrium and the equations that require the number of gram atoms of the elements in the system be conserved. If a single phase system contains c components, the equilibrium state is specified by assigning values to the mole fractions of each of these c components. If the components are built from e elements, then there are e conservation of atoms equations, one for each element, of the form

$$m_j = \sum_{k=1}^{c} b_{jk} n_k \qquad (j = 1, 2, \ldots, e) \qquad (11.56)$$

where the coefficient b_{jk} is the number of atoms of element j contained in a molecule of k. (If k doesn't contain the element j, then $b_{jk} = 0$ for that component.) Note that these equations are linear in the unknown n_k values. There are $r = (c - e)$ independent reactions in this system, each with its condition for equilibrium, Eq. (11.55), and corresponding value for its equilibrium constant. Each of these equilibrium constants is related to the composition of the system through an equation like Eq. (11.33). These relationships are normally not linear because the activity values involved have exponents. Adding these $r = (c - e)$ equilibrium constant relations to the set of e conservation of atoms equations yields a total of $[e + (c - e)] = c$ equations in c unknowns and the state of the system is therefore fully determined mathematically.

Example 11.3. A gas mixture has following composition:

Component	H_2	O_2	H_2O	CO	CO_2	CH_4
Mole Fraction	0.05	0.05	0.15	0.25	0.40	0.10

Find the equilibrium composition of this mixture at 600°C.

The set of conservation equations corresponding to Eq. (11.56) for this system is

$$m_C = (0)n_{H_2} + (0)n_{O_2} + (0)n_{H_2O} + (1)n_{CO} + (1)n_{CO_2} + (1)n_{CH_4}$$

$$m_O = (0)n_{H_2} + (2)n_{O_2} + (1)n_{H_2O} + (1)n_{CO} + (2)n_{CO_2} + (0)n_{CH_4}$$

$$m_H = (2)n_{H_2} + (0)n_{O_2} + (2)n_{H_2O} + (0)n_{CO} + (0)n_{CO_2} + (4)n_{CH_4}$$

The total number of moles in the system at any instant is the sum of the number of moles of each of the components.

$$n_T = n_{H_2} + n_{O_2} + n_{H_2O} + n_{CO} + n_{CO_2} + n_{CH_4}$$

For this example, $c = 6$ and $e = 3$; there are three independent reactions. Find three nonredundant reactions that incorporate all six components.

$$2H_2 + O_2 = 2H_2O \tag{1}$$

$$2CO + O_2 = 2CO_2 \tag{2}$$

$$CH_4 + 2O_2 = 2H_2O + CO_2 \tag{3}$$

The standard free energy changes for these reactions at 600°C can be found from Appendix G:

$$\Delta G^o_{[1]} = -406,200 \qquad \Delta G^o_{[2]} = -414,500 \qquad \Delta G^o_{[3]} = -797,900$$

The corresponding equilibrium constants, assuming the gas mixture is ideal so that the activities are equal to the mole fractions, are

$$K_{[1]} = \frac{X^2_{H_2O}}{X^2_{H_2} X_{O_2}} = 2.02 \times 10^{24}$$

$$K_{[2]} = \frac{X^2_{CO_2}}{X^2_{CO} X_{O_2}} = 6.34 \times 10^{24}$$

$$K_{[3]} = \frac{X^2_{H_2O} X_{CO_2}}{X_{CH_4} X^2_{O_2}} = 5.53 \times 10^{47}$$

since for each component $X_k = n_k/n_T$. Thus there are seven equations relating the seven variables, (six n_k values and n_T). This system of equations can be solved using an iterative solver from a mathematics applications software program to yield the final equilibrium composition for a system with the initial composition given at the beginning of the example.

The equilibrium composition of the gas phase computed from these seven simultaneous equations is

Component	H_2	H_2O	CO	CO_2	CH_4	O_2
Mole Fraction	0.136	0.144	0.231	0.445	0.052	1.6×10^{-24}

The problem posed in Example 11.3 is typical of this class of problems. An initial composition of the system is known which permits calculation of the number of gram atoms m_j of each of the elements. The molecular formulas of the components give the coefficients in the set of equations corresponding to the conservation of the elements in the system. A set of $r = (c - e)$ independent chemical reactions can then be formulated. Values for ΔG^o are then obtained for these reactions. For many common molecular components, values for the reaction that forms the component from its elements ΔH_f^o and ΔS_f^o are tabulated (see, for example, references 11.1 to 11.4, or Appendices G, H and I of this text). Since $\Delta G^o = \Delta H^o - T \Delta S^o$, this information permits computation of ΔG_f^o at any temperature, for any formation reaction involving these components. The equilibrium constant K can be computed for each of these reactions from the condition for equilibrium, Eq. (11.55). Of the resulting c equations in c unknowns, e are linear, derived from the conservation of the elements. The remaining $(c - e)$ equations come from the equilibrium constants and are nonlinear equations. Mathematics applications packages, such as TK-Solver^{TM} and MathCad^{TM}, provide a framework for solving such sets of equations.

More explicit applications software, such as the program SOLGASMIX [11.5], has been developed to permit solution of this class of problem for a system with an arbitrary number of components. In its most complete form, this software incorporates a data base with enthalpies and entropies of formation of a large number of molecular components that normally form in the gas phase. Input to the program includes a list of the elements in the system and a range of temperatures that are of interest in the problem. The software typically executes the following steps:

1. Develops a list of the molecular components in its data base that form from the elements that are input.
2. Generates a list of independent balanced chemical reactions among these components.
3. Looks up enthalpies and entropies for these reactions in its data base.
4. Computes ΔG_j^o values for each reaction at a temperature of interest.
5. Computes the corresponding equilibrium constant, K_j.
6. Writes the e conservation of element equations.
7. Writes the $(r - e)$ equilibrium constant relations.

8. Solves this set of equations for the compositions of each of the components at equilibrium.

Output includes a list of the concentrations of all the components considered in the program. Typically concentrations of perhaps 20 to 30 components can be included in the output. For practical purposes the vast majority of these are small enough to be neglected and the composition of the gas at equilibrium can be usefully specified in terms of a handful of components. Variation of the gas composition with temperature can be explored by incrementing T and repeating the algorithm. Figure 11.2 shows a typical output of the SOLGASMIX program.

TEMPERATURE=1000 K;
TOTAL PRESSURE=1.000e+00 ATM;

INITIAL COMPOSITION:

CO(g)	Co$_2$(g)	SO2
3.12E-01	5.09E-01	1.79E-01

EQUILIBRIUM COMPOSITION:

1 COS (g)	p=9.25E-03
2 S$_2$O (g)	p=2.34E-04
3 SO$_3$ (g)	p=6.03E-10
4 SO$_2$ (g)	p=3.22E-02
5 SO (g)	p=5.19E-06
6 CS$_2$ (g)	p=1.13E-05
7 CS (g)	p=6.23E-10
8 S$_8$ (g)	p=3.44E-08
9 S$_7$ (g)	p=4.47E-07
10 S$_6$ (g)	p=6.84E-06
11 S$_5$ (g)	p=2.75E-05
12 S$_4$ (g)	p=6.60E-06
13 S$_3$ (g)	p=2.15E-03
14 S$_2$ (g)	p=7.30E-02
15 S (g)	p=1.39E-09
16 C$_3$O$_2$ (g)	p=4.68E-23
17 CO$_2$ (g)	p=8.78E-01
18 CO (g)	p=5.30E-03
19 O$_3$ (g)	p=8.78E-36
20 O$_2$ (g)	p=1.00E-16
21 O (g)	p=1.56E-18
22 C$_5$ (g)	p=3.39E-64
23 C$_4$ (g)	p=1.75E-59
24 C$_3$ (g)	p=2.39E-46
25 C$_2$ (g)	p=1.37E-43

FIGURE 11.2

Example of the equilibrium composition of a gas mixture with a given initial composition of 1000 K computed with GASMIX [11.5].

11.2 REACTIONS IN MULTIPHASE SYSTEMS

It is not unusual in materials science for a system of practical interest to be multi-component, multiphase, *and* capable of exhibiting chemical reactions. For example, the oxidation of a metal involves three phases: the metal, the ceramic oxide, and the gas phase, which is the source of the oxygen. The description of such a system begins with a list of the components in each phase. The construction of a valid list is a non-trivial task. Within a given phase, the components are those chemical species whose compositions can be varied independently when the system is not at equilibrium.

In the gas phase the list includes in principle all the known molecular species that the elements in the system can combine to form. If electrical effects are important in the problem at hand, this list must be extended to include ionic species as well (see Chap. 15). In metallic phases the list normally incorporates the elements in the system since molecular combinations separate into their elemental components upon dissolution in metals. Ceramic phases include the vast range of ionic, covalent, and polar solid phases normally described in chemistry as "chemical compounds" like oxides, carbides, nitrides, sulfides, sulfates, etc. If variations from the stoichiometric composition are of interest in the problem, then variations in the number of moles of the *elemental* components must be included in the description of a compound phase. If electrical fields are involved, the components can be usefully taken as the ionic species that make up the ceramic phase. Ionic species also play a key role in behavior if one of the phases is an electrolyte.

As an example, consider the oxidation of copper metal. Three phases normally participate: the metallic phase (α), the gas phase (g), and the ceramic phase, copper oxide (ϵ), nominally Cu_2O. In a restricted range of oxygen potential and temperature, CuO may also form. At equilibrium all three of these phases are solutions: some oxygen atoms dissolve in copper metal, some copper vapor exists in the gas along with the oxygen, the oxide phase departs from its stoichiometric composition. Some components exist in one phase but not in others. The gas phase contains the molecular form of oxygen, O_2. The oxygen in the copper metal is monatomic; in order to dissolve oxygen in copper, the oxygen molecule must dissociate into atoms at the metal/gas interface. The oxygen in the oxide is present as oxygen ions, the copper as copper ions. These ions cannot exist either in the metal or the vapor. Other ionic species can exist as defects in the oxide. The components can be taken as: Cu^α, $Cu^{++\epsilon}$, $O^{--\epsilon}$, Cu^g, and O_2^g. Even this list is not exhaustive. Spectroscopic analysis of the gas phase reveals monatomic oxygen O_1, ozone O_3, and a variety of copper and oxygen complexes, some of which are ions, present in very small but detectable quantities.

In most practical applications the list of components considered sufficient to describe the behavior of the system is much shorter than the exhaustive list contemplated in the last paragraph. Components may be deleted from the description of the behavior of a system if it is known from independent information that the variation in their number of moles is negligible in the context of the problem. The simplest treatment of the copper-oxygen system visualizes only three components: $Cu(\alpha)$, $CuO(\epsilon)$, and $O_2(g)$. Each of the *phases* in the system (α, ϵ, g) is also viewed as a *component* (Cu, CuO, O_2) because the compositions of the phases are assumed to be invariant.

At this level of sophistication there are two elements and three components in this three phase system. Since $r = (c - e) = (3 - 2) = 1$, it is a *univariant* system.

The strategy for finding the conditions for equilibrium can now be applied to this three phase, three component system. The number of copper atoms in the system at any instant in time is

$$m_{Cu} = n_{Cu}^{\alpha} + n_{CuO}^{\epsilon} \tag{11.57}$$

The number of oxygen atoms is

$$m_O = 2n_{O_2}^g + n_{CuO}^{\epsilon} \tag{11.58}$$

In an isolated system the number of atoms of each element is conserved.

$$dm_{Cu} = 0 = dn_{Cu}^{\alpha} + dn_{CuO}^{\epsilon} \rightarrow dn_{Cu}^{\alpha} = -dn_{CuO}^{\epsilon}$$

$$dm_O = 0 = 2dn_{O_2}^g + dn_{CuO}^{\epsilon} \rightarrow dn_{O_2}^g = -\frac{1}{2}dn_{CuO}^{\epsilon}$$

The expression for the change in entropy of this three phase system involves terms in U' and V' for each of the phases. When these terms are combined with the internal energy and volume isolation constraints, they yield the conditions for thermal ($T^{\alpha} = T^{\epsilon} = T^g$) and mechanical ($P^{\alpha} = P^{\epsilon} = P^g$) equilibrium. Since the current focus is on the chemical effects in the system, the entropy can be written as

$$dS'_{iso} = [[\quad]] - \frac{1}{T}\left[\mu_{Cu}^{\alpha}dn_{Cu}^{\alpha} + \mu_{CuO}^{\epsilon}dn_{CuO}^{\epsilon} + \mu_{O_2}^g dn_{O_2}^g \right] \tag{11.59}$$

where the notation $[[\ ]]$ signifies the four noncompositional terms in the expression. Applying the conservation equations

$$dS'_{iso} = [[\quad]] - \frac{1}{T}\left[\mu_{Cu}^{\alpha}(-dn_{CuO}^{\epsilon}) + \mu_{CuO}^{\epsilon}dn_{CuO}^{\epsilon} + \mu_{O_2}^g \left(-\frac{1}{2}dn_{CuO}^{\epsilon} \right) \right]$$

$$dS'_{iso} = [[\quad]] - \frac{1}{T}\left[\mu_{CuO}^{\epsilon} - \left(\mu_{Cu}^{\alpha} + \frac{1}{2}\mu_{O_2}^g \right) \right](dn_{CuO}^{\epsilon}) \tag{11.60}$$

To find the conditions for equilibrium, set the coefficients of the differentials in this expression equal to zero. Applying that tactic to the terms in $[[\ ]]$ yields the conditions for thermal and mechanical equilibrium as previously noted. The coefficient of the compositional variable can be recognized as the affinity for the reaction

$$Cu(\alpha) + \frac{1}{2}O_2(g) = CuO(\epsilon)$$

defined to be

$$\mathscr{A}_{CuO} = \mu_{CuO}^{\epsilon} - \left(\mu_{Cu}^{\alpha} + \frac{1}{2}\mu_{O_2}^g \right) \tag{11.61}$$

The condition for chemical equilibrium in this system requires that the coefficient of dn_{CuO} in Eq. (11.60) be equal to zero.

$$\mathscr{A}_{CuO} = 0 \tag{11.62}$$

Express the chemical potentials in terms of the corresponding activities and collect like terms to derive

$$\Delta G^o_{CuO} = -RT \ln K_{CuO} \tag{11.63}$$

where

$$\Delta G^o_{CuO} = G^o_{CuO} - \left(G^o_{Cu} + \frac{1}{2} G^o_{O_2} \right) \tag{11.64}$$

and

$$K_{CuO} = \frac{a_{CuO}}{a_{Cu} a^{1/2}_{O_2}} \tag{11.65}$$

Thus, the condition for equilibrium in this *multiphase* reacting system is formally identical with that derived for reactions in the gas phase.

Example 11.4. Find the partial pressure of oxygen that exists in a system in which pure copper is equilibrated with CuO at 900°C. The standard free energy for the formation of CuO from copper and oxygen at 900°C is

$$\Delta G^o_{900C} = -184 \text{ KJ}$$

The equilibrium constant for the reaction is

$$K = e^{-(-184,000/8.314 \cdot 1173)} = e^{+18.9} = 1.56 \times 10^8$$

The proper quotient of activities can be computed by noting that $a_{Cu} = 1$ and $a_{CuO} = 1$ since they are in their reference states for the given conditions. The activity of oxygen is the ratio of its partial pressure in the system to the pressure in the reference state.

$$a_{O_2} = \frac{P_{O_2}}{P^o_{O_2}}$$

Since the reference state used in evaluating ΔG^o for the reaction is 1 atmosphere, the activity of oxygen is numerically equal to its partial pressure P_{O_2}. The equilibrium constant, Eq. (11.64), is thus

$$K_{CuO} = \frac{a_{CuO}}{a_{Cu} a^{1/2}_{O_2}} = \frac{1}{1 \cdot P^{1/2}_{O_2}} = 1.56 \times 10^8$$

Solving for the partial pressure of oxygen

$$P_{O_2} = \left(\frac{1}{1.56 \times 10^8} \right)^2 = 4.2 \times 10^{-17} \text{atm}$$

A reacting system is *multivariant* if $r = c - e > 1$. In multiphase systems, the extension from univariant to multivariant reactions follows the same logical sequence developed for the single gas phase in Sec. 11.1.2. The result is the same and the conditions for chemical equilibrium correspond to a set of equations of the form

$$\mathscr{A}_{[j]} = 0 \qquad [j = 1, 2, \ldots, r] \tag{11.66}$$

and the equivalent working equations

$$\Delta G^o_{[j]} = -RT \ln K_{[j]} \qquad [j = 1, 2, \ldots, r] \tag{11.67}$$

where r is the number of independent chemical reactions for the system. It is further true that, since any chemical reaction among the components in the system can be constructed from a linear combination of the independent reactions, Eqs. (11.66) and (11.67) apply to *every chemical reaction that may be written among the components in the system.*

In treating practical problems involving multicomponent, multiphase reacting systems it is frequently convenient to limit consideration to a single equilibrium that exists between a few of the components that happen to be of interest in a particular application. This tactic is possible because an equation of the form of Eq. (11.67) holds for every reaction. Caution must be applied in making such a simplification since the equilibrium situation thus deduced may be *metastable*. One or more of the components left out of consideration in making such a simplification may in fact be the predominant species in the actual stable equilibrium state. Deductions based upon the simplified analysis may therefore be misleading. A representation of multivariant equilibria that incorporates all the components in a system, called the *predominance diagram* for the system, is presented in Sec. 11.4.

11.3 PATTERNS OF BEHAVIOR IN COMMON REACTING SYSTEMS

It is clear that evaluation of the equilibrium constant for a reaction is the key to the solution of practical problems in reacting systems. The equilibrium constant is in turn most frequently computed from information about the standard free energy change for the reaction, ΔG^o, through Eq. (11.67). The variation of the equilibrium state with temperature and pressure may be traced to the variation of ΔG^o. Since the standard free energy change is defined in terms of the free energy of the components in their reference states and these are normally defined to be the pure components, computation of the variation of ΔG^o with temperature and pressure employs the familiar thermodynamic relationships derived in earlier chapters for simple, one component systems.

11.3.1 Richardson-Ellingham Charts for Oxidation

The standard free energy change for any reaction can be expressed in terms of the standard enthalpy and entropy of the reaction through the definitional relationship

$$\Delta G^o = \Delta H^o - T \Delta S^o \tag{11.68}$$

In most studies of reacting systems in materials science the variation of the behavior of the system with temperature at one atmosphere pressure is most important. The variation of the enthalpy with temperature at constant pressure is given by

$$\Delta H^o(T) = \Delta H^o(T_0) + \int_{T_0}^{T} \Delta C_P^o(T)dT \tag{11.69}$$

where ΔC_p^o is the heat capacity of the pure components of the products minus the heat capacity of the reactants. On the scale of energies associated with thermodynamic processes, heats of reactions are very large numbers, typically a few hundred kilojoules. Heat exchanges associated with altering the temperature of reactants or products more typically range below a few tens of kilojoules. The *difference* in these heating effects for reactants and products is yet an order of magnitude smaller. The second term in Eq. (11.69) therefore is negligible for most practical purposes and the heat of reaction is independent of temperature. A similar observation holds for the standard entropy of a reaction. In the relation

$$\Delta S^o(T) = \Delta S^o(T_0) + \int_{T_0}^{T} \frac{\Delta C_P^o(T)}{T} dT \qquad (11.70)$$

the second term may be neglected. Thus the enthalpy and entropy terms in Eq. (11.68) may be treated as constants.

With these approximations a plot of ΔG^o for a reaction versus temperature is based simply on Eq. (11.68) with ΔH^o and ΔS^o as constants. Accordingly, $\Delta G^o(T)$ is expected to be linear with a slope equal to $(-\Delta S^o)$ and an intercept at $T = 0$ K equal to the standard enthalpy change for the reaction, ΔH^o. Figure 11.3 shows such a plot for the formation of nickel oxide. Deviations from linearity are measurable but not important in most practical applications. The curve has three discontinuities in its slope. These are associated with phase changes of the components in the reaction. For example, at 1450°C nickel metal melts. Below that temperature the reference state for nickel is pure crystalline (FCC) nickel, with the reaction

$$2Ni(c) + O_2 = 2NiO(c)$$

Above the melting point of nickel the stable form, and thus the reference state for the component in the reaction, is pure liquid nickel. The ΔG^o plot represents the reaction

$$2Ni(l) + O_2 = 2NiO(c)$$

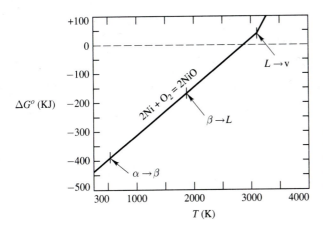

FIGURE 11.3
Standard free energy change for the formation of nickel oxide (NiO) plotted as a function of temperature illustrates that the temperature dependence of ΔH^o and ΔS^o can be neglected.

The enthalpies and entropies of these reactions differ by amounts that correspond to the difference in reference states for the nickel as expressed in the reaction

$$2Ni(c) = 2Ni(l)$$

showing that they differ by twice the heat and entropy of fusion of nickel. Thus at the melting point the slope of the ΔG^o plot changes by $2\Delta S_f^o$ for pure nickel. The intercept at $T = 0$ K changes by $2\Delta H_f^o$. The more drastic change in slope and intercept that occurs at 3380°C is associated with the vaporization of nickel. Entropies of vaporization are typically an order of magnitude larger than entropies of fusion. A break in the curve, which decreases its slope, can be traced to a phase change for the product, in this case NiO.

Thus a plot of ΔG^o versus temperature is typically a broken line with straight segments with the discontinuities in slope corresponding to phase changes that occur in the components of the reaction.

It is frequently useful to compare the stabilities of families of compounds in seeking answers to practical questions such as "Which of these metals is more resistant to oxidation at 1000°C?" or "Which of these nitrides is more stable, or less likely to dissociate, at 1400°C?" Such comparisons are facilitated by the existence of patterns of behavior within classes of similar compounds. In developing a comparison of the oxides, nitrides, or the carbides, it is useful to formulate the problem in terms of reactions written to involve the same number of moles of the component common in all the reactions. Thus to compare oxidation behavior, write and balance all the reactions considered in terms of the consumption of one mole of oxygen.

Figure 11.4 shows a plot formulated in this way, known as a *Richardson Elling-ham Chart*, introduced independently in the 1940s by these scientists. Standard free energy changes for a collection of oxidation reactions are plotted as a function of temperature on this chart. The curve representing a given reaction is a broken line, with changes in slope occurring at phase changes for the metal or oxide. Each line segment is described by Eq. (11.68) and has a slope equal to $-\Delta S^o$ for the reaction and an intercept *at* $T = 0$ K equal to ΔH^o. At low temperatures where both metal and oxide are solid phases, the lines all have essentially the same slope. Evidently the standard entropy changes for these reactions are nearly identical. This reflects the fact that the primary contribution to the change in entropy for these reactions comes from the change in volume associated with the contraction of one mole of the gas phase for each unit of reaction that occurs. Differences between entropies of the solid phases exist but are small in comparison with this effect. In support of this point note that the reaction

$$C + O_2 = CO_2$$

is essentially horizontal on this plot. In this case one mole of gas forms one mole of gas, the volume change is essentially zero, and the change in entropy associated with the volume change is zero. Similarly, the reaction

$$2C + O_2 = 2CO$$

has a *negative* slope because the system expands from one to two moles of gas for each unit of reaction that occurs.

Since the entropies of the reactions that transform a metal to its oxide have similar values, the major differences between the curves are contained in the heats of reaction displayed as the intercept for each curve on the $T = 0$ K ordinate. Since the temperature scale plotted on the diagram is in degrees Celsius, the intercept at 0 K corresponds to an intercept on the line at $-273°C$ to the left of the ordinate. The order of the reaction lines from the top to the bottom of the chart is primarily determined by the corresponding heat of reaction per mole of oxygen consumed.

Each of the oxidation reactions represented in this chart is written on the basis of one mole of oxygen consumed. If the oxide has the formula M_uO_v, the balanced equation for the reaction reads

$$\frac{2u}{v}M + O_2 = \frac{2}{v}M_uO_v \tag{11.71}$$

One consequence of this tactic is that the equilibrium constant for all of the reactions has the form

$$K = \frac{a_{M_uO_v}^{(2/v)}}{a_M^{(2u/v)} P_{O_2}} \tag{11.72}$$

In most applications the departure of the composition of the oxide from its reference state is negligible and the activity of the oxide can be taken as 1. If the metal in the problem is pure, or the solvent in a dilute solution, its activity may also be taken as 1. With these two assumptions, Eq. (11.72) may be written

$$K = \frac{1}{P_{O_2}} \tag{11.73}$$

In this case the value of P_{O_2} determined from the chart is the partial pressure of oxygen that is in equilibrium with the pure metal and its oxide. This value of P_{O_2} is frequently referred to as the *dissociation pressure* of the oxide because it reports the limit of stability of the oxide in question. If the value of P_{O_2} for a given atmosphere under consideration lies below the pressure in equilibrium with the oxide, the oxide spontaneously decomposes or dissociates into the metal and oxygen.

It is very useful to construct some additional scales on this chart that make it possible to read equilibrium constants and dissociation pressures graphically. Because for any reaction $\Delta G^o = -RT \cdot \ln K = (-R \cdot \ln K) \cdot T$, any combination of values $(\Delta G^o, T)$, which represents any point on the chart, also represents a particular value of K. The locus of points in $(\Delta G^o, T)$ space that all share the *same* value of K is a straight line through the origin with a slope equal to $(-R \cdot \ln K)$. The value of the equilibrium constant corresponding to any point on the chart could be read directly if an envelope of such labelled straight lines were superimposed as shown in Fig. 11.5 The same information could be obtained simply by constructing the scale at the right and bottom perimeter of this graph with points labelled that correspond to the intercept of a line with a fixed K value on that scale. With this scale the value of K corresponding to the combination of ΔG^o and T labelled P can be obtained simply by overlaying a

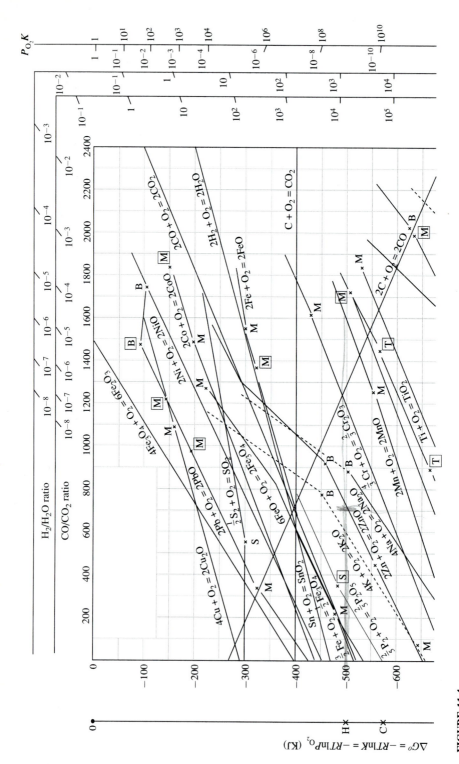

FIGURE 11.4
Richardson–Ellingham for the formation of oxides.

326

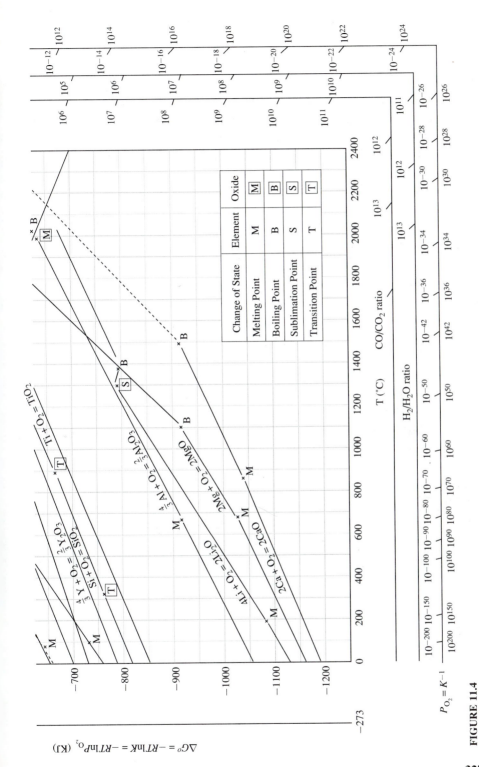

FIGURE 11.4

327

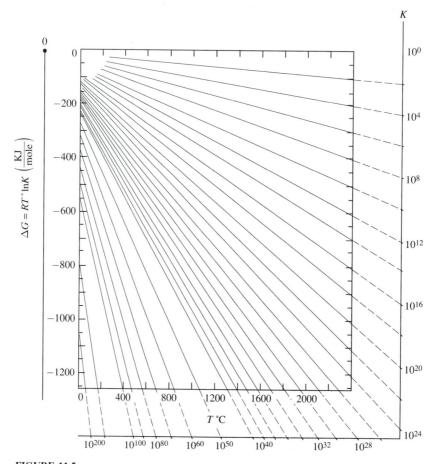

FIGURE 11.5
Lines that represent constant values of the equilibrium constant, K, plotted in ΔG^o-T space corresponding to the Richardson-Ellingham chart.

straightedge from the origin at O through P. The intercept on the K-scale, properly interpolated, gives the corresponding value for the equilibrium constant.

Example 11.5. Find the equilibrium constant for the oxidation of zinc at 700°C.
Find the zinc oxide reaction line on the chart. Read the value of ΔG^o for this reaction at 700°C,

$$\Delta G^o_{700C} = -500 \text{ KJ}$$

The equilibrium constant corresponding to this condition can be calculated algebraically from this value.

$$K = e^{-(-500,000/8.314\cdot973)} = 7 \times 10^{26}$$

To solve the problem using the K scale on the chart, find the point on the zinc oxide line at 700°C. With a straightedge, lay a line between this point and the origin O. Read the intercept of this line on the K-scale and interpolate, visualizing a log scale, between 10^{26} and 10^{27}.

If the approximations that lead to Eq. (11.73) are valid, specifically that the metal and oxide activities are approximately 1, then the value of K read from the K-scale is simply the reciprocal of the dissociation pressure for the oxide in the reaction under consideration. An oxygen pressure scale constructed in this manner is incorporated into the Richardson-Ellingham chart, Fig. 11.4. Use of this scale allows quick answers to a variety of questions that may be asked about the reactions represented in the chart. Points on the K-scale are given by the reciprocal of the value on the P_{O_2} scale.

Example 11.6. Evaluate the dissociation pressure of zinc oxide at $700°C$.
    The equilibrium constant for this reaction at $700°C$ is computed in Example 11.5 to be 7×10^{26}. According to Eq. (11.73),

$$P_{O_2} = \frac{1}{K} = 1.4 \times 10^{-27} \text{atm}$$

The problem can be solved graphically with the P_{O_2} scale in Fig. 11.4. On the zinc oxide reaction line mark the point at $700°C$. Lay a straightedge from this point through the origin. Read the intercept on the P_{O_2} scale, interpolating logarithmically between 10^{-27} and 10^{-26}.

If the assumptions leading to Eq. (11.73) are not valid, then the activity of the metal in the system, such as a solute in a dilute solution, must be measured or modelled and input to the calculation. If the solvent in this binary metallic system is noble, meaning that it does not form an oxide spontaneously, then the K-scale can be used to compute the equilibrium constant and the more general form, Eq. (11.72), can be applied with the activity of the oxide taken as 1. If the second element in the alloy can also react to form an oxide, then a more sophisticated treatment is required.

Example 11.7. Compute the partial pressure of oxygen in an atmosphere that is equilibrated at $700°C$ with zinc oxide and a gold-zinc alloy with $X_{Zn} = 0.005$. Assume the Henry's law constant for this dilute solution is 8.5 at $700°C$. Except for its effect upon the activity of zinc in solution, the gold can be treated as inert in this system.
    The activity of zinc in the alloy is

$$a_{Zn} = \gamma_{Zn}^o X_{Zn} = 8.5 \cdot (0.005) = 0.0425$$

Insert this value in the equilibrium constant read from the K-scale on the chart:

$$K = \frac{1}{a_{Zn}^2 P_{O_2}} = 7 \times 10^{26}$$

Solving for P_{O_2}

$$P_{O_2} = \frac{1}{(0.0425)^2 \cdot 7 \times 10^{26}} = 7.9 \times 10^{-25} \text{atm}$$

The partial pressure of oxygen in a system is frequently referred to as its *oxygen potential*. This term derives from the fact that the chemical potential of oxygen in the system is functionally determined by the oxygen partial pressure since

$$\mu_{O_2} = \mu_{O_2}^o + RT \ln P_{O_2} \tag{11.74}$$

If the oxygen potential in a system lies above the equilibrium value at any temperature, then the metal oxidizes and the oxide is stable. If the potential lies below the equilibrium value at the temperature of the system, the oxide is unstable and dissociates. This situation is illustrated in Fig. 11.6. The standard free energy change for the oxidation of zinc at 700°C is given by the point B on this diagram and the corresponding equilibrium oxygen potential is labelled C. Consider a system with the oxygen potential labelled E on the P_{O_2} scale. From the construction that produced the P_{O_2} scale, it can be deduced that the point labelled D has a y value, in units of ΔG, given by $RT \cdot \ln(P_{O_2})_E$. The length $(B - D)$ on the graph is thus:

$$B - D = RT \ln \left(P_{O_2} \right)_{eq} - RT \ln \left(P_{O_2} \right)_E = RT \ln \frac{\left(P_{O_2} \right)_{eq}}{\left(P_{O_2} \right)_E}$$

This difference is precisely the *affinity* for the reaction when the partial pressure is $(P_{O_2})_E$. Recall Eq. (11.35):

$$\mathscr{A} = RT \ln \frac{Q}{K}$$

If the activity of metal and oxide are taken to be 1, then Q is $1/(P_{O_2})_E$ and K is $1/(P_{O_2})_{eq}$ and

$$\mathscr{A} = RT \ln \frac{\left(P_{O_2} \right)_{eq}}{\left(P_{O_2} \right)_E} = B - D \tag{11.75}$$

Thus, the vertical distance from a point on the chart representing any given nonequilibrium oxygen potential to the equilibrium line is equal to the affinity for the reaction for that composition. If D lies above the line, the affinity $(B - D)$ is negative, and by arguments previously developed, products (the oxide) spontaneously form. In the context of the Richardson-Ellingham chart, points in the domain above the equilibrium line have affinities that correspond to oxidation of the metal or the formation of a stable oxide. Points lying below the line yield a positive affinity and products decompose; in this region the oxide is unstable and dissociates.

Example 11.8. Find the affinities for the oxidation reactions for copper, nickel, zinc, titanium, and aluminum in a system with an oxygen potential of 10^{-16} atm at 1000°C.
    The standard free energies for these reactions at 1000°C can be read from the chart:

Component	CuO	NiO	ZnO	TiO$_2$	Al$_2$O$_3$
ΔG_{1000}^o (KJ)	−176	−255	−427	−682	−853

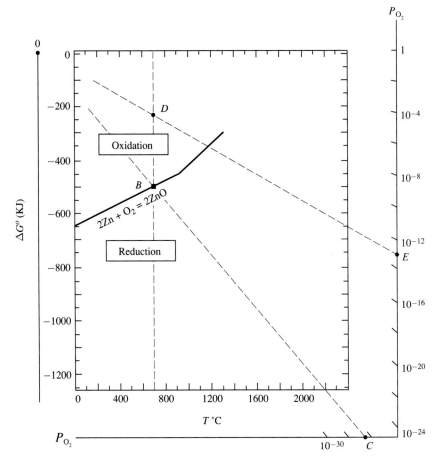

FIGURE 11.6
The line representing equilibrium states between Zn, ZnO, and O_2 divides the graph into two domains. Combinations of oxygen potential and temperature that lie above this line oxidize the metal; combination below the line reduce the oxide.

The value of $RT \cdot \ln P_{O_2}$ corresponding to 10^{-16} atm at 1000°C is

$$D = (8.314) \cdot (1273) \cdot \ln(10^{-16}) = -390,000 \text{ J} = -390 \text{ KJ}$$

The values for the affinities are obtained by subtracting D from ΔG^o for each reaction:

Component	CuO	NiO	ZnO	TiO$_2$	Al$_2$O$_3$
A_{1000} (KJ)	+214	+135	−37	−292	−463

Recall that the affinity may be thought of as $\mu_{\text{products}} - \mu_{\text{reactants}}$. If $A > 0$, $\mu_{\text{products}} > \mu_{\text{reactants}}$ and products decompose. For $A < 0$, $\mu_{\text{products}} < \mu_{\text{reactants}}$ and products form. It is concluded that this atmosphere reduces CuO and NiO, while ZnO, TiO$_2$, and Al$_2$O$_3$ are stable in this system.

The demonstration that the affinity for a reaction for any given oxygen potential can be visualized as the vertical distance from the equilibrium line in the chart provides a convenient tool for deducing patterns of behavior. For example, consider a system with a fixed oxygen pressure containing a metal M and its oxide $M_u O_v$. Figure 11.6 illustrates this situation for Zn and ZnO but the strategy is general. The point E locates the line through the origin that represents the locus of points on the chart that have the oxygen potential given by E. As the temperature of the system is changed for this fixed atmosphere, the point D moves along the line DE. This line crosses the equilibrium reaction line for M at some temperature T_{eq} for which P_{O_2} is the dissociation pressure. At lower temperatures, D lies above the line and the M oxidizes. Above T_{eq} this atmosphere is reducing to $M_u O_v$. Thus if the oxide is heated in this atmosphere it is stable up to T_{eq} and then begins to decompose with further increase in temperature.

A comparison of the relative stabilities of the oxides can be visualized with the same construction. Consider any temperature T on the chart. The oxygen potential for any atmosphere represented by a point D located on that constant temperature line can be obtained from the P_{O_2} scale. Oxides with equilibrium lines that lie above D are reduced, or are unstable, in that atmosphere; those below are stable. Evidently the order of stability of the oxides at any given temperature is given by the order in which their equilibrium lines cross that temperature line.

This observation is the basis for one strategy for preventing the oxidation of a particular metal at high temperatures in the laboratory. For example, the oxidation of a nickel sample can be prevented by encapsulating it with titanium chips, usually prepared by simply turning a titanium rod in a lathe and collecting the chips. The formation of TiO_2, by far the more stable oxide in this system, establishes the oxygen potential in the system. This potential is well below the dissociation pressure of NiO and a nickel sample remains clean.

The arguments just presented for evaluating the affinity extend in a straightforward way to systems in which the metal and oxide are not in their pure reference states so that their activities differ from 1. In this case, the point B in Fig. 11.6 has an equilibrium constant K, which can be read from the K-scale on the chart. A system not in equilibrium that has an activity quotient Q above K on that scale can be represented by the corresponding point D at the temperature T. The affinity for the reaction in this state is given by Eq. (11.35) and again corresponds to the vertical distance $(B - D)$ on the chart. If D lies above B, the metal oxidizes and oxide is stable. If D is below B, the oxide dissociates.

11.3.2   Oxidation in CO/CO_2 and H_2/H_2O Mixtures

In Sect. 11.3.1 it is assumed that the only component present in the gas phase is oxygen. Thus the only means of controlling the oxygen partial pressure is through a reduction in the total pressure of the system. Since the best vacuum attainable in the laboratory is about 10^{-10} atm, which is well above most dissociation pressures, this approach to controlling oxidation is both inflexible and very limited. The most convenient means for controlling the oxygen potential of an atmosphere and thus the oxidation reactions that occur in its presence is through control of the chemical composition of the gas phase. The simplest atmospheres that provide this control incorporate mixtures of CO/CO_2, H_2/H_2O, or both. The development required to understand the behavior of these two classes of atmospheres is identical.

Focus on oxidation behavior in an atmosphere containing CO and CO_2. At any temperature the oxidation potential, which is the partial pressure of oxygen in the equilibrated atmosphere, is controlled by the ratio of the partial pressures of CO and CO_2. This relationship derives from the condition for equilibrium that applies to an atmosphere containing these three components. The reaction that describes this univariant system is

$$2CO + O_2 = 2CO_2$$

The standard free energy change for this reaction is plotted as a function of temperature on the Richardson-Ellingham chart. The equilibrium constant for this reaction, assuming the gas mixture to be ideal, is

$$K = \frac{X_{CO_2}^2}{X_{CO}^2} \cdot \frac{1}{X_{O_2}} = \frac{P_{CO_2}^2}{P_{CO}^2} \cdot \frac{1}{P_{O_2}} \tag{11.76}$$

assuming the total pressure in the system is one atmosphere. At any temperature ΔG^o for this reaction can be read from the chart and the numerical value of K can be computed or read from the K-scale. At that temperature, for any value of the partial pressure of oxygen, there is a unique value for the ratio (P_{CO_2}/P_{CO}):

$$P_{O_2} = \frac{P_{CO_2}^2}{P_{CO}^2} \cdot \frac{1}{K} \tag{11.77}$$

A high ratio of CO_2 to CO in the atmosphere yields a high oxygen potential. To decrease the oxygen potential, increase the relative concentration of the reducing gas, CO.

The construction in Fig. 11.5 demonstrates that the ordinate on the chart can be interpreted as an $[RT \cdot \ln P_{O_2}]$ scale. With this interpretation in mind, consider the locus of points in this graph that correspond to a *fixed ratio of* $[P_{CO_2}/P_{CO}]$. The connection is established through the condition for equilibrium related to Eq. (11.76):

$$\Delta G_{[CO_2]}^o = \Delta H_{[CO_2]}^o - T \Delta S_{[CO_2]}^o = -RT \ln \left(\frac{P_{CO_2}^2}{P_{CO}^2} \frac{1}{P_{O_2}} \right)$$

which can be written

$$\Delta H_{[CO_2]}^o - T \Delta S_{[CO_2]}^o = -RT \ln \left(\frac{P_{CO_2}}{P_{CO}} \right)^2 - RT \ln \frac{1}{P_{O_2}}$$

$$RT \ln P_{O_2} = \Delta H_{[CO_2]}^o + T \left[R \ln \left(\frac{P_{CO_2}}{P_{CO}} \right)^2 - \Delta S_{[CO_2]}^o \right] \tag{11.78}$$

For a constant ratio of $[P_{CO2}/P_{CO}]$ a plot of this relationship on an $(RT \cdot \ln P_{O_2})$ versus T chart is a straight line with an intercept at $T = 0$ K of $\Delta H_{CO_2}^o$ and a slope equal to the quantity in brackets in Eq. (11.78). Figure 11.7 shows the envelope of straight

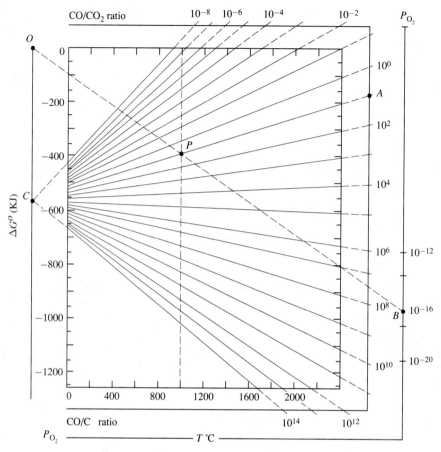

FIGURE 11.7
Locus of points on the Richardson-Ellingham chart that correspond to fixed values of CO/CO_2 in the gas phase.

lines, all passing through the point C on the $T = 0$ K line, that represent constant (CO/CO_2) ratios. This envelope can be replaced by constructing a (CO/CO_2) scale around the perimeter of the chart in the same manner as the K and P_{O_2} scales already discussed.

To calculate the oxygen potential of any given CO/CO_2 mixture at a particular temperature, use the K-scale to determine the equilibrium constant for the CO/CO_2 reaction, substitute into Eq. (11.77), and compute the partial pressure of oxygen in equilibrium in the given atmosphere and temperature. To determine the oxygen potential graphically using Fig. 11.7, lay a straightedge on the chart connecting the C point with the given value on the (CO/CO_2) scale, labelled A in the example in Fig. 11.7. Mark the point P at which this line intersects the temperature of interest, 1000°C in this example. The oxygen partial pressure corresponding to the point P is found as before. Lay a straightedge from O through P and read its intersection on the P_{O_2}

scale, labelled B in Fig. 11.7. This construction can be used to read oxygen potentials for any combination of (CO/CO_2) ratio and temperature on the chart.

The procedure for relating (H_2/H_2O) ratios to oxygen partial pressures is identical except that the intercept of the envelope of radiating lines is $\Delta H^o_{H_2O}$, labelled H on the chart, and the (H_2/H_2O) scale applies. All four scales, K, P_{O_2}, (CO/CO_2), and (H_2/H_2O), are constructed in Fig. 11.4.

Example 11.9. Estimate the oxygen potential for a gas mixture with a CO_2/CO ratio of 10^{-4} at $1200°C$.

Solution. The standard free energy change for the CO-CO_2 reaction at $1200°C$ can be read from Fig. 11.4: $\Delta G^o = -310$ KJ. The corresponding equilibrium constant is

$$K = e^{-(-310,000/8.314 \cdot 1473)} = 9.8 \times 10^{10}$$

The corresponding oxygen partial pressure is given by Eq. (11.77)

$$P_{O_2} = \left(\frac{P_{CO_2}}{P_{CO}}\right)^2 \cdot \frac{1}{K} = \left(10^{-4}\right)^2 \cdot \frac{1}{9.8 \times 10^{10}} = 1.0 \times 10^{-19} \text{atm}$$

To obtain the same result graphically, lay a straightedge on the chart connecting the point C on the ordinate and the value 10^{+4} on the (CO/CO_2) scale. (Caution: the scale is constructed in terms of the ratio of CO to CO_2 and the input information in this example gives the value for the ratio of CO_2 to CO.) Mark the point at which this line crosses the $1200°C$ temperature line; the coordinates of this point are $(1200, -530)$. Now lay a straightedge from the O point on the chart through this point and read to P_{O_2} scale:

$$P_{O_2} \cong 10^{-19} \text{atm}$$

Example 11.10. What is the maximum water content that could be tolerated in a hydrogen atmosphere used to prevent the oxidation of copper samples annealed at $900°C$?

Solution. Use the H_2/H_2O scale to find the ratio in equilibrium with copper and its oxide at $900°C$. Lay a straightedge from the H point on the ordinate through the point on the diagram where the ΔG^o curve for Cu and Cu_2O crosses the $900°C$ line; the coordinates of this point are $(900, -184)$. Read the corresponding equilibrium ratio on the (H_2/H_2O) scale:

$$\frac{P_{H_2}}{P_{H_2O}} \cong \frac{1}{10^3} \cong 10^{-3}$$

The corresponding oxygen partial pressure is about 10^{-8} atm and can be neglected in computing the composition of the atmosphere. Thus, as long as the partial pressure of water vapor is kept below 10^{-3}, or about 0.1%, the formation of copper oxide at $900°C$ is prevented. The composition of water vapor can be monitored by measuring the dew point of the atmosphere.

The arguments presented in Sect. 11.3.1 that demonstrate the application of the Richardson-Ellingham chart to compute the affinity for a system not at equilibrium can also be applied to the (CO/CO_2) and (H_2/H_2O) scales on the chart. For example, any combination of an atmosphere composition given by a (CO/CO_2) ratio and a temperature plots as a point D on the chart, as shown in Fig. 11.6. Every such point has a corresponding oxygen potential. The vertical distance from that point D to a point B on an equilibrium line that represents oxidation of some metal M of interest is the affinity for the reaction

$$\frac{2u}{v}M + 2CO_2 = \frac{2}{v}M_uO_v + 2CO \qquad (11.79)$$

If D lies above B, the atmosphere oxidizes M and if D is below B, the oxide dissociates.

11.4   PREDOMINANCE DIAGRAMS AND MULTIVARIANT EQUILIBRIA

Consider a system with c components made from e elements and exhibiting $r = (c-e)$ independent reactions. The conditions for chemical equilibrium in this system can be described by a system of equations that have the form given by Eq. (11.67):

$$\Delta G^o_{[j]} = -RT \ln K_{[j]} \qquad [j = 1, 2, \ldots, r] \qquad (11.67)$$

It was emphasized that an equation of this form can be written for every possible chemical reaction among the components in the system, although only r of these equations are independent. The rest of the equations describing the system can be deduced from these independent equations. As the number of components increases, the calculation and representation of the equilibrium state becomes progressively less tractable.

A convenient representation of the behavior of a multivariant system is embodied in a *predominance diagram* that can be constructed to represent its complex behavior as a reacting system. Of the collection of independent potentials necessary to describe the state of the system, two that are of particular interest in a given application are chosen as variables to be explored in describing the behavior of the system. All but one of the remaining potentials are fixed. This permits display of the rivalry between the components that are competing for predominance in the system on a two dimensional graph. Because the variables involved are all thermodynamic potentials, such *predominance diagrams* take the form of a cell structure with areas that represent domains of predominance of a particular component separated by lines that are *limits of predominance* of competing components. These lines meet at triple points. Predominance diagrams have an appearance similar to phase diagrams plotted in potential space (see Secs. 10.3 and 10.4). However, as is demonstrated, while a predominance diagram may provide an approximate representation of such a phase diagram, it is not identical with a phase diagram. *Domains of predominance* displayed in these diagrams are not identical with the *domains of stability* normally represented in phase diagrams.

11.4.1   Pourbaix High Temperature Oxidation Diagrams

This form of predominance diagram chooses as independent variables the oxygen potential and temperature and displays the domains of predominance of the various oxides of a metal that exhibits more than one stable oxide. Consider a metal M that is found to exist in two different oxidation states yielding oxides with the formulas MO and MO_2. This system of four components (M, O_2, MO, and MO_2) and two elements (M and O) has $r = 4 - 2 = 2$ independent reactions; it is *bivariant*. A total of three reactions can be written among the components. As in the development of the Richardson-Ellingham chart, it is convenient to write these reactions in terms of one mole of O_2 consumed in each case.

$$2M + O_2 = 2MO \tag{11.9}$$

$$M + O_2 = MO_2 \tag{11.10}$$

$$2MO + O_2 = 2MO_2 \tag{11.11}$$

Let R denote the component other than O_2 on the reactant side of each equation and P denote the product component. Then each of these reactions has the generic form

$$xR + O_2 = yP \tag{11.12}$$

The condition for equilibrium associated with each reaction has the form

$$\Delta G^o = -RT \ln K = -RT \ln \frac{a_P^y}{a_R^x} \cdot \frac{1}{P_{O_2}} \tag{11.80}$$

which can also be written

$$\Delta H^o - T\Delta S^o = -RT \ln f + RT \ln P_{O_2}$$

where f is defined as the *predominance ratio* for the reaction

$$f \equiv \frac{a_P^y}{a_R^x} \tag{11.81}$$

If $f \gg 1$, the product component in the reaction predominates at equilibrium; if $f \ll 1$, the reactant predominates. Solve for

$$\ln P_{O_2} = \frac{\Delta H^o}{RT} + \left[\ln f - \frac{\Delta S^o}{R} \right] \tag{11.82}$$

In the practical application of this equation it is convenient to convert from natural to common logarithms by dividing through by $\ln 10 = 2.303$.

$$\log P_{O_2} = \frac{\Delta H^o}{R'T} + \left[\log f - \frac{\Delta S^o}{R'} \right] \tag{11.83}$$

where $R' = 2.303R = 19.147$ J/mole-K. This is the working equation involved in constructing these predominance diagrams. It applies to all oxidation reactions written so that the coefficient of O_2 is 1. It is a rewritten form of Eq. (11.80), the condition for equilibrium.

Assume that ΔH^o and ΔS^o are independent of temperature, so that the integrals involving ΔC_P for the reaction are negligible, [see Eq. (11.69) and (11.70)]. For a fixed value of f, the predominance ratio, a plot of $\log P_{O_2}$ versus $(1/T)$ for any given equilibrium is a straight line with slope equal to $(\Delta H^o/R')$ and intercept given by $[\log f - \Delta S^o/R']$. It is conventional to graph the $(1/T)$ scale in reverse so that the temperature qualitatively increases from left to right; this convention changes the sign of the slope on the graph to $(-\Delta H^o/R')$. Since oxidation reactions are exothermic (heat is given off in the reaction so that the heat absorbed by the system is negative) slopes of lines on these plots are positive.

Figure 11.8 sketches a plot of Eq. (11.82) for a reaction like [11.9]. In this representation the predominance ratio f is treated as a parameter and allowed to vary from 10^{-4} to 10^{+4}. Because the scale is logarithmic, each order of magnitude change in f displaces the intercept by 1 unit. Below the line, where $f \ll 1$, the reactants predominate at equilibrium. Above the line where $f \gg 1$, the product component is the predominant species at equilibrium. In the range of values of f near $f = 1$ neither reactants nor products are said to predominate. It is convenient to define the condition $f = 1$ to be the *limit of predominance* with respect to the competing components. The line drawn with the condition $f = 1$ serves as a boundary between domains of predominance of the components involved.

The competition between three of the components, such as that represented by Eqs. [11.9], [11.10] and [11.11] at the beginning of this section, can be represented by three envelopes of parallel lines like those shown for a single reaction in Fig. 11.8. Note that the values of ΔH^o and ΔS^o for these three reactions are not independent;

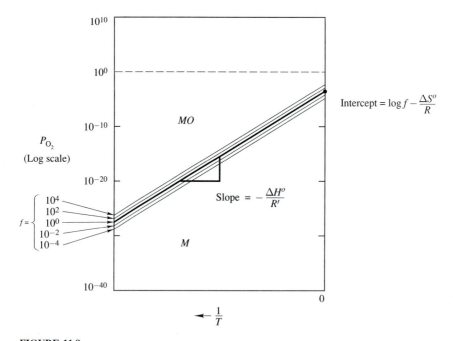

FIGURE 11.8
Plot of equilibrium oxygen potential for reacton [11.9] as a function of temperature for a set of values of the predominance ratio, f.

reaction [11.11] is a linear combination of reactions [11.9] and [11.10]. For the particular stoichiometry in this example, [11.11] = 2 · [11.10] − [11.9]. Figure 11.9*a* illustrates the juxtaposition of three competing reaction lines. Figure 11.9*b* shows details of the structure in the region around their common intersection. Focus on the solid lines representing the *limits of predominance*, $f = 1$, in each case. That these three lines intersect at a single point can be derived algebraically from the relationships between their slopes and intercepts dictated by the linear relation among the three chemical reactions they represent. Segments of these three intersecting lines bound domains of predominance of each of the competing components. Extensions of these lines beyond the triple point are dashed because in each case they represent competition between two components in a domain in which the third component has independently been shown to be predominant. Below the line labelled *OA* in Figure 11.9*a*, M predominates with respect to MO; above the line *OC*, MO_2 predominates at equilibrium relative to MO. Thus the dashed segment *OB* represents the competition between M and MO_2 in a region in which MO has been shown to be the predominant component with respect to *both* M and MO_2. The same argument can be applied to the extensions labelled *OD* and *OF*.

To generate a high temperature oxidation predominance diagram it is necessary to list all the oxide forms known to exist for the given metal. Heats and entropies of formation of each of the oxide forms must be obtained as input. Next an exhaustive list of all the possible reactions between metal and oxides, as well as between the oxides, is assembled. Values of ΔH^o and ΔS^o can be computed for all these reactions from the information about the formation reactions and no additional information is required. The generic form of an oxidation reaction on these diagrams is

$$x M_a O_b + O_2 = y M_u O_v$$

It is easy to show that the coefficients required to balance these equations depend on the chemical formulas of the components as

$$x = \frac{2u}{va - ub} \qquad y = \frac{2a}{va - ub} \tag{11.84}$$

Reactions involving the pure metal M can be included in this compilation simply by setting $b = 0$. Enthalpies and entropies of all reactions are related to the corresponding properties in the formation reactions

$$\Delta H^o_{[f]} = y \Delta H^o_{f, M_u O_v} - x \Delta H^o_{f, M_a O_b} \tag{11.85}$$

and similarly for the entropy change.

An efficient approach for generating such a diagram focuses on computation of the triple points formed by the intersection of triplets of predominance limit lines. Begin with a list of the components other than oxygen, specifically the metal and all of its oxides. Make an exhaustive list of the arrangements of these components taken three at a time. This represents the possible triple points that may exist in the system. The coordinates of any triple point can be computed from the intersection

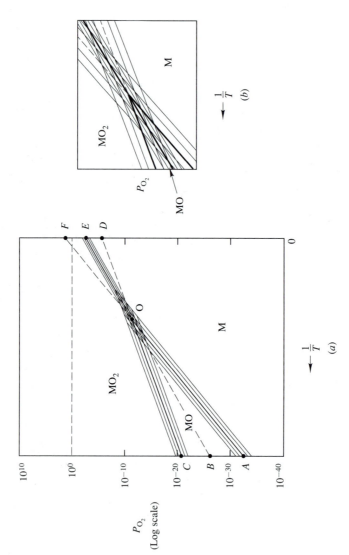

FIGURE 11.9

(a) Predominance diagram for a system with three competing components, M, MO and MO_2, (b) shows the pattern of predominance ratios near the triple point. solid lines correspond to the *limits of predominance*, for which $f = 1$.

of any two of the three lines that meet at that point. Consider for example, the triple point formed in the limits of predominance for three oxides, M_2O, MO, and MO_2. The corresponding oxidation reactions, based on one mole of oxygen, can be written with coefficients determined by inspection or computed from Eq. (11.84) as

$$2M_2O + O_2 = 4MO \qquad\qquad\qquad [A]$$

$$\frac{2}{3}M_2O + O_2 = \frac{4}{3}MO_2 \qquad\qquad\qquad [B]$$

$$2MO + O_2 = 2MO_2 \qquad\qquad\qquad [C]$$

From heats and entropies of formation of these compounds, compute ΔH^o and ΔS^o for each of these reactions. Note that [C] is a linear combination of [A] and [B]. Expressions based on Eq. (11.83) give equations for the limits of predominance for reactions [A] and [B]:

$$\log P_{O_2} = \frac{\Delta H^o_{[A]}}{R'T} - \frac{\Delta S^o_{[A]}}{R'}$$

$$\log P_{O_2} = \frac{\Delta H^o_{[B]}}{R'T} - \frac{\Delta S^o_{[B]}}{R'}$$

$$\log P_{O_2} = \frac{\Delta H^o_{[C]}}{R'T} - \frac{\Delta S^o_{[C]}}{R'}$$

Because the slopes and intercepts of these lines are related, it can be shown that these lines intersect at a single point. The coordinates of this triple point in $(1/T, \log P_{O_2})$ space can be found by solving any two of these equations simultaneously. The resulting relations are

$$\frac{1}{T} = \frac{\Delta S^o_{[A]} - \Delta S^o_{[B]}}{\Delta H^o_{[A]} - \Delta H^o_{[B]}} \qquad\qquad (11.86)$$

$$\log P_{O_2} = \frac{\Delta H^o_{[B]}\Delta S^o_{[A]} - \Delta H^o_{[A]}\Delta S^o_{[B]}}{R'(\Delta H^o_{[A]} - \Delta H^o_{[B]})} \qquad\qquad (11.87)$$

Coordinates of all the triple points that exist among the components taken three at a time can be computed from equations like these. Triple points that lie outside the domain of the diagram [such as cases for which $(1/T)$ is negative or $P_{O_2} > 10^{10}$] are deleted from the list. A further subset of triple points can be deleted from consideration because the triple points lie in a region of predominance of a component not involved in forming the triple point. This subset can be determined from an analysis of the predominance ratios for all the reactions that do not contain the three components in a given triple point. The remaining triple points all appear on the finished predominance diagram. To complete the diagram, construct straight lines representing limits of predominance between pairs of triple points that share a common pair of components. Finally, label the fields thus enclosed with the component that is predominant in that area.

Example 11.11. Construct a predominance diagram for manganese and its oxides:

Component	H_f^o (KJ)	S_f^o (J/K)
Mn	0	0
MnO	-385	-73
Mn_3O_4	-1387	-357
MnO_2	-521	-184

(Mn_2O_3 may be observed under some conditions but is neglected in this example.)

Solution. First write down all of the balanced reactions that combine these oxides with each other or with the metal Mn. There are 4!/(2!2!) or six such reactions. Compute the values of ΔH^o and ΔS^o for all six reactions; they are linear combinations of the formation quantities given above. Use these values to compute slopes and intercepts and plot the limits of predominance given by setting $f = 1$ in Eq. (11.83). Next write out the possible combinations of these four components taken three at a time, 4!/(3!2!) or four possible triple points. Use Eqs. (11.86) and (11.87) to compute the triple points in the diagram:

Triple Point	$(1/T)(K^{-1})$	log P_{O_2}
[Mn, MnO, Mn_3O_4]	-4.24×10^{-4}	24.21
[Mn, MnO, MnO_2]	-1.53×10^{-4}	13.76
[Mn, Mn_3O_4, MnO_2]	-3.19×10^{-4}	10.48
[MnO, Mn_3O_4, MnO_2]	2.81×10^{-4}	7.60

In this case three of the triple points yield negative values for $1/T$. This is mathematically valid, but not thermodynamically so. Only the triple point [MnO, Mn_3O_4, MnO_2] lies within the diagram; however even this point occurs at a very high oxygen potential corresponding to a partial pressure of oxygen of about 10^8 atm. Connect the dots to form the diagram, Fig. 11.10.

11.4.2 Predominance Diagrams with Two Compositional Axes

Consider the reaction of a metal M with an atmosphere that contains two independent compositional variables. The most common example of this class of diagram displays domains of predominance of the various compounds that may be formed in the presence of an atmosphere containing both oxygen and sulfur, O_2 and S_2. The axes of the diagram are taken to be the potentials log P_{O_2} and log P_{S_2}. In order to plot such a diagram the temperature must be fixed. Competing components now involve oxides, sulfides, and components with M, S, and O in their formulas. Three types of generic

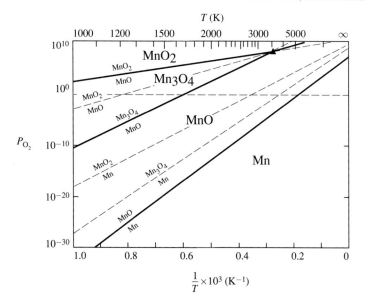

FIGURE 11.10
Predominance diagram for the manganese-oxygen system constructed by computing triple points and connecting them appropriately.

equations can be constructed:

No S_2: $\qquad\qquad x_1 M_a S_b O_c + O_2 = y_1 M_u S_v O_w$ [11.13]

No O_2: $\qquad\qquad x_2 M_a S_b O_c + S_2 = y_2 M_u S_v O_w$ [11.14]

Both S_2 and O_2: $\qquad x_3 M_a S_b O_c + O_2 + m S_2 = y_3 M_u S_v O_w$ [11.15]

Normally equations of class [11.13] are simple oxidation equations, generically identical with that which leads to Eq. (11.84) for the stoichiometric coefficients. However, the components may contain sulfur provided that $b/a = v/u$. A similar statement can be made for the sulfidation equation of class [11.14]. Equation (11.84) also yields the stoichiometric coefficients required to balance this class of equations. The coefficients for equations in class [11.15] are

$$ x_3 = \frac{2u}{aw - cu} \qquad y_3 = \frac{2a}{aw - cu} \qquad m = \frac{av - ub}{aw - cu} \qquad (11.88) $$

Example 11.12. Construct a balanced equation with the components MS_2, O_2, S_2, and $M_2 S_3 O_5$.

Comparison with the generic form given in equation [11.15] above gives

$$ M_a S_b O_c \text{ is } MS_2 \rightarrow a = 1 \qquad b = 2 \qquad c = 0 $$

$$ M_u S_v O_w \text{ is } M_2 S_3 O_5 \rightarrow u = 2 \qquad v = 3 \qquad w = 5 $$

so that

$$ (aw - cu) = (1 \cdot 5 - 0 \cdot 2) = 5 $$

and

$$(av - ub) = (1 \cdot 3 - 2 \cdot 2) = -1$$

The corresponding coefficients are

$$x_3 = \frac{(2 \cdot 2)}{5} = \frac{4}{5} \qquad y_3 = \frac{(2 \cdot 1)}{5} = \frac{2}{5} \qquad m = -\frac{1}{5}$$

The balanced equation reads

$$\frac{4}{5}MS_2 + O_2 = \frac{1}{5}S_2 + \frac{2}{5}M_2S_3O_5$$

The negative result obtained for m simply places S_2 on the right side of the equation. To check the balance note that there are 4/5 moles of M atoms, 2 moles of O and 8/5 moles of S on both sides of the equation.

A list of the components considered to exist in the system yields an exhaustive compilation of the reactions between competing components. The condition for equilibrium for any of these reactions can be written as

$$\Delta G^o_{[j]} = -RT \ln K_{[j]} = -R'T \log K_{[j]} \qquad [j = 1, 2, \ldots, r] \tag{11.89}$$

For reactions of type [11.13] that do not involve S_2

$$K_{[11.13]} = \frac{f_{[11.13]}}{P_{O_2}}$$

where $f_{[11.13]}$ is the predominance ratio for the non-oxygen components. Equation (11.89) becomes

$$\Delta G^o_{[11.13]} = -R'T \log f_{[11.13]} + R'T \log P_{O_2}$$

At the limit of predominance $f_{[11.13]} = 1$ and $\log f_{[11.13]} = 0$. Thus the equation defining the limit of predominance for this class of reactions is

$$\log P_{O_2} = \frac{\Delta G^o_{[11.13]}}{R'T} \tag{11.90}$$

A similar argument applied to equations of type [11.14] yields

$$\log P_{S_2} = \frac{\Delta G^o_{[11.14]}}{R'T} \tag{11.91}$$

for the limit of predominance. The equilibrium constant for reactions that involve both S_2 and O_2 is

$$K_{[11.15]} = \frac{f_{[11.15]}}{P_{O_2} P_{S_2}^m}$$

Equation (11.89) becomes

$$\Delta G^o_{[11.15]} = -R'T \left[\log f_{[11.15]} - \log P_{O_2} - m \log P_{S_2} \right]$$

The limit of predominance for this class of reactions is obtained by setting $f_{[11.15]} = 1$ or $\log f_{[11.15]} = 0$. Then

$$\log P_{O_2} = -m \log P_{S_2} + \frac{\Delta G^o_{[11.15]}}{R'T} \tag{11.92}$$

Expressions (11.90), (11.91), and (11.92) are the equations for the limits of predominance for each of the three classes of reactions. When plotted on coordinates of log P_{O_2} versus log P_{S_2} as in Fig. 11.11, these equations yield simple results. On this scale Eq. (11.90) simply reads "log P_{O_2} is a constant" and plots as a horizontal line. This reflects the fact that S_2 is not involved in class [11.13] reactions so that their competition is independent of the S_2 content in the atmosphere. Equation (11.91) reads "log P_{S_2} is a constant and plots as a vertical line." Competitions between components that involve both O_2 and S_2, class [11.15], described by Eq. (11.92), plot as a straight line with slope equal to $-m$ [which is determined by the formulas for the compounds involved and is given by Eq. (11.88)] and an intercept (at log $P_{S_2} = 0$ or $P_{S_2} = 1$) on the log P_{O_2} scale at $\Delta G^o_{[11.15]}/R'T$. An example of the predominance diagram for copper in oxygen-sulfur atmospheres at 1000 K is shown in Fig. 11.12.

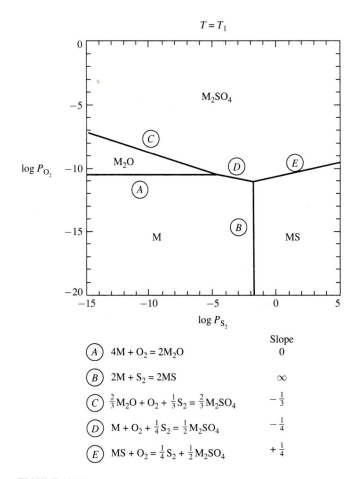

FIGURE 11.11
Predominance diagram for a system describing reactions with a gas containing both oxygen and sulfur.

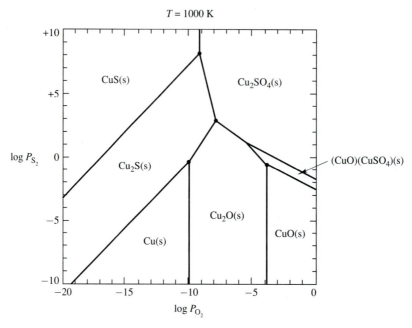

T = 1000 K

FIGURE 11.12
Predominance diagram calculated for the Cu-O-S system at 1000K

11.4.3   Interpretation of Predominance Diagrams

Predominance diagrams provide a convenient means for displaying the behavior of multivariant reacting systems. Their construction assumes that the system has reached equilibrium and also that heats and entropies of reactions are essentially independent of temperature. It is important that the distinction be established between a predominance diagram for a system and the phase diagram for the same system. Figure 11.13 compares the predominance diagram and the phase diagram for the iron-oxygen system. The phases shown in Figure 11.13a are: α and γ iron-rich terminal phases, the liquid, and the three intermediate phases, two line compounds (at least at low temperatures), Fe_2O_3 and Fe_3O_4, and wustite, which has a variable composition near FeO. At the composition and temperature labelled P in Fig. 11.13a, the system consists of a single phase, which is wustite, at equilibrium.

In contrast, the predominance diagram treats the three intermediate compounds, Fe_2O_3, Fe_3O_4, and FeO, as *components* rather than phases. At the corresponding point P in the predominance diagram in Fig. 11.13b, the system is considered to consist of a mixture of all the components Fe, FeO, Fe_2O_3, and Fe_3O_4. In this view the component designated by the molecular formula FeO predominates over all the remaining components at the point P. At equilibrium the system is predicted to consist *mostly* of the chemical component FeO.

From a fundamental point of view, these descriptions are distinctly different. However, it can be demonstrated rigorously that with careful manipulation these dif-

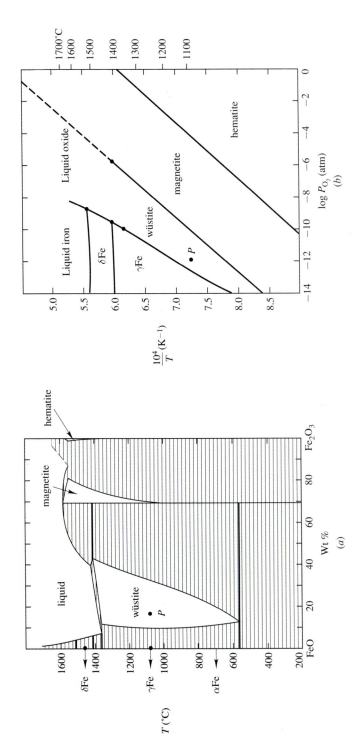

FIGURE 11.13

Comparison of (*a*) a phase diagram with (*b*) the analogous predominance diagram plotted in (log P_{O_2} − 1/*T*) space for the Fe–O system.

ferent descriptions of system behavior can be transformed from one to another. Specifically, the true phase diagram for a system can be derived from a description of the system using components that involve assumed intermediate molecular compositions like M_2O or M_2O_5, and not just M and O. However, in this transformed version of predominance diagrams, the limits of stability of the phases do not correspond precisely with the limits of predominance which, after all, are given by the condition $f_{[j]} = 1$, corresponding to the assumption that the participarts are in their reference states, with $a_j = 1$. Nonetheless, predominance diagrams provide a useful basis for displaying the pattern of compound formation as a function of the applied thermodynamic potentials. They may be used to place limits on tolerable atmosphere compositions, form a basis for the design of atmospheres, support suggestions for candidate materials for use in a given atmosphere, and suggest possible chemical and thermal histories that a failed material may have experienced.

11.5 COMPOUNDS AS COMPONENTS IN PHASE DIAGRAMS

It is common practice in the chemical literature, particularly in ceramic materials, to choose chemical compound compositions as the components in displaying a phase diagram. Where this is justified the description of behavior of the systems may be greatly simplified. It is not always justified.

Figure 11.14 presents the phase diagram for the alumina-silica system. The components in this diagram are taken to be Al_2O_3 and SiO_2. The four phases that exist in this system (alumina, mullite, silica, and liquid) are solutions. These solutions are viewed as mixtures, not of the components Al, Si, and O, but of the components

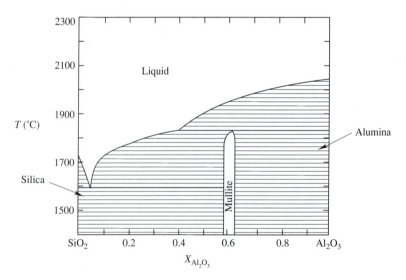

FIGURE 11.14
Alumina-silica phase diagram in which molecular compounds are treated as components.

Al_2O_3 and SiO_2. The free energies of mixing of these phases are each expressed in terms of the mole fractions of the components Al_2O_3 and SiO_2. Chemical potentials and activities in the system are properties of the components Al_2O_3 and SiO_2.

$$\mu_{Al_2O_3} \equiv \left(\frac{\partial G^\alpha}{\partial n_{Al_2O_3}} \right)_{T,P,n_{SiO_2}}$$

The conditions for equilibrium that determine the positions of the phase boundaries are also familiar. For example, for the $(\alpha + \epsilon)$ [alumina + mullite] two phase equilibrium,

$$\mu^\alpha_{Al_2O_3} = \mu^\epsilon_{Al_2O_3} \qquad \mu^\alpha_{SiO_2} = \mu^\epsilon_{SiO_2}$$

The calculation of this phase diagram can be based on solution models for the phases, viewed as solutions of Al_2O_3 and SiO_2, and on information about the relative stability of the phase forms for pure alumina and silica as presented in detail in Chap.10. For many purposes this representation is adequate to understand the behavior of the system.

On a more sophisticated level, it is understood that this diagram is actually a pseudobinary section through the ternary Al-Si-O phase diagram. The sequence of compositions used to designate the abscissa in Fig. 11.14 lies along the straight line connecting the compositions corresponding to Al_2O_3 on the Al-O binary side of the diagram with the composition SiO_2 on the Si-O binary side. The single phase regions that are intersected by that line may exhibit small but finite composition ranges, departing from stoichiometric ratios in directions that lie outside the pseudobinary composition line. Since these small deviations are associated with defects in the structure of the phases, they may exert influences on properties that are significant. Whether the system can be treated as a binary alumina-silica system or as the ternary Al-Si-O system depends upon the application.

In many cases it is not possible to represent the phase relationships in a system made from two compounds on a binary phase diagram. Figure 11.15 shows the ternary isotherm for a system A-B-C. This system has intermediate compounds on the binaries at AB, AC and BC. The compounds AB and AC lie at ends of a two phase field in the ternary and, at least at the temperature shown, can be treated as forming a binary system. A binary system can also be constructed using the compounds AC and BC as components. However, an attempt to construct a binary phase diagram using the two compounds AB and BC would be disastrous. Equilibria that are sampled by the sequence of compositions along the line joining these compound compositions involve tie lines and tie triangles connecting compositions that are far outside the line. The behavior of the system in this region cannot be described by a binary phase diagram involving the compound components AB and BC.

11.6 SUMMARY OF CHAPTER 11

A system is classified as *reacting* if the number of components c is larger than the number of elements e. The number of independent chemical reactions r a system may exhibit is equal to $c - e$.

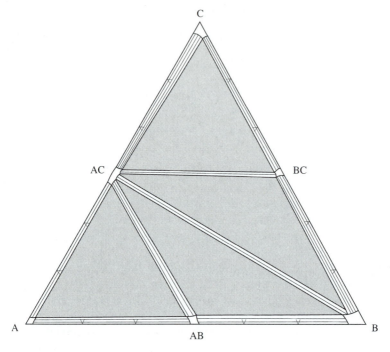

FIGURE 11.15

The ternary system ABC has three line compounds: AB, AC, BC. These compounds may be used as components in plotting the binaries; $AB - AC$, and $AC - BC$. However, the $AB - AC$ section cuts across two and three phase fields; a binary phase diagram with AB and AC as components cannot be constructed.

In an isolated reacting system the equations conserving the number of moles of each component must be replaced by constraints that conserve the number of gram atoms of each element. These constraints give rise to relations between changes in molar concentration that can occur which in turn produce relations between the chemical potentials of those components when the system attains equilibrium.

The *affinity* for a reaction is defined to be the difference between the chemical potential of the products and reactants. For the reaction

$$l\text{L} + m\text{M} = r\text{R} + s\text{S}$$

the affinity is expressed as

$$\mathscr{A} \equiv (r\mu_\text{R} + s\mu_\text{s}) - (l\mu_\text{L} + m\mu_\text{M})$$

The affinity is zero when the system reaches equilibrium. For a given nonequilibrium state,

$$\mathscr{A} < 0 \qquad \text{Products Form}$$

$$\mathscr{A} > 0 \qquad \text{Products Decompose}$$

The condition for equilibrium in a reacting system is

$$\mathcal{A} = 0$$

Substitution of the definition of activity for each chemical potential converts this equation to

$$\Delta G^o = -RT \ln K$$

where ΔG^o is the *standard free energy for the reaction,*

$$\Delta G^o \equiv (r G_R^o + s G_S^o) - (l G_L^o + m G_M^o)$$

and K is the *equilibrium constant*

$$K \equiv \left(\frac{a_R^r \cdot a_S^s}{a_L^l \cdot a_M^m} \right)_{\text{equil}}$$

For a system that has not reached equilibrium the *proper quotient of activities* Q has the same form as K but the activities in this ratio correspond to some state of interest other than the equilibrium state. The affinity can be written as

$$\mathcal{A} = RT \ln \frac{Q}{K}$$

and the direction of spontaneous change for the given state depends upon the value of Q/K:

$$\frac{Q}{K} < 1 \rightarrow \mathcal{A} < 0 \qquad \text{Products Form}$$

$$\frac{Q}{K} > 1 \rightarrow \mathcal{A} > 0 \qquad \text{Products Decompose}$$

For a *multivariate* reacting system for which $r = (c - e) > 1$, there are $(c - e)$ independent conditions for equilibrium that have the form

$$\Delta G_{[j]}^o = -RT \ln K_{[j]} \qquad [j = 1, 2, \ldots, (c - e)]$$

Since it can be shown that all possible reactions among the components are linear combinations of this set of independent reactions, an equation of this form must be satisfied for *every chemical reaction* that can be written for the system.

If the system involves condensed phases then some components may exhibit departures from the simple stoichiometric ratios (of the elements) characteristic of the gas phase. To the extent that departures from stoichiometry can be neglected, multiphase systems can be treated as reacting systems and the conditions for equilibrium described above can be applied. Where departures from stoichiometry are important in the application at hand, it is necessary to recognize these compositional variations and treat the problem as a competition for stability between phases rather than as a competition for predominance between components.

The Richardson-Ellingham chart for the oxides is a succinct summary of the thermodynamic behavior of oxide forming reactions. Specially constructed scales on the diagram permit easy estimation of the equilibrium oxygen potential and the equivalent CO/CO_2 ratio and H_2/H_2O ratio for any metal and its oxide presented on the chart. Similar charts can be constructed for sulfides, chlorides, nitrides and carbides.

If departures from stoichiometry are not critical, intermediate phases can be treated as components in a multivariate reacting system. The equilibrium condition of the system can then be adequately represented on a *predominance diagram*. The typical predominance diagram is plotted in potential space, either temperature or the chemical potential of governing components used as axes. The lines on the diagram are *limits of predominance* that arise in the comparison of the reacting components two at a time. Domains enclosed within the lines are regions in which, at equilibrium, one component has a concentration larger than that of every other component in the competition.

PROBLEMS

11.1. Consider a gas mixture at a high temperature with the following components:

$$H_2, H_2O, O_2, SiO(g), SiH_4, Si(g)$$

(*a*) List the elements in the system.
(*b*) How many *independent* chemical reactions does this system have?
(*c*) Make a complete list of all the possible reactions.
(*d*) From this list choose a set of independent reactions.
(*e*) Select one of the remaining reactions and show that it is a linear combination of independent reactions.

11.2. A gas mixture at 1200 K has the following composition:

Component	CO	CO_2	O_2
Mole fraction	0.25	0.60	0.15

Compute the affinity for the reaction

$$CO + \tfrac{1}{2}O_2 = CO_2$$

for this mixture. In what direction does the reaction go?

11.3. Consider the following gas mixture at 900K:

Component	H_2	H_2O	O_2	CH_4	CO	CO_2
Mole fraction	0.19	0.05	1×10^{-5}	0.30	0.05	0.50

(*a*) Write a set of independent reactions for this system.
(*b*) Obtain equilibrium constants for these reactions.
(*c*) Evaluate the affinity for each of these reactions.
(*d*) What can you conclude about the direction of spontaneous change for this system?

11.4. A gas mixture initially has the composition:

Component	CO	O_2	CO_2
Mole fraction	0.50	0.12	0.38

(a) Write out the conservation equations for atoms of carbon and oxygen in terms of the three n_k values.

(b) Write an expression for the total number of moles n_T in the system.

(c) Write the expression for the equilibrium constant in terms of the n_k values and n_T.

(d) Solve this set of equations to find the equilibrium composition for this reaction at 1000 K.

11.5. A gas mixture of nitrogen and oxygen may contain the following components:

$$N_2, O_2, NO, NO_2, N_2O_4, N_2O, N_2O_5$$

(a) Write out the set of formation reactions for this mixture.

(b) Write all the reactions of the form

$$nA + O_2 = mB$$

that can be constructed for this system.

(c) From this list, how many sets of three equations that each share in three components (other than oxygen) can be written?

(d) List the triplets of components other than O_2 and N_2 involved.

11.6. Nickel forms NiO upon exposure to an atmosphere containing oxygen.

(a) Find the standard free energy of formation of NiO as a function of temperature.

(b) Evaluate and plot the equilibrium constant for this reaction as a function of temperature.

(c) Compute the equilibrium oxygen partial pressure at 962 K, assuming the nickel and NiO are pure.

(d) Compute the affinity for a system with pure nickel and pure NiO in air at 962 K.

11.7. Consider the oxidation of silicon with water vapor at 800°C.

(a) Find the standard free energy of formation of SiO_2 and H_2O.

(b) Combine them to compute ΔG^o for the reaction

$$Si + 2H_2O = SiO_2 + 2H_2$$

(c) Evaluate the equilibrium constant and the (H_2/H_2O) ratio in equilibrium with silicon and its oxide at 800°C.

(d) Compute the equilibrium partial pressure of oxygen in a silicon-silica system.

(e) Compute the equilibrium partial pressure of oxygen for a gas mixture at 800°C with an (H_2/H_2O) ratio computed in (c).

(f) Compare oxygen pressure values in (d) and (e).

11.8. Use the Richardson-Ellingham chart to verify the results of the calculation in Problem 11.7.

11.9. Use the Richardson-Ellingham chart for oxides to find the following:

(a) The dissociation pressure of CoO at 1000°C.

(b) The equilibrium constant for the formation of SiO_2 at 800°C.

(c) The (CO/CO_2) ratio in equilibrium with Na and Na_2O at 600°C.

(d) Will an atmosphere with an H_2/H_2O ratio of $10^5/1$ prevent the oxidation of chromium at 1200°C?

(e) Find the composition of an H_2/H_2O atmosphere that has the same oxygen potential as an atmosphere with a CO/CO_2 ratio of 10^3 at 900°C.

(f) Find the oxygen potential in (e).

11.10. From the Richardson-Ellingham chart:

(a) Does aluminum reduce Fe_2O_3 to iron at 1200°C?

(b) Can an atmosphere with a CO/CO_2 ratio of 0.06 oxidize nickel at 800°C?

(c) Does silica oxidize liquid zinc at 700°C?

(d) Does sodium reduce water to oxygen and hydrogen at 500°C?

11.11. Compute the affinities for the following reactions under the following conditions. Assume non-gaseous components are pure.

Reaction	T (°C)	P_k (atm)
$4Cu + O_2 = 2Cu_2O$	800	$P_{O_2} = 2 \times 10^{-8}$
$2Ca + O_2 = 2CaO$	1200	$P_{O_2} = 2 \times 10^{-16}$
$2Ni + S_2 = 2NiS$	700	$P_{S_2} = 2 \times 10^{-2}$
$Ca + Cl_2 = CaCl_2$	600	$P_{Cl_2} = 3 \times 10^{-6}$

11.12. Compute and plot a Pourbaix high temperature diagram for copper and its oxides, CuO and Cu_2O.

11.13. Write and balance chemical equations relating the following components:

(a) Na_2O and Na_2SO_3.

(b) Nb_2O_5 and NbS_2O_5.

(c) Cu_2O and $CuSiO_4$.

(d) Cu_2S and $CuSO_4$.

11.14. List the possible triple points that may exist in a $\log P_{O_2}$–$\log P_{S_2}$ predominance diagram involving the following components:

$$Fe, FeO, Fe_2O_3, Fe_3O_4, FeS_2, FeS, FeSO_4$$

11.15. Can you find a CO/CO_2 atmosphere that both prevents the oxidation of an iron-carbon alloy with $a_C = 0.48$ at 1200 K and at the same time carburizes it?

11.16. Initially the composition of a system is

Component	H_2	H_2O	CO	CO_2	O_2
Mole fraction	0.40	0.10	0.40	0.08	0.02

(a) Find the equilibrium composition for this system at 800°C.

(b) Compute the affinities for the reactions from this starting point at 800°C.

REFERENCES

1. *CRC Handbook of Physics and Chemistry*, 71st edition, D.R. Lide, Editor in Chief, CRC Press, Boca Raton, Fla., 1990.

2. Cox, J.D., D.D. Wagman, and V.A. Medvedev: *CODATA Key Values of Thermodynamics*, Hempshire Pub. Corp., New York, N.Y., 1989.

3. Barin, I., O. Knacke and O. Kubaschewski: *Thermochemical Properties of Inorganic Substances Supplement*, Springer-Verlag, New York, N.Y. 1977.

4. Chase, M.W., et.al.: *JANAF Thermodynamic Tables*, Third edition, J. Phys. Chem. Ref. Data, 11, supp. 2, 1982.

5. On line access to the SOLGASMIX program is available through F*A*C*T* (Facility for the Analysis of Chemical Thermodynamics), Ecole Polytechnique, CRCT, Montreal, Quebec, Canada.

CHAPTER 12

CAPILLARITY
EFFECTS IN
THERMODYNAMICS

Thermodynamic systems that consist of more than one phase necessarily have internal interfaces, that is, regions in which the transition from the intensive properties of one phase to those of a neighboring phase is accommodated. Since this transition is accommodated over distances that are of the order of a few atoms, these regions are nearly two dimensional in their geometry and are commonly called surfaces or interfaces between the phases.

Effects associated with these zones of transition and their geometry are unusually important in materials science. Solid materials have an internal *microstructure* and the geometry of microstructure plays a key role in the behavior of materials. Properties are altered and controlled by processing the material, by heat treatment, hot pressing, and extrusion for example, under conditions that change the internal microstructure. Since the features in the microstructure are defined by the interfaces that bound them, the thermodynamic behavior of interfaces plays a central role in both the processing of materials and in their behavior in service.

Atoms residing in the transition zone between two phases necessarily have patterns of neighbors that are not characteristic of either phase. Thus, each element of an interface has an extra energy associated with it that can be called its "surface" or "interfacial" energy. A rigorous definition of this property is given in Sec. 12.2.

This surface energy operates to influence properties of the adjacent phases through the geometry of the surface. The geometric property that operates in these relationships is the *local curvature* of the surface or interface. Section 12.1 develops the geometric concepts associated with the curvature of surfaces and the relationships between curvature and other geometric properties of the interface.

Effects upon thermodynamic properties that derive from the curvature of interfaces in a system are commonly called *capillarity* effects because the initial documentation of these relationships was studied experimentally with fine glass tubes, called capillaries, which could be made with reproducible internal radii or curvature. These capillarity effects derive from the conditions for equilibrium in a multicomponent two phase system with curved interfaces. Accordingly, the general strategy for finding conditions for equilibrium is applied to this class of systems in Sec. 12.3. It is shown that, in the presence of curved interfaces, the condition for *mechanical equilibrium* is altered. A purely mechanical development of this result is presented in Sec. 12.4. The consequences of this result are deduced in Sec. 12.5.

A second geometric feature found in most microstructures results from the intersection of three interfaces to form a line, called a *triple line* in the structure. The conditions for equilibrium at a triple line are developed in Sec. 12.6. Application to the wetting of one phase by another, penetration of a liquid into a solid along its internal grain boundaries, and the development of the geometry of microstructures illustrate these results.

Because the arrangement of atoms in the transition zone between phases is different from either phase, the properties of that region are different from either adjacent phase. In particular, the composition of the components may be significantly different in the transition zone than in either adjacent phase when the system attains equilibrium. This phenomenon, called *adsorption*, is treated in Sec. 12.7.

12.1 THE GEOMETRY OF SURFACES

To develop the relationships between the thermodynamic properties of surfaces and the properties of the adjacent phases, it is necessary first to introduce the concepts required to describe the geometry in the vicinity of a point P on a surface. The quantities required, called the *principal normal curvatures* at P, are defined in this section. In addition it is shown that if a curved surface moves in space, say as the result of kinetic processes that occur in the adjacent phases, the change in area of the surface is related to the volume swept through by this displacement through the local curvature of the surface.

Consider first the definition of the curvature at a point P on a curved line in two dimensional space, Fig. 12.1. In this construction, two points A and B on the curve are selected. A circle can be constructed through any three points in the plane; construct the circle that passes through the points A, P, and B. This circle has a center and a radius. Now let A and B approach P. In the limit as A and B arrive at P, a unique circle is approached in this construction. This circle is called the "osculating circle at P" because it just kisses the curve at this point. The circle has its center at O in Fig. 12.1, called the center of curvature for P, and has a radius r, called the radius of

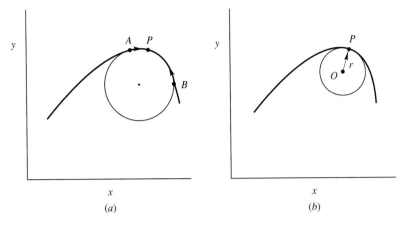

FIGURE 12.1
Definition of the curvature at a point P on a line in two dimensions.

curvature at P. The reciprocal of the radius,

$$k = \frac{1}{4}$$ (12.1)

is called the local curvature of the curve at the point P. In general k varies continuously as P moves along the curve.

Now consider an element of a smoothly curved surface embedded in three dimensional space, Fig. 12.2. There exists a tangent plane at any point P and a unit vector $\hat{N}$ perpendicular to the tangent plane called the *normal* vector at P. A

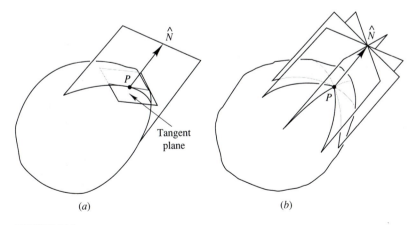

FIGURE 12.2
(*a*) A plane through the point P on a surface and containing the surface normal, N, intersects the surface with a plane curve that can be analyzed as in Figure 12.1. (*b*) The local curvature of these curves of intersection varies with the orientation of the intersecting plane.

plane that intersects this surface at P and contains the normal, a plane that is locally perpendicular to the tangent plane, Fig. 12.2a, produces a curve of intersection with the surface conceptually identical to the plane curve shown in Fig. 12.1. Thus the curvature of this line of intersection through P has a value given by the construction outlined for Fig. 12.1.

If the intersecting plane is rotated to another orientation that still contains the normal, the intersecting curve in general will be different. The local curvature at P varies as the intersecting plane rotates about the normal through P, Fig. 12.2b. In *differential geometry* it is shown that there are two orientations of the normal plane for which the curvature of the intersecting line has a maximum and a minimum value. These are called the principal directions at P on the surface. It can be shown that these directions are perpendicular to each other.

These observations lead to the construction shown in Fig. 12.3. The vector $\hat{N}$ is the normal vector at P, and $\hat{u}$ and $\hat{v}$ are unit vectors in the principal directions. The radii r_1 and r_2 are the principal radii of curvature at P; $\kappa_1 = 1/r_1$ and $\kappa_2 = 1/r_2$ are the *principal normal curvatures* at P. The latter two quantities are the most convenient descriptors of the local surface geometry.

Two geometric properties related to the local curvatures find useful application:

$$H \equiv \frac{1}{2}(\kappa_1 + \kappa_2) \ . \tag{12.2}$$

is the local *mean curvature* of the surface at P. The product of the curvatures,

$$K \equiv \kappa_1\kappa_2 \tag{12.3}$$

is sometimes called the *total curvature* at P. It is shown that thermodynamics operates in the geometry of a microstructure through H, the local mean curvature.

Consider a smooth closed surface, that is, one that encloses a microstructural feature such as a particle of the β phase. Define the curvature to be positive if the

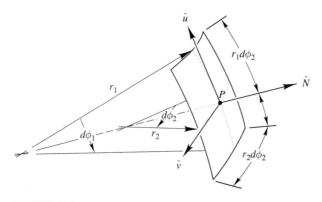

FIGURE 12.3
The normal $\hat{N}$ and principal directions $\hat{u}$ and $\hat{v}$ define two planes normal to the surface at P. The lengths, r_1 and r_2, are radii of curvature in these planes. The reciprocals of these radii are the *principal normal curvatures* of the surface at P.

vector at P that points towards its center of curvature points *in to the β phase* and negative if it points *outward*. Three kinds of surface elements may be found on a smooth surface that encloses a feature or a phase, Fig. 12.4:

1. *Convex surface*, for which both vectors point inward.
2. *Concave surface*, with both vectors pointing outward.
3. *Saddle surface*, for which the two vectors point in opposite directions.

The local mean curvature H is always positive for convex surface elements, negative for concave elements, and may be either positive or negative for saddle surface, depending upon the relative values of the curvatures.

Fig. 12.3 shows an element of surface near P with principal radii of curvature r_1 and r_2 generating the element by rotating through infinitesimal arcs $d\phi_1$ and $d\phi_2$. The dimensions of the element are $r_1 d\phi_1$ and $r_2 d\phi_2$. The area of the element is

$$dA_0 = r_1 d\phi_1 \cdot r_2 d\phi_2 \tag{12.4}$$

Now suppose as a result of processes that occur in the adjacent phases the interface moves a distance δn along its normal, Fig. 12.5. The volume swept out by this displacement is

$$\delta V = \delta n \cdot dA_0 = \delta n \cdot r_1 d\phi_1 r_2 d\phi_2 \tag{12.5}$$

The radii of curvature change to $(r_1 + \delta n)$ and $(r_2 + \delta n)$. The area of the element after displacement is

$$dA_1 = (r_1 + \delta n)(r_2 + \delta n)d\phi_1 d\phi_2$$

$$= \left[r_1 r_2 + (r_1 + r_2)\delta n + \delta n^2 \right] d\phi_1 d\phi_2$$

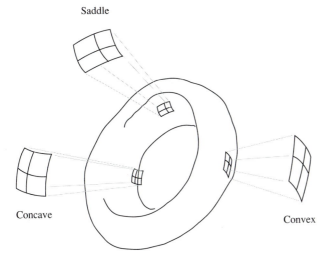

FIGURE 12.4
Three classes of surface elements are illustrated for a bowl (*a*) convex, (*b*) concave, and (*c*) saddle surface.

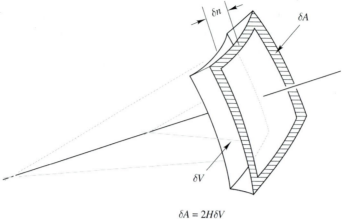

$$\delta A = 2H\delta V$$

FIGURE 12.5
The displacement of a surface element at distance δn along its normal sweeps through a volume δV and changes the area of the element by an amount δA.

The term δn^2 can be neglected as a higher order differential. The change in area associated with the displacement of the interfacial element is

$$\delta A = dA_1 - dA_0 = (r_1 + r_2)\delta n \cdot d\phi_1 d\phi_2$$

Multiply and divide by $r_1 r_2$

$$\delta A = \frac{(r_1 + r_2)}{r_1 r_2} \cdot \delta n \cdot r_1 r_2 d\phi_1 d\phi_2 \qquad (12.6)$$

Note that the ratio can be written

$$\frac{r_1 + r_2}{r_1 r_2} = \frac{1}{r_1} + \frac{1}{r_2} = \kappa_1 + \kappa_2 = 2H$$

Further, the remaining product of factors in Eq. (12.6), $\delta n r_1 r_2 d\phi_1 d\phi_2$, can be recognized as the volume swept through, δV, Eq. (12.5). Thus the change in surface area of a curved surface as it migrates is related to the volume swept through and to the local mean curvature by

$$\delta A = 2H\delta V \qquad (12.7)$$

This relationship constitutes a constraint upon the variables that describe the system and it plays a key role in the derivation of the conditions for equilibrium in systems with curved internal interfaces.

12.2 SURFACE EXCESS PROPERTIES

This section presents the apparatus developed by J. Willard Gibbs for ascribing an appropriate part of the total value of an extensive property of a system to the presence of internal interfaces. Gibbs realized that the zone of transition between two phases is too thin to admit direct measurement of its properties. He proposed a strategy for evaluating the contribution due to surfaces in terms of the measurable properties of the macroscopic system. This strategy leads to the definition of *surface excess* thermodynamic properties.

Figure 12.6 illustrates a system composed of two phases α and β and the transition zone σ between them. The boundaries of the transition zone need not be precisely defined as long as they enclose the region in which occur all the changes in intensive properties from their value in α to their value in β. Gibbs referred to the transition zone σ as the "physical surface of discontinuity," recognizing that it is not truly a surface but is a slab of small but finite thickness.

Let B' be any extensive thermodynamic property. A corresponding local *density* of B can be defined at any point P in the system. Consider a small volume V' that has the value B' for the property of interest. The local density of the property B at the point P is given by

$$\lim_{V' \to 0} \frac{B'}{V'} \equiv b_v \tag{12.8}$$

Because it has a value defined at each point in the system, b_v is an *intensive* property. The value of the property B' contained in a volume element dv neighboring P is $b_v dv$. The total value of B' for the system is evidently the integral of the density b_v over the volume of the system:

$$B' = \iiint_{V'} b_v \, dv \tag{12.9}$$

These concepts form the basis for describing *nonuniform systems* discussed in detail in Chap. 14. If the system is in internal equilibrium and no external fields operate, the

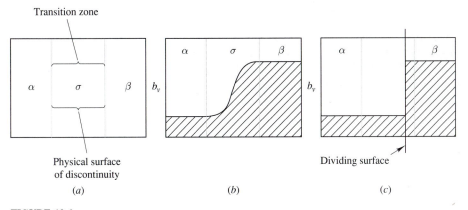

FIGURE 12.6

(*a*) Visualization of the zone of transition, σ, between two phases defining the physical surface of discontinuity. (*b*) Sketch of the variation of the density of a thermodynamic property b_v through the zone of transition. (*c*) The hypothetical system introducing the *dividing surface* with the step function variation for the density b_v.

intensive properties within each phase do not vary with position outside the physical surface of discontinuity. Figure 12.6b sketches the variation of b_v with position through the physical surface of discontinuity. Suppose the system under consideration has a constant cross-sectional area. Then the total value of B'_{sys} for the system can be visualized in Fig. 12.6b as proportional to the shaded area under the curve.

Gibbs next introduced a true two dimensional geometric surface, which he called the "dividing surface." The location of the dividing surface is arbitrary except that it lies within the physical surface of discontinuity, Fig. 12.6c. It is shown that some choices for the actual location in a given application are more convenient than others. Once the location is chosen in conjunction with one of the properties, it must be retained for all of the others.

The surface excess of B, that is, the part of the total value of B' for the system that may be associated with the presence of interface, is assigned by imagining a hypothetical system in which the densities b_v^α and b_v^β in α and β are maintained up to the dividing surface, at which there is a discontinuity in b_v, Fig. 12.6c. The total value of B'_{hyp} for this hypothetical system can be visualized as being proportional to the shaded area under the density curve in Fig. 12.6c. This quantity can be calculated from the known densities b_v^α and b_v^β in the parts of the system where these quantities are constant and from the location of the dividing surface.

The surface excess of B for the system as a whole is defined to be

$$B'^s \equiv B'_{sys} - B'_{hyp} \tag{12.10}$$

That is, B'^s is the difference in B evaluated for the actual system and that computed from the intensive properties b_v^α and b_v^β and the location of the dividing surface. This quantity can be visualized as being proportional to the difference between the area under the curve in Fig. 12.6b and under the step function in Fig. 12.6c. Let A be the area of the interface in the system. Then the "specific surface excess of B" is given by

$$B^s \equiv \frac{B'^s}{A} \tag{12.11}$$

This quantity, which has units of "B per unit area," can be defined for any extensive thermodynamic property. Thus S^s is the specific surface excess entropy for the α/β interface, G^s is the specific surface excess Gibbs free energy, and so on. Of particular interest in Sec. 12.6 is the "specific surface excess number of moles of component k," which is the central notion in the description of adsorption phenomena. Specific surface excess properties are *intensive variables*. This observation is important because it implies that they are functions of the intensive properties of the system but not the extensive properties.

Equations (12.10) and (12.11) can be combined with minor rearrangement to give

$$B'_{sys} = B'_{hyp} + B'^s = B'_{hyp} + B^s A \tag{12.12}$$

Since the hypothetical system consists of an α and a β part, this expression can be written more explicitly as

$$B'_{sys} = B'^\alpha_{hyp} + B'^\beta_{hyp} + B^s A \tag{12.13}$$

An arbitrary change in the state of this system must now explicitly include the possibility that the area of the interface may change. Thus, for an arbitrary change,

$$dB'_{\text{sys}} = dB'^{\alpha}_{\text{hyp}} + dB'^{\beta}_{\text{hyp}} + B^s dA \tag{12.14}$$

12.3 CONDITIONS FOR EQUILIBRIUM IN SYSTEMS WITH CURVED INTERFACES

The general strategy for finding conditions for equilibrium is applied to a system consisting of α and β phases, each containing c components and separated by curved interfaces. First, write an expression for the change in entropy of the system. According to Eq. (12.14), this includes contributions from two phases and the surface excess term:

$$dS'_{\text{sys}} = dS'^{\alpha}_{\text{hyp}} + dS'^{\beta}_{\text{hyp}} + S^s dA \tag{12.15}$$

where S^s is the specific superficial excess entropy for the system. The hypothetical subsystems, with entropy densities of the phases extending to the dividing surface, have intensive properties identical with those of the α and β phases outside the zone of transition. Thus,

$$
dS'_{\text{sys}} = \left[\frac{1}{T^{\alpha}} dU'^{\alpha}_{\text{hyp}} + \frac{P^{\alpha}}{T^{\alpha}} dV'^{\alpha}_{\text{hyp}} - \sum_{k=1}^{c} \frac{\mu_k^{\alpha}}{T^{\alpha}} dn^{\alpha}_{k,\text{hyp}} \right]
$$
$$
+ \left[\frac{1}{T^{\beta}} dU'^{\beta}_{\text{hyp}} + \frac{P^{\beta}}{T^{\beta}} dV'^{\beta}_{\text{hyp}} - \sum_{k=1}^{c} \frac{\mu_k^{\beta}}{T^{\beta}} dn^{\beta}_{k,\text{hyp}} \right] + S^s dA \tag{12.16}
$$

The extensive properties in the right-hand side of this relation are understood to be those of the hypothetical system.

Next, write out the isolation constraints. The constant internal energy constraint in an isolated system with internal interfaces can be written

$$dU'_{\text{sys}} = 0 = dU'^{\alpha}_{\text{hyp}} + dU'^{\beta}_{\text{hyp}} + U^s dA$$

or

$$dU'^{\beta}_{\text{hyp}} = - \left(dU'^{\alpha}_{\text{hyp}} + U^s dA \right) \tag{12.17}$$

The constant volume constraint is:

$$dV'_{\text{sys}} = 0 = dV'^{\alpha}_{\text{hyp}} + dV'^{\beta}_{\text{hyp}} + V^s dA$$

The superficial excess volume V^s is *zero* since the construction that leads to the definition of superficial excess properties assumes that the total volume of the hypothetical system is the same as the actual system. Thus this constraint yields simply

$$dV'^{\beta}_{\text{hyp}} = -dV^{\alpha}_{\text{hyp}} \tag{12.18}$$

The total number of moles of each component is constant

$$dn_{k,\text{sys}} = 0 = dn^{\alpha}_{k,\text{hyp}} + dn^{\beta}_{k,\text{hyp}} + \Gamma_k dA$$

or Γ_k is defined to be the specific interfacial excess of component K by

$$\Gamma_k = \frac{n_k^s}{A}$$

$$dn_{k,\text{hyp}}^\beta = -\left(dn_{k,\text{hyp}}^\alpha + \Gamma_k dA\right) \tag{12.19}$$

Substitute the isolation constraints, Eqs. (12.17)–(12.19) into the entropy expression, Eq. (12.16), eliminating the dependent variables in an isolated system

$$dS'_{\text{sys,iso}} = \frac{1}{T^\alpha}dU_{\text{hyp}}^{'\alpha} + \frac{P^\alpha}{T^\alpha}dV_{\text{hyp}}^{'\alpha} - \sum_{k=1}^{c}\frac{\mu_k^\alpha}{T^\alpha}dn_{k,\text{hyp}}^\alpha$$

$$+ \frac{1}{T^\beta}(-dU_{\text{hyp}}^{'\alpha} - U^s dA) + \frac{P^\beta}{T^\beta}(-dV_{\text{hyp}}^{'\alpha})$$

$$- \sum_{k=1}^{c}\frac{\mu_k^\beta}{T^\beta}(-dn_{k,\text{hyp}}^\alpha - \Gamma_k dA) + S^s dA$$

Collecting like terms

$$dS'_{\text{sys,iso}} = \left[\frac{1}{T^\alpha} - \frac{1}{T^\beta}\right]dU_{\text{hyp}}^{'\alpha} + \left[\frac{P^\alpha}{T^\alpha} - \frac{P^\beta}{T^\beta}\right]dV_{\text{hyp}}^{'\alpha}$$

$$- \sum_{k=1}^{c}\left[\frac{\mu_k^\alpha}{T^\alpha} - \frac{\mu_k^\beta}{T^\beta}\right]dn_{k,\text{hyp}}^\alpha \tag{12.20}$$

$$+ \left[S^s - \frac{1}{T^\beta}U^s + \sum_{k=1}^{c}\frac{\mu_k^\beta}{T^\beta}\Gamma_k\right]dA$$

The variables in this equation are still not all independent since there exists a *geometric* relationship between the area and volume variables derived in Eq. (12.7). To apply this relationship it is first necessary to choose a convention for the sign of the surface curvature. Let the curvature be defined positive when the surface is convex relative to the β phase. Then the volume enclosed by the surface is that of the β phase, and Eq. (12.7) must be written

$$dA = 2H \cdot dV_{\text{hyp}}^\beta = -2H \cdot dV_{\text{hyp}}^\alpha \tag{12.21}$$

Substitute this result into Eq. (12.20) to eliminate dA

$$dS'_{\text{sys,iso}} = \left[\frac{1}{T^\alpha} - \frac{1}{T^\beta}\right]dU_{\text{hyp}}^{'\alpha}$$

$$+ \left[\left(\frac{P^\alpha}{T^\alpha} - \frac{P^\beta}{T^\beta}\right) - \left(S^s - \frac{1}{T^\beta}U^s + \sum_{k=1}^{c}\frac{\mu_k^\beta}{T^\beta}\Gamma_k\right) \cdot 2H\right]dV_{\text{hyp}}^{'\alpha} \tag{12.22}$$

$$- \sum_{k=1}^{c}\left[\frac{\mu_k^\alpha}{T^\alpha} - \frac{\mu_k^\beta}{T^\beta}\right]dn_{k,\text{hyp}}^\alpha$$

The conditions for equilibrium in this system can now be determined by setting all the coefficients of the differentials in this equation simultaneously equal to zero. The coefficient of the internal energy yields the condition for thermal equilibrium,

$$\frac{1}{T^\alpha} = \frac{1}{T^\beta} = 0 \qquad \rightarrow \qquad T^\alpha = T^\beta \tag{12.23}$$

The coefficient of dn_k yields the condition for chemical equilibrium,

$$\frac{\mu_k^\alpha}{T^\alpha} - \frac{\mu_k^\beta}{T^\beta} = 0 \qquad \rightarrow \qquad \mu_k^\alpha = \mu_k^\beta \quad (k = 1, 2, \ldots, c) \tag{12.24}$$

These results are identical with those obtained for a system without curved interfaces. The condition for *mechanical equilibrium* is not the same:

$$\left(\frac{P^\alpha}{T^\alpha} - \frac{P^\beta}{T^\beta} \right) - \left(S^s - \frac{1}{T^\beta} U^s + \sum_{k=1}^{c} \frac{\mu_k^\beta}{T^\beta} \Gamma_k \right) \cdot 2H = 0$$

Since by equation (12.23) the temperatures are equal, they may be eliminated from this equation by multiplying by T. The condition for mechanical equilibrium may be written:

$$P^\alpha - P^\beta + \left(U^s - T S^s - \sum_{k=1}^{c} \mu_k \Gamma_k \right) \cdot 2H = 0 \tag{12.25}$$

To simplify the application of this result, introduce the definition,

$$\gamma \equiv U^s - T S^s - \sum_{k=1}^{c} \mu_k \Gamma_k \tag{12.26}$$

It will subsequently be shown that, in simple systems, γ is the *specific interfacial free energy*. Then the condition for mechanical equilibrium in Eq. (12.25) may be written

$$P^\beta - P^\alpha = 2\gamma H \tag{12.27}$$

Thus, at equilibrium in a two phase system with curved interfaces, the pressure in the two phases are not equal. It can be shown that γ is always positive. Whether the pressure in the β phase is larger or smaller than in the α phase depends upon the sign of the mean curvature, H. Since, by the choice of convention made earlier, H is positive for surface elements that are convex relative to the β phase, the pressure inside a simple closed particle of the β phase is higher than that outside it at equilibrium. The magnitude of the pressure difference increases as the particle size gets smaller or, more generally, as H increases. The influences exerted by curved surfaces upon the thermodynamic behavior of systems developed in the remainder of this chapter are all directly traceable to this condition for mechanical equilibrium.

Alternate visualizations of γ can be obtained by developing the combined statement of the first and second laws for two phase systems with curved interfaces. From the definition of surface excess quantities, the change in internal energy experienced by such a system can be written

$$dU'_{\text{sys}} = dU'^\alpha_{\text{hyp}} + dU'^\beta_{\text{hyp}} + U^s dA \tag{12.28}$$

Application of the combined statements of the first and second laws to the uniform hypothetical systems gives

$$
dU'_{sys} = \left[T^\alpha dS'^\alpha_{hyp} - P^\alpha dV'^\alpha_{hyp} + \sum_{k=1}^{c} \mu_k^\alpha dn^\alpha_{k,hyp} \right]
$$

$$
+ \left[T^\beta dS'^\beta_{hyp} - P^\beta dV'^\beta_{hyp} + \sum_{k=1}^{c} \mu_k^\beta dn^\beta_{k,hyp} \right] + U^s dA \qquad (12.29)
$$

If this two phase system remains in equilibrium as its state is changed, Eqs. (12.23) and (12.24) permit replacement of T^α and T^β with a single system temperature T and replacement of μ_k^α and μ_k^β with μ_k for the system. Equation (12.29) can be written

$$
dU'_{sys} = T(dS'^\alpha_{hyp} + dS'^\beta_{hyp}) - P^\alpha dV'^\alpha_{hyp} - P^\beta dV'^\beta_{hyp}
$$

$$
+ \sum_{k=1}^{c} \mu_k(dn^\alpha_{k,hyp} + dn^\beta_{k,hyp}) + U^s dA \qquad (12.30)
$$

Use Eq. (12.14) to rewrite the terms in brackets in terms of the superficial excess properties,

$$
dU'_{sys} = T(dS'_{sys} - S^s dA) - P^\alpha dV'^\alpha_{hyp} - P^\beta dV'^\beta_{hyp}
$$

$$
+ \sum_{k=1}^{c} \mu_k(dn_{sys} - \Gamma_k dA) + U^s dA \qquad (12.31)
$$

Collecting like terms

$$
dU'_{sys} = T dS'_{sys} - P^\alpha dV'^\alpha_{hyp} - P^\beta dV'^\beta_{hyp} + \sum_{k=1}^{c} \mu_k dn_{sys}
$$

$$
+ \left(U^s - T S^s - \sum_{k=1}^{c} \mu_k \Gamma_k \right) dA
$$

The coefficient of dA is precisely γ; thus

$$
dU'_{sys} = T dS'_{sys} - P^\alpha dV'^\alpha_{hyp} - P^\beta dV'^\beta_{hyp} + \sum_{k=1}^{c} \mu_k dn_{sys} + \gamma dA \qquad (12.32)
$$

Evidently γ can be appropriately interpreted in terms of the change in internal energy of a two phase system with change in internal area,

$$
\gamma = \left(\frac{\partial U'}{\partial A} \right)_{S', V'^\alpha, V'^\beta, n_k} \qquad (12.33)
$$

For a unary system ($c = 1$) it is possible to choose the location of the dividing surface so that Γ is zero. For this case Eq. (12.26) becomes simply

$$
\gamma = U^s - T S^s = F^s = G^s \qquad (12.34)
$$

That is, γ is the *specific interfacial free energy*. (The Gibbs and Helmholtz functions are identical in this case because V^s is zero.) This property of the interface has units of energy per unit area and is usually reported in units of J/m^2. The value of γ depends on the nature of both phases involved. For crystalline materials the value of γ also depends on the orientation of the crystal planes at the interface. Table 12.1 presents typical values of γ measured for a variety of surfaces and interfaces. Values of γ for additional systems are collected in Appendix F.

Example 12.1. Compute the pressure in a one micron diameter water droplet formed in supercooled water vapor at 80°C. The vapor pressure of water at 80°C is 0.52 atm. The specific interfacial free energy of water may be taken as 80 ergs/cm^2.

For a spherical particle the mean curvature $H = 1/r$. From Eq. (12.27),

$$P^L = P^V + 2\gamma \left(\frac{1}{r}\right)$$

$$= 0.52 \ (\text{atm}) + 2\left(80 \ \frac{\text{ergs}}{\text{cm}^2}\right)\left(\frac{1}{0.5 \times 10^{-4} \ \text{cm}}\right)\left(\frac{1 \ \text{J}}{10^7 \ \text{ergs}}\right)\left(\frac{82.06 \ \text{cc-atm}}{8.314 \ \text{J}}\right)$$

$$P^L = 0.52(\text{atm}) + 3.16(\text{atm}) = 3.68(\text{atm})$$

TABLE 12.1
Typical values for surface free energies of selected interfaces [12.3,12.4]

Material	γ (ergs/cm^2)($T°C$)*	Material	γ (ergs/cm^2)($T°C$)*
	Liquid metals at their melting point		
Cesium	60 (mp)	Gold	1140 (mp)
Lead	450 (mp)	Copper	1300 (mp)
Aluminum	866 (mp)	Nickel	1780 (mp)
Silicon	730 (mp)	Rhenium	2700 (mp)
	Solid-vapor surfaces of pure metals		
Bismuth	550 (250)	Copper	1780 (925)
Aluminum	980 (450)	Nickel	2280 (1060)
Gold	1400 (1100)	Tungsten	2800 (2000)
	Solid-liquid interfaces for pure metals		
Sodium	20	Gold	132
Lithium	30	Copper	177
Lead	33	Platinum	240
	Grain boundaries in pure metals		
Aluminum	324 (450)	Copper	625 (925)
Iron (δ phase)	468 (1450)	Nickel	866 (1060)
Iron (γ phase)	756 (1350)	Tungsten	1080 (2000)
	Compounds		
Water (liquid)	72 (25)	MgO	1000 (25)
NaCl (100) face	300 (25)	TiC	1190 (1100)
Be$_2$O$_3$(liquid)	80 (900)	CaF$_2$ (111)	450 (25)
Al$_2$O$_3$ (liquid)	700 (2080)	CaCO$_3$ (1010)	230 (25)
Al$_2$O$_3$ (solid)	905 (1850)	LiF (100)	340 (25)

*Multiply by 10^{-3} to convert to J/m^2.

12.4 SURFACE TENSION: THE MECHANICAL ANALOG OF SURFACE FREE ENERGY

The mechanical effects derived from the surface free energy can also be interpreted as arising from a force acting tangentially in the physical surface of discontinuity. Figure 12.7 shows an element of area with arbitrary principal curvatures. If this element were cut out of the surface, the element would shrink and disappear. To prevent the shrinkage of the element it is necessary to apply a tensile force tangentially around its perimeter.

Define the *surface tension* σ of the surface to be a normalized expression of this force with units of force per unit length, usually reported in units of dynes/cm. The force acting on the edge AB arising from surface tension is $F_1 = \sigma r_1 d\phi_1$ since $r_1 d\phi_1$ is the length over which the normalized force σ acts. Similarly, the force acting over the line CD is also $F_1 = \sigma r_1 d\phi_1$. The forces acting on the segment BC and DA are each $F_2 = \sigma r_2 d\phi_2$. These force vectors act in the surface, Fig. 12.7. Because the surface is curved, these forces are not quite perpendicular to the local surface normal $\hat{N}$, but make small angles $(1/2)d\phi_1$ and $(1/2)d\phi_2$ with vectors perpendicular to the normal. The component of the force acting along the segment AB that acts normal to the surface is

$$F_1 \sin(\tfrac{1}{2}d\phi_2) \cong F_1(\tfrac{1}{2}d\phi_2) = \sigma r_1 d\phi_1 \cdot (\tfrac{1}{2}d\phi_2) \tag{12.35}$$

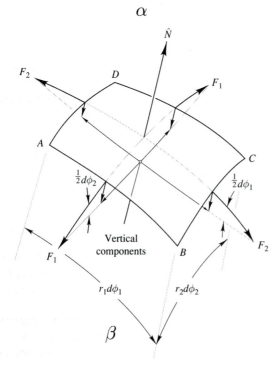

FIGURE 12.7
Forces acting on a surface element derived from the surface tension. If the surface is curved the forces have components that act in the normal direction.

since for small angles $\sin(x) \rightarrow x$. The normal component derived from the force acting along CD has the same value. The normal component due the force acting along BC is the same as that along DA and is given by

$$F_2 \sin(\tfrac{1}{2}d\phi_1) \cong F_1(\tfrac{1}{2}d\phi_1) = \sigma r_2 d\phi_2 \cdot (\tfrac{1}{2}d\phi_1) \tag{12.36}$$

The total force acting in the direction normal to the surface derived from the surface tension is the sum of these four components,

$$F_{\text{surf}} = 2\left[\sigma r_1 d\phi_1 \cdot (\tfrac{1}{2}d\phi_2)\right] + 2\left[\sigma r_2 d\phi_2 \cdot (\tfrac{1}{2}d\phi_1)\right]$$
$$F_{\text{surf}} = \sigma[r_1 + r_2]d\phi_1 d\phi_2 \tag{12.37}$$

The normal force acting upon the β phase adjacent to the element of surface in Fig. 12.7 is the sum of the force exerted by the α phase and that exerted by the surface tension acting through the curvature of the local surface element,

$$F^\beta = F^\alpha + F_{\text{surf}} \tag{12.38}$$

The force exerted by the α phase can be expressed in terms of the pressure in the α phase acting over the area of the surface element,

$$F^\alpha = P^\alpha r_1 r_2 d\phi_1 d\phi_2 \tag{12.39}$$

The force acting on the β phase can be expressed in terms of the pressure in the β phase,

$$F^\beta = P^\beta r_1 r_2 d\phi_2 d\phi_2 \tag{12.40}$$

Substitute Eqs. (12.37), (12.39), and (12.40) into Eq. (12.38):

$$P^\beta r_1 r_2 d\phi_1 d\phi_2 = P^\alpha r_1 r_2 d\phi_1 d\phi_2 + \sigma[r_1 + r_2]d\phi_1 d\phi_2$$

Divide through by $r_1 r_2 d\phi_1 d\phi_2$:

$$P^\beta = P^\alpha + \sigma \frac{r_1 + r_2}{r_1 r_2} \tag{12.41}$$

Note that

$$\frac{r_1 + r_2}{r_1 r_2} = \frac{1}{r_1} + \frac{1}{r_2} = 2H \tag{12.42}$$

so that Eq. (12.41) can be written

$$P^\beta = P^\alpha + 2\sigma H \tag{12.43}$$

This equation is identical with Eq. (12.27) with the specific interfacial free energy γ replaced by the surface tension σ.

Thus, for the simplest case of a unary two phase system in which both phases are isotropic fluids, the surface tension, which is a mechanical force acting in the surface, and the surface free energy, which is an excess energy associated with the presence of the interface, are equal. Their units (dyne/cm and ergs/cm^2 = (dyne-cm)/cm^2) are convertible. However, the situation increases significantly in complexity if one or both phases is crystalline. The specific interfacial free energy is a scalar, that is, a property of the system that has a value at each point on the surface, and varies with the orientation(s) of the crystal(s) involved. The surface tension passes to a surface stress in this more general case and is a tensor, as are all other local mechanical conditions in the solid state. Treatment of this more sophisticated case is beyond the scope of this introductory text. Comprehensive and understandable presentations of these ideas can be found in references 12.1 and 12.2.

12.5 CAPILLARITY EFFECTS ON PHASE DIAGRAMS

The influence of curved interfaces upon the behavior of materials systems is manifested primarily through the shift of phase boundaries on phase diagrams derived from the altered condition for mechanical equilibrium. The melting point of a phase is altered if the solid phase is present as fine particles. The vapor pressure in equilibrium with liquid is raised if the liquid is present as a dispersion of droplets. Phase boundaries in a binary system are slightly shifted if the minor phase is finely divided. Although these effects are usually small unless the particulate phase is extremely fine, they play a profound role in determining the course of many microstructural processes that occur in materials science.

12.5.1 Phase Boundary Shifts in Unary Systems

Consider a unary two phase system $(\alpha + \beta)$ in which it is explicitly recognized that the curvature of the $\alpha\beta$ boundary plays a role in the conditions for equilibrium:

$$T^\beta = T^\alpha \tag{12.44}$$

$$P^\beta = P^\alpha + 2\gamma H \tag{12.45}$$

$$\mu^\beta = \mu^\alpha \tag{12.46}$$

Equation (12.45) contains the convention that H is positive for surface elements that are convex relative to the β phase. The following development is an incremental generalization of the Clausius-Clapeyron equation derived in Chap. 7.

In considering possible variations in the state of this system it is necessary to include not only temperature and pressure but also to explicitly include variations that can arise from changes in the state of subdivision of the system reported in the geometric factor H. If an element of volume of the α phase is taken through an arbitrary change in state, the change in chemical potential is related to any changes in temperature and pressure it experiences by the equation

$$d\mu^\alpha = -S^\alpha dT^\alpha + V^\alpha dP^\alpha \tag{12.47}$$

Similarly, for a sample of the β phase,

$$d\mu^\beta = -S^\beta dT^\beta + V^\beta dP^\beta \tag{12.48}$$

If throughout these changes in state, the α and β phases are maintained in thermodynamic equilibrium, then Eqs. (12.44)–(12.46) must hold. Then

$$dT^\beta = dT^\alpha = dT \qquad (12.49)$$

$$dP^\beta = dP^\alpha + 2\gamma dH \qquad (12.50)$$

$$d\mu^\beta = d\mu^\alpha = d\mu \qquad (12.51)$$

Equation (12.50) recognizes that the pressure in the β phase can change as a result of two independent influences: the pressure in the α phase *and* the curvature of the $\alpha\beta$ interface.

These last five equations can be combined into a single expression,

$$d\mu^\beta = -S^\beta dT + V^\beta(dP^\alpha + 2\gamma dH) = d\mu^\alpha = -S^\alpha dT + V^\alpha dP^\alpha$$

Combining like terms

$$(S^\alpha - S^\beta)dT - (V^\alpha - V^\beta)dP^\alpha + 2\gamma V^\beta dH = 0 \qquad (12.52)$$

Define, as in Chap. 7:

$$\Delta S \equiv S^\alpha - S^\beta$$

$$\Delta V \equiv V^\alpha - V^\beta$$

Since by convention, the β phase is the reference phase for the definition of sign of the curvatures in the system, these definitions define changes from the reference (β) phase to the α phase. This convention is maintained throughout this development and must be used in applying the results to practical problems. With these definitions Eq. (12.52) can be written

$$\Delta S dT - \Delta V dP^\alpha + 2\gamma V^\beta dH = 0 \qquad (12.53)$$

The first two terms are identical with those obtained in the derivation of the Clausius-Clapeyron, Eq. (7.14), in which variations due to geometric effects are not considered; the third term contains the effect of capillarity influences upon the equilibrium conditions.

12.5.2 Vapor Pressure in Equilibrium with Curved Surfaces

Systematic exploration of Eq. (12.53) is most readily carried out by comparing equilibrium states in systems in which one of the variables is fixed. To derive an expression for the effect of curvature on vapor pressure, compare two liquid-vapor systems that are at the same temperature but differ in their state of subdivision reported by a difference in mean curvature H, Fig. 12.8. In this comparison the β phase is the liquid, α is the vapor and $dT = 0$; Eq. (12.53) becomes

$$-\Delta V dP + 2\gamma V^\beta dH = 0 \qquad (12.54)$$

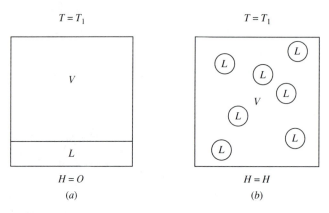

$T = T_1$ $T = T_1$

FIGURE 12.8
Compare two unary liquid plus vapor systems that have the same temperature; (*a*) a bulk system, with a flat interface and $H = 0$; (*b*) a system in which the liquid is finely divided in droplet form with a mean curvature everywhere equal to H.

If it is assumed that the vapor behaves like an ideal gas,

$$\Delta V = V^G - V^L \cong V^G = \frac{RT}{P}$$

since the volume of one mole of liquid is very small compared to that occupied by one mole of gas. Equation (12.54) can be written

$$-\frac{RT}{P}dP + 2\gamma V^L dH = 0$$

Or

$$d(\ln P) = \frac{2\gamma V^L}{RT}dH \qquad (12.55)$$

Since γ and V^L are not significant functions of the curvature of the interface in the system, this equation can be integrated from conditions corresponding to the bulk liquid with a flat interface and $H = 0$ to a system with the liquid surface exhibiting an arbitrary value of H,

$$\int_{P(H=0)}^{P(H)} d\ln P = \int_{H=0}^{H} \frac{2\gamma V^L}{RT}dH = \frac{2\gamma V^L}{RT}\int_{H=0}^{H} dH$$

$$\ln\left(\frac{P(H)}{P(H=0)}\right) = \frac{2\gamma V^L}{RT}H$$

$$P(H) = P(H=0)e^{(2\gamma V^L/RT)H} \qquad (12.56)$$

where $P(H)$ is the vapor pressure in a system with liquid droplets that have surface mean curvature H and $P(H = 0)$ is the vapor pressure over the bulk liquid at the same temperature. Since H has units of length^{-1}, the coefficient of H in the exponent must have units of length. This quantity can be called the *capillarity length scale* for liquid-vapor systems, written λ_v:

$$\lambda_v \equiv \frac{2\gamma V^L}{RT} \tag{12.57}$$

With this definition Eq. (12.56) can be written as

$$P(H) = P(H = 0)e^{\lambda_v H} \tag{12.58}$$

This equation is frequently used in an approximate form obtained by expanding the exponential and neglecting terms beyond the linear term,

$$P(H) = P(H = 0)[1 + \lambda_v H] \tag{12.59}$$

This approximation is valid as long as $\lambda_v H \ll 1$. It is concluded that the vapor pressure in a system containing liquid droplets is *higher* than the value found over a bulk liquid at the same temperature. For the effect of curvature to be significant, the product $\lambda_v H$ must be larger than about 0.01. Table 12.2 gives typical values of the capillarity length scale for a variety of liquids at their melting points. Evidently the effect of capillarity becomes significant for droplets with a radius of curvature smaller than about one micron (10^{-6} m or 10^{-4} cm).

Example 12.2. Calculate and plot as a function of radius the vapor pressure in a system with liquid zinc droplets suspended in zinc vapor at 900 K. The vapor pressure of zinc over bulk liquid at this temperature is 1×10^{-2} atm; the molar volume of the liquid is 9.5 cc/mole and $\gamma = 380$ ergs/cm^2.

Solution. The capillarity length scale for this system is

$$\lambda_v = \frac{2\left(380\ \frac{\text{ergs}}{\text{cm}^2}\right)\left(9.5\ \frac{\text{cc}}{\text{mole}}\right)}{\left(8.314\ \frac{\text{J}}{\text{mol K}}\right)\left(10^7\ \frac{\text{ergs}}{\text{J}}\,900\ \text{K}\right)} = 9.6 \times 10^{-8}\ \text{cm}$$

Insert this value into Eq. (12.58) to compute the vapor pressure in this system

$$P(r) = (1 \times 10^{-2}\ \text{atm})e^{\left[\frac{9.6\times10^{-8}\ (\text{cm})}{r\ (\text{cm})}\right]}$$

TABLE 12.2
Typical values for capillary length scales

	liquid-vapor λ_v (nm)	solid-liquid λ_m (nm)
Water	0.8	0.2
Metals	1	0.4
Salts	2	—
Oxides	0.8	—

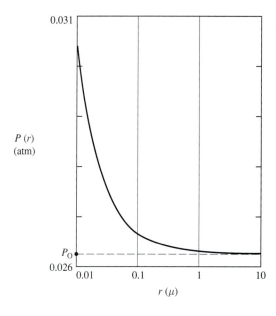

FIGURE 12.9
Plot of the vapor pressure of zinc as a function of droplet radius at 900 K.

This result is plotted in Fig. 12.9. The computed vapor pressure begins to deviate measurably from the bulk value for droplets below one micron in size.

Equation (12.59) describes the shift in vapor pressure as a function of curvature at each temperature along the liquid-vapor curve. For a given value of H this effect can be visualized as a shift in the vapor pressure curve on the (P,T) diagram, Fig. 12.10. Note that this capillarity shift can also be visualized as a depression of the vaporization temperature at a fixed pressure. This latter result could have been derived from Eq. (12.53) by comparing the equilibrium *temperature* of two systems that are constrained to have the *same vapor pressure*, but different states of subdivision. This application of the strategy is explored for the solid-liquid equilibrium in the next section.

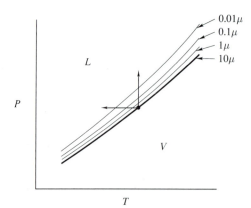

FIGURE 12.10
Shift in the vapor pressure curve with size of the liquid droplets in the $L + V$ system.

12.5.3 Effect of Curvature upon the Melting Temperature

To examine the effect of curvature upon solid-liquid equilibria, compare the systems shown in Fig. 12.11. In this comparison the external pressure, which is also the pressure in the liquid phase, is the same in the two systems; they differ in the value of H that is characteristic of each system. Equation (12.53) forms the basis for determining the difference in temperature that these two systems exhibit when they come to equilibrium. In the convention used in deriving Eq. (12.53) the β phase is the finely divided solid phase S and α is the liquid. Since in this comparison $dP = 0$, Eq. (12.53) can be written

$$\Delta S dT + 2\gamma V^S dH = 0 \tag{12.60}$$

Rearranging

$$dT = -\frac{2\gamma V^S}{\Delta S} dH$$

Integrating

$$\int_{T(H=0)}^{T(H)} dT = -\int_{H=0}^{H} \frac{2\gamma V^S}{\Delta S} dH = -\frac{2\gamma V^S}{\Delta S} H$$

since the terms in the integrand are not functions of curvature. Thus,

$$T(H) = T(H = 0) - \frac{2\gamma V^S}{\Delta S} H \tag{12.61}$$

where $T(H)$ is the equilibrium melting temperature in the system with solid particles that have mean curvature H and $T(H = 0)$ is the bulk melting temperature at the

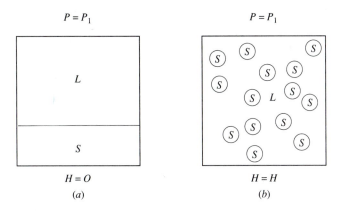

$P = P_1$ $P = P_1$

L S

$H = O$ $H = H$

(a)   (b)

FIGURE 12.11
Compare two unary solid plus liquid systems that have the same pressure; (a) a bulk system, with a flat interface and $H = 0$; (b) a system in which the solid is finely divided, with a mean curvature everywhere equal to H.

same pressure. This equation can also be written

$$T(H) = T(H = 0)[1 - \lambda_m H] \tag{12.62}$$

where the coefficient λ_m has units of length and can be viewed as a capillarity length scale for melting, given by

$$\lambda_m = \frac{2\gamma V^S}{T(H = 0)\Delta S} = \frac{2\gamma V^S}{\Delta H} \tag{12.63}$$

since the heat of fusion, $\Delta H = T(H = 0)\Delta S$. Because $\Delta S = S^L - S^S$, the entropy of fusion in the system is positive; it is concluded that increasing the curvature of the solid particles in a solid-liquid system *lowers* the melting point. Application of this result to each value of P along the melting curve produces a shift in that phase boundary to lower temperatures, Fig. 12.12. Typical values of capillarity length scales for solid-liquid systems are given in Table 12.2.

Example 12.3. When a liquid is cooled below its bulk melting point the solid phase forms and grows as dendrites. The scale of the resulting microstructure, indicating whether it is fine grained or a coarse structure and the rate at which it solidifies, is strongly influenced by the size of the tips of these tree-like structures as they grow. The tip radius is in turn determined by the temperature difference between the liquid adjacent to the tip and that of the supercooled liquid surrounding it. Calculate this temperature difference for a silicon dendrite with a tip radius of 0.1 microns growing into a surrounding liquid that is undercooled 5°C below the bulk melting point of silicon. The surface energy of the liquid-solid interface may be taken as 150 ergs/cm².

Solution. The melting point of bulk silicon is 1683 K, the heat of fusion is 46.5 KJ/mole, and the molar volume of solid silicon is 11.2 cc/mole. The capillarity length scale for this system can be calculated from Eq. (12.63) as

$$\lambda_m = \frac{2\left(150 \; \frac{\text{ergs}}{\text{cm}^2}\right)\left(10^{-7} \; \frac{\text{J}}{\text{erg}}\right)\left(11.2 \; \frac{\text{cc}}{\text{mol}}\right)}{\left(46500 \; \frac{\text{J}}{\text{mol}}\right)} = 7.2 \times 10^{-9} \; \text{cm}$$

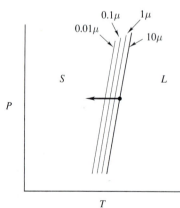

FIGURE 12.12
Shift in the melting curve with the size of the particles of the solid phase suspended in the liquid.

Assuming that the liquid and solid in a volume element at the tip of the dendrite are in *local equilibrium*, the temperature of the liquid at the interface is given by Eq. (12.62)

$$T = 1683 \text{ K} \left[1 - (7.2 \times 10^{-9} \text{ cm}) \frac{1}{0.1 \text{ microns} \left(\frac{10^{-4} \text{ cm}}{1 \text{ micron}} \right)} \right]$$

$$T = 1681.8 \text{ K}$$

The temperature of the surrounding supercooled liquid is

$$T^L = 1683 \text{ K} - 5 \text{ K} = 1678 \text{ K}$$

Thus the temperature at the tip lies above that of the surrounding liquid. Heat flows from the tip into the liquid, promoting solidification, and the solid silicon dendrite grows in these surroundings.

Example 12.4. Assume an array of dendrites with a distribution of tip sizes exists in the supercooled liquid silicon described in Example 12.3. If the temperature of a tip lies above that of the surrounding supercooled liquid, heat flows away from the tip into the surroundings and the tip grows. Conversely, if a tip has a temperature below T^L, heat flows from the surrounding liquid toward the tip, melting it. Compute the radius of the tip that neither grows nor melts.

Solution. If the tip radius is such that its equilibrium melting temperature equals that of the supercooled surroundings, 1678 K in this case, the tip remains stationary because it is in thermal equilibrium with its surroundings. The tip radius at which this occurs is found by setting $T(H)$ equal to 1678 K and solving for $H = 1/r$:

$$1678 \text{ K} = 1638 \text{ K} \left[1 - (7.2 \times 10^{-9} \text{ cm}) \cdot \frac{1}{r} \right]$$

Solving for the tip radius

$$r = \frac{(7.2 \times 10^{-9} \text{ cm})(1683 \text{ K})}{1683 \text{ K} - 1678 \text{ K}} = 2.4 \times 10^{-6} \text{ cm} = 0.024 \text{ microns}$$

This value for r is the *critical radius for growth*; tips smaller than 0.17 microns melt under these conditions while those that are larger than this radius grow.

12.5.4 Phase Boundary Shifts in Binary Systems

The altered condition for mechanical equilibrium that characterizes capillarity effects also operates on the boundaries of an $(\alpha + \beta)$ field in a binary system, producing shifts similar to those just derived for unary systems. To derive expressions for these capillarity shifts, it is necessary to follow a strategy analogous to that leading to Eq. (12.53) for a unary system. The following development paraphrases that strategy.

Equations (12.23), (12.24), and (12.27) present the conditions for equilibrium in a two phase multicomponent system. For a binary system $c = 2$ and these equations can be written

$$T^\beta = T^\alpha \qquad \mu_1^\beta = \mu_1^\alpha \qquad \mu_2^\beta = \mu_2^\alpha \qquad P^\beta = P^\alpha + 2\gamma H$$

Focus for the moment on the behavior of component 1 in the α phase. Because the chemical potential is an intensive property, it is a function of the temperature, pressure, and composition of the α phase:

$$d\mu_1^\alpha = -\overline{S}_1^\alpha dT^\alpha + \overline{V}_1^\alpha dP^\alpha + \mu_{12}^\alpha dX_2^\alpha \tag{12.64}$$

where the mole fraction of component 2 is chosen as the independent composition variable. $\overline{S}_1^\alpha$ and $\overline{V}_1^\alpha$ are the partial molal entropy and volume for component 1. The coefficient

$$\mu_{12}^\alpha \equiv \left(\frac{\partial \mu_1}{\partial X_2} \right)_{T,P}^\alpha \tag{12.65}$$

reports the change in chemical potential of component 1 with composition in the α phase. Analogous expressions can be written for changes with composition of the other three chemical potentials in the system, μ_{22}^α, μ_{12}^β, and μ_{22}^β.

If α and β are maintained in equilibrium as the state of the $(\alpha + \beta)$ system is altered, the conditions for equilibrium require that

$$dT^\beta = dT^\alpha = dT \tag{12.66}$$

$$dP^\beta = dP^\alpha + 2\gamma dH \tag{12.67}$$

$$d\mu_1^\beta = d\mu_1^\alpha \tag{12.68}$$

$$d\mu_2^\beta = d\mu_2^\alpha \tag{12.69}$$

Expressions like Eq. (12.64) can be substituted into Eq. (12.68) to give

$$d\mu_1^\alpha = -\overline{S}_1^\alpha dT^\alpha + \overline{V}_1^\alpha dP^\alpha + \mu_{12}^\alpha dX_2^\alpha$$
$$= d\mu_1^\beta = -\overline{S}_1^\beta dT^\beta + \overline{V}_1^\beta dP^\beta + \mu_{12}^\beta dX_2^\beta$$

Use the condition for thermal equilibrium to replace dT^α and dT^β with dT. Write dP^β in terms of dP^α and dH with the condition for mechanical equilibrium,

$$d\mu_1^\alpha = -\overline{S}_1^\alpha dT + \overline{V}_1^\alpha dP^\alpha + \mu_{12}^\alpha dX_2^\alpha$$
$$= d\mu_1^\beta = -\overline{S}_1^\beta dT + \overline{V}_1^\beta [dP^\alpha + 2\gamma dH] + \mu_{12}^\beta dX_2^\beta$$

Substitution for dP^β introduces the mean curvature into the relationship. Collecting like terms

$$-(\overline{S}_1^\alpha - \overline{S}_1^\beta)dT + (\overline{V}_1^\alpha - \overline{V}_1^\beta)dP^\alpha - 2\gamma \overline{V}_1^\beta dH + \mu_{12}^\alpha dX_2^\alpha - \mu_{12}^\beta dX_2^\beta = 0$$

Define

$$\Delta \overline{S}_1 \equiv \overline{S}_1^\alpha - \overline{S}_1^\beta \tag{12.70}$$

$$\Delta \overline{V}_1 \equiv \overline{V}_1^\alpha - \overline{V}_1^\beta \tag{12.71}$$

Note that these quantities are differences in *partial molal properties* for component 1 as distinguished from the corresponding total properties that appear in Eq. (12.53) for a unary system. With these definitions,

$$-\Delta\overline{S}_1 dT + \Delta\overline{V}_1 dP^\alpha - 2\gamma\overline{V}_1^\beta dH + \mu_{12}^\alpha dX_2^\alpha - \mu_{12}^\beta dX_2^\beta = 0 \tag{12.72}$$

The same derivation applied to Eq. (12.69) gives the corresponding equation for component 2 as

$$-\Delta\overline{S}_2 dT + \Delta\overline{V}_2 dP^\alpha - 2\gamma\overline{V}_2^\beta dH + \mu_{22}^\alpha dX_2^\alpha - \mu_{22}^\beta dX_2^\beta = 0 \tag{12.73}$$

Equations (12.72) and (12.73) are generalizations of Eq. (12.53) derived for a unary system. They represent two equations relating the five variables T, P, H, X_2^α, and X_2^β. To explore the effect of H upon phase boundary compositions, it is necessary to fix temperature and pressure. With $dT = 0$ and $dP^\alpha = 0$, the analysis focuses on the variation of the compositions of the phases with curvature. With these constraints, Eq. (12.72) and (12.73) become

$$-2\gamma\overline{V}_1^\beta dH + \mu_{12}^\alpha dX_2^\alpha - \mu_{12}^\beta dX_2^\beta = 0 \tag{12.74}$$

$$-2\gamma\overline{V}_2^\beta dH + \mu_{22}^\alpha dX_2^\alpha - \mu_{22}^\beta dX_2^\beta = 0 \tag{12.75}$$

The chemical potential derivatives that appear as coefficients in these equations, the μ_{kj} terms, are related within each phase by the Gibbs-Duhem equation. For any given phase,

$$X_1 d\mu_1 + X_2 d\mu_2 = 0$$

For a fixed temperature and pressure $d\mu_1$ can be evaluated from Eq. (12.64):

$$(d\mu_1)_{T,P} = \mu_{12} dX_2$$

Similarly, for μ_2

$$(d\mu_2)_{T,P} = \mu_{22} dX_2$$

Insert these evaluations into the Gibbs-Duhem equation:

$$X_1 \cdot \mu_{12} dX_2 + X_2 \cdot \mu_{22} dX_2 = 0$$

Thus, the compositional derivative of μ_2 is related to that for μ_1

$$\mu_{22} = -\frac{X_1}{X_2}\mu_{12} \tag{12.76}$$

This equation applies separately to the α or β phase.

With these relationships Eq. (12.74) and (12.75) can be rewritten

$$\mu_{12}^\alpha \left(\frac{dX_2^\alpha}{dH}\right) - \mu_{12}^\beta \left(\frac{dX_2^\beta}{dH}\right) = 2\gamma\overline{V}_1^\beta \tag{12.77}$$

$$-\frac{X_1^\alpha}{X_2^\alpha}\mu_{12}^\alpha \left(\frac{dX_2^\alpha}{dH}\right) + \frac{X_1^\beta}{X_2^\beta}\mu_{12}^\beta \left(\frac{dX_2^\beta}{dH}\right) = 2\gamma\overline{V}_2^\beta \tag{12.78}$$

The derivatives in these equations are the rates of change of the $(\alpha+\beta)$ phase boundary compositions with curvature. These equations form two simultaneous linear equations in these derivatives. They can be solved by substitution or determinants to give

$$\left(\frac{dX_2^\alpha}{dH}\right) = 2\gamma(X_1^\beta \overline{V}_1^\beta + X_2^\beta \overline{V}_2^\beta) \cdot \frac{X_2^\alpha}{\mu_{12}^\alpha(X_2^\alpha - X_2^\beta)} \tag{12.79}$$

$$\left(\frac{dX_2^\beta}{dH}\right) = 2\gamma(X_1^\alpha \overline{V}_1^\beta + X_2^\alpha \overline{V}_2^\beta) \cdot \frac{X_2^\beta}{\mu_{12}^\beta(X_2^\alpha - X_2^\beta)} \tag{12.80}$$

Note that the quantity in brackets in Eq. (12.79) is simply the molar volume of the β phase V^β; the corresponding quantity in Eq. (12.80) has units of volume per mole but is not equal to the molar volume of either phase.

The chemical potential of component 1 generally decreases with increasing component 2 (except within a miscibility gap) so that the μ_{12} terms are negative in these equations. Choose component 2 so that it is rich in the β phase and the solute in α, that is, so that $X_2^\beta > X_2^\alpha$. Then both derivatives are positive. This shift in the compositions at the phase boundaries of an $(\alpha + \beta)$ field is sketched in Fig. 12.13. With increasing curvature of the $\alpha\beta$ boundary both phases become richer in component 2. The magnitude of the shift is particularly sensitive to the values of the μ_{12} factors and to the width of the two phase field given by $(X_2^\beta - X_2^\alpha)$. Evidently, the narrower the two phase field, the larger the magnitude of the shift.

These results, though general, are most frequently applied to the case in which both the α and β phases are dilute solutions. The α phase is dilute in component 2 and the β phase is dilute in component 1. The chemical potential of component 1 in the α phase is given by Raoult's Law for the solvent:

$$\mu_1^\alpha = \mu_1^{o\alpha} + RT \ln X_1^\alpha$$

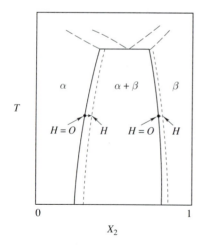

FIGURE 12.13
Effect of curvature of the $\alpha\beta$ interface upon the equilibrium composition of the α and β phases in a binary system.

Evaluate

$$\mu_{12}^{\alpha} = \left(\frac{d\mu_1^{\alpha}}{dX_2^{\alpha}}\right)_{T,P} = \left(\frac{d\mu_1^{\alpha}}{dX_1^{\alpha}}\right)_{T,P} \cdot \frac{dX_1^{\alpha}}{dX_2^{\alpha}} = -\frac{RT}{X_1^{\alpha}} \cong -RT$$

since in dilute α, $X_1^{\alpha} \cong 1$. In the β phase, where component 1 is the solute, Henry's law describes its chemical potential as

$$\mu_1^{\beta} = \mu_1^{o\beta} + RT \ln \gamma_1^{o} + RT \ln X_1^{\beta}$$

where γ_1^{o}, the Henry's law coefficient, is independent of composition. The required derivative is

$$\mu_{12}^{\beta} = \left(\frac{d\mu_1^{\beta}}{dX_2^{\beta}}\right)_{T,P} = \left(\frac{d\mu_1^{\beta}}{dX_1^{\beta}}\right)_{T,P} \frac{dX_1^{\beta}}{dX_2^{\beta}} = -\frac{RT}{X_1^{\beta}}$$

Further, since both α and β are dilute solutions $(X_2^{\alpha} - X_2^{\beta}) \cong (-X_2^{\beta}) \cong -1$. With these evaluations, Eq. (12.79) and (12.80) become

$$\left(\frac{dX_1^{\alpha}}{dH}\right) = 2\gamma V^{\beta} \frac{X_2^{\alpha}}{-RT \cdot (-1)} = \frac{2\gamma \overline{V}_1^{\beta}}{RT} X_2^{\alpha} \tag{12.81}$$

$$\left(\frac{dX_2^{\beta}}{dH}\right) = 2\gamma \overline{V}_1^{\beta} \frac{1}{-\left(\frac{RT}{X_1^{\beta}}\right) \cdot (-1)} = \frac{2\gamma \overline{V}_1^{\beta}}{RT} X_1^{\beta} \tag{12.82}$$

Note that the volume factor in Eq. (12.80) simplifies to the partial molal volume of component 1 in the β phase.

Equation (12.81) can be integrated after separating the variables:

$$\int_{X_2^{\alpha}(H=0)}^{X_2^{\alpha}(H)} \frac{dX_2^{\alpha}}{X_2^{\alpha}} = \int_{H=0}^{H} \frac{2\gamma V^{\beta}}{RT} dH$$

where $X_2^{\alpha}(H = 0)$ is the composition of the α side of the two phase field in a system with flat interfaces (a bulk system) and $X_2^{\alpha}(H)$ is the composition of α in equilibrium with β when the $\alpha\beta$ interface has a mean curvature H. Since the integrand on the right side is independent of the curvature of the interface, integration gives

$$\ln \left[\frac{X_2^{\alpha}(H)}{X_2^{\alpha}(H = 0)}\right] = \frac{2\gamma V^{\beta}}{RT} H \tag{12.83}$$

The coefficient of H has the same form as that obtained for the shift in vapor pressure in a unary system, Eq. (12.56). It can be defined as a capillarity length scale for the composition shift as

$$\lambda^{\alpha} \equiv \frac{2\gamma V^{\beta}}{RT} \tag{12.84}$$

With this substitution and some rearrangement Eq. (12.83) can be written

$$X_2^{\alpha}(H) = X_2^{\alpha}(H = 0)e^{\lambda^{\alpha} H} \tag{12.85}$$

An approximate version of this equation is usually applied in practical situations, obtained by expanding the exponential and neglecting terms beyond the first order in H,

$$X_2^\alpha(H) = X_2^\alpha(H = 0)[1 + \lambda^\alpha H] \tag{12.86}$$

Thus, the concentration of component 2 in the α phase, the *solute* in this case, which is in equilibrium with the β phase in a system in which the $\alpha\beta$ interface has mean curvature H, is larger than the concentration in the analogous bulk system, that is, one for which the interface is flat. This increase in solubility is proportional to the mean curvature H of the interface.

A similar result can be obtained for the composition of the solute in the β phase in this system by integrating Eq. (12.82):

$$X_1^\beta(H) = X_1^\beta(H = 0)e^{-\lambda^\beta H} \tag{12.87}$$

Note that the solute is component 1 in the β phase. Note also the negative sign in the exponent. The capillarity length scale for the β phase is given by

$$\lambda_c^\beta \equiv \frac{2\gamma \overline{V}_1^\beta}{RT} \tag{12.88}$$

If $\lambda_c^\beta H$ is small compared to 1,

$$X_1^\beta(H) = X_1^\beta(H = 0)[1 - \lambda^\beta H] \tag{12.89}$$

Thus, in the β phase the equilibrium concentration of component 1, the *solute* in this case, is decreased in proportion to the mean curvature of the interface.

Recall that the convention for choosing the sign of H assumes that the mean curvature is positive for surface elements that are convex relative to the β phase. If, in the system under consideration, the α phase is present as particles in a matrix of β, then, by this convention, H is *negative* in this system. The equations describing the capillarity shift in such a system are exactly those given here, unchanged. Within these equations, however, H is negative. Thus the phase boundary shifts are in the opposite direction on the composition axis.

The shifts in composition at constant temperature and pressure given in Eqs. (12.86) and (12.89) can be computed at each temperature along a phase boundary. The result can be viewed as a shift in the phase boundaries of the α and β phases, Fig. 12.13. The effects are qualitatively similar if the α and β phases are not dilute so that the general Eqs. (12.79) and (12.80) must be applied.

12.5.5 Local Equilibrium and the Application of Capillarity Shifts

The development presented in Sec. 12.5 visualizes a system composed of two phases with an interface on which the mean curvature is constant. The simplest example of such a microstructure consists of a collection of spherical particles of the β phase that all have the same radius r, so that $H = 1/r$. In real microstructures there always exists a distribution of mean curvature values over the interfaces in the system. Even in the simplest case of spherical particles, real microstructures exhibit a distribution of sizes. Thus the structures visualized in Figs. 12.8 and 12.11 never exist in real structures.

The results derived in that development are generally used in practice by applying the *principle of local equilibrium* at an element of interface in the two phase structure. Each interfacial element has its value of mean curvature H. It is assumed that volume elements in the α and β phases adjacent to the interface are *locally* in equilibrium. Explicitly, it is assumed that the conditions for equilibrium, equality of temperature and chemical potentials, and the condition for mechanical equilibrium that gives rise to a pressure difference determined by the *local* curvature of the interface, all hold for the α and β volume elements in contact at the interface. The results derived from these conditions, contained, for example, in Eqs. (12.86) and (12.89), also hold for these volume elements.

If the principle of local equilibrium is applied at each interfacial element in a microstructure, the deductions from that assumption give rise to effects that can be used to understand microstructure changes that may be ongoing. Figure 12.14a sketches a microstructure consisting of a dispersion of β particles in an α matrix. Prolonged exposure to high temperature causes such a microstructure to coarsen to the structures shown in Fig. 12.14b and c. This process is called *coarsening* or *Ostwald ripening*. Small particles shrink and disappear while large particles grow in this process; at a later time the same volume of β phase is distributed over fewer larger particles. This process can be understood by focusing on the interaction between two particles, Fig. 12.15. According to Eq. (12.86) the concentration of solute in α adjacent to the small particle is shifted more by capillarity than that adjacent to the large particle, Fig. 12.15. Thus the concentration difference created in the α phase produces a diffusion flux of solute from the region of high concentration by the small particle, to the region of lower concentration adjacent to the large particle. As a result the small particle shrinks as it supplies the solute to this diffusion process and the large particle grows as it

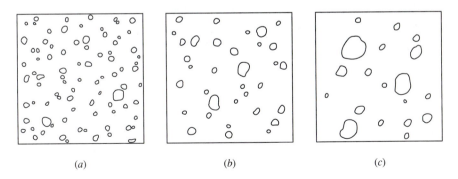

(a)　　　　　　　　　　(b)　　　　　　　　　　(c)

FIGURE 12.14
Three stages of the coarsening process in which a dispersion of fine β particles evolves toward microstructures with fewer, larger particles.

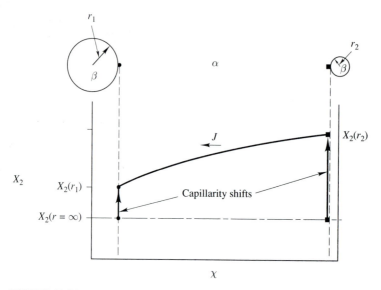

FIGURE 12.15

The principle of local equilibrium implies capillarity shifts in matrix compositions required to produce diffusional flow from small particles (which shrink and disappear) to large particles (which grow).

accumulates solute. Repetition of this interaction between pairs of interfacial elements throughout the microstructure yields the coarsening process.

Application of capillarity shift effects on phase diagrams through the principle of local equilibrium is invoked in the analysis of a variety of processes that produce changes in microstructure, including coarsening, dendrite growth, eutectic and eutectoid transformations, powder processing in ceramics and powder metallurgy, grain growth, and nucleation. This section provides the fundamental thermodynamic background underlying these applications.

12.6   THE EQUILIBRIUM SHAPE OF CRYSTALS: THE GIBBS-WULFF CONSTRUCTION

The development in Sec. 12.5 assumes that the value of the specific interfacial free energy is *isotropic*; that is, it does not depend on the orientation of the element of interface under consideration. If one or both of the phases involved is crystalline, then this assumption may be a poor approximation. It can be readily visualized that different crystal faces exposed to their equilibrium vapor will exhibit differences in surface energy. The atoms thus exposed in a close packed plane have a configuration of neighbors more like that in the bulk crystal than atoms lying on an irrational crystal plane at the surface. In general the value of γ varies significantly with the orientation of the crystal plane exposed.

Visualize a plot of $\gamma(\theta, \phi)$ over the sphere of orientation, Fig. 12.16a. Orientations corresponding to close packed planes may exhibit local minima in the form

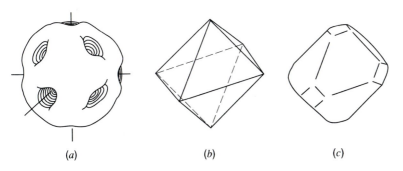

(a)   (b)   (c)

FIGURE 12.16
A plot of surface energy in spherical coordinates (a) may correspond to an equilibrium crystal shape composed entirely of facets (b), or a faceted shape with rounded edges and corners (c).

of cusps in this γ plot because their surface energies are significantly smaller than neighboring orientations. The crystal may minimize its surface energy by favoring the formation of surface facets with these low energy orientations. The resulting equilibrium shape may be a polyhedron completely bounded by facets, or may consist of facets connected by smoothly curved surfaces, Fig. 12.16b and Fig. 12.6 c. The shape of the crystal at equilibrium can be modelled from such a γ plot by applying once more the strategy for finding conditions for equilibrium. In order to develop this strategy it is first necessary to review some aspects of the geometry of polyhedra.

The volume of a polyhedron can be related to its surface area and its "radius" in much the same way as a sphere. Label the F faces on the polyhedron shown in Fig. 12.17a from 1 to F. The point O is the centroid (center of mass) of the polyhedron. Let A_j be the area of the jth face. Define λ_j to be the *perpendicular* distance from the centroid to the jth face; this property is called the *pedal function* for the jth face. Draw lines from the centroid O to the corners of the jth face. This construction establishes a pyramid with the jth face as its base, O as its apex, and λ_j, as its altitude. The volume

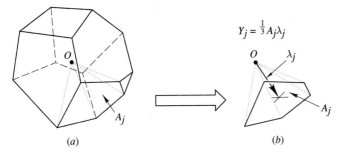

$$Y_j = \tfrac{1}{3} A_j \lambda_j$$

(a)   (b)

FIGURE 12.17
The volume of a flat faced polyhedra (a) can be computed from the areas A_j of the individual faces and (b) from their pedal functions, λ_j.

of this pyramid is $(1/3)A_j\lambda_j$. The volume of the polyhedron is

$$V^c = \sum_{j=1}^{F} \frac{1}{3} A_j \lambda_j \tag{12.90}$$

Now compare the geometry of this polyhedron with one obtained from it by displacing each of the facets a distance $d\lambda_j$, which may be different for each face. The jth face sweeps through a volume given by $A_j d\lambda_j$. The total change in volume of the polyhedron is the sum of the contributions from the faces

$$dV^c = \sum_{j=1}^{F} A_j d\lambda_j \tag{12.91}$$

In this evaluation of dV^c contributions along the edges and at the corners are neglected. However, since these are of a higher order in the differential $d\lambda_j$, in the limit they are in fact, negligible.

This volume change can also be computed by differentiating Eq. (12.90)

$$dV^c = \sum_{j=1}^{F} \frac{1}{3} [A_j d\lambda_j + \lambda_j dA_j] \tag{12.92}$$

Set these two expressions for dV^c equal to each other and simplify,

$$\sum_{j=1}^{F} A_j d\lambda_j = \frac{1}{2} \sum_{j=1}^{F} \lambda_j dA_j \tag{12.93}$$

Substitute this result into Eq. (12.91) obtaining a relation between the volume change of a faceted crystal and the changes in areas of its facets as

$$dV^c = \frac{1}{2} \sum_{j=1}^{F} \lambda_j dA_j \tag{12.94}$$

This expression is the analog to Eq. (12.7) for smooth surfaces, connecting the change in volume of the crystal to changes in its area.

Next apply the general strategy for finding conditions for equilibrium. The sequence of relationships that result in this application is essentially equivalent to that developed in Sec. 12.3 for curved interfaces except that each term involving the change in area in Eqs. (12.15) through (12.20) is replaced by the corresponding summation over the facets on the crystal surface. With α designated as the vapor (v) and β the crystal (c), Eg. (12.21) becomes

$$dS'_{\text{sys,iso}} = \left(\frac{1}{T^v} - \frac{1}{T^c} \right) dU^{/v}_{\text{hyp}} + \left(\frac{P^v}{T^v} - \frac{P^c}{T^c} \right) dV^{/v}_{\text{hyp}}$$

$$- \sum_{k} \left(\frac{\mu_k^v}{T^v} - \frac{\mu_k^c}{T^c} \right) dn_{k,\text{hyp}}^v - \sum_{j=1}^{F} \frac{\gamma_j}{T^c} dA_j \tag{12.95}$$

where

$$\gamma_j = U_j^s - T S_j^s - \sum_k \mu_k \Gamma_{kj} \tag{12.96}$$

for the jth face. The differential in the second term in Eq. (12.95) can be converted to $dV_{\text{hyp}}^{\prime c}$ since, in this isolated system, $dV_{\text{hyp}}^{\prime v} = -dV_{\text{hyp}}^{\prime c}$. Apply Eq. (12.94) to relate this volume change to the changes in facet areas:

$$dS'_{\text{sys,iso}} = \left(\frac{1}{T^v} - \frac{1}{T^c}\right) dU^{\prime v} + \left(\frac{P^v}{T^v} - \frac{P^c}{T^c}\right)\left[-\frac{1}{2}\sum_{j=1}^{F}\lambda_j dA_j\right]$$

$$- \sum_k \left(\frac{\mu_k^v}{T^v} - \frac{\mu_k^c}{T^c}\right) dn_k^v - \sum_{j=1}^{F} \frac{\gamma_j}{T^c} dA_j \tag{12.97}$$

Collect like terms:

$$dS'_{\text{sys,iso}} = \left(\frac{1}{T^v} - \frac{1}{T^c}\right) dV_{\text{hyp}}^{\prime v} - \sum_k \left[\frac{\mu_k^v}{T^v} - \frac{\mu_k^c}{T^c}\right] dn_{k,\text{hyp}}^v$$

$$+ \sum_{j=1}^{F}\left[\left(\frac{P^c}{T^c} - \frac{P^v}{T^v}\right)\frac{1}{2}\lambda_j - \frac{\gamma_j}{T^c}\right] dA_j$$

With the entropy of the isolated system thus expressed in terms of independent variables, the conditions for equilibrium are found by setting the coefficients equal to zero. The first $(c + 1)$ terms in this equation yield the usual conditions for thermal and chemical equilibrium. The remaining F terms are each of the form

$$\left(\frac{P^c}{T^c} - \frac{P^v}{T^v}\right)\frac{1}{2}\lambda_j - \frac{\gamma_j}{T} = 0$$

which, given that $T^c = T^v = T$, can be rearranged as

$$P^c = P^v + 2\frac{\gamma_j}{\lambda_j} \qquad (j = 1, 2, \ldots, F) \tag{12.98}$$

This condition for mechanical equilibrium is analogous to Eq. (12.27) for smoothly curved surfaces.

The *shape* of the equilibrium crystal can be inferred from the observation that in Eq. (12.98) the pressures are values for the crystal and vapor phases and thus are the same for all the exposed faces. Consequently, at equilibrium the ratio

$$\frac{\gamma_j}{\lambda_j} = \text{constant} \tag{12.99}$$

for all the faces. This implies that the pedal function for each facet, the perpendicular distance from the centroid of the crystal to each facet, is proportional to the surface energy of that facet. Facets with a low surface energy take on a position closer to the centroid of the crystal than those with higher surface energy. As a consequence these facets occupy more of the surface area of the resulting crystal.

The relationship given in Eq. (12.99) provides the basis for determining the quantitative shape of a crystal at equilibrium embodied in the *Gibbs-Wulff* construction. Figure 12.18*a* shows a two dimensional polar plot of γ versus orientation illustrating this construction. For any given orientation the radius vector r of the polar plot is proportional to the surface energy. In simple terms, Eq. (12.99) states that the pedal function for each face is proportional to the value of γ for that face and thus the radius vector on the polar plot. This geometric relationship can be visualized by constructing the plane perpendicular to r at the tip of each radius vector. Repetition of this construction for the set of all radius vectors over the sphere of orientation yields a collection of overlapping planes that envelop and delineate the shape that satisfies the conditions for mechanical equilibrium for the crystal that has the given polar plot. Equilibrium shapes for a variety of polar plots are sketched in Fig. 12.19.

This condition for mechanical equilibrium for fine crystals can be translated into capillarity shifts of vapor pressure and equilibrium compositions by paraphrasing the strategies that yield Eqs. (12.56), (12.62), (12.79), and (12.80), in each case replacing the product γH with the ratio γ_j / λ_j.

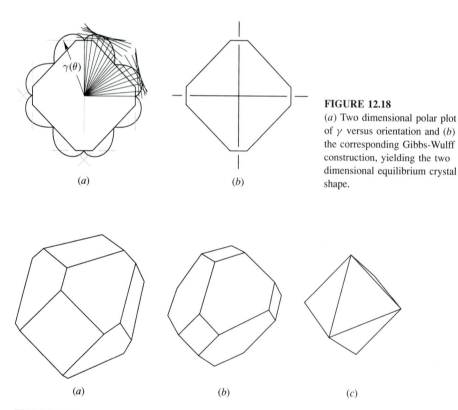

(*a*)   (*b*)

FIGURE 12.18
(*a*) Two dimensional polar plot of γ versus orientation and (*b*) the corresponding Gibbs-Wulff construction, yielding the two dimensional equilibrium crystal shape.

(*a*)   (*b*)   (*c*)

FIGURE 12.19
Equilibrium crystal shapes derived from a variety of γ polar plots.

12.7   EQUILIBRIUM AT TRIPLE LINES

Consider the rudimentary three phase $(\alpha + \beta + \epsilon)$ system shown in Fig. 12.20. These phases meet pair-wise at $\alpha\beta, \alpha\epsilon$, and $\beta\epsilon$ interfaces; each interface has its value of specific interfacial free energy designated, respectively, $\gamma_{\alpha\beta}, \gamma_{\beta\epsilon}$, and $\gamma_{\alpha\epsilon}$. In the development presented in this section it is assumed that these surface energies are isotropic. The three phases meet along a line designated $\alpha\beta\epsilon$ and called a *triple line*. The three interfaces also meet along this triple line. In general the triple line is not straight nor even a plane curve; it is a space curve. Figure 12.21 shows a plane section through this system at the point P, constructed so that the section is locally perpendicular to the triple line. The traces of the three interfaces form interior angles ϕ_α, ϕ_β, and ϕ_ϵ, as shown in this figure.

The condition for equilibrium in this system can be found by applying the general strategy used repeatedly in this presentation. Development of this application is cumbersome because the system consists of three phases and three interfaces. The change in entropy is given by

$$dS'_{\text{sys}} = dS'^{\alpha}_{\text{hyp}} + dS'^{\beta}_{\text{hyp}} + dS'^{\epsilon}_{\text{hyp}} + S^s_{\alpha\beta}dA_{\alpha\beta} + S^s_{\beta\epsilon}dA_{\beta\epsilon} + S^s_{\alpha\epsilon}dA_{\alpha\epsilon} \qquad (12.100)$$

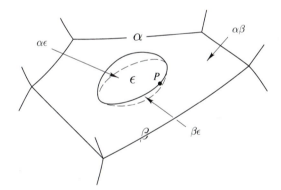

FIGURE 12.20
A typical triple line in a three phase microstructure. A particle of the ϵ phase lies on the interface between α and β phases. The line in the $\alpha\beta$ interface where all three phases meet is an $\alpha\beta\epsilon$ triple line.

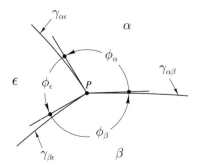

FIGURE 12.21
View of the local microstructure on a plane perpendicular to the triple line at a point P.

The isolation constraints are

$$dU'_{sys} = 0$$

$$= dU'^{\alpha}_{hyp} + dU'^{\beta}_{hyp} + dU'^{\epsilon}_{hyp} + U^s_{\alpha\beta}dA_{\alpha\beta} + U^s_{\beta\epsilon}dA_{\beta\epsilon} + U^s_{\alpha\epsilon}dA_{\alpha\epsilon} \quad (12.101)$$

$$dV'_{sys} = 0 = dV'^{\alpha}_{hyp} + dV'^{\beta}_{hyp} + dV'^{\epsilon}_{hyp} \quad (12.102)$$

$$dn_{k,sys} = 0 = dn^{\alpha}_{k,hyp} + dn^{\beta}_{k,hyp} + dn^{\epsilon}_{k,hyp}$$

$$+ \Gamma_{k,\alpha\beta}dA_{\alpha\beta} + \Gamma_{k,\beta\epsilon}dA_{\beta\epsilon} + \Gamma_{k\alpha\epsilon}dA_{\alpha\epsilon} \quad (k = 1, 2, \ldots, c) \quad (12.103)$$

Use of the isolation constraints to eliminate dependent variables in the expression for the entropy yields the familiar terms that lead to conditions for thermal, mechanical, and chemical equilibrium among the three phases, along with three additional terms:

$$dS'_{sys,iso} = [\ldots 2(c + 2) \text{ terms} \ldots] - \frac{\gamma_{\alpha\beta}}{T}dA_{\alpha\beta} - \frac{\gamma_{\beta\epsilon}}{T}dA_{\beta\epsilon} - \frac{\gamma_{\alpha\epsilon}}{T}dA_{\alpha\epsilon} \quad (12.104)$$

These three changes in area of the interfaces are not independent. It is necessary to derive the relationships between these area changes associated with an arbitrary displacement of the triple line.

A displacement dl of a triple line that is arbitrary in both magnitude and direction is shown in Fig. 12.22. The area of each of the three interfaces is changed as a result of this displacement by amounts that are determined by the length changes $dl_{\alpha\beta}, dl_{\beta\epsilon}$ and $dl_{\alpha\epsilon}$. For an element of length dL *along* the triple line, the change in area of each

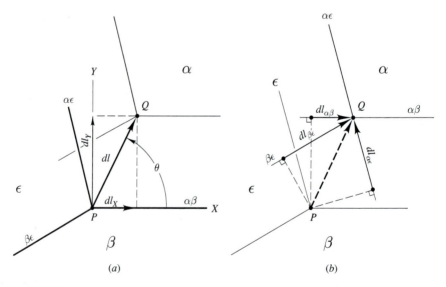

FIGURE 12.22
(a) The arbitrary displacement of a triple point P to a point Q can be described in terms of its components dl_x and dl_y. (b) As a result the lengths of the traces of the $\alpha\beta$, $\alpha\epsilon$ and $\beta\epsilon$ interfaces change by amounts $dl_{\alpha\beta}, dl_{\alpha\epsilon}$ and $dl_{\beta\epsilon}$.

of the boundaries is:

$$dA_{\alpha\beta} = dL \cdot dl[\cos\theta\cos\phi_\epsilon + \sin\theta\sin\phi_\epsilon]$$

$$dA_{\beta\epsilon} = -dL \cdot dl[\cos\theta\cos\phi_\alpha - \sin\theta\sin\phi_\alpha]$$

$$dA_{\alpha\epsilon} = -dL \cdot dl[\cos\theta]$$

where θ gives the arbitrary direction of displacement of the triple line. Substitute these results into the expression for the entropy, Eq. (12.104):

$$dS'_{\text{sys,iso}} = [\ldots\ldots] + \frac{1}{T}(dL \cdot dl) \times [\gamma_{\alpha\beta}(\cos\theta\cos\phi_\epsilon + \sin\theta\sin\phi_\epsilon)$$

$$+ \gamma_{\beta\epsilon}(\cos\theta\cos\phi_\alpha - \sin\theta\sin\phi_\alpha) + \gamma_{\alpha\epsilon}(\cos\theta)]$$

Collecting like terms

$$dS'_{\text{sys,iso}} = [\ldots\ldots]$$

$$+ \frac{1}{T}[-dL] \cdot [\gamma_{\alpha\beta}\cos\phi_\epsilon + \gamma_{\beta\epsilon}\cos\phi_\alpha + \gamma_{\alpha\epsilon}] \cdot dl\cos\theta$$

$$+ \frac{1}{T}[-dL] \cdot [\gamma_{\alpha\beta}\sin\phi_\epsilon - \gamma_{\beta\epsilon}\sin\phi_\alpha] \cdot dl\sin\theta$$

$$dS'_{\text{sys,iso}} = [\ldots\ldots]$$

$$+ \frac{1}{T}[-dL] \cdot [\gamma_{\alpha\beta}\cos\phi_\epsilon + \gamma_{\beta\epsilon}\cos\phi_\alpha + \gamma_{\alpha\epsilon}] \cdot dl_x \qquad (12.105)$$

$$+ \frac{1}{T}[-dL] \cdot [\gamma_{\alpha\beta}\sin\phi_\epsilon - \gamma_{\beta\epsilon}\sin\phi_\alpha] \cdot dl_y$$

where dl_x and dl_y are the independent components of the displacement of the triple line.

Equation (12.105) thus expresses the change in entropy for an isolated three phase system with interfaces and triple line in terms of variables that are now independent. The conditions for equilibrium are found by setting all the coefficients of the independent differentials in this equation equal to zero. The terms included in the brackets, [......], give the usual conditions for thermal, mechanical, and chemical equilibrium. The remaining two coefficients, treated explicitly in Eq. (12.105), yield

$$\gamma_{\alpha\beta}\cos\phi_\epsilon + \gamma_{\beta\epsilon}\cos_\alpha + \gamma_{\alpha\epsilon} = 0 \qquad (12.106)$$

and

$$\gamma_{\alpha\beta}\sin\phi_\epsilon - \gamma_{\beta\epsilon}\sin\phi_\alpha = 0 \qquad (12.107)$$

These equations are identical to those derived from a mechanical balance of three force vectors of magnitudes $\gamma_{\alpha\beta}$, $\gamma_{\beta\epsilon}$, and $\gamma_{\alpha\epsilon}$ acting at the triple point. In this force balance, Eq. (12.106) represents the sum of the forces in the x direction and Eq. (12.107) is the sum in the y direction.

These two equations can also be written

$$\frac{\gamma_{\alpha\beta}}{\sin\phi_\epsilon} = \frac{\gamma_{\beta\epsilon}}{\sin\phi_\alpha} = \frac{\gamma_{\alpha\epsilon}}{\sin\phi_\beta} \tag{12.108}$$

by applying some trigonometric identities (see Problem 12.13). These conditions for equilibrium at a triple line form the basis for determining *relative* interfacial energies of the intersecting surfaces since, with minor rearrangement,

$$\frac{\gamma_{\beta\epsilon}}{\gamma_{\alpha\beta}} = \frac{\sin\phi_\alpha}{\sin\phi_\epsilon} \tag{12.109}$$

$$\frac{\gamma_{\alpha\epsilon}}{\gamma_{\alpha\beta}} = \frac{\sin\phi_\beta}{\sin\theta_\epsilon} \tag{12.110}$$

Thus, if the absolute value of any one of the three surface energies has been measured independently, the values of the other two can be determined simply by measuring the angles between the traces of the surfaces on a section perpendicular to the triple line.

Example 12.5. A nickel bicrystal is equilibrated with its vapor. Where the grain boundary intersects the free surface, a groove is formed, as shown in Fig. 12.23. The surface energy of a nickel-vapor interface is nearly isotropic at 1400 K and can be taken as 1780 ergs/cm^2 = 1.78 J/m^2. Determine the interfacial energy of this grain boundary.

Solution. Measure the dihedral angle (the angle between the surface normals) at the root of the groove. Since the grain boundary energy is significantly smaller than the surface energy, this angle is near 0°. The most accurate method for determining such angles is based upon sighting vertically down on the groove with an interference microscope and analyzing the interference fringes. For this illustrative example, measurement of the angle between *tangent lines* drawn at the root of the groove yields 168°. The remaining angles are equal to each other and are thus equal to $(360° - \phi_v)/2$. Equation (12.109) becomes

$$\frac{\gamma_{gb}}{\gamma_{sv}} = \frac{\sin\phi_v}{\frac{1}{2}\sin(360 - \phi_v)} = \frac{\sin(168°)}{\sin(96°)} = \frac{0.208}{0.994} = 0.209$$

$$\gamma_{gb} = 0.209 \cdot \gamma_{sv} = 0.209 \cdot 1.78 \ \frac{J}{m^2} = 0.37 \ \frac{J}{m^2}$$

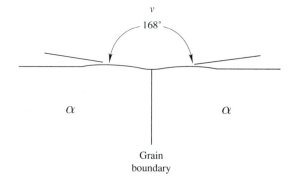

v

168°

α α

Grain
boundary

FIGURE 12.23
Configuration commonly observed where a grain boundary in a solid phase meets a free surface. The groove angle yields the ratio of the grain boundary energy to the surface energy.

Equations (12.106) and (12.107) assume that the surface energies of the three interfaces involved are not functions of orientation. If one or more of the three phases is crystalline, this assumption is not valid. A more general set of conditions for equilibrium, which considers variations in the energies of the boundaries due to rotations as well as translations, introduces additional "torque" terms into these equations that encompass the changes in the γ values with orientation. While such effects have been shown to be important in the analysis of low angle grain boundaries, it is usually not necessary to appeal to this level of sophistication in the description of the behavior of materials systems. See reference 12.4 for more details.

The *sessile drop* experiment is sometimes used to assess surface energies, Fig. 12.24. A droplet of a liquid is placed on an inert substrate and allowed to equilibrate with its vapor. For metals, glasses, and ceramics this experiment is carried out in a muffle furnace since the system must be above the melting point of the droplet. A telescope is sighted on the droplet and a shadow image of the droplet is recorded. An analysis of the droplet shape that incorporates both gravitational and capillarity effects yields the liquid-vapor surface energy and the angle θ shown in Fig. 12.24. Since the substrate is inert, it remains flat and equilibrium is not obtained in the vertical direction such that Eq. (12.107) is not realized. Equation (12.106) holds, with $\phi_\epsilon = \pi$ and $\phi_\alpha = (\pi - \theta)$ and α, β, and ϵ becoming the vapor (V), liquid (L) and solid (S) phases. Equation (12.106) becomes

$$\gamma_{VL} \cos(\pi - \theta) + \gamma_{IS} \cos(\pi) + \gamma_{VS} = 0 \qquad (12.111)$$

so that

$$\gamma_{VL} \cos\theta - \gamma_{LS} + \gamma_{VS} = 0$$

or

$$\cos\theta = \frac{\gamma_{VS} - \gamma_{LS}}{\gamma_{VL}} \qquad (12.112)$$

The dihedral angle θ is taken as a measure of the tendency of the liquid to *wet* the substrate. In Fig. 12.24 it is shown that for values of θ larger than 90° the liquid

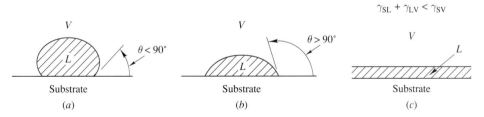

FIGURE 12.24
The dihedral angle measurement in a sessile drop experiment reports the relative surface energies of the three interfaces that meet at the triple line, and provides a measure of the tendency for the liquid phase to wet the substrate.

equilibrates as a bead resting on the substrate. As θ approaches $0°$ a given volume of droplet spreads over an increasing area of the substrate. The limiting case, $\theta = 0°$, or $\cos\theta = 1$, corresponds to the condition

$$\gamma_{VS} = \gamma_{LS} + \gamma_{VL} \qquad (12.113)$$

If the liquid-substrate and the liquid-vapor interfaces have a combined energy lower than the substrate-vapor interface energy, the system can minimize its surface energy by replacing the high energy (SV) interface completely with the low energy (LS) interface, Fig. 12.24c. The liquid completely covers the solid substrate with a film and is said to *wet* the solid.

These same phenomena operate to influence the geometry of microstructures of materials, sometimes with disastrous effects. Figure 12.25 shows a particle of the β phase at a grain boundary in the matrix phase α. If it is assumed that $\gamma_{\alpha\beta}$ is isotropic, two of the interfaces at this $\alpha\beta\alpha$ triple line have the same energy. The third, $\alpha\alpha$, is a grain boundary in the matrix phase. Equation (12.106) becomes

$$\gamma_{\alpha\alpha} = 2\gamma_{\alpha\beta}\cos\frac{\theta}{2} \qquad (12.114)$$

If $\theta = 0°$, corresponding to $\gamma_{\alpha\alpha} = 2\gamma_{\alpha\beta}$, the β phase wets the grain boundary completely. For values of $\gamma_{\alpha\beta}$ smaller than this limiting value, the system can minimize its surface energy by completely replacing the relatively high energy grain boundary with two $\alpha\beta$ interfaces; at equilibrium the β phase exists as a grain boundary film, Fig. 12.25b. If, during processing of this material, the system is taken to a temperature in which the β phase is liquid, all the grain boundaries in the structure are replaced by a liquid film and the material disintegrates into separate grains. This phenomenon is occasionally used in research to disintegrate a polycrystal into separate grains so that the distribution of shapes and sizes of grains can be examined. If the same phenomenon occurs during hot working of an alloy ingot, the disintegration of the grain structure can be disastrous.

The equilibrium angle at triple lines, determined by the relative energies of the three interfaces that meet there, plays a role in a significant number of phenomena that influence the evolution of microstructures in materials science. The presence of a small amount of liquid phase accelerates sintering in the processing of some important ceramics. If the liquid is retained as a thin, glassy film in the fi-

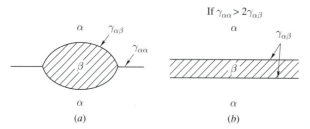

FIGURE 12.25
(a) A particle of the β phase may form at a grain boundary in the matrix. If $\gamma_{\alpha\alpha} > 2\gamma_{\alpha\beta}$, β wets the grain boundary and may form a grain boundary film at equilibrium.

nal product, the mechanical properties may be significantly degraded. The shape of porosity in the late stages of sintering of ceramics may be significantly influenced by the dihedral angle developed at grain boundaries that intersect the surface of internal porosity. Grain growth in polycrystalline metallic and ceramic systems results from a continuing attempt by the system to establish the equilibrium angles at the grain edges and corners. Large grain sizes are detrimental to mechanical properties. Failures in thin film stripes that connect electronically active devices on integrated circuits may be associated with the development of pinholes initiated at triple lines in the grain boundary network of the film. Wetting tendencies are crucial in the spreading of adhesives and solders in joining of materials as well as the application of coatings that protect them. The nucleation of particles of a newly forming phase is strongly favored if the particles form *heterogeneously* on a substrate that the new phase tends to wet; the energy required to form the surface of the new small particle, which is the primary barrier to nucleation, is thus significantly reduced. The distribution of second phase particles in a microstructure at grain faces, triple lines, or quadruple points in the grain boundary network, or dispersed within the grains, is largely determined by the competing energetics of formation of nuclei at these various sites.

12.8 ADSORPTION AT SURFACES

Every extensive thermodynamic property has a specific surface excess contribution associated with interfaces in the system. In particular, the number of moles of any component can be expected to exhibit such a surface excess, which may be positive or negative. Component k is said to be *adsorbed* at the surface or interface in an equilibrated system. The general strategy underlying the description of adsorption of components at interfaces is presented in this section. The treatment is confined to liquid-vapor and solid-vapor surfaces because this permits certain simplifications but still allows development of the strategy. For a more general treatment of the thermodynamics of adsorption see reference 12.5.

12.8.1 Measures of Adsorption

Figure 12.26a illustrates the variation of concentration of component k through a system containing a flat surface between the α and vapor phases, where α may be liquid or solid. In general, except near the critical point, the concentration of any component in the vapor phase can be treated as negligible in comparison with the α phase because the molar volume of the vapor is very large in comparison with V^α. Thus, if β is the vapor phase in Fig. 12.26, then $c_k^\beta \approx 0$. Figure 12.26b illustrates the corresponding hypothetical system used in defining surface excess quantities with C_k^α constant up to the position of the dividing surface at x_s. It is shown in Sec. 12.2 that the superficial excess of any extensive property can be visualized as the difference between the area under the curve in Fig. 12.26a and the area under the step function defined by the hypothetical system. For component k, this difference is the shaded

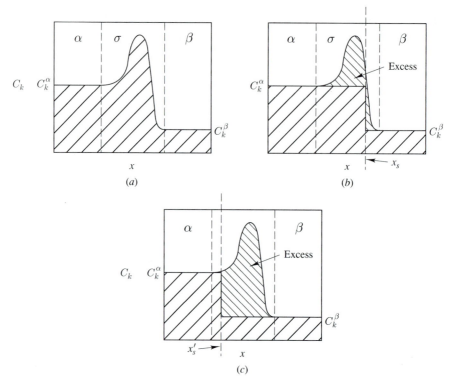

FIGURE 12.26

(*a*) Illustration of the distribution of component k through the physical surface of discontinuity between α and β phases. (*b*) The corresponding hypothetical system with the dividing surface at x_r(*c*). An alternate choice for x_r.

area in Fig. 12.26*b*. Mathematically,

$$n_k^s = n_k - C_k^\alpha \cdot A x_s = \int_a^b C_k \cdot A \ dx - C_k^\alpha \cdot A \ x_s \qquad (12.115)$$

Figure 12.26*c* shows an alternate choice for the position of the dividing surface at x_s' with the corresponding shaded area visualizing the surface excess of component k. This shaded area is roughly twice that shown in Fig. 12.26*b*. Evidently the value of the surface excess of k can be very sensitive to the choice of the position of the dividing surface. Since the total thickness of the surface of discontinuity is about one atom, displacement of the dividing surface by a fraction of an Ångstrom may produce a significant change in the computed value for the surface excess of component k or any other surface excess property.

This impractical situation can be avoided by following yet another strategy due to Gibbs; defining *reduced* measures of adsorption that are independent of the choice of the position of the dividing surface. Consider the surface excess of the total number of moles of all the components defined for the dividing surface at x_s:

$$n_T^s = n_T - C_T^\alpha \cdot A \, x_s = \int_a^b C_T \cdot A dx - C_T^\alpha \cdot A x_s \qquad (12.116)$$

where

$$C_T = \sum_{k=1}^c C_k$$

at each point in the system. Eliminate the position of the dividing surface in Eqs. (12.115) and (12.116). First, solve Eq. (12.116) for the product $A x_s$

$$A x_s = \frac{1}{C_T^\alpha} \left[\int_a^b C_T(x) \cdot A dx - n_T^s \right]$$

Substituting this result into Eq. (12.115)

$$n_k^s = \int_a^b C_k \cdot A dx - C_k^\alpha \cdot \frac{1}{C_T^\alpha} \left[\int_a^b C_T(x) \cdot A dx - n_T^s \right]$$

Collecting the surface excess terms on the left side of the equation

$$n_k^s - \frac{C_k^\alpha}{C_T^\alpha} \cdot n_T^s = \int_a^b C_k \cdot A dx - \frac{C_k^\alpha}{C_T^\alpha} \int_a^b C_T(x) \cdot A dx \qquad (12.117)$$

Evaluation of the integrals on the right side of this equation does not require a definition of the dividing surface since x_s has been eliminated from the equation. Thus the quantity on the left side of this equation is independent of the position of the dividing surface. These integrals are illustrated in Fig. 12.27. The first is the total area under the curve for component k, Fig. 12.27a. The second integral is the area under the curve for the total number of moles of all the components, Fig. 12.27b. The second term on the right-hand side of Eq. (12.117) multiplies this term by $C_k^\alpha / C_T^\alpha = X_k^\alpha$, the mole fraction of k in the α phase; the lower line in Fig. 12.27b represents this term. Thus, the cross-hatched area in Fig. 12.27c is a representation of the quantity defined on the left-hand side of Eq. (12.117). This area reports the total number of atoms of component k in the surface of discontinuity, reduced in proportion to the fraction of the total number of atoms in this region that would be component k if this component were present in the same proportion as in the α phase. The quantity defined on the left side of Eq. (12.117) is called the *reduced surface excess of component k*.

The corresponding *specific* reduced surface excess of component k, written $\overline{\Gamma}_k$, can be obtained by dividing the surface excess by the area of the dividing surface

$$\overline{\Gamma}_k \equiv \frac{n_k^s - X_k^\alpha \cdot n_T^s}{A} \qquad (12.118)$$

Evidently this property is also independent of the choice of location of the dividing surface.

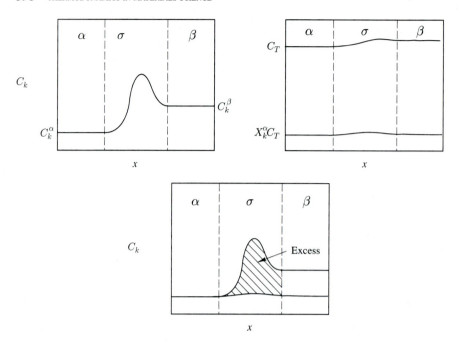

FIGURE 12.27

(a) Sketch of the variation of concentration of k through the physical surface of discontinuity. (b) Variation of total concentration with position; dashed line is $X_k^\alpha C_T$. (c) The shaded area is the *reduced surface excess* of component k.

12.8.2   The Gibbs Adsorption Equation

A strategy based upon a generalization of the Gibbs-Duhem equation leads to the working equation for interpreting adsorption phenomena. Consider a two phase multi-component system with flat internal interfaces. If the system is in internal equilibrium, then the temperature, pressure, and chemical potentials each have a single value for the system. The combined statement of the first and second laws for a two phase system with flat internal interfaces may be written

$$dU'_{\text{sys}} = T dS'_{\text{sys}} - P dV'_{\text{sys}} + \sum_{k=1}^{c} \mu_k dn_{k,\text{sys}} + \gamma dA \qquad (12.119)$$

It is not necessary to write out separate terms for the two phases since the intensive properties are the same in both phases. Consider the formation of this two phase system by adding quantities of the components to an initially empty system at constant T, P, and composition. The total internal energy of this system can computed by integrating Eq. (12.119) for this process:

$$U'_{\text{sys}} = T S'_{\text{sys}} - P V'_{\text{sys}} + \sum_{k=1}^{c} \mu_k n_{k,\text{sys}} + \gamma A \qquad (12.120)$$

Take the total differential of this expression for an arbitrary change in state of the system

$$dU'_{sys} = TdS'_{sys} + S'_{sys}dT - PdV'_{sys} - V'_{sys}dP$$

$$+ \sum_{k=1}^{c} \mu_k dn_{k,sys} + \sum_{k=1}^{c} n_{k,sys}d\mu_k + \gamma dA + Ad\gamma \qquad (12.121)$$

Comparison of this result with Eq. (12.119) yields a generalized form of the Gibbs-Duhem equation

$$S'_{sys}dT - V'_{sys}dP + \sum_{k=1}^{c} n_{k,sys}d\mu_k + Ad\gamma = 0 \qquad (12.122)$$

Each of the system properties can be expressed in terms of the corresponding surface excess property by applying Eq. (12.13)

$$[S_{hyp}^{'\alpha} + S_{hyp}^{'\beta} + S^{'s}]dT - [V_{hyp}^{'\alpha} + V_{hyp}^{'\beta}]dP + \sum_{k=1}^{c}[n_{k,hyp}^{\alpha} + n_{k,hyp}^{\beta} + n_k^s]d\mu_k + Ad\gamma = 0$$
$$(12.123)$$

A separate Gibbs-Duhem equation holds for each of the uniform hypothetical parts of the system; for the α and β parts,

$$S_{hyp}^{'J}dT - V_{hyp}^{'J}dP + \sum_{k=1}^{c} n_{k,hyp}^{j}d\mu_k = 0 \qquad (J = \alpha, \beta) \qquad (12.124)$$

Removal of these terms from Eq. (12.123) leaves the relation

$$S^s dT + \sum_{k=1}^{c} n_k^s d\mu_k + Ad\gamma = 0 \qquad (12.125)$$

Solving for $d\gamma$

$$d\gamma = -\frac{S^s}{A} - \sum_{k=1}^{c} \frac{n_k^s}{A}d\mu_k = -s^s dT - \sum_{k=1}^{c} \Gamma_k d\mu_k \qquad (12.126)$$

where s^s and Γ_k are *specific interfacial excess properties*. This result is one form of the Gibbs adsorption equation.

With some manipulation the Gibbs adsorption equation can be expressed in terms of *reduced* specific interfacial excess properties:

$$d\gamma = -\bar{s}^s dT - \sum_{k=1}^{c} \overline{\Gamma_k} d\mu_k \qquad (12.127)$$

It is most useful in this form because the quantities contained are independent of the choice of the dividing surface. This equation is usually applied at constant temperature to the evaluation of the variation of the adsorption of a particular component with composition of the α phase. With $dT = 0$, Eq. (12.127) reads

$$d\gamma = -\sum_{k=1}^{c} \overline{\Gamma_k} d\mu_k \qquad (12.128)$$

For a binary system the summation in this equation has two terms. With proper choice of the dividing surface the surface excess of component 1 can be set to zero. Then Eq. (12.128) can be written

$$d\gamma = -\overline{\Gamma_2} d\mu_2$$

and the reduced specific surface excess can be computed from

$$\overline{\Gamma_2} = -\frac{d\gamma}{d\mu_2} \qquad (12.129)$$

If component 2 is the solute in a dilute solution, Henry's law relates μ_2 to X_2, the solute concentration and

$$d\mu_2 = RTd\ln X_2$$

The adsorption equation becomes:

$$\overline{\Gamma_2} = -\frac{d\gamma}{RTd\ln X_2} = -\frac{X_2}{RT} \cdot \frac{d\gamma}{dX_2} \qquad (12.130)$$

In a dilute binary solution if the addition of solute lowers the surface free energy [i.e., the derivative in Eq. (12.130) is negative], then the surface excess for that component is positive and solute adsorbs on the surface. If the surface energy is raised by the addition of the solute, then the solute is depleted at the surface at equilibrium.

12.9   SUMMARY OF CHAPTER 12

The superficial excess contribution to any extensive thermodynamic property can be visualized by comparing the value for a real system to that obtained for a hypothetical system in which intensive properties of each phase have a discontinuity at a *dividing surface*.

Application of the general strategy for finding conditions for equilibrium to a multicomponent, two phase system with a curved interface yields the usual conditions for thermal and chemical equilibrium, but the condition for mechanical equilibrium is found to be

$$P^\beta = P^\alpha + 2\gamma H$$

where H is the local mean curvature of the surface and

$$\gamma = U^s - TS^s - \sum_{k=1}^{c} \mu_k \Gamma_k$$

is, in simple cases, the *specific interfacial free energy* of the interface. A mechanical analog yields the same relationship between the pressures if both phases are fluids and demonstrates that for such cases the surface tension and specific interfacial free energy are identical.

The altered condition for mechanical equilibrium associated with curved interfaces can be combined with the conditions for thermal and mechanical equilibrium to compute shifts of phase boundaries associated with the curvature of internal interfaces. For a unary system, the vapor pressure is shifted by

$$P(H) = P(H = 0)e^{(2\gamma V^L/RT)H} = P(H = 0)e^{\lambda v H}$$

where λ_v is the *capillarity length scale* for the liquid-vapor equilibrium. The melting point is decreased in proportion to the curvature

$$T(H) = T(H = 0) - \frac{2\gamma V^L}{\Delta S} H$$

where ΔS is the entropy of fusion.

The phase boundaries of a two phase field in a binary system are also shifted if the interface between the phases is curved. For dilute terminal phases,

$$X_2^\alpha(H) = X_2^\alpha(H = 0)e^{(2\gamma V^\beta/RT)H}$$

$$X_1^\beta(H) = X_1^\beta(H = 0)e^{(2\gamma \overline{V_1}^\beta/RT)H}$$

These concepts are frequently applied *locally* in microstructures in which the curvature varies with position, invoking the assumption of *local equilibrium*.

For crystalline solids, γ varies with orientation. The condition for equilibrium yields an equilibrium shape for the crystal defined by the condition

$$P^c = P^v + 2\frac{\gamma_j}{\lambda_j}$$

where γ_j is the interfacial energy of the jth face and λ_j is its *pedal function*. The Gibbs-Wulff construction yields the equilibrium polyhedral shape for any crystal, given a polar plot of γ over the sphere of orientation.

Where three phases meet at a triple line, the equilibrium configuration of the mating surfaces is given by

$$\frac{\gamma_{\alpha\beta}}{\sin\phi_\epsilon} = \frac{\gamma_{\beta\epsilon}}{\sin\phi_\alpha} = \frac{\gamma_{\alpha\epsilon}}{\sin\phi_\beta}$$

This condition is equivalent to a mechanical force balance at the triple line. Relative values of these angles may play a key role in wetting of one phase on another, microstructural shapes, nucleation processes, and thin film device failures.

Components tend to adsorb at interfaces at equilibrium. In a dilute binary system the reduced surface excess of the solute is given by

$$\overline{\Gamma}_2 = -\frac{X_2}{RT} \cdot \frac{d\gamma}{dX_2}$$

Adsorption is positive if component 2 lowers the surface energy; component 2 is reduced at the surface if it increases γ.

PROBLEMS

12.1. Sketch a familar object that has convex, concave and saddle surface elements. Draw an exploded view of the object that illustrates each of these classes of features.

12.2. Write out an explicit expression for the change in internal energy for a two phase, two component system, including surface terms.

12.3. It has been asserted that V^s, the "specific superficial excess volume," is zero for any interface. Prove this assertion.

12.4. Paraphrase the development leading to Eq. (12.33) to show that

$$\gamma = \left(\frac{\partial F'}{\partial A} \right)_{T, V^\alpha, V^\beta, n_k}$$

12.5. A thin wire of radius r and length l tends to shorten under the action of surface tension, unless a weight F is hung on it sufficient to balance the contracting forces.
(a) Write a force balance that relates the weight F to the surface tension σ and the dimensions of the wire.
(b) Write an expression for the change in energy associated with an incremental lengthening dl of the rod in terms of γ and the dimensions of the rod.
(c) Interpret the energy change in terms of the work of displacing the force F through a distance dl.
(d) Equate the two alternate evaluations of F to show that $\sigma = \gamma$.

12.6. Use Eq. (12.53) to sketch a surface that represents the variation of vapor pressure with temperature and curvature.

12.7. Compute the vapor pressure of liquid copper over a flat surface at 1400 K. Compute the equilibrium vapor pressure inside an 0.5 micron diameter bubble of copper vapor suspended in liquid copper at 1400 K.

12.8. It is asserted that the capillarity shift of the melting point of a material is significantly smaller than the shift expected for an allotropic transformation. Justify this conjecture.

12.9. Below 892 K the phase diagram for the system A-B consists of dilute terminal solid solutions α and β with no intermediate phases. At 680 K the solubility limits are $X_2^\alpha = 0.025$ and $X_2^\beta = 0.967$. The molar volume of β is 9.5 cc/mole and the partial molal volume of component 1 in β is 11.2 cc/mole.
(a) Compute the capillarity length scales for the α and β phases.
(b) Compute and plot the equilibrium interface compositions as a function of particle size. Assume the particles are spheres.

12.10. The ϵ to β transformation occurs at 1155 K in pure titanium. The element B is a β stabilizer when added to titanium. Assume for this problem that the β and ϵ phases are ideal solutions. The properties of the system are:

Component	$T_k^{\epsilon \to \beta}$ (K)	$\Delta S_k^{\alpha \to \beta}$ (J/mol-K)	V^β (cc/mol)
Titanium	1155	3.4	11.5
B	830	5.2	9.7

$$\gamma = 470 \text{ ergs/cm}^2$$

(a) Compute the bulk compositions X_B^ϵ and X_B^β at 1100 K.

(*b*) An alloy with $X_B = 0.12$ is quenched from the β phase to 1100 K where ϵ nucleates and grows. Compute and plot the interface compostiion in the β phase as a function of particle radius.

(*c*) Sketch the capillarity shift on a phase diagram.

12.11. Sketch the microstructure at the advancing ($\alpha\beta$/L) interface in a solidifying eutectic. How does the capillarity shift alter the composition in the liquid at the curved interface

(*a*) In front of the α phase?

(*b*) In front of the β phase?

Illustrate your answer by sketching shifted liquidus curves on the phase diagram near the eutectic point.

12.12. The tetrakiadecahedron is a polyhedron with six {100} faces and eight {111} faces. (*Tetrakiadeca* is Greek for fourteen.) On a *regular* tetrakiadecahedron all the edges have the same length. Calculate the ratio of $\gamma_{\{111\}}$ to $\gamma_{\{100\}}$ that would be required to produce this shape.

12.13. Show that Eq. (12.108) is consistent with Eqs. (12.106) and (12.107).

12.14. The surface energy of the interface between nickel and its vapor is estimated to be 1580 ergs/cm^2 at 1100 K. The average dihedral angle measured for grain boundaries intersecting the free surface is 168°. Thoria dispersed nickel alloys are made by dispersing fine particles of ThO$_2$ in nickel powder and consolidating the aggregate. The particles are left at the grain boundaries in the nickel matrix. Prolonged heating at elevated temperatures gives the particles their equilibrium shape. The average dihedral angle measured inside the particle is found to be 145°. Estimate the specific interfacial energy of the thoria-nickel interface.

12.15. Nuclear reactors used in submarine power plants circulate liquid sodium as a coolant in the reactor core. Stainless steel piping is proposed for use in the heat exchangers. Will liquid sodium significantly penetrate the grain boundaries in stainless steel?

12.16. Visualize a gold thin film stripe as a connector in a microelectronic chip. Suppose the stripe is 0.1 micron thick and has a bamboo grain structure. (This means the grain boundaries run laterally completely across the stripe.) Take the grain boundary energy of gold at 600 K to be 420 ergs/cm^2; its surface energy is 1440 ergs/cm^2.

(*a*) Compute the dihedral angle where a grain boundary meets the external surface.

(*b*) Find a critical grain boundary spacing s_c for which the equilibrium grain shape produces a hole in the film.

12.17. Show that the specific interfacial excess properties associated with a grain boundary are *independent* of the choice of the position of the dividing surface.

12.18. Sketch a concentration profile crossing an interface between α and β phases. Sketch a plot of the variation of the specific interfacial excess of the component as the choice of the position of the dividing surface moves from one side of the physical surface of discontinuity to the other. Pay particular attention to the form of the curve.

12.19. Compute the specific interfacial excess of iron adsorbed at the liquid-solid interface in dilute nickel-iron alloys, given that

$$\gamma = 1600 - 150 X_{Fe}$$

Repeat the calculation for tin absorbed at a nickel-tin liquid-solid interface, given that

$$\gamma = 1250 e^{-8.2 X_{Sn}} + 600$$

REFERENCES

1. Herring, C.: in *Structure and Properties of Solid Surfaces,* R. Gomer and C.S. Smith, eds., University of Chicago Press, Chicago, Ill., 1952.
2. Murr, L. E.: *Interfacial Phenomena in Metals and Alloys,* Addison-Wesley, Reading, Mass., 1975.
3. Kingery, W. D., H. K. Bowen and D. R. Uhlmann: *Introduction to Ceramics,* second edition, John Wiley & Sons, New York, N.Y., p.183, 1976.
4. Shewmon, P. G., and W. M. Robertson: in *Metal Surfaces, Structure, Energetics and Kinetics*, ASM, Materials Park, Ohio, 1963.
5. Lupis, C. H. P., *Chemical Thermodynamics of Materials*, Elsevier Science Pub. Co., New York, N.Y., pp. 389–430, 1983.

CHAPTER

13

DEFECTS IN
CRYSTALS

Stable solids are generally crystalline. Atoms of the components vibrate about well defined positions in space relative to one another. This local pattern repeats indefinitely in three dimensions. This characteristic periodic structure plays a key role in determining most of the physical, electronic, mechanical, and chemical properties of solids.

Solid crystals are not perfect. These imperfections in the arrangement of the atoms in space occur as isolated points, along lines, or as surfaces in the structure. The thermodynamics of surfaces is treated in Chap. 12. Line defects, known as *dislocations*, dominate the mechanical behavior of ductile crystals and are important defects in electronic materials. However, an understanding of their behavior is not traditionally formulated in thermodynamic terms. In contrast, most of the elements of the behavior of *point defects* are formulated thermodynamically.

Perhaps the most important role played by point defects in the behavior of solids is in *diffusion*, that is, the atom-by-atom transport of components through the crystal lattice. Most processes in materials science that produce changes in microstructure involve diffusion. The elemental step in diffusion is the motion of an atom from a normal crystal site into an adjacent point defect. Processes such as precipitation, phase changes, sintering, oxidation, solid state bonding, and some forms of creep depend on the presence of point defects in the system. Point defects influence the resistivity of conductors, the losses in insulators, and the conductivity of semiconductors. Their influence is particularly important in stoichiometric and nonstoichiometric ceramic and

405

intermetallic compounds where the distribution of a variety of defect types controls performance.

This chapter begins by developing a thermodynamic description of the behavior of point defects in elemental crystals, applying one more time, the general strategy for finding conditions for equilibrium. This approach is then extended to stoichiometric binary compounds, which by definition are constrained to a fixed ratio of the two elements comprising the system. These results set the groundwork for the treatment of the more general case of nonstoichiometric compounds.

13.1 POINT DEFECTS IN ELEMENTAL CRYSTALS

The concept of a point defect presumes the existence of a periodic lattice of sites that are normally occupied by atoms in a crystal. The two primary classes of point defects found to exist in elemental crystals are *vacancies* and *interstitials*.[1] A vacancy exists in a crystal where a normal lattice site is unoccupied, Fig. 13.1a. An interstitial defect occurs when an atom occupies a position in the crystal other than a normal lattice site Fig. 13.1b. While each defect contributes an increase to the energy of the crystal, each defect also increases the entropy of the crystal. These effects combine to guarantee that at equilibrium a crystal contains some point defects: a crystal is not perfect at equilibrium. The concentration of point defects in an elemental crystal is normally very small. Even in the extreme, near the melting point, defects occur at only about one site in 10,000. Concentrations of interstitials in elemental crystals are expected to be much smaller than vacancies under the same conditions. Nonetheless, this small fraction of defect sites plays a crucial role in materials science.

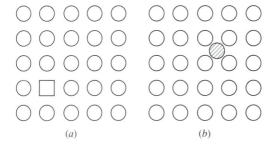

FIGURE 13.1
Two classes of point defects in an elemental crystal: (*a*) vacancy, and (*b*) an interstitial.

(*a*)   (*b*)

---

[1]Every crystal lattice has a set of normal sites and a set of interstitial sites which lie between the normal sites. If an added component atoms is small enough, it tends to occupy interstitial sites; hydrogen, carbon, nitrogen, and sometimes oxygen dissolve in most metals as such *interstitial components*. Component atoms that normally are found in normal lattice sites are called *substitutional components*. A small atom of an interstitial component occupying an interstitial site is not normally considered to be a defect in the crystal. However, if an atom of a substitutional component occupies an interstitial site, the configuration is called an *interstitial defect*.

13.1.1 Conditions for Equilibrium in a Crystal with Vacant Lattice Sites

To find the concentration of defects at equilibrium, apply the familiar strategy. Consider a system composed of a homogeneous crystalline phase (α) and its vapor (g). The combined statement of the first and second laws must explicitly recognize that the crystal can change its internal energy not only by changing entropy, volume, and the number of moles of each component, but also by changing the number of vacancies in the system.

$$U'^{\alpha} = U'^{\alpha}(S'^{\alpha}, V'^{\alpha}, n_1^{\alpha}, n_2^{\alpha}, \ldots, n_c^{\alpha}, n_v^{\alpha}) \tag{13.1}$$

where n_v is the number of vacant lattice sites in the crystal. The change in internal energy for the crystal when taken through an arbitrary change in state is

$$dU'^{\alpha} = T^{\alpha}dS'^{\alpha} - P^{\alpha}dV'^{\alpha} + \sum_{k=1}^{c} \mu_k^{\alpha}dn_k^{\alpha} + \mu_v^{\alpha}dn_v^{\alpha} \tag{13.2}$$

This equation implicitly contains as a coefficient relationship the definition of μ_v, the chemical potential of a vacancy in the crystal. The analogous expression for the vapor phase is familiar

$$dU'^{g} = T^{g}dS'^{g} - P^{g}dV'^{g} + \sum_{k=1}^{c} \mu_k^{g}dn_k^{g} \tag{13.3}$$

Rearrange each of these equations to solve for the change in entropy for each phase. Then combine them to yield an expression for the change in entropy of the system:

$$dS_{\text{sys}} = \frac{1}{T^{\alpha}}dU'^{\alpha} + \frac{P^{\alpha}}{T^{\alpha}}dV'^{\alpha} - \frac{1}{T^{\alpha}}\sum_{k=1}^{c}\mu_k^{\alpha}dn_k^{\alpha} - \frac{1}{T^{\alpha}}\mu_v^{\alpha}dn_v^{\alpha}$$

$$+ \frac{1}{T^{g}}dU'^{g} + \frac{P^{g}}{T^{g}}dV'^{g} - \frac{1}{T^{g}}\sum_{k=1}^{c}\mu_k^{g}dn_k^{g} \tag{13.4}$$

In deriving the conditions for equilibrium, attention is focused on an isolated system. Because the system is isolated from its surroundings,

$$dU'_{\text{sys}} = 0 = dU'^{\alpha} + dU'^{g} \rightarrow dU'^{g} = -dU'^{\alpha} \tag{13.5}$$

$$dV'_{\text{sys}} = 0 = dV'^{\alpha} + dV'^{g} \rightarrow dV'^{g} = -dV'^{\alpha} \tag{13.6}$$

$$dn_{k,\text{sys}} = 0 = dn_k^{\alpha} + dn_k^{g} \rightarrow dn_k^{g} = -dn_k^{\alpha} \tag{13.7}$$

Isolation of the system from its surroundings does not put any limitation on the number of lattice sites in the crystal. Lattice sites can be created or destroyed by moving an atom from the interior of the crystal to its surface and vice versa. Thus the number of vacancies n_v is not constrained in an isolated system. Insert these isolation constraints into Eq. (13.4) and collect terms:

$$dS_{\text{sys,iso}} = \left(\frac{1}{T^\alpha} - \frac{1}{T^g} \right) dU'^\alpha + \left(\frac{P^\alpha}{T^\alpha} - \frac{P^g}{T^g} \right) dV'^\alpha$$

$$- \sum_{k=1}^{c} \left(\frac{\mu_k^\alpha}{T^\alpha} - \frac{\mu_k^g}{T^g} \right) dn_k^\alpha - \frac{\mu_v^\alpha}{T^\alpha} dn_v^\alpha \qquad (13.8)$$

In this equation, n_v is an *independent* variable: n_v varies independently in an isolated system because lattice sites may be created or annihilated with no other changes in the system. The maximum in entropy that corresponds to the equilibrium state for the system is found by setting all the coefficients of the differentials of the independent variables in the system equal to zero. This yields the familiar conditions for thermal, mechanical, and chemical equilibrium in this solid-vapor system. In addition, it yields the condition

$$\mu_v^\alpha = 0 \qquad (13.9)$$

Thus in a crystal at equilibrium, the chemical potential of vacancies is zero.

13.1.2 The Concentration of Vacancies in a Crystal at Equilibrium

A mixture of vacant sites with sites occupied by normal atoms can be considered to be a dilute solution of vacancies and normal atoms. The chemical potential of any component, including vacancies, is identical with the partial molal Gibbs free energy. In Chap. 8 it is demonstrated that the partial molal Gibbs free energy can always be decomposed into ideal and excess parts. Thus for vacancies,

$$\Delta \mu_v = \Delta \overline{G_v} = \Delta \overline{G_v}^{\text{xs}} + \Delta \overline{G_v}^{\text{id}} \qquad (13.10)$$

These contributions can be written

$$\mu_v^\alpha - \mu_v^{o\alpha} = \left[\Delta \overline{H_v} - T \Delta \overline{S_v}^{\text{xs}} \right] + kT \ln X_v^\alpha \qquad (13.11)$$

where X_v^α is the atom fraction of vacant sites in the crystal and $\Delta \overline{H_v}$ and $\Delta \overline{S_v}^{\text{xs}}$ are partial molal properties associated with vacancies. Let the properties in this equation be reported for the formation of the defect crystal from a perfect crystal. Since there are no defects in a perfect crystal, $\mu_v^{o\alpha}$ is zero and the chemical potential of vacancies in the solution is

$$\mu_v^\alpha = \left[\Delta \overline{H_v} - T \Delta \overline{S_v}^{\text{xs}} \right] + kT \ln X_v^\alpha \qquad (13.12)$$

where k is Boltzmann's constant. According to Eq. (13.9), at equilibrium this quantity is zero,

$$\mu_v^\alpha = 0 = \left[\Delta \overline{H_v} - T \Delta \overline{S_v}^{\text{xs}} \right] + kT \ln X_v^\alpha \qquad (13.13)$$

Thus the equilibrium fraction of vacant sites in a crystal is:

$$X_v = e^{(\Delta \overline{S_v}^{xs}/k)} \cdot e^{(-\Delta \overline{H_v}/kT)} \tag{13.14}$$

In this equation, $\Delta \overline{H_v}$ can be thought of as the enthalpy of formation of a vacancy from a perfect crystal. $\Delta \overline{S_v}^{xs}$ is the excess entropy associated with this process, physically associated with changes in the vibrational behavior of atoms surrounding the vacant site. In many texts this quantity is written $\Delta \overline{S_v}^{vib}$. Since the enthalpy of formation of a vacancy is positive, Eq. (13.14) demonstrates that the concentration of vacancies increases with temperature.

Experimental tests of these relations are usually indirect. For example, defects act as scattering centers in the flow of electrons through a conductor. In a dilute solution of vacancies each vacancy makes the same contribution to scattering the flow of electrons; hence the electrical resistivity is proportional to the vacancy concentration. Changes in electrical resistivity may thus be used to monitor changes in vacancy concentrations, provided that other influences are eliminated [13.2]. Careful thermal analysis can also be used to monitor changes in vacancy concentration. Measurement of the power required to heat two samples identical except for defect concentrations has been used to follow interstitial and vacancy annihilation processes during annealing after deformation or neutron irradiation.

In a few cases direct measurements of the equilibrium vacancy concentration as a function of temperature have been carried out, Fig. 13.2. In one sample the length was measured carefully as a function of temperature in a sensitive dilatometer. These length measurements could be converted to changes in the sample volume. The

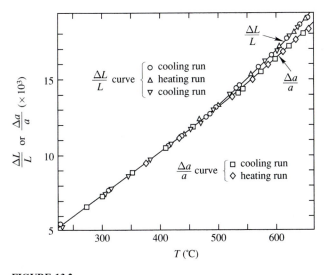

FIGURE 13.2
Comparison of the molar volume of a crystal computed from dilatometric measurements and X-ray lattice parameter measurements. The difference between these curves at any temperature is the volume of vacancies in the crystal [13.1].

volume increases partly because the average distance between atoms expands with temperature but also as a result of an increase in the number of vacant lattice sites with temperature. X-ray measurements on the same sample at a set of temperatures permit evaluation of the lattice parameters of the crystal and thus the mean interatomic distance. The volume computed for one mole of atoms from this X-ray measurement is that of a defect free crystal. Subtraction of the volume calculated in the X-ray measurements from the total volume measured directly gives the volume of sites in the crystal that are vacant and thus the number of vacancies. Equation (13.14) suggests that a plot of the logarithm of the measured vacancy concentration versus $1/T$ should be linear with a slope equal to $\Delta \overline{H}_v/k$ and an intercept (at $1/T = 0$) equal to $\Delta \overline{S}_v/k$. Such a plot is shown in Fig. 13.3. Table 13.1 presents properties of vacancies for a few metallic systems [13.3].

13.1.3   Interstitial Defects and Divacancies

These arguments easily extend to other defects in elemental crystals with analogous results. At equilibrium, the chemical potential of an interstitial defect is zero, implying

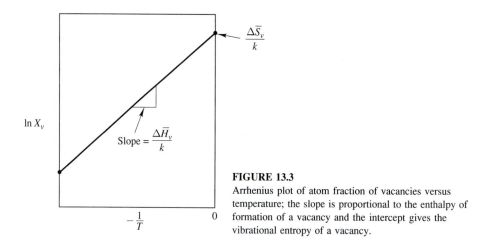

FIGURE 13.3
Arrhenius plot of atom fraction of vacancies versus temperature; the slope is proportional to the enthalpy of formation of a vacancy and the intercept gives the vibrational entropy of a vacancy.

TABLE 13.1
Properties of vacancies for some typical metals [13.3]

Metal	$X_v(T_m)$*	$\Delta \overline{S}_v$ (J/gm-atom K)	$\Delta \overline{H}_v$ (KJ/gm-atom)
Aluminum	9.0×10^{-4}	19	74
Copper	1.9×10^{-4}	12	113
Gold	7.2×10^{-4}	10	92
Lead	2.0×10^{-4}	21	56
Platinum	6.0×10^{-4}	9	135
Silver	1.7×10^{-4}	12	101

a relation between defect concentration and temperature of the general form

$$X_D = f_D e^{(\Delta \overline{S_d}^{xs}/k)} \cdot e^{-(\Delta \overline{H_D}/kT)} \tag{13.15}$$

where f_D is the ratio the number of interstitial sites to the number of normal lattice sites in the crystal.

Theoretical calculations have demonstrated that the enthalpy to form a vacancy is significantly lower than that associated with the formation of an interstitial defect. At the same temperature it can be expected that the concentration of interstitial defects is very much smaller (usually by several orders of magnitude) than that of vacancies at equilibrium. Thus interstitial defects are not expected to participate in processes that occur near equilibrium. However, they may play an important role in crystals that are far from equilibrium. For example, in neutron irradiation, collisions of high energy neutrons with atoms on normal sites may displace atoms over large distances. These displaced atoms come to rest in regions in which the normal sites are all occupied and thus these atoms reside in interstitial sites. The empty sites they leave behind are vacancies. During subsequent annealing these collections of equal numbers of vacancies and interstitial defects may recombine to form normal, occupied lattice sites. However, interstitial defects do not play an important role in most applications.

Defects may occur in combinations in an elemental crystal. The most common of these is the *divacancy*, which is a pair of adjacent vacant lattice sites. Analysis shows that such a defect would be significantly more mobile than a single vacancy. However, the equilibrium concentration of such defects in a crystal is much smaller than that of single vacancies. Adapting the result in Eq. (13.14) to this case at equilibrium,

$$X_{vv} = e^{(\Delta \overline{S_{vv}}^{xs}/k)} \cdot e^{-(\Delta \overline{H_{vv}}/kT)} \tag{13.16}$$

where the subscript *vv* denotes properties of divacancies.

It is useful to visualize the formation of a divacancy in two steps:

(1) the separate formation of two vacancies from a perfect crystal
(2) the formation of the divacancy configuration from two separated single vacancies

The enthalpy change associated with the first process is simply $2\Delta \overline{H_v}$. Let $\Delta \overline{H_{\text{int}}}$, the *interaction enthalpy*, be the change in enthalpy associated with the second process. Thus, the enthalpy of formation of a divacancy can be written

$$\Delta \overline{H_{vv}} = 2\Delta \overline{H_v} + \Delta \overline{H_{\text{int}}} \tag{13.17}$$

The same argument can be used to write the excess entropy

$$\Delta \overline{S_{vv}} = 2\Delta \overline{S_v} + \Delta \overline{S_{\text{int}}} \tag{13.18}$$

$\Delta \overline{H_{\text{int}}}$ and $\Delta \overline{S_{\text{int}}}$ are the interaction enthalpy and entropy for the pair of vacancies, both of which can be expected to be negative since the property values for a divacancy can be expected to be smaller than for a pair of isolated vacancies. Equation (13.16) can be written

$$X_{vv} = e^{(1/k)[2\Delta \overline{S_v} + \Delta \overline{S_{\text{int}}}]} \cdot e^{-(1/kT)[2\Delta \overline{H_v} + \Delta \overline{H_{\text{int}}}]}$$

$$X_{vv} = \left[e^{(1/k)[\Delta \overline{S_v}]} \cdot e^{-(1/kT)[\Delta \overline{H_v}]} \right]^2 \cdot e^{(\Delta \overline{S_{\text{int}}}/k)} \cdot e^{-(\Delta \overline{H_{\text{int}}}/kT)}$$

$$X_{vv} = (X_v)^2 \cdot e^{(\Delta \overline{S_{\text{int}}}/k)} e^{-(\Delta \overline{H_{\text{int}}}/kT)} \tag{13.19}$$

It is concluded that at equilibrium the concentration of divacancies is somewhat larger than the square of the concentration of single vacancies. Since the square of a small fraction is a much smaller fraction, equilibrium concentrations of divacancies probably do not play a significant role in most processes. However, it is possible to introduce a supersaturated concentration of vacancies into a crystal by quenching from a high temperature where vacancy concentrations are high, by neutron irradiation, or by ion bombardment such as is routinely done in ion implantation of thin films for microelectronic devices or in surface treatments of bulk materials. Under these conditions, far from equilibrium, significant divacancy concentrations can be developed and these vacancies can play a role in enhancing diffusion rates and in electronic processes.

13.2 POINT DEFECTS IN STOICHIOMETRIC COMPOUND CRYSTALS

Chapter 8 deals with the thermodynamic description of binary and higher order solutions. In the context of that chapter the crystal structure of solid phases is not explicitly treated because the phenomenological approach does not require it. It is assumed more or less explicitly that such a crystal consisted of a single class of lattice sites which could be readily occupied by atoms of any of the components in the system. Many intermediate phases and line compounds have crystal structures that have two (or more) distinct classes of lattice sites, Fig. 13.4, called *sublattices*. In ionic crystals, one set of sites typically contains the cations, (positively charged ions) and the other, the anions (negatively charged ions). In such crystals, one set of sites is termed the cation sites and the other the anion sites. This characteristic holds even if the bonding in the crystal is not purely ionic. In this case the more electronegative atom occupies the anion sites and the less electronegative atom the cation sites.

The anion (or more electronegative) component in a compound crystal is typically (though not always) a nonmetallic element like oxygen, nitrogen, carbon, sulfur, chlorine, etc. In the following description this element is designated generically as X. The other more metallic (cation) element in the system is designated with an M. A vacant lattice site is designated with a V. These components (atoms or ions) can occupy cation (M) sites, anion (X) sites, or interstitial (i) sites. Finally, each entity visualized has an associated electronic charge.

A broad variety of defects can be visualized, even in a simple crystal with only two types of normal lattice sites. To facilitate the explicit description of this variety

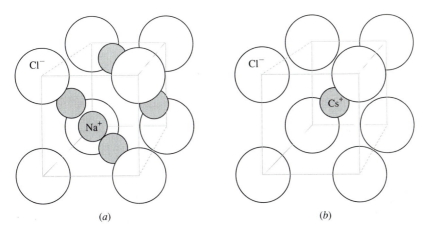

FIGURE 13.4
Common crystal structures with two classes of lattice sites: (*a*) NaCl; (*b*) CsCl.

of entities, a widely accepted notation was devised by Kröger and Vink [13.4]. This notation exhibits three important elements in the identification of a particular defect:

1. The entity occupying the defect site (M, X, V, or substitutional elements)
2. A subscript for the type of site occupied (M, X, or i)
3. A superscript for the *excess* charge associated with the site, $(^{x}),(\cdot)$, or $(')$

   Designations 1 and 2 are self evident. The notation used to describe electrical charge associated with the defect requires some discussion. It is convenient to describe the charge in comparison with that normally associated with a particular site, the local *excess* charge. If the entity occupying the site carries the charge of the species normally occupying that site, then a superscript $(^{x})$ is used. Thus in an alumina crystal (Al_2O_3) normal sites would be designated Al^x_{Al} and O^x_O. Similarly, a trivalent chromium ion on a cation site in alumina would be designated Cr^x_{Al} because it carries the same charge as the aluminum ions in the structure. A superscript $(\cdot)$ designates one unit of excess positive charge. Thus $Al^{\cdots}_i$ describes a trivalent aluminum ion in an interstitial site where its three positive charges are all excess charge; $Ca^{\cdot}_K$ represents a divalent calcium ion on a normally monovalent potassium site in KCl. A superscript $(')$ represents an excess negative charge. An oxygen ion in an interstitial site is represented by O''_i; Mg'_{Al} describes a divalent magnesium ion on a (normally trivalent) aluminum cation site. Some generic examples of this notation are reviewed in Table 13.2.

   Vacant lattice sites on either sublattice have an associated excess charge of equal magnitude but opposite sign to the ion that normally occupies the site. For example, the removal of a cation from a cation site to produce a cation vacancy leaves an excess negative charge associated with the surrounding anions, which is no longer balanced. Thus a cation vacancy in potassium chloride is designated V'_K; in alumina, the designation is V'''_{Al}. An anion vacancy in MgO is represented by $V^{\cdot\cdot}_O$. Cation

TABLE 13.2
Defect designations using the Kröger-Vink[13.4] notation applied to a compound with nominal composition MX and normal valence of M as +2, X as −2

Defect	Excess charge	Symbol
Vacancy on M sublattice	−2	V''_M
Vacancy on X sublattice	+2	$V^{\cdot\cdot}_X$
M atom in interstitial site	+2	$M^{\cdot\cdot}_i$
X atom in interstitial site	−2	X''_i
M atom on X site	+4	$M^{\cdot\cdot\cdot\cdot}_X$
X aton on M site	−4	X''''_M
Divacancy on M and X sites	0	$(V_M V_X)$
M interstitial paired with M on X site	+6	$(M_i M_x)^{\cdot\cdot\cdot\cdot\cdot\cdot}$
Solute cation L with +3 charge on M site	+1	$L^{\cdot}_M$
Solute anion Y with −1 charge on X site	+1	$Y^{\cdot}_X$
Free (unattached) electron	−1	e'
Electron hole	+1	$h^{\cdot}$

vacancies carry an excess negative charge and anion vacancies an excess positive charge.

13.2.1 Frenkel Defects

The equilibrium concentration of defects in a compound crystal can be derived from the general strategy with careful attention paid to the conservation equations that form the isolation constraints. The results take a form analogous to the conditions for equilibrium in a multivariate reacting system derived in Sec. 11.4. Indeed, these conditions can be formulated in terms of defect "reactions," affinities, formation energies of the defects, and corresponding equilibrium constants. These results are illustrated in this section for the simplest form of defect in a binary compound, known as a *Frenkel defect*. A Frenkel defect is formed on the cation sublattice by removing an M ion from a normal M site and placing it in an interstitial site, Fig. 13.5. It is also possible to form a Frenkel defect on the anion sublattice. A Frenkel defect is called an *intrinsic* defect because it can be formed without any interaction with the surroundings of the crystal.

Consider a crystal MX in which the normal valance of M is +2 and that of X is −2. If this crystal contains Frenkel defects derived from cation sites, four distinct entities exist in such a crystal: M^x_M, X^x_X, V''_M and $M^{\cdot\cdot}_i$ (see Fig. 13.5). In words, this notation describes: M ions on cation (M) sites; X ions on anion (X) sites; vacant sites on the cation (M) sublattice; and M ions in an interstitial (i) site. The number of each of these entities can be varied in the crystal; however, these variations are not independent. It is possible to define a chemical potential for each of these entities as the rate of change of the Gibbs free energy of the crystal with respect to the number of each particular entity at constant temperature and pressure.

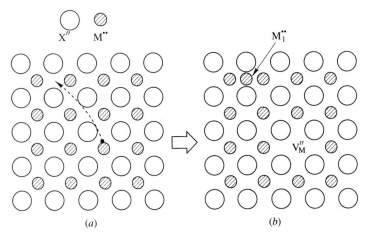

FIGURE 13.5
A Frenkel defect is formed on the cation sublattice by removing an M atom from a normal cation lattice
site and placing it in an interstitial site.

To find the conditions for equilibrium in such a system, apply the familiar
strategy. First, write an expression for the entropy of this homogeneous crystalline
phase incorporating changes in the numbers of each possible entity.

$$dS'_{\text{sys}} = \frac{1}{T}dU' + \frac{P}{T}dV' - \frac{1}{T}\left[\mu_{M_M}dn_{M_M} + \mu_{X_X}dn_{X_X} + \mu_{V_x}dn_{V_x} + \mu_{M_i}dn_{M_i}\right] \quad (13.20)$$

In an isolated system, $dU' = 0$ and $dV' = 0$. Conservation of atoms of M and X
requires

$$dm_X = dn_{X_X} = 0 \quad (13.21)$$

$$dm_M = 0 = dn_{M_M} + dn_{M_i} \rightarrow dn_{M_M} = -dn_{M_i} \quad (13.22)$$

since all X atoms remain on anion sites while the M atoms are distributed over cation
and interstitial sites. Further, each Frenkel defect consists of one vacancy and one
interstitial atom. In the process of creating Frenkel defects the change in the number
of vacancies and interstitials therefore must be the same:

$$dn_{V_M} = dn_{M_i} \quad (13.23)$$

Insertion of these constraints into Eq. (13.20) simplifies to

$$dS'_{\text{sys,iso}} = -\frac{1}{T}[\mu_{M_M}(-dn_{M_i}) + \mu_{X_X}(0) + \mu_{V_M}(dn_{M_i}) + \mu_{M_i}dn_{M_i}]$$

$$dS'_{\text{sys,iso}} = -\frac{1}{T}[\mu_{M_i} + \mu_{V_M} - \mu_{M_M}]dn_{M_i} \quad (13.24)$$

The quantity in brackets can be thought of as the affinity for the defect reaction

$$M_M = V_M + M_i \quad [13.1]$$

which represents the unit process in the formation of a Frenkel pair from an M atom on a normal M site.

The condition for equilibrium in the crystal can be obtained by setting the affinity equal to zero:

$$\mu_{M_i} + \mu_{V_M} - \mu_{M_M} = 0 \tag{13.25}$$

Each of the chemical potentials in this expression can be written in the form

$$\mu_k = \mu_k^o + \overline{\Delta G_k}^{xs} + kT \ln X_k$$
$$= G_k^o + (\overline{G_k}^{xs} - G_k^o) + kT \ln X_k$$
$$\mu_k = \overline{G_k}^{xs} + kT \ln X_k \qquad (k = M_i, V_M, M_M) \tag{13.26}$$

Substituting these expressions into the condition for equilibrium, Eq. (13.25)

$$\left[\overline{G_{V_M}}^{xs} + \overline{G_{M_i}}^{xs} - \overline{G_{M_M}}^{xs} \right] + kT \left[\ln X_{V_M} + \ln X_{M_i} - \ln X_{M_M} \right] = 0 \tag{13.27}$$

This result can be cast in a form very similar to the law of mass action for a reacting system, Eq. (11.34).

$$\overline{\Delta G_{fd}}^{xs} = -kT \ln K_{fd} \tag{13.28}$$

where

$$\overline{\Delta G_{fd}}^{xs} \equiv \overline{G_{V_M}}^{xs} + \overline{G_{M_i}}^{xs} - \overline{G_{M_M}}^{xs} \tag{13.29}$$

and

$$K_{fd} \equiv \frac{X_{V_M} X_{M_i}}{X_{M_M}} \tag{13.30}$$

There are important differences between this relation and Eq. (11.34). $\overline{\Delta G_{fd}}^{xs}$ is not the "standard free energy change for the reaction;" it is the difference in *partial molal excess free energies* of the components at equilibrium. Similarly, K_{fd} in Eq. (13.28) is not the "proper quotient of activities" defined for reactions to be the equilibrium constant [see Eq. (11.33)]. Rather it is the analogous ratio expressed in terms of the atom fractions of the entities in the crystal, not their activities. Nonetheless, the result has the same form as the familiar law of mass action and is clearly related to it.

Since the solution is very dilute, Raoult's law applies to the solvent M_M:

$$\overline{\Delta G_{M_M}}^{xs} = 0 = \overline{G_{M_M}}^{xs} - G_{M_M}^o \rightarrow \overline{G_{M_M}}^{xs} = G_{M_M}^o \tag{13.31}$$

Further, X_{M_M} is nearly one. Equation (13.27) can be rewritten as

$$\left[\overline{G_{V_M}}^{xs} + \overline{G_{M_i}}^{xs} - G_{M_M}^o \right] + kT \ln \left(X_{V_M} X_{M_i} \right) = 0$$
$$X_{V_M} X_{M_i} = e^{-\frac{1}{kT} \left[\overline{G_{V_M}}^{xs} + \overline{G_{M_i}}^{xs} - G_{M_M}^o \right]} \tag{13.32}$$

Applying the usual definitional relation permits the substitution

$$\left[\overline{G_{V_M}}^{xs} + \overline{G_{M_i}}^{xs} - G_{M_M}^o\right] = \left[\overline{H_{V_M}} + \overline{H_{M_i}} - H_{M_M}^o\right] - T\left[\overline{S_{V_M}}^{xs} + \overline{S_{M_i}}^{xs} - S_{mM}^o\right]$$

$$= \Delta\overline{H_{fd}} - T\Delta\overline{S_{fd}} \tag{13.33}$$

where $\Delta\overline{H_{fd}}$ and $\Delta\overline{S_{fd}}^{xs}$ are implicitly defined in Eq. (13.33). Note that $\Delta\overline{H_{fd}}$ is the enthalpy of formation of the defect, described by reaction [13.1], since the excess enthalpy is the total enthalpy. The condition for equilibrium, Eq. (13.32), can now be expressed as

$$X_{V_M} \cdot X_{M_i} = e^{(\Delta\overline{S_{fd}}^{xs}/k)} \cdot e^{-(\Delta\overline{H_{fd}}/kT)} \tag{13.34}$$

Finally, since the number of vacancies and interstitials are constrained to be the same, Eq. (13.23), $X_{V_M} = X_{M_i}$ so that $(X_{V_M})(X_{M_i}) = (X_{V_M})^2 = (X_{fd})^2$, where X_{fd} is the atom fraction of Frenkel defects in the structure. Thus,

$$X_{V_M} \cdot X_{M_i} = \left(X_{V_M}\right)^2 = (X_{fd})^2 = e^{(\Delta\overline{S_{fd}}^{xs}/k)} \cdot e^{-(\Delta\overline{H_{fd}}/kT)}$$

or

$$X_{fd} = e^{(\Delta\overline{S_{fd}}^{xs}/2k)} \cdot e^{-(\Delta\overline{H_{fd}}/2kT)} \tag{13.35}$$

If $\Delta\overline{H_{fd}}$ and $\Delta\overline{S_{fd}}^{xs}$ are insensitive to temperature, the equilibrium concentration of Frenkel defects is expected increase with temperature with the typical Arrhenius functional form.

13.2.2 Schottky Defects

In the simplest case of a crystal with formula MX the anions and cations carry the same number of charges. In an MX crystal a Schottky defect consists of a vacant cation site and a vacant anion site, Fig. 13.6. In this simple case the formation of Schottky defects does not disturb the electrical neutrality of the crystal. Like the Frenkel defect, this structural imperfection is *intrinsic* since it can be formed in a perfect crystal without adding or subtracting atoms or charges to the crystal.

The equilibrium concentration of Schottky defects in a crystal with formula MX begins with an expression for the change in entropy allowing in this case for changes in the number of cation and anion vacancies:

$$dS_{sys}' = \frac{1}{T}dU' + \frac{P}{T}dV' - \frac{1}{T}\left[\mu_{M_M}dn_{M_M} + \mu_{X_X}dn_{X_X} + \mu_{V_M}dn_{V_M} + \mu_{V_X}dn_{V_x}\right] \tag{13.36}$$

The isolation constraints in this case are:

$$dU' = 0 \qquad dV' = 0$$

$$dm_M = dn_{M_M} = 0 \tag{13.37}$$

$$dm_X = dn_{X_X} = 0$$

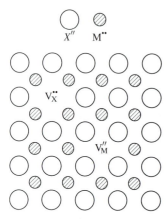

FIGURE 13.6
A Schottky defect consists of a vacant cation site and a vacant anion site.

The latter two equations report the conservation of atoms of M and X, respectively. The condition that the ratio of anion to cation sites is constrained to be 1/1 requires that the equal numbers of each type of vacancies be formed:

$$dn_{V_X} = dn_{V_M} \tag{13.38}$$

Put the constraints into the expression for the entropy

$$dS'_{\text{sys,iso}} = -\frac{1}{T}\left[\mu_{V_X}dn_{V_X} + \mu_{V_M}dn_{V_M}\right] = -\frac{1}{T}\left[\mu_{V_X} + \mu_{V_M}\right]dn_{V_M} \tag{13.39}$$

The coefficient of dn_{V_M} in this expression can be thought of as the affinity for the defect reaction

$$\text{null} = V_X + V_M \tag{13.2}$$

which describes the formation of two vacancies in a region that is initially a perfect crystal. The notation "null" in this context means the initially defect-free crystal. The condition for equilibrium is obtained by setting the coefficient in Eq. (13.39) equal to zero,

$$\mu_{V_X} + \mu_{V_M} = 0 \tag{13.40}$$

Following an argument to connect Eq. (13.25) with Eq. (13.35) yields an expression for the equilibrium concentration of Schottky defects at any temperature T:

$$X_{\text{sd}} = e^{(\Delta \overline{S_{\text{sd}}}^{\text{xs}}/2k)} \cdot e^{-(\Delta \overline{H_{\text{sd}}}/2kT)} \tag{13.41}$$

where $\Delta \overline{S_{\text{sd}}}^{\text{xs}}$ is the excess entropy associated with the formation of the pair of vacancies from a perfect crystal and $\Delta \overline{H}_{\text{sd}}$ is the corresponding enthalpy change.

In a crystal with formula $M_u X_v$, in which the ratio of M to X sites is u/v, conservation of the ratio of anion to cation sites dictated by the geometry of the crystalline arrangement requires that a Shottky defect be made up of u cation vacancies and v anion vacancies. This condition also preserves charge neutrality with the formation of a Schottky defect. The condition for equilibrium reflects the constraint on the ratio of lattice sites and is described by a defect equation containing these coefficients:

$$\text{null} = u V_M + v V_X \qquad [13.3]$$

13.2.3 Combined Defects in Binary Compounds

A cation site is normally occupied by an M ion carrying a positive charge ez_M, where z_M is the valence of the cation and e is the magnitude of the charge carried by an electron. If the site is vacant, meaning that there is no M ion occupying it, then there exists an uncompensated *negative* charge equal to $(-ez_M)$ associated with surrounding anions. Similarly, a vacant anion site carries a charge of $(-ez_X)$ where z_X is the normal valence of the anion. Since z_X is negative, this excess charge is *positive*. These oppositely charged entities in a crystal can be expected to attract each other to form a cation-anion vacancy pair, Fig. 13.7. If z_X and z_M are equal, this vacancy complex has zero excess charge. For the more general case for which z_M and z_X are not equal, the complex carries a net charge. This relationship can be represented by the defect reaction,

$$V_M + V_X = (V_M V_X) \qquad [13.4]$$

where the notation $(V_M V_X)$ represents the complex defect consisting of the oppositely charged pair of vacancies. The condition for equilibrium with respect to this interaction is determined by setting the corresponding affinity equal to zero

$$\mu_{(V_M V_X)} - \left(\mu_{V_M} + \mu_{V_X}\right) = 0 \qquad (13.42)$$

so that

$$\frac{X_{(V_M V_X)}}{X_{V_M} X_{V_X}} = e^{(\Delta \overline{S_{mx}}^{xs}/k)} \cdot e^{-(\Delta \overline{H_{mx}}/kT)} \qquad (13.43)$$

where $\Delta \overline{S_{mx}}^{xs}$ is the difference in excess entropy and $\Delta \overline{H_{mx}}$ is the enthalpy difference for the process expressed by Eq. [13.4]. The entropy difference is arguably small.

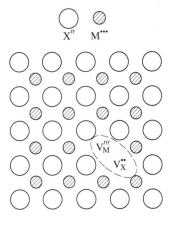

FIGURE 13.7
Vacancies on the two sublattices tend to attract each other to form complex vacancies, $V_M V_X$. The net charge on this complex is $e(z_x - z_m)$.

The enthalpy difference is negative since the oppositely charged vacancies are spontaneously attracted. Thus the concentration of complexes is larger than the product of the two single vacancy concentrations by an amount that depends upon the energy of the attraction. The enthalpy difference can be estimated as the difference in energy between two charged particles at the equilibrium separation distance in the vacancy pair and that of a pair of infinitely separated vacancies. This estimate gives reasonable accuracy for ionic crystals in which the excess charges are strongly associated with the sites. More sophisticated models are required to describe the behavior of compound crystal structures in which other types of bonding play a significant role.

13.2.4 Multivariate Equilibrium Among Defects in a Stoichiometric Compound Crystal

This section introduces the strategy for determining defect concentrations in a crystal of the compound $M_u X_v$ that is isolated from its surroundings. Thus, the collections of defects that can form may be considered *intrinsic*, since no interaction with the surroundings is involved in their formation. Sections 13.3.1 and 13.4 treat more general problems in which interactions with surroundings are explicitly included. The entities that may exist in this crystal when it comes to equilibrium include M_M, M_i, V_M, X_X, X_i, and V_X. Defect complexes may also exist but are neglected in this development. In applying the general strategy for finding conditions for equilibrium, the expression for the change in entropy contains a term of the form $\mu_k dn_k$ for each of the six entities listed above.

$$dS'_{sys} = \frac{1}{T}dU' + \frac{P}{T}dV' - \frac{1}{T}\left[\mu_{M_M}dn_{M_M} + \mu_{M_i}dn_{M_i} + \mu_{V_M}dn_{V_M}\right.$$

$$\left. + \mu_{X_X}dn_{X_X} + \mu_{X_i}dn_{X_i} + \mu_{V_X}dn_{V_X}\right] \tag{13.44}$$

In addition to the usual isolation constraints on changes in internal energy and volume, three constraining equations operate on the changes in the numbers of each entity in this isolated crystal:

1. Conservation of M atoms

$$dm_M = dn_{M_M} + dn_{M_i} = 0 \rightarrow dn_{M_M} = -dn_{M_i} \tag{13.45}$$

2. Conservation of X atoms

$$dm_X = dn_{X_X} + dn_{X_i} = 0 \rightarrow dn_{X_X} = -dn_{X_i} \tag{13.46}$$

3. Conservation of the ratio of sites in the two sublattices

$$v\, dn_{S_M} = u\, dn_{S_X} \tag{13.47}$$

where n_S denotes the number of each kind of site in the lattice. Since sites are either occupied or vacant,

$$v\left[dn_{M_M} + dn_{V_M}\right] = u\left[dn_{X_X} + dn_{V_X}\right] \tag{13.48}$$

Thus three of the six dn_k terms are dependent upon the other three. Choose dn_{M_i}, dn_{X_i}, and dn_{V_X} as independent variables. The third equation permits expression of dn_{V_M} in terms of these variables:

$$dn_{V_M} = \frac{u}{v}\left[dn_{X_X} + dn_{V_X}\right] - dn_{M_M} \tag{13.49}$$

Substitute these constraints into Eq. (13.44) and collect terms:

$$dS'_{\text{sys,iso}} = -\frac{1}{T}\left[(\mu_{M_i} + \mu_{V_M} - \mu_{M_M})dn_{M_i} + (\mu_{X_i} + \mu_{V_X} - \mu_{X_X})dn_{X_i}\right.$$
$$\left. + \left(\frac{u}{v}\mu_{V_M} + \mu_{V_X}\right)dn_{V_X}\right] \tag{13.50}$$

To find the conditions for equilibrium, set the coefficients equal to zero. These three coefficients are the affinities for the three defect chemistry equations:

$$M_M = M_i + V_M \tag{13.5}$$

$$X_X = X_i + V_X \tag{13.6}$$

$$\text{null} = uV_M + vV_X \tag{13.7}$$

The first two equations correspond to the formation of Frenkel defects on the cation and anion sublattices respectively and the last equation is the Schottky reaction for an $M_u X_v$ crystal.

It is customary to describe the concentrations of each of the entities as a fraction of the sites in the corresponding sublattice. These measures of composition are designated with brackets. Thus $[V_M]$ is the ratio of the number of vacant cation sites to the total number of cation sites: n_{V_M}/n_{S_M}. Similarly, $[X_i]$ is the ratio of the number of interstitial X ions to the number of anion sites in the crystal: n_{X_i}/n_{S_X}. Focus on one mole of the compound $M_u X_v$. More explicitly, focus on a quantity of the crystal that contains $(u + v)N_o$ lattice sites, where N_o is Avogadro's number. The number of M sites is $u(N_o)$ and the number of X sites is $v(N_o)$. Thus, the total number of cation vacancies in one mole of compound can be written $n_{V_M} = uN_o[V_M]$, the number of cation interstitials is $n_{M_i} = uN_o[M_i]$, the number of anion vacancies is $n_{V_X} = vN_o[V_X]$, and the number of anion interstitials can be written $n_{X_i} = vN_o[X_i]$.

The condition for equilibrium corresponding to the formation of a Frenkel defect on the cation lattice represented in Eq. [13.5] can be written:

$$\frac{[M_i][V_M]}{[M_M]} = K_{\text{fd,c}} = K^o_{\text{fd,c}}e^{-(\Delta\overline{H_{\text{fd,c}}}/kT)}$$

where $K^o_{\text{fd,c}}$ is an entropy term and $\Delta\overline{H_{\text{fd,c}}}$ is the enthalpy of formation of a Frenkel defect on the cation lattice. Since the defect concentrations are generally very small, $[M_M]$ can be taken as 1. A similar approximation is also valid for $[X_X]$. The conditions for equilibrium corresponding to the defect Eq. [13.5] through [13.7] are:

$$[M_i][V_M] = K^o_{\text{fd,c}}e^{-(\Delta\overline{H_{\text{fd,c}}}/kT)} \tag{13.51}$$

$$[X_i][V_X] = K^o_{\text{fd,a}}e^{-(\Delta\overline{H_{\text{fd,a}}}/kT)} \tag{13.52}$$

$$[V_M]^u [V_X]^v = K_{sd}^o e^{-(\Delta \overline{H_{sd}}/kT)} \tag{13.53}$$

The K_r^o coefficients contain the entropy factors in these equations. These results represent three equations among the four variables, $[V_M]$, $[M_i]$, $[V_X]$, and $[X_i]$. A fourth equation is supplied by the condition for charge neutrality in the crystal.

Each of the four defect entities has an excess charge associated with it. If z is the normal charge on an occupied cation site, then $-(u/v)z$ is the charge on a normal anion site. For example, for the crystal Al_2O_3, $z = +3$ and the charge on the anion site is $-(2/3)(+3) = -2$. The excess charge associated with each entity is thus

1. Cation vacancy, V_M: $-z$
2. Cation interstitial, M_i: $+z$
3. Anion vacancy, V_X: $+(u/v)z$
4. Anion interstitial, X_i: $-(u/v)z$

Charge neutrality requires that the total excess charge sum to zero. Mathematically

$$-zn_{V_M} + zn_{M_i} + \frac{u}{v}zn_{V_X} - \frac{u}{v}zn_{X_i} = 0$$

Substitute the expressions for the defect concentrations and simplify

$$-[V_M] + [M_i] + \frac{u}{v}[V_X] - \frac{u}{v}[V_i] = 0 \tag{13.54}$$

If the entropy and energy factors are known, then eqs.(13.51) through (13.54) provide four equations among four variables and the concentrations of all four defects can be computed for the crystal at any given temperature. Unfortunately, well documented and consistent information of this type is not yet available even for thoroughly studied ceramic systems [13.2]. A critical obstacle to the development of such information is the crucial role associated with impurity atoms in compound crystals (see Sec. 13.4). However, these equations can provide reasonable relative values for the four types of defects. For example, in sapphire (undoped alumina) in the range from 1200 to 1500°C the dominant defect is V_{Al}''' with the concentration of anion defects about ten orders of magnitude lower and aluminum interstitials more than twenty orders of magnitude smaller [13.2].

13.3 NONSTOICHIOMETRIC COMPOUND CRYSTALS

The notion that chemical substances are stoichiometric, specifically that the number of atoms of the elements that compose them occur in ratios of small whole numbers, can be traced to the discovery that many substances exist as molecules. H_2O, CO_2, and CH_4 are familiar examples of stoichiometric compounds. When elements M and X combine to form a crystalline compound $M_u X_v$, the resulting structure forms in two sublattices, one normally occupied by M atoms and the other by X atoms. The geometry of the crystal structure dictates that the numbers of the two kinds of sites occur in the ratio (v/u) and the ratio of sites is *stoichiometric*. In a perfect crystal, with all M sites occupied by M atoms and X sites by X atoms, the *composition* of the system is also stoichiometric, and the ratio of X to M atoms is (v/u). If the crystal contains *only intrinsic* defects, such as those described in Sec. 13.2, the *composition* is still stoichiometric.

The defect structure of real crystals is not confined to intrinsic defects. Interactions with the surroundings may produce defects that are not simply equivalent to an internal rearrangement of atoms on sites in the crystal. In an oxidizing atmosphere the compound M_uO_v may dissolve more oxygen atoms than the number given by the stoichiometric ratio (v/u). The composition of the compound departs from its stoichiometric ratio and becomes *oxygen rich*. However, the geometric requirements of the crystal lattice for the compound require that the ratio of sites remains at (u/v). Thus the excess oxygen atoms must be accommodated either by

Placing oxygen ions in interstitial positions, or

Placing the oxygen ions on normal anion sublattice sites and simultaneously creating vacancies in the *cation* sublattice.

In the first case the compound is said to be "oxygen excess"; this case can be represented by the formula, $M_uO_{v+\delta}$, where δ reports the quantity of excess oxygen in the system. The second case is said to be "metal deficient" and can be represented by $M_{u-\delta}O_v$. Both of these conditions describe a system in which the ratio of oxygen to metal is greater than (v/u).

For example, an oxide with the stoichiometric formula MO_2 may analyze at the composition $MO_{2.05}$. This departure from stoichiometry corresponds to either an "oxygen excess" compound, $MO_{2+0.05}$, or a "metal deficient" compound, $M_{1-0.024}O_2$. Note that the value of δ corresponding to the O/M ratio, 2.05/1, is not the same in these two cases. In either case the corresponding number of defects can be computed from the measured composition. In the case of oxygen excess, δ is the ratio of interstitial oxygen atoms to anion sites in the system. For the metal deficient model, δ is the fraction of cation sites that are vacant.

If, in contrast, the ratio of metal to oxygen atoms in the system is less than (v/u), the compound is *metal rich*. This excess of metal relative to oxygen can be accommodated by

Placing metal ions in interstitial sites (metal excess),

Placing metal ions on normal cation sites and creating an equal number of vacancies on the *anion* sublattice (oxygen deficient).

The "metal excess" case can be represented by $M_{u+\delta}O_v$ and alternately the "oxygen deficient" case by $M_uO_{v-\delta}$. In the first case δ is the ratio of interstitial metal atoms to cation sites while for the oxygen deficient description, δ is the fraction of anion sites that are vacant.

In a real M_uO_v crystal all four of the defects listed above can be expected to exist simultaneously in the system. Indeed, in principle all the possible defects and defect combinations that can be visualized for a system can be expected to exist at thermodynamic equilibrium. However, the *relative* concentrations of defect types can vary by many orders of magnitude so that some defects are present in concentrations that are negligibly small while others dominate. The identification of which defects are important in a given crystal under a given set of conditions depends upon their thermodynamic properties.

13.3.1   Equilibrium in Compound Crystals with a Variety of Defects

The problem of identifying those defects present in sufficient concentrations to play a role in the behavior of the system is closely analogous to the description of homogeneous multivariant reacting systems developed in Sec. 11.1.2. All the defects that can be visualized can be expected to exist in any real compound crystal in equilibrium with its surrounding atmosphere. The internal energy and entropy of a crystal is a function of the numbers of each of these entities; each has a chemical potential derived from coefficient relations in the expression for dU' or dG'. The change in entropy of such a crystal for an arbitrary change in its state may be written

$$dS'^\alpha = \frac{1}{T^\alpha}dU'^\alpha + \frac{P^\alpha}{T^\alpha}dV'^\alpha - \frac{1}{T^\alpha}\sum_{k=1}^{c^\alpha}\mu_k^\alpha dn_k^\alpha \qquad (13.55)$$

Here the summation covers the complete list of separately countable entities presumed to exist in the crystal, including:

> ions or atoms on their normal sublattice sites
> vacant sites on each sublattice
> ions or atoms on interstitial sites
> impurity ions or atoms on each type of sublattice site
> unassociated electrons and holes in the system
> combinations of these entities

If the total number of distinguishable types of these enumerated atoms, ions, and defects is c^α, then there are $[c^\alpha + 2]$ terms in this expression for the entropy.

A departure from stoichiometry requires an exchange of components with the surroundings. In deriving the conditions that define the final state for a compound crystal, in equilibrium with its surroundings, it is also necessary to include the change in entropy of the gas surroundings with its components. In the following it is assumed that the surroundings is a gas phase.

$$dS'^g = \frac{1}{T^g}dU'^g + \frac{P^g}{T^g}dV'^g - \frac{1}{T^g}\sum_{k=1}^{c^g}\mu_k^g dn_k^g \qquad (13.56)$$

where c^g is the number of components in the gas phase. There are $[c^g + 2]$ terms in this equation. The change in entropy of the system is the sum of the two expressions given in Eqs. (13.55) and (13.56) and it contains $[c^\alpha + c^g + 4]$ terms.

The general strategy for finding the conditions for equilibrium focuses on an isolated system. Some of the isolation constraints are familiar in this two phase system:

$$dU'_{sys} = dU'^\alpha + dU'^g = 0 \rightarrow dU'^g = -dU'^\alpha \tag{13.57}$$

$$dV'_{sys} = dV'^\alpha + dV'^g = 0 \rightarrow dV'^g = -dV'^\alpha \tag{13.58}$$

The numbers of gram atoms of each of the elements in the system is conserved

$$dm_j = 0 = \sum_{k=1}^{c^\alpha} v_{kj} dn_k^\alpha + \sum_{k=1}^{c^g} v_{kj} dn_k^g \tag{13.59}$$

Here, as in Sec. 11.1.2, the coefficients v_{kj} represent the number of atoms of the element j contained in each enumerated entity k, including normally occupied sites, defects, and components in both phases. Many of these coefficients are often zero; most entities contain either M but not X or vice versa. The numbers of vacant sites also have zero coefficients in these equations since they do not contain atoms.

Two new constraints arise from the nature of the system. The ratio of the number n_{SX} of X sites to the number n_{SM} of M sites is a characteristic of the crystal lattice structure and must be maintained. If the compound has the nominal formula $M_u X_v$, this ratio is (v/u). Thus,

$$v n_{S_M} = u n_{S_X} \rightarrow v dn_{S_M} = u dn_{S_X} \tag{13.60}$$

Here n_{SM} includes all the entities in the list that occupy a site on the M sublattice and n_{SX} incorporates those on the X sublattice. This includes atoms or ions and vacancies on such sites, but does not include interstitials, electrons, or holes, which do not reside on sublattice sites.

Yet another constraint arises because most of the entities enumerated carry an electric charge. Because the crystal is electrically neutral, whatever changes in the numbers of these entities occur, the change in total electrical charge in the crystal is zero. This charge balance relationship is conveniently expressed in terms of the excess charges on each of the entities present, assuming an M site normally carries a charge of $-e(z_M)$ and an X site carries a charge of $e(z_X)$:

$$dq = 0 = \sum_{k=1}^{c^\alpha} z_k e dn_k^\alpha \tag{13.61}$$

where z_k is the excess number of charge units carried by entity k in the crystal. The value of z_k may be positive, negative, or zero, in agreement with the assigned nature of the entity k.

Examination of the constraining equations shows that there are $[2+e+1+1] = [4+e]$ equations relating the $[c^\alpha + c^g + 4]$ variables in the expression for the entropy. Thus $[4+e]$ variables can be eliminated, leaving $[(c^\alpha + c^g + 4) - (4+e)]$ or $[c^\alpha + c^g - e]$ independent variables. The conditions for equilibrium are obtained by setting the coefficients of each of these independent variables in the expression for the change in entropy of the system equal to zero.

Two of the resulting equations are the familiar conditions for thermal and mechanical equilibrium. The remaining $[c^\alpha + c^g - e - 2]$ equations are linear combinations of the chemical potentials of the entities involved. Each of these independent equations has the form of an *affinity*, as is found to be characteristic of multivariate reacting systems in Sec. 11.1.2. Each affinity expression can be represented by a "chemical equation" with reactants, products and coefficients appropriate to "balance" the equation (see Eq. [13.1], for example). However, in addition to balancing the number of atoms of each element on both sides of the equation, these defect chemistry reactions also balance the electric charge in order to satisfy Eq. (13.61) and the ratio of X to M sites required to satisfy Eq. (13.60).

The construction of balanced defect chemistry equations is a nontrivial exercise because all three of the above conditions must be satisfied. In simple cases it is possible to balance such an equation by inspection or by trial and error. However, a proper set of coefficients can be deduced for any assumed reaction by applying the associated conservation equations. To illustrate this procedure, again focus on a crystal with the nominal formula $M_u X_v$. The normal valence of M is z and that of X is $-(u/v)z$. Consider the reaction with its surroundings that adds M to a normal site on the cation lattice. This incorporation reaction must be accompanied by the formation of vacant sites on the anion lattice, each of which carries a charge of $+(u/v)z$, and may also require formation of unassociated electrons to balance the charge. The corresponding incorporation reaction might be written

$$aM(g) \xrightarrow{M_n X_v} bM_M + cV_X^{(u/v)z} + de^- \qquad [13.8]$$

To find the coefficients, apply the conservation equations:

1. Conservation of M: $a = b$
2. X/M site ratio is $v/u : c/b = v/u$
3. Balance of excess charge: $b(0) + c[(u/v)z] - d = 0$

These three conditions yield values for the coefficients in Eq. [13.8]:

$$b = a \qquad c = (v/u)a \qquad d = az$$

Write the reaction for one atom of M; thus $a = 1$. The defect Eq. [13.8] can be written

$$M(g) \xrightarrow{M_n X_v} M_M + \frac{v}{u}V_X + ze^- \qquad [13.9]$$

Example 13.1. Consider the incorporation of Si into SiO_2. For this case: $u = 1$, $v = 2$ and $z = 4$. Take a to be 1. Then Eq. [13.9] can be adapted to read:

$$Si(g) \xrightarrow{SiO_2} Si_{Si} + 2V_O'' + 4_e' \qquad [13.10]$$

Solution. It can be verified by inspection that all three conditions are satisfied.

Example 13.2. Write a balanced defect equation for the incorporation of aluminum into alumina. The constants are: $u = 2$; $v = 3$; and $z = 3$. Apply Eq. [13.9], again taking $a = 1$

$$Al(g) \overset{Al_2O_3}{\rightarrow} Al_{Al} + \frac{3}{2}V''_o + 3'_e$$

or

$$2Al(g) \overset{Al_2O_3}{\rightarrow} 2Al_{Al} + 3V''_o + 6'_e \qquad [13.11]$$

Alternate reactions can be written for the incorporation of M into the crystal. For example, M may enter as an interstitial. A generic equation analogous to Eq. [13.9] could also be derived for this case by incorporating the balance equations as above.

In further analogy with the results for ordinary chemical equilibria obtained in Chap. 11, it can also be concluded that every other reaction (balanced equation) that can be written among the enumerated entities is a linear combination of these independent equations. Thus at equilibrium, the affinity is equal to zero for *every reaction* that can be written among the participants.

Finally, by following steps similar to those connecting Eq. (13.25) with (13.30) for the simple case of the Frenkel defect, each of these affinity expressions can be converted to a form reminiscent of the law of mass action

$$\Delta \overline{G_r}^{xs} = \Delta \overline{H_r} - T \Delta \overline{S_r}^{xs} = -kT \ln K_r \qquad (13.62)$$

where $\Delta \overline{H_r}$ is the change in partial molal enthalpy associated with the defect reaction and $\Delta \overline{S_r}^{xs}$ is the corresponding excess entropy change. The "equilibrium constant" is the ratio of atom fractions of the entities on the product side of the reaction to those on the reactant side.

13.3.2 Illustration of the Conditions for Equilibrium for Alumina

Alumina has the nominal formula Al_2O_3, with $z_{Al} = +3$ and $z_O = -2$. In this example assume that an alumina crystal is in equilibrium with its vapor and neglect the presence of impurities. The components in the gas phase are $Al(g)$ and $O_2(g)$. The list of entities that may exist in the alumina crystal includes Al^x_{Al}, O^x_O, $Al^{\cdots}_i$, O''_i, V'''_{Al}, and $V^{\cdots}_O$. Neglect combinations of defects. The expression for the change in entropy that this two phase system may exhibit contains terms for each of this list of components. The internal energy and volume terms give rise to the usual conditions for thermal and mechanical equilibrium. This treatment focuses on the defect chemistry and other chemical changes for the system. Thus Eqs. (13.55) and (13.56) can be combined to read

$$dS'_{sys} = [\![\quad]\!] - \frac{1}{T^g} \left[\mu^g_{Al} dn^g_{Al} + \mu^g_{O_2} dn^g_{O_2} \right]$$

$$- \frac{1}{T^\alpha} \left[\mu_{Al_{Al}} dn_{Al_{Al}} + \mu_{Al_i} dn_{Al_i} + \mu_{V_{Al}} dn_{V_{Al}} \right] \qquad (13.63)$$

$$- \frac{1}{T^\alpha} \left[\mu_{O_o} dn_{O_o} + \mu_{O_i} dn_{O_i} + \mu_{V_o} dn_{V_o} + \mu_e dn_e \right]$$

Conservation of aluminum and oxygen atoms in this isolated system requires

$$dm_{Al} = 0 = dn^g_{Al} + dn_{Al_{Al}} + dn_{Al_i} \rightarrow dn^g_{Al} = - \left[dn_{Al_{Al}} + dn_{Al_i} \right] \quad (13.64)$$

$$dm_O = 0 = 2dn^g_{O_2} + dn_{O_o} + dn_{O_i} \rightarrow dn^g_{O_2} = -\frac{1}{2} \left[dn_{O_o} + dn_{O_i} \right] \quad (13.65)$$

Maintenance of the ratio of cation to anion sites at 2/3 requires

$$3 \left[dn_{Al_{Al}} + dn_{V_{Al}} \right] = 2 \left[dn_{O_o} + dn_{V_o} \right] \quad (13.66)$$

The requirement for charge balance can be written

$$\delta q^{xs} = +3dn_{Al_i} - 3dn_{V_{Al}} - 2dn_{O_i} + 2dn_{V_o} - dn_e = 0 \quad (13.67)$$

Substitute these constraints into equation (13.63) and collect terms

$$
\begin{aligned}
dS'_{sys,iso} = [\![\quad]\!] &- \frac{1}{T} \left[\mu_{Al_{Al}} + 3\mu_e - \mu^g_{Al} - \mu_{V_{Al}} \right] dn_{Al_{Al}} \\
&- \frac{1}{T} \left[\mu_{Al_i} + 3\mu_e - \mu^g_{Al} \right] dn_{Al_i} \\
&- \frac{1}{T} \left[\mu_{O_o} + \frac{2}{3}\mu_{V_{Al}} - \frac{1}{2}\mu^g_{O_2} - 2\mu_e \right] dn_{O_o} \\
&- \frac{1}{T} \left[\mu_{O_i} - \frac{1}{2}\mu^g_{O_2} - 2\mu_e \right] dn_{O_i} - \frac{1}{T} \left[\frac{2}{3}\mu_{V_{Al}} + \mu_{V_o} \right] dn_{V_o}
\end{aligned}
\quad (13.68)
$$

The coefficients in these equations are the affinities for the following balanced defect reactions:

$$Al(g) + V'''_{Al} = Al^x_{Al} + 3'_e \qquad [13.12]$$

$$Al(g) = Al_i^{\cdots} + 3'_e \qquad [13.13]$$

$$3O_2(g) + 12'_e = 6O_o{}^x + 4V'''_{Al} \qquad [13.14]$$

$$O_2(g) + 4'_e = 2O''_i \qquad [13.15]$$

$$null = 2V'''_{Al} + 3V_o^{\cdots} \qquad [13.16]$$

Other reactions that can be written for this system can all be expressed as linear combinations of these five independent reactions. For example, the Frenkel reaction,

$$Al^x_{Al} = Al_i^{\cdots} + V'''_{Al}$$

can be obtained by subtracting reaction [13.12] from reaction [13.13].

13.4 IMPURITIES IN NONSTOICHIOMETRIC COMPOUNDS

Up to this point the description of the defect structure in a compound $M_u X_v$ has been limited to systems in which the only elements existing are M and X. No real crystalline compound is pure; additional elements are always incorporated to some extent. Indeed, in some systems it is possible to replace one element, say M, completely with another, L. The crystalline compound phase is capable of forming a complete series of solid solutions from M_uX_v to L_uX_v. In this case the compounds L_uX_v and M_uX_v form an isomorphous phase diagram. The system NiO-CoO is an example of such an isomorphous system (see Fig. 9.15). For significant solubility to exist in such cases, the elements involved must have the same valence, nearly the same atom size, and very similar electronegativity values. Where these three conditions are not met, which corresponds to the vast majority of combinations of the elements, the solubility of L in M_uX_v is limited to impurity levels.

As might be expected, the conditions for equilibrium in these more complex cases are derived from the general strategy and can be represented by defect energetics and equilibrium constants that correspond to defect chemistry equations. The construction of balanced reaction equations must conserve the atoms of all the elements involved, preserve the ratio of anion to cation sites, and maintain electrical neutrality. Some examples follow.

Magnesium oxide (MgO) is widely used as a sintering aid in the consolidation of alumina powders. MgO can be incorporated into the alumina lattice in a variety of reactions [13.2].

$$2MgO \xrightarrow{Al_2O_3} 2Mg_{Al}' + 2O_O^x + V_O^{\cdot\cdot} \qquad [13.17]$$

Addition of Mg (which is divalent) at a trivalent cation sublattice site leaves an uncompensated charge of (-1). In addition to adding oxygen ions at anion sites, it is necessary to create an anion vacancy in order to preserve the ratio of anion to cation sites of $3/2$. The charges are thus balanced. Alternatively,

$$3MgO + Al_{Al}^x \xrightarrow{Al_2O_3} 3Mg_{Al}' + 3O_O^x + Al_i^{\cdot\cdot\cdot} \qquad [13.17]$$

has the net effect of increasing the number of cation sites by 2 while creating 3 anion sites and an aluminum atom in an interstitial site. A third possibility,

$$3MgO \xrightarrow{Al_2O_3} 2Mg_{Al}' + 3O_O^x + Mg_i^{\cdot\cdot} \qquad [13.17]$$

produces a magnesium interstitial while creating no vacancies on either sublattice. Estimates of the energetics of these reactions remain speculative. Calculations based upon such estimates show that the dominant defect in MgO doped sapphire is $V_O^{\cdot\cdot}$, vacancies on the oxygen sublattice. Equilibrium concentrations of V_{Al}''', estimated to be the dominant defect in undoped sapphire, are about ten orders of magnitude lower in sapphire doped with 500 ppm of MgO [13.2]. Clearly, impurities may have a dominant effect on the distribution of defects in compound crystals.

These results demonstrate the complexity of thermodynamic relationships involved in determining the distribution of defects in compound crystals. Most undoped ceramic materials contain very significant concentrations of impurities. Impurity additions can completely alter the equilibrium distribution of defects. The thermodynamic treatment of defect chemistry presented in this chapter provides a foundation for understanding physical phenomena such as ionic conductivity, diffusion, oxide layer growth, and dielectric and optical behavior in crystalline compounds. The application of these principles to the interpretation of experimental observations in real ceramic structures constitutes a challenging area of study for the foreseeable future.

13.5   SUMMARY OF CHAPTER 13

When it attains its equilibrium state, a crystal contains defects in its lattice structure. These defects usually take the form of vacant lattice sites or atoms in interstitial sites. Associations of defects are also common in crystals.

Application of the general strategy for finding conditions for equilibrium in crystals with defects yields results formally identical to those obtained for multicomponent reacting systems. Interactions of defects with normal lattice sites and with the surroundings can be described by defect chemistry equations, with accompanying enthalpies and excess entropies of reaction and equilibrium constants.

In compound crystals, balanced defect chemistry reactions must conserve the elements involved, the ratio of lattice sites and charge neutrality.

Intrinsic defects formed in stoichiometric crystals form as Frenkel pairs or as Schottky clusters. Departures from stoichiometry always involves the formation of extrinsic defects.

Impurity atoms can be incorporated into crystals in a variety of ways, each requiring the formation of a specific set of defects. Since the number of defects can be simply related to the number of impurity atoms, small concentrations of impurity atoms can produce defect concentrations far in excess of those that form intrinsically. Thus, it is essential to treat explicitly impurity effects in considering those aspects of the behavior of compound crystals that are governed by their defect structure.

PROBLEMS

13.1. Careful dilatometric and high temperature diffraction measurements of pure face centered cubic A give the following information:

	a_0 (cm)	V (cm^3/mole)
At $T_1 = 1040$ K	3.7021×10^{-8}	7.6404
At $T_2 = 1340$ K	3.7113×10^{-8}	7.6992

Derive a relationship for the temperature dependence of the vacancy concentration in pure A.

13.2. The enthalpy of formation of an interstitial in an FCC material is found to be 188 KJ/mole; the excess entropy is 7.6 J/mole-K.

(a) Calculate and plot the equilibrium concentration of interstitials as a function of temperature in the range from 600 K to 1300 K.

(b) Compare this result to that found for vacancies in this material in Problem 13.1.

13.3. Suppose that the interaction parameters for divacancies are about 10 percent of the values of the corresponding single defect parameters:

$$\Delta \overline{H_{\text{int}}} = -0.1 \Delta \overline{H_v}$$

$$\Delta \overline{S_{\text{int}}} = -0.1 \Delta \overline{S_v}$$

Using the results of Problem 13.1, compute and plot the mole fraction of divacancies as a function of temperature for pure A in the range from 600 K to 1300 K.

13.4. Enthalpies of formation of Frenkel defects range from 350 to 550 KJ/mole and entropies from 1 to 10 J/mole K. Use these ranges to compute and plot limits of the defect concentrations expected as a function of T. Could very careful measurements of volume be used to estimate values of formation energies of Frenkel defects?

13.5. Frenkel and Schottky defects are *intrinsic* to the crystal in which they occur. While both of these classes of defects may form in compound crystals, only one may form in a pure elemental crystal. Identify which of these defects may form in an elemental crystal and explain why this is not possible for the other defect.

13.6. Consider the stoichiometric crystal silicon nitride, Si_3N_4. The following defects may form in this system: V_{Si}, Si_i, V_N, N_i.

   (a) Write out the conditions for equilibrium that permit the computation of these defect concentrations at any temperature.

   (b) What information is required to calculate actual equilibrium defect concentrations as a function of temperature?

13.7. The system M-O forms the oxides M_2O^γ, $M_2O_3^\epsilon$ and MO_2^η at 1600 K. The composition of the ϵ phase (M_2O_3) in equilibrium with $\gamma(M_2O)$ is $X_O = 0.5994$. At the other phase boundary of ϵ the composition in equilibrium with $\eta(MO_2)$ is $X_O = 0.6003$. Describe the composition of the ϵ phase when equilibrated with γ

   (a) As a metal excess oxide;

   (b) As an oxygen deficient oxide.

   Describe the composition of the ϵ phase when equilibrated with η

   (c) As a metal deficient oxide;

   (d) As an oxygen excess oxide.

   Explain briefly what each of these descriptions implies for the nature of the defect structure.

13.8. Write a balanced equation that represents the incorporation of silicon into a silicon sublattice site in silicon nitride, Si_3N_4.

13.9. Develop a series of balanced defect equations necessary to relate defect concentrations in boron nitride (BN) to the composition of the atmosphere in which it is equilibrated.

13.10. Construct a set of equations that describe the incorporation of yttria (Y_2O_3) into zirconia (ZrO_2). Describe *in words* what defects are created in each case.

REFERENCES

1. Simmons, R., and R. Balluffi: *Physical Reviews*, vol. 117, p. 52, 1960.
2. Lagerlof, K. P. D., T. E. Mitchell, and A. H. Heuer: "Lattice Diffusion Kinetics in Undoped and Impurity-Doped Sapphire(α-Al_2O_3): A Dislocation Loop Annealing Study," *J. Am. Ceramic Soc.*, vol. 72, p. 2159, 1989.
3. Mukherjee, K.: *Trans. AIME*, vol. 6, 1324, 1966.
4. Kröger, F. A.: *The Chemistry of Imperfect Crystals*, North-Holland Publishing Co., Amsterdam, 1964.

CHAPTER
14

EQUILIBRIUM IN CONTINUOUS SYSTEMS: THERMODYNAMIC EFFECTS OF EXTERNAL FIELDS

In treating equilibria between phases in earlier chapters it has been tacitly assumed that the intensive properties *within a phase* are uniform. In assigning a value T^α to the α phase, it is assumed that the α phase can be described by a single value of the temperature that is characteristic of that part of the system. This same assumption has been made for pressure P^α, composition X_k^α, chemical potential μ_k^α, and all the partial molal properties. In an equilibrium between two phases some of these properties have different values in the two phases. For example, if the phases are separated by a curved interface, the pressures in the two phases are different. However, it has been assumed that the pressure *within each phase* is uniform.

If external fields act on a system the intensive properties are found to *vary with position* within a phase when it comes to equilibrium. For example, in the gravitational field of the earth, at equilibrium the pressure varies with height in the atmosphere. In a centrifugal field, such as may be maintained in an ultracentrifuge rotating at constant velocity, the composition varies along the radius. This nonuniform distribution may be used to separate components within a single phase. In an electrostatic field it is found that the chemical potential of charged components varies with position within a phase when the system attains equilibrium.

To describe the variation of thermodynamic properties in such homogeneous but *nonuniform* systems, it is necessary to define and derive *densities* of thermodynamic extensive properties, such as the entropy density or Gibbs free energy density. Densities are local intensive properties that can be associated with each point in the system. This notion is introduced in Sec. 12.2 in defining surface excess properties. Formulation of the combined statement of the first and second laws of thermodynamics for a volume element in terms of the local densities lays the foundation for applying the general strategy for finding conditions for equilibrium to nonuniform systems. This strategy is first used for a system without external fields to demonstrate the approach and to show that the assumption that the intensive properties are uniform in such a system is indeed valid. Subsequently, conditions for equilibrium in the presence of external fields are developed. The presentation in Secs. 14.1 through 14.3 rests heavily on the foundation laid down by Haase [14.1] in his brief but comprehensive treatment.

It is also important to explore the thermodynamic description of nonuniform systems as a basis for describing the behavior of systems that are *not at equilibrium*. Much of materials science examines the processes that occur in materials as they experience spontaneous changes on their path toward equilibrium. The evolution of the system is generally viewed as being driven by precisely the nonuniformities that represent departures from equilibrium. Heat flow is a response to temperature gradients; diffusion responds to concentration or chemical potential gradients; electrical flow arises from electrical potential gradients. The description of such processes cannot begin without an apparatus for describing nonuniformities in the properties of the system.

In the description of ordinary nonuniform systems, it is generally assumed that the local properties at a point depend upon each other through conventional thermodynamic relationships. For example, the chemical potentials of the components at a point are assumed to be determined by the temperature, pressure, and composition at that point with the same functional relationship that would exist for a uniform large system with the same intensive properties. It has been found that this simple picture is inadequate for the description of nonuniform systems in which the intensive properties vary significantly over short distances. In very fine structures, such as one in which the composition varies periodically with position and the wavelength of the variation is of the order of a few tens of atoms, the Gibbs free energy is larger than the value that would be calculated from the local intensive properties. This extra energy, associated with property *variations* as opposed to property *values*, has been called the *gradient energy* of the system. The concept of the gradient energy and some of its applications are presented in Sec. 14.4. The classic papers introducing this concept are due to Cahn [14.2] and Cahn and Hilliard [14.3]. Hilliard [14.4] provides an excellent review of the concepts and pertinent experimental observations.

14.1 THERMODYNAMIC DENSITIES AND THE DESCRIPTION OF NONUNIFORM SYSTEMS

Consider any extensive thermodynamic property B'; B' may thus be S', V', U', H', F', G', or n_k. Each of these is a property associated with the system as a whole. A local density of each of these properties can be defined as in Sec. 12.2, Eq. (12.8):

$$b_v \equiv \lim_{V' \to 0} \frac{B'}{V'} \tag{14.1}$$

Familiar examples of this concept include the local mass density, in which B' is m, the mass of the volume V', and the limiting process yields ρ_m, the mass density, which may vary from point to point in the system. Alternatively, if B', is taken to be n_k, the number of moles of component k, then the limit of the ratio of n_k/V is the local *concentration* of component k, c_k.

Focus on a volume element δv in the vicinity of a point P located at coordinates (x, y, z) in the system. If $b_v(x, y, z)$ is the local density of B at P, then $b_v(x, y, z)\delta v$ is the total value of the extensive property B' in the volume element. The total value B' for the system is obtained by summing over all the volume elements:

$$B' = \iiint_{V'} b_v(x, y, z)\delta v \tag{14.2}$$

The change in B' when the system is taken through a process is obtained by taking the differential of both sides of Eq. (14.2)

$$dB' = d \iiint_{V'} b_v(x, y, z)\delta v = \iiint_{V'} db_v \delta v \tag{14.3}$$

The order of the differential and integral operations can be interchanged because the volume elements can be considered fixed during the changes in local density values. The differential of the density of any extensive property can be obtained by taking the differential of the definition, Eq. (14.1)

$$db_v = d\left(\frac{B'}{V'}\right) = \frac{1}{V'}dB' - \frac{B'}{V'^2}dV' \tag{14.4}$$

It is possible to formulate a *local version* of the combined statement of the first and second laws in terms of the corresponding density functions. Apply Eq. (14.4) to describe the change in the internal energy density

$$du_v = d\left(\frac{U'}{V'}\right) = \frac{1}{V'}dU' - \frac{U'}{V'^2}dV' \tag{14.5}$$

The first term can be evaluated from the combined statement for the first and second laws for the macroscopic system since

$$dU' = TdS' - PdV' + \sum_{k=1}^{c} \mu_k dn_k \tag{14.6}$$

Integration of this equation at constant intensive properties (T, P, and μ_k) gives an expression for the internal energy

$$U' = TS' - PV' + \sum_{k=1}^{c} \mu_k n_k \tag{14.7}$$

Substituting Eqs. (14.6) and (14.7) into (14.5)

$$du_v = \frac{1}{V'}\left[TdS' - PdV' + \sum_{k=1}^{c} \mu_k dn_k\right] - \frac{1}{V'^2}\left[TS' - PV' + \sum_{k=1}^{c} \mu_k n_k\right]dV'$$

Grouping like terms

$$du_v = T\left[\frac{1}{V'}dS' - \frac{S'}{V'^2}dV'\right] - P\left[\frac{1}{V'}dV' - \frac{V'}{V'^2}dV'\right] + \sum_{k=1}^{c} \mu_k\left[\frac{1}{V'}dn_k - \frac{n_k}{V'^2}dV'\right] \tag{14.8}$$

The first term in brackets is precisely the differential of the entropy density ds_v. The second term in brackets is zero which arises because V' is the normalization factor in the definition of the density. The term in brackets inside the summation is the differential of the density of number of moles of component k, generally referred to as the *molar concentration* of component k, dc_k. Thus Eq. (14.8) can be written

$$du_v = Tds_v + \sum_{k=1}^{c} \mu_k dc_k \tag{14.9}$$

This is the combined statement of the first and second laws applied to the densities of thermodynamic functions in a volume element. Rearrangement of this equation gives the corresponding expression for the change in entropy density,

$$ds_v = \frac{1}{T}du_v - \frac{1}{T}\sum_{k=1}^{c} \mu_k dc_k \tag{14.10}$$

This expression is required in the application of the general strategy to find the conditions for equilibrium within a single phase continuous system.

14.2 CONDITIONS FOR EQUILIBRIUM IN THE ABSENCE OF EXTERNAL FIELDS

Consider a *nonuniform* multicomponent, single phase, nonreacting, otherwise simple system. The steps in deriving the conditions for equilibrium are familiar. To evaluate the change in entropy for an infinitesimal alteration of the distribution of densities in this nonuniform system, insert Eq. (14.10) into the generic Eq. (14.4)

$$dS' = \iiint_{V'}\left[\frac{1}{T}du_v - \frac{1}{T}\sum_{k=1}^{c} \mu_k dc_k\right]\delta v \tag{14.11}$$

If during its displacement the system is isolated from its surroundings, then its internal energy cannot change and

$$dU' = \iiint_{V'} du_v \delta v = 0 \tag{14.12}$$

Isolation prevents exchange of components with the surroundings. Since the system being considered is also nonreacting, the number of moles of each component is conserved and

$$dn_k = \iiint_{V'} dc_k \delta v = 0 \tag{14.13}$$

The constant volume isolation constraint is implicit in this treatment because volume is used as the normalization factor.

Equilibrium is attained in an isolated system when its entropy reaches a maximum. To find this extremum in this case it is necessary to use the method of Lagrange multipliers used in Sec. 6.2.3 to derive the Boltzmann distribution function in statistical thermodynamics. Multiply each of the $(c + 1)$ constraining equations by an arbitrary constant and add the result to the expression for the function to be maximized, Eq. (14.11):

$$dS' + \alpha dU' + \sum_{k=1}^{c} \beta_k dn_k = 0 \tag{14.14}$$

where α and the set of β_k values are constants in the system, which means explicitly that they are *not functions of position*. Write this equation in terms of the corresponding integrals,

$$\iiint_{V'} \left[\frac{1}{T} du_v - \frac{1}{T} \sum_{k=1}^{c} \mu_k dc_k \right] \delta v + \alpha \iiint_{V'} du_v \delta v + \sum_{k=1}^{c} \beta_k \iiint_{V'} dc_k \delta v = 0$$

Collect terms

$$\iiint_{V'} \left[\left(\frac{1}{T} + \alpha \right) du_v - \frac{1}{T} \sum_{k=1}^{c} (\mu_k - \beta_k) dc_k \right] \delta v = 0 \tag{14.15}$$

For this integral to equal zero for arbitrary changes in the differentials of the densities, the coefficients of these differentials must simultaneously vanish. Thus for the extremum,

$$\frac{1}{T} = -\alpha \rightarrow T = \text{constant} \tag{14.16}$$

$$\mu_k = \beta_k \rightarrow \mu_k = \text{constant} \qquad (k = 1, 2, \ldots, c) \tag{14.17}$$

where the notation " = constant" in this context means "is not a function of position" in the system. A more explicit statement of these conditions uses the concept of the gradient developed in vector calculus. For any function of position in space, $f(x,y,z)$, the *gradient* is a vector given by

$$\text{grad} f \equiv \left(\frac{\partial f}{\partial x} \right) \hat{i} + \left(\frac{\partial f}{\partial y} \right) \hat{j} + \left(\frac{\partial f}{\partial z} \right) \hat{k} \tag{14.18}$$

where $\hat{\mathbf{i}}$, $\hat{\mathbf{j}}$ and $\hat{\mathbf{k}}$ are unit vectors in the x, y and z directions. The gradient vector at any position (x, y, z) gives the magnitude and direction of the most rapid change in the function f at that point. Using this concept the conditions for equilibrium in a continuous system can be stated:

$$\text{grad } T = 0 \qquad \text{(Thermal Equilibrium)} \qquad (14.19)$$

$$\text{grad } \mu_k = 0 \quad (k = 1, 2, \ldots, c) \qquad \text{(Chemical Equilibrium)} \qquad (14.20)$$

The condition for mechanical equilibrium can be derived from these results. Recall the variation of the chemical potential of a component with temperature, pressure, and composition, Eq. (12.57):

$$d\mu_k = -\overline{S}_k dT + \overline{V}_k dP + \sum_{j=2}^{c} \mu_{kj} dX_j \qquad (14.21)$$

where the summation is over $(c - 1)$ independent mole fractions. In the context of a nonuniform system, the variations considered in this equation are variations with *position*. Accordingly, each of the differentials can be replaced by the corresponding *gradient* of the intensive property

$$\text{grad } \mu_k = -\overline{S}_k \text{ grad } T + \overline{V}_k \text{ grad } P + \sum_{j=2}^{c} \mu_{kj} \text{ grad } X_j \qquad (14.22)$$

The conditions for thermal and chemical equilibrium, Eqs. (14.19) and (14.20), eliminate two of the terms in this expression

$$0 = \overline{V}_k \text{ grad } P + \sum_{j=2}^{c} \mu_{kj} \text{ grad } X_j$$

Multiply by X_k and sum over all of the components

$$0 = \sum_{k=1}^{c} X_k \overline{V}_k \text{ grad } P + \sum_{k=1}^{c} X_k \sum_{j=2}^{c} \mu_{kj} \text{ grad } X_j$$

Note that the summation in the first term is the molar volume. Interchange the order of the summations in the second term

$$0 = V \text{ grad } P + \sum_{j=2}^{c} \left[\sum_{k=1}^{c} X_k \mu_{kj} \right] \text{ grad } X_j$$

Recognize that the inner summation is a form of the Gibbs-Duhem equation (see Sec. 12.5.4) so that

$$\sum_{k=1}^{c} X_k \mu_{kj} = 0$$

Thus the condition for mechanical equilibrium is as expected

$$\text{grad } P = 0 \qquad (14.23)$$

This result completes the demonstration that temperature, pressure, chemical potential, and hence composition are uniform within any single phase region provided there are no external fields acting on the system. Thus this presumption, used extensively if implicitly in earlier chapters, is justified.

14.3   CONDITIONS FOR EQUILIBRIUM IN THE PRESENCE OF EXTERNAL FIELDS

This section considers the influence that time-invariant gravitational, electrical, and centrifugal fields have upon the conditions for equilibrium in a multicomponent, single phase system. It is essential to limit consideration to time-invariant fields because if the fields and their influence on the properties of the system were changing with time, then it would not be possible for the system to come to equilibrium, which is a time-invariant condition.

In classical physics the influence these phenomena have upon the behavior of a system can be formulated on the basis of the concept of potential energy. Each is visualized to have a corresponding *force field*, the strength of which varies from point to point in space. The force field at some point P in space is a representation of the force that would act upon a particle placed at P. Gravitational and centrifugal fields interact with the *mass* of the particle while an electrical field interacts with its *charge*.

14.3.1   Potential Energy of a Continuous System

An alternate formulation of the same phenomena is based upon the idea of the *potential field*, which describes the *potential energy* of a particle in a force field. Introduce the concept of the *specific potential function* $\psi(x, y, z)$, defined to be the potential energy *per unit mass* at the point (x, y, z) derived from an electrical, gravitational, or centrifugal field. The force acting on a particle depends upon its mass m and the gradient of the potential function:

$$\vec{F} = -m \text{ grad } \psi \tag{14.24}$$

The work done on the particle in moving it through some infinitesimal displacement dx is

$$\delta W_{PF} = \vec{F} \cdot d\vec{x}$$

$$= -m \left[\left(\frac{\partial \psi}{\partial x} \right) \hat{i} + \left(\frac{\partial \psi}{\partial y} \right) \hat{j} + \left(\frac{\partial \psi}{\partial z} \right) \hat{k} \right] \cdot [dx\hat{i} + dy\hat{j} + dz\hat{k}]$$

$$\delta W_{PF} = -m \left[\left(\frac{\partial \psi}{\partial x} \right) dx + \left(\frac{\partial \psi}{\partial y} \right) dy + \left(\frac{\partial \psi}{\partial z} \right) dz \right] = -m \, d\psi \tag{14.25}$$

since the quantity in brackets in the last equation is precisely the differential of the specific potential function ψ. For any finite change in position of the particle in the field from point a to point b, the work done on the particle is

$$W_{PF} = \int_a^b \vec{F} \cdot d\vec{x} = \int_a^b -m d\psi = -m[\psi(b) - \psi(a)] \tag{14.26}$$

In a *conservative field*, the work done depends only upon the potential at the initial and final positions in the field and is independent of the path by which the particle moves between these points. Since gravitational, centrifugal, and electrostatic fields are conservative, the specific potential function gives the potential energy per unit mass of a particle situated in the field.

In applying these concepts to thermodynamic systems, the focus is on the changes that are associated not with displacements of a particle in the field, but with the influence of the potential field on an element of volume in a system that may be experiencing a variety of internal changes that could affect its mass. Focus on an element of volume δv in a multicomponent single phase system. If c_k is the molar concentration of component k in the element, then the number of moles of k is $c_k \delta v$. The *mass* of component k in the element is $M_k c_k \delta v$ where M_k is the molecular weight (gram atomic weight for an element) of component k. The total mass δm contained in the volume element δv is given by

$$\delta m = \sum_{k=1}^{c} M_k c_k \delta v \tag{14.27}$$

Since ψ is the potential energy per unit mass, the potential energy of this volume element is $\psi \delta m$. The total potential energy of the system is obtained by summing the contributions from each volume element

$$E'_{\text{pot}} = \iiint_{V'} \psi \sum_{k=1}^{c} M_k c_k \delta v \tag{14.28}$$

Suppose the system experiences an infinitesimal alteration of its internal state in which all its intensive properties are allowed to vary. Since ψ is fixed in time (the field is assumed to be time invariant) and the molecular weight of each component is constant, the change in potential energy of each volume element can only arise from changes in the concentrations of the components in that element. Thus the change in potential energy of the system is given by

$$dE'_{\text{pot}} = \iiint_{V'} \psi \sum_{k=1}^{c} M_k dc_k \delta v \tag{14.29}$$

The total energy of the system is the sum of its internal energy, derived from its temperature, pressure, composition, and the potential energy associated with the external field:

$$E'_{\text{tot}} = U' + E'_{\text{pot}} = \iiint_{V'} u_v \delta v + \iiint_{V'} \psi \sum_{k=1}^{c} M_k c_k \delta v \tag{14.30}$$

The change in total energy of the system accompanying an arbitrary change in its state is given by

$$dE'_{\text{tot}} = \iiint_{V'} \left[du_v + \psi \sum_{k=1}^{c} M_k dc_k \right] \delta v \tag{14.31}$$

All the effects of external force fields upon the conditions for equilibrium arise from this relationship.

14.3.2   Conditions for Equilibrium

The entropy of the system is not influenced by the external field. In applying the general strategy, the change in entropy experienced by the system is given by Eq. (14.11). The influence of the external field operates through the *isolation constraints*. In an isolated system which cannot exchange energy with its surroundings, the total energy of the system cannot change. Equation (14.12) is replaced by the more general condition

$$dE'_{\text{tot}} = 0 = \iiint_{V'} \left[du_v + \psi \sum_{k=1}^{c} M_k dc_k \right] \delta v \tag{14.32}$$

Conservation of the components, described by Eq. (14.13), is unaffected by the external field. To find the extremum in the entropy, apply the method of Lagrange multipliers, slightly generalized from Eq. (14.14)

$$dS' + \alpha dE'_{\text{tot}} + \sum_{k=1}^{c} \beta_k dn_k = 0 \tag{14.33}$$

Substitute from Eqs. 14.11, 14.13, and 14.32:

$$\iiint_{V'} \left[\frac{1}{T} du_v - \frac{1}{T} \sum_{k=1}^{c} \mu_k dc_k \right] \delta v + \alpha \iiint_{V'} \left[du_v + \psi \sum_{k=1}^{c} M_k dc_k \right] \delta v$$

$$+ \sum_{k=1}^{c} \beta_k \iiint_{V'} dc_k \delta v = 0$$

Combining like terms

$$\iiint_{V'} \left[\left[\frac{1}{T} + \alpha \right] du_v + \sum_{k=1}^{c} \left[-\frac{1}{T} \mu_k + \alpha \psi M_k + \beta_k \right] dc_k \right] \delta v = 0 \tag{14.34}$$

Set the coefficients of each of the differentials in the integrand equal to zero to find the conditions for equilibrium. Keep in mind that α and the set of β_k values are constants, independent of position in the system. The coefficient of du_v gives the usual condition for thermal equilibrium:

$$\frac{1}{T} = -\alpha \;\rightarrow\; \text{grad } T = 0 \tag{14.35}$$

Use this result to evaluate α in the coefficients of each of the dc_k terms

$$-\frac{1}{T} \mu_k - \frac{1}{T} \psi M_k = \beta_k \;\rightarrow\; \mu_k + \psi M_k = \text{ constant} \tag{14.36}$$

Express this result in terms of the gradients of the variables

$$\text{grad } (\mu_k + \psi M_k) = 0$$

Since the gradient of a sum is the sum of the gradients, this result may also be written

$$\text{grad } \mu_k = -M_k \text{ grad } \psi \qquad (k = 1, 2, \ldots, c) \qquad (14.37)$$

Note that M_k, the molecular weight of k, is a constant. Thus in the presence of an external force field, the chemical potential of each of the components in a homogeneous system is *not uniform*.

The condition for mechanical equilibrium is derived by paraphrasing Eqs. (14.21) through (14.23). The left-hand side of Eq. (14.22) is not zero in the presence of an electrical field but is given by Eq. (14.37)

$$-M_k \text{ grad}\psi = \overline{V}_k \text{ grad}P + \sum_{j=2}^{c} \mu_{kj} \text{ grad } X_j \qquad (14.38)$$

Multiply by X_k and sum

$$-\sum_{k=1}^{c} X_k M_k \text{ grad } \psi = \sum_{k=1}^{c} X_k \overline{V}_k \text{ grad } P + \sum_{k=1}^{c} X_k \sum_{j=2}^{c} \mu_{kj} \text{ grad } X_j$$

The Gibbs-Duhem relation demonstrates that the double summation term is zero. The coefficient of $\text{grad}\psi$ can be recognized as M, the mass per mole of solution at the position (x, y, z), while the coefficient of $\text{grad}P$ is V, the volume per mole. Their ratio M/V is equal to ρ, the *mass density* (gm/cc) at the point P. The condition for mechanical equilibrium is:

$$\text{grad } P = -\rho \text{ grad } \psi \qquad (14.39)$$

It is concluded that when a system comes to equilibrium in the presence of an external force field, the temperature does not vary with position. In contrast, both the pressure and the chemical potential of each component vary with position in a manner determined by the positional variation of the potential field.

14.3.3 Equilibrium in a Gravitational Field

In a gravitational field, the force acting on a mass m is given by Newton's second law:

$$\vec{F} = m\vec{a} = m\vec{g} \qquad (14.40)$$

where g is the acceleration engendered by the field. Comparison of this equation with the general definition of the specific potential, Eq. (14.24), yields

$$\vec{g} = - \text{ grad } \psi \qquad (14.41)$$

The conditions for equilibrium stated in Eqs. (14.37) and (14.39) for the case of a gravitational field are:

$$\text{grad } \mu_k = M_k\vec{g} \qquad (k = 1, 2, \ldots, c) \qquad (14.42)$$

$$\text{grad } P = \rho\vec{g} \qquad (14.43)$$

The latter equation is a general form of the familiar *barometric* equation relating pressure to altitude in the earth's atmosphere. The following examples illustrate these results.

Example 14.1. Derive an expression for the variation of pressure with altitude for a unary system.

Solution. The condition for mechanical equilibrium, Eq. (14.43), provides the basis for the relationship. In a Cartesian coordinate system with $\hat{k}$ the unit vector in the vertical direction, $g = -g\hat{k}$, where g is the magnitude of the gravitational acceleration. The negative sign is required because the acceleration vector is directed downward while $\hat{k}$ by convention points upward. The vector Eq. (14.43) can be written explicitly as

$$\left(\frac{\partial P}{\partial x}\right)\hat{i} + \left(\frac{\partial P}{\partial y}\right)\hat{j} + \left(\frac{\partial P}{\partial z}\right)\hat{k} = \rho(-g\hat{k})$$

A vector equation represents three equations: corresponding coefficients of $\hat{i}$, $\hat{j}$ and $\hat{k}$ on the two sides of the equation must be separately equal. Since the $\hat{i}$ and $\hat{j}$ components on the right side are zero, the variation of μ_k in the horizontal directions is zero. Equating coefficients in the $\hat{k}$ direction yields

$$\frac{dP}{dz} = -\rho g \qquad (14.44)$$

Recall that ρ is the mass density of the system, which for a unary system is the ratio of the molecular weight M_1 (mass/mole) to the molar volume V:

$$\rho = \frac{M_1}{V} = \frac{M_1}{\frac{RT}{P}}$$

assuming the atmosphere is an ideal gas. The mass density thus varies with height z since the pressure varies with height. Insert this result into the condition for mechanical equilibrium, Eq. (14.44):

$$\frac{dP}{dz} = -M_1\left(\frac{P}{RT}\right)g$$

Separating the variables

$$\frac{dP}{P} = -\frac{M_1 g}{RT}dz$$

Integrating

$$\ln\left(\frac{P(z)}{P(z_0)}\right) = -\frac{M_1 g}{RT}(z - z_0) = \frac{M_1 g}{RT}h$$

$$P(z) = P(z_0)e^{(M_1 g/RT)h} \qquad (14.45)$$

This is the barometric equation, which demonstrates that at equilibrium in a unary system the pressure decreases exponentially with altitude. The analogous result is more complex for a multicomponent system because the molar mass varies with composition, which also varies with altitude.

Example 14.2. Planet X has an atmosphere that consists of a binary mixture of hydrogen and nitrogen. Derive an expression for the variation of composition of this atmosphere with altitude. Assume the atmosphere is an ideal gas mixture. At the planetary surface the mole fraction of H_2 is 0.35. Planet X has a radius of 10,000 km and a mass that yields a gravitational acceleration of 1000 cm/sec^2. Take the temperature of the atmosphere to be 800 K. Compute the composition of the atmosphere at 100 km above the surface.

Solution. For this one dimensional field the vector equation describing the condition for chemical equilibrium in this atmosphere, Eq. (14.42), simplifies to

$$\frac{d\mu_k}{dz} = -M_k g$$

again noting that the $\vec{g}$ is $-g\hat{\mathbf{k}}$. For an ideal gas mixture,

$$\mu_k = \mu_k^o + RT \ln X_k$$

The reference state for component k assumed in this expression is "pure k at the temperature and pressure of the solution." Since the pressure of the solution varies with height, it is essential to note that the chemical potential of the reference state depends upon its pressure. Substitute the expression for the chemical potential into the condition for equilibrium

$$\frac{d\mu_k}{dz} = \frac{d\mu_k^o}{dP} \cdot \frac{dP}{dz} + RT \frac{d \ln X_k(z)}{dz} = -M_k g$$

The variation of chemical potential of the reference state with pressure may be obtained from a coefficient relationship in the relation

$$d\mu_k^o = -S_k^o dT + V_k^o dP$$

$$\left(\frac{\partial \mu_k^o}{\partial P} \right)_T = V_k^o = \frac{RT}{P}$$

The derivative of pressure with height in a multicomponent system is given by

$$\frac{dP}{dz} = -\rho\, g = -\frac{M}{V} g = -\frac{M}{\frac{RT}{P}} g = -\frac{P}{RT} \cdot Mg$$

where $M = \sum_k M_k X_k$ is the mass per mole of solution. The condition for chemical equilibrium becomes:

$$\frac{RT}{P} \cdot \left(-\frac{P}{RT} \cdot Mg \right) + RT \frac{d \ln X_k(z)}{dz} = -M_k g$$

or

$$RT \frac{d \ln X_k(z)}{dz} = Mg - M_k g$$

Rearrange

$$\frac{d \ln X_k(z)}{dz} = \frac{g}{RT} [M - M_k] \tag{14.46}$$

Thus, if the molecular weight of component k is larger than the average, the right side of the equation is negative and the mole fraction of that component *decreases* with altitude. If M_k is smaller than the average, the right side is positive and the mole fraction of that component *increases* with altitude. These results are in agreement with intuition, which suggests heavier components are enriched at lower altitudes.

For a binary system the average molar mass is

$$\sum_{j=1}^{2} X_j M_j = X_1 M_1 + X_2 M_2 = (1 - X_2)M_1 + X_2 M_2$$

Substitute this result into Eq. (14.46)

$$\frac{d \ln X_2(z)}{dz} = \frac{g}{RT}[(1 - X_2)M_1 + X_2 M_2 - M_2]$$

$$= \frac{g}{RT}(1 - X_2) \cdot (M_1 - M_2)$$

Separating the variables

$$\frac{d \ln X_2}{1 - X_2} = \frac{g(M_1 - M_2)}{RT} \cdot dz$$

Integrating

$$\frac{X_2(z)}{1 - X_2(z)} = \frac{X_2(z_0)}{1 - X_2(z_0)} \cdot e^{(g/RT)(M_1 - M_2)(z - z_0)} \tag{14.47}$$

Choose component 2 to be H_2 and insert the values given in the problem:

$$\frac{X_2(z)}{1 - X_2(z)} = \frac{0.35}{0.65} \exp\left[\frac{1000 \, \frac{cm}{sec^2}}{8.314 \, \frac{J}{mol-K} \, 800 \, K} \cdot (28 - 2)\frac{gm}{mol} \cdot 100 \, km\right]$$

Convert gm-cm^2/sec^2 to Joules to make the units compatible:

$$\frac{X_2(z)}{1 - X_2(z)} = \frac{0.35}{0.65} \cdot e^{3.91} = 26.8$$

Solve for the composition of hydrogen at 100 km:

$$X_{H_2}(100 \, km) = 0.964$$

The atmosphere is significantly enriched in the lighter element at 100 km.

14.3.4 Equilibrium in a Centrifugal Field

Consider now a system that is rotating with an angular velocity of ω radians/sec. The acceleration vector for an element of volume contained a distance r from the axis of rotation is given by

$$\vec{a} = \omega^2 \vec{r} \tag{14.48}$$

where the vector r is directed toward the center of rotation. The acceleration is directed inward. The force acting upon the mass dm contained in the volume element is

$$d\vec{F} = dm \, \vec{a} = dm \, \omega^2 \vec{r}$$

Comparison with Eq. (14.24) yields the specific potential field for a centrifugal force field as

$$\text{grad } \psi = -\omega^2 \vec{r} \tag{14.49}$$

For this case the conditions for chemical equilibrium [Eq. (14.37)] and mechanical equilibrium [Eq. (14.39)] become

$$\text{grad } \mu_k = M_k \omega^2 \vec{r} \tag{14.50}$$

$$\text{grad } P = \rho \omega^2 \vec{r} \tag{14.51}$$

Example 14.3. Estimate the maximum pressure achieved in a turbine blade in a jet turbine engine rotating at 22,000 rpm. The root of the blade is 3 cm from the axis of rotation and the blade is 8 cm long. Take the density of the blade to be 9 gm/cc. Assume the state of stress is a simple hydrostatic pressure.

Solution. Compute the angular velocity in radians per second

$$\omega = \frac{2\pi \cdot 22,000}{60} = 2,300 \; \frac{\text{radians}}{\text{sec}}$$

The condition for mechanical equilibrium for a radial centrifugal field is

$$\frac{dP}{dr} = \rho \omega^2 r$$

Integrating

$$P(r) = \frac{1}{2} \rho \omega^2 r^2$$

Insert the properties stated in the problem.

$$P(r) = \frac{1}{2} \left(9 \; \frac{\text{gm}}{\text{cc}} \right) \left(2300 \; \frac{\text{radians}}{\text{sec}} \right)^2 (11 \text{ cm})^2 \left(\frac{1 \text{ J}}{10^7 \text{ ergs}} \right) \left(\frac{82.06 \text{ cc-atm}}{8.314 \text{ J}} \right)$$

Calculating

$$P(r) = 2850 \text{ atm} = 41,900 \text{ psi}$$

Thus centrifugal stresses of the order of a few kilobars are not uncommon in an operating turbine blade. Actual stresses are not hydrostatic and the deviations from this simple equilibrium condition play a much more significant role in determining the performance of a turbine blade.

14.3.5 Equilibrium in an Electrostatic Field

The description of forces that operate in an electrostatic field is formulated in terms slightly different from those used for gravitational and centrifugal fields. An electrical potential field $\phi(x, y, z)$ is defined such that the force acting on a particle carrying a *charge q* is given by

$$\vec{F} = -q \text{ grad } \phi \tag{14.52}$$

In this view, $-\text{grad}\phi$ is the force per unit charge. This contrasts with the definition of the specific potential field ψ defined in Eq. (14.24) such that $-\text{grad}\phi$ is the force per unit *mass*.

These two potentials are related through the charge per unit mass that a system may possess. Focus on a volume element δv. Within that volume element, focus on the atoms of component k. Let z_k be the charge number associated with a unit of component k. Then, z_k is a small whole number; it is zero, positive, or negative depending upon the nature of the component. The mass of the atoms of component k in δv,

$$\delta m_k = M_k c_k \delta v \qquad (14.53)$$

The charge carried by these atoms is

$$\delta q_k = z + k(N_0 e) c_k \delta v \qquad (14.54)$$

where e is the magnitude of the charge carried by an electron and N_o is Avogadro's number. The product $N_o e$ is the magnitude of the electric charge carried by one mole of electrons. This quantity is called the *Faraday*, written $\mathcal{F}$, equal to 96512 Coulombs/mole, after the genius who pioneered the study of electrochemical effects early in the last century. The force acting upon this subset of atoms in the volume element δv can be formulated in terms of either Eq. (14.24) or (14.52).

$$\delta \vec{F}_k = -\delta m_k \text{ grad } \psi = -\delta q_k \text{ grad } \phi$$

$$-(M_k c_k \delta v) \text{ grad } \psi = -(z_k \mathcal{F} c_k \delta v) \text{ grad } \phi$$

$$M_k \text{ grad } \psi = z_k \mathcal{F} \text{ grad } \phi \qquad (14.55)$$

This establishes the relationship between the specific potential ψ and the electric potential ϕ. The conditions for equilibrium, Eqs. (14.37) and (14.39), can now be written in terms of the electrical potential. In particular, for chemical equilibrium,

$$\text{grad } \mu_k = -z_k \mathcal{F} \text{ grad } \phi \qquad (k = 1, 2, \dots, c) \qquad (14.56)$$

Introduce the following definition:

$$\eta_k \equiv \mu_k + z_k \mathcal{F} \phi \qquad (14.57)$$

The property η_k is called the *electrochemical potential* of component k. With this definition, the condition for equilibrium in the presence of an electrostatic field can be written

$$\text{grad } \eta_k = 0 \qquad (k = 1, 2, \dots, c) \qquad (14.58)$$

Thus in an electrostatic field the electrochemical potential is uniform at equilibrium. Note that the gradient in chemical potential, and hence qualitatively the gradient in composition, depends upon the sign of z_k. For positively charged components the chemical potential decreases with increasing electrical potential; for negatively charged components the chemical potential and electric field gradients are in the same direction.

For mechanical equilibrium,

$$\text{grad } P = -\rho \text{ grad } \psi = -\frac{1}{V} \left(\sum_{k=1}^{c} M_k X_k \right) \text{ grad } \psi$$

Distribute gradψ throughout the sum

$$\text{grad } P = -\frac{1}{V} \sum_{k=1}^{c} M_k X_k \text{ grad } \psi$$

Substitute in each term from Eq. (14.55)

$$\text{grad } P = -\frac{1}{V} \sum_{k=1}^{c} X_k (z_k \mathcal{F} \text{ grad } \phi)$$

Note that in each term $X_k / V = c_k$, the molar concentration. Finally,

$$\text{grad } P = -\mathcal{F} \text{ grad } \phi \sum_{k=1}^{c} z_k c_k \tag{14.59}$$

The quantity in the summation can be thought of as the average charge density at some position (x, y, z) in the system. Since in general, the right side of Eq. (14.59) differs from zero for a system with charged particles in an electrical field, a pressure gradient is expected to develop in such a system when it comes to equilibrium.

Example 14.4.   Given an electrostatic field that decays exponentially from the surface of the system toward its interior, describe the variation of composition of the components with position when this system comes to equilibrium. Let $X_k(\infty)$ be the composition well away from the surface of the system corresponding to the chemical potential $\mu_k(\infty)$.

Solution. Since at equilibrium, the electrochemical potential [Eq. (14.57)] is constant [Eq. (14.58)], then

$$\mu_k + z_k \mathcal{F} \phi = \text{const} = \mu_k(\infty)$$

$$\mu_k - \mu_k(\infty) = -z_k \mathcal{F} \phi_0 e^{-ax}$$

Components for which z_k is negative exhibit a chemical potential distribution that mimics the electric field function, with μ_k and thus X_k high near the surface and decaying down to $X_k(\infty)$. The distribution for positively charged components is the opposite, rising from the lowest value of concentration adjacent to the surface, toward their far field value. This result is consistent with the notion that negatively charged particles move toward regions of high electrical potential while positively charged ions are repelled from those regions.

14.4   THE GRADIENT ENERGY IN NONUNIFORM SYSTEMS

The preparation of microstructures that have extremely fine characteristic length scales is an important trend in materials science. Microelectronic devices are prepared with controlled composition distributions that span a few nanometers. Sol gel materials have pore structures in the nanometer size range. Rapidly solidified materials yield very fine scale microstructures. Nanocomposite materials are being designed and fabricated. Solute rich clusters that form prior to precipitation in age hardening materials have composition gradients on a fine scale. Spinodal decomposition and ordering transformations also occur at this level of microstructural scale. Nucleation processes also may involve large changes in intensive properties over short distances.

The description of the thermodynamics of nonuniform systems presented in Secs. 14.1 through 14.3 is founded on the concept of the local density of thermodynamic properties, Eq. (14.1). The total value of any extensive property is then taken to be the integral of the local density function over the volume of the system, Eq. (14.2). It has been found that this description is *inadequate* if the properties of the system vary significantly over very short distances. Such a system is characterized as having extremely large values of *gradients* of its intensive properties.

To illustrate this idea consider the concentration gradient in a typical interdiffusion study. In an experimental diffusion couple joining A to B, the atom fraction of B may change from 0 to 1 over a range of perhaps 10 microns (10^{-6} m). The average gradient in atom fraction in this structure is thus $1/10^{-6}$ or 10^6 m^{-1}. In a microstructure undergoing spinodal decomposition, the atom fraction of B may change from, say, 0.8 to 0.2 in a few nanometers. The gradient in this system is thus ($0.6/3 \times 10^{-9}$ m^{-1} = 2×10^9 m^{-1}) or about three orders of magnitude larger than in the diffusion couple.

In the presence of large gradients it is found that the value of the local density of a thermodynamic function b_v depends not only upon the values of the intensive properties in the volume element but also on their rates of change, that is, their local *gradients*. A rigorous formulation of this contribution to the properties of systems was first presented by Cahn and Hilliard in 1959 [14.3].

Consider as an example the Helmholtz free energy of a binary system, F'. In the traditional description the local density of Helmholtz free energy is a function of the local temperature, pressure (in solids, the state of stress), and composition:

$$f_v = f_v(T, P, X_2) = f_v(T, P, c_2) \tag{14.60}$$

To simplify the problem, suppose the temperature and pressure are uniform and the system exhibits a nonuniformity only in composition. Then the Helmholtz free energy of the system is the integral

$$F = \iiint_{V'} f_v[c_2(x, y, z)]\delta v \tag{14.61}$$

If the variation of composition with position is known, together with a solution model that permits calculation of the Helmholtz free energy from the composition, then F' can be computed.

Cahn and Hilliard assumed that f_v depends not only upon c_2 but also on all measures of the variation of c_2 with position including c_2', $(c_2')^2$, $(c_2')^3$, and c_2'', $(c_2'')^2$, etc. Here the (') and ('') denote the first and second derivatives with respect to position. Thus, it is assumed that f_v depends on the composition, the gradient, the square of the gradient, the second derivative, and so on. A generalization of the Taylor expansion of a function about a point permits this dependence to be expressed as

$$f_v = f_v^o[c_2] + L[c_2'] + K_1[c_2''] + K_2[c_2'^2] + \cdots \tag{14.62}$$

where the coefficients have explicit but unfamiliar meaning:

$$L = \left(\frac{\partial f_v}{\partial c_2'} \right) \tag{14.63}$$

$$K_1 = \left(\frac{\partial f_v}{\partial c_2''} \right) \tag{14.64}$$

$$K_2 = \left(\frac{\partial f_v}{\partial c_2'^2} \right) \tag{14.65}$$

Terms beyond these second order quantities are neglected in the development.

In crystals with a center of symmetry in the unit cell, the value of f_v must be independent of the choice of direction for the coordinate system describing the variation of composition. Since altering the direction of the x-axis changes the sign of the first derivative c_2', the coefficient of this term, L, must be zero. Further mathematical manipulation demonstrates that the remaining two terms can be combined into

$$K_1 \frac{d^2 c_2}{dx^2} + K_2 \left(\frac{dc_2}{dx} \right)^2 = K \left(\frac{dc_2}{dx} \right)^2$$

where

$$K = K_2 - \frac{dK_1}{dc_2}$$

Thus the local free energy density can be written

$$f_v = f_v^o(c_2) + K \left(\frac{dc_2}{dx} \right)^2 \tag{14.66}$$

The total Helmholtz free energy of the system is then

$$F' = \iiint_{V'} \left[f_v^o(c_2) + K (\nabla c_2)^2 \right] \delta v \tag{14.67}$$

where the symbol ∇ is shorthand notation for the *gradient* of the composition function [see Eq. (14.18)]. Comparison of this result with Eq. (14.61) identifies the additional term in the Helmholtz free energy, which is proportional to the *square of the gradient* of the composition distribution. Since it can be shown that the coefficient K must be positive, it follows that this gradient energy contribution is always positive.

It may be useful to view the gradient energy as a distributed or smeared out surface energy. In Gibbs' development of the thermodynamics of surfaces, the effects of the nonuniformity associated with the transition from the properties of one phase to those of its neighbor are assigned to a discontinuity located at the dividing surface, Sec. 12.2. The gradient energy formalism recognizes the diffuse nature of this distribution in many real systems, characterizing them with a smooth variation in intensive properties, or a *gradient*, yet realizing an excess energy in the system associated with such nonuniformities.

Like surface energy, the gradient energy may provide a barrier to be overcome by a process, as in nucleation, or it may contribute to the driving force, as in coarsening processes. Consider for example, the spinodal decomposition process described in connection with miscibility gaps in Sec. 10.1.4. If a uniform solution is quenched into the spinodal region it may spontaneously unmix. In more detail, the as-quenched structure contains an array of random statistical fluctuations in composition, Fig. 14.1a. The spontaneous unmixing process develops as an increasing variation in the composition range in the system, Fig. 14.1b. This can be viewed as an increase in the *amplitude* of components of the composition wave. The process evidently requires the motion of atoms of the components through the solution, a diffusion process. Components of the composition fluctuation that have the shortest wavelength have the shortest diffusion distance and largest concentration gradient. Amplitudes of these shortest waves therefore grow most rapidly, ultimately resulting in a modulated structure with a wavelength equal to the distance between atoms.

This is not observed experimentally. The structure that results has a characteristic wavelength in the nanometer range but large compared with the interatomic distance. The value of this characteristic wavelength varies in a predictable way with temperature and the starting composition of the solution. The atomically fine compositional variation does not form because the energy associated with the extremely large gradients in this system prevents it. Rather, the system chooses a length scale that optimizes the competition between shortened diffusion distances and increased gradient energy as the wavelength shortens. These observations could not be explained without invoking the gradient energy concept.

The gradient energy can add to the driving force for diffusion for a process that requires the spontaneous elimination of gradients artificially introduced into a system. Hilliard and Philofsky [14.5] prepared a series of diffusion couples by vapor depositing thin alternating layers of gold and palladium on a substrate, Fig. 14.2. They

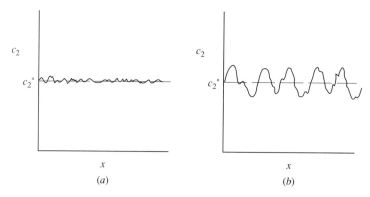

FIGURE 14.1
A random composition fluctuation about the average composition c_2^o in a solution quenched into a miscibility gap (a) begins to unmix spontaneously (b) by amplifying some of the harmonic components in the composition wave form.

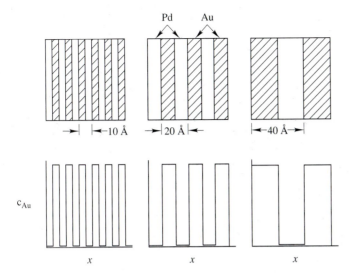

FIGURE 14.2
Thin film diffusion couples prepared by alternately vapor depositing layers of gold and palladium.

were able to prepare these layered structures with different layer thicknesses ranging from about one to four nanometers. The gradients in the one nanometer composite were thus about four times larger than those in the coarser, four nanometer structure. Since the gradient energy term is proportional to the *square* of the gradient, the ratio of gradient energies in the two extreme cases was about 16.

In spinodal decomposition the spontaneous process is unmixing, or more specifically, the spontaneous amplification of compositional fluctuations. In these initially layered structures the spontaneous process is the elimination of these compositional variations with the equilibrium structure exhibiting a uniform composition. In this *homogenization* process the system has gradients initially; in its final state the gradients are eliminated. Thus in this case the gradient energy provides an additional driving force that increases the diffusion rate. In the experiments of Hilliard and Philofsky, this showed up as a variation of the diffusion coefficient with the initial spacing between the layers. Their one nanometer structure with the largest gradient energy gave the lowest diffusion coefficient, Fig. 14.3. The diffusion coefficient reports the proportionality between the flux and the driving force. Since the driving force is increased by the gradient energy term for the finer structures, a lower value of the diffusion coefficient is required to obtain the same flux.

The sharp drop in diffusion coefficient that appears near 3 nanometers is associated with a loss of mechanical coherency in these modulated structures. Elastic effects are found to play an important role in ultrafine structures, and may even eclipse the gradient energy effects in the system. For example, if the elastic strains associated with the compositional variations that develop in spinodal decomposition are sufficiently large, the process can be completely suppressed. For an excellent review of the theory and associated experimental observations, see Hilliard [14.4].

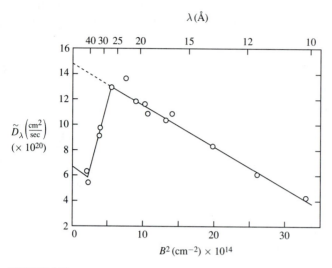

FIGURE 14.3
Variation of diffusion coefficient with wave length of the starting layered structure demonstrates that the gradient energy contribution is real [14.5].

14.5 SUMMARY OF CHAPTER 14

In systems of ordinary dimensions, the thermodynamics of nonuniform systems can be described by the variation of the local density of any of its extensive properties b_v with position.

The global value of a property is the integral over the volume of the system of its local density.

Application of the general strategy for finding conditions for equilibrium in continuous systems showed that

1. In the absence of external fields, at equilibrium,

$$\text{grad } T = 0 \tag{14.19}$$

$$\text{grad } P = 0 \tag{14.23}$$

$$\text{grad } \mu_k = 0 \quad (k = 1, 2, \ldots, c) \tag{14.20}$$

2. In the presence of external fields, described by the specific potential energy function ψ, at equilibrium,

$$\text{grad } T = 0 \tag{14.35}$$

$$\text{grad } P = -\rho \text{ grad } \psi \tag{14.39}$$

$$\text{grad } \mu_k = -M_k \text{ grad} \psi \quad (k = 1, 2, \ldots, c) \tag{14.37}$$

The specific potential energy function is related to familiar physical quantities:

1. In a gravitational field, $\text{grad}\psi = -g$, the gravitational acceleration vector.
2. In a centrifugal field, $\text{grad}\psi = -\omega^2 r$, where ω is the angular velocity and r is the distance from the axis of rotation.
3. In an electrostatic field, $M_k \text{grad}\psi = z_k \mathcal{F} \text{grad}\phi$, where M_k is the molecular weight of component k, z_k is its charge number, $\mathcal{F}$ is Faraday's constant, and ϕ is the electric potential.

If the length scales of variation of properties is in the nanometer range, then it is necessary to include a gradient energy term in the description of the thermodynamic state of each volume element. For example, the density of the Helmholtz free energy must be written

$$F' = \iiint_{V'} \left[f_v^o(c_2) + K(\nabla c_2)^2 \right] \delta v \tag{14.67}$$

where $f_v(c_2)$ is the free energy density associated with an element of the same temperature, pressure, and composition in an homogeneous system. Phenomena that occur in fine scale structures cannot be adequately explained without including this gradient energy contribution.

PROBLEMS

14.1. A bar of copper is insulated along its length L. One end is held at 498 K and the other end fixed at 298 K until reaching the steady state temperature profile

$$T(x) = T_2 - \frac{T_2 - T_1}{L} \cdot x$$

The ends of the bar are then insulated from the heat source and sink (the system is isolated) If $L = 2$ cm, and the diameter, $d = 1$ cm.
(a) Find the equilibrium temperature distribution.
(b) Compute the change in entropy for the system during its transition to equilibrium. Begin by writing an expression for the change in entropy density of the system as a function of position.

14.2. Explain why there is no PdV type term in the local formulation of the combined statement of the first and second laws.

14.3. Use the conditions for equilibrium in a single phase system to prove that, in the absence of external fields, the *composition* X_k of each component is uniform in the system.

14.4. Potential field effects are incorporated into the conditions for equilibrium through Eq. (14.31), which expresses one of the constraints that operates in an isolated system. Justify the contention that the energy function, $E'_{\text{tot}} = U' + E'_{\text{pot}}$, is constrained to be constant, rather than just the internal energy U', in developing the description of an isolated system.

14.5. Assume that the mixing behavior of the A-B system is described by the familiar equation

$$\Delta G_{\text{mix}} = a_0 X_A X_B + \Delta G_{\text{mix}}^{\text{id}}$$

where a_0 is independent of temperature and pressure.

(a) Write an expression for the chemical potential of component B as a function of T, P, and X_B.

(b) Derive an expression for the variation of composition with altitude for this system.

(c) Compare this result with that obtained for an ideal solution.

14.6. One strategy for separating isotopes of uranium (U^{235} and U^{238}) in the Manhattan Project was to form a dilute solution containing these isotopes and centrifuge it. Examine the thermodynamics of this situation and assess its feasibility.

14.7. Describe an electrical field that generates the same pressure distribution as a centrifugal field developed in an ultracentrifuge rotating at 20,000 rpm. State *all* your assumptions. (Several essential elements have been left out of the statement of this problem.)

14.8. Develop an expression for the variation of composition with distance from the surface of a system for a dilute NaCl aqueous solution given that the electrical potential decays exponentially from the surface as in Example 14.4.

14.9. Consider a nonuniform single phase system in which the composition varies periodically in one direction:

$$X_2(x) = 0.5 + 0.05 \cos \frac{2\pi x}{\lambda} \qquad 0 \leq x \leq L$$

where $\lambda = 5$ nm (nanometers). Assume

$$\Delta G_{mix} = \Delta F_{mix} = 12200 X_A X_B + \Delta G_{mix}^{id}$$

and $V = 7.1$ cc/mole, independent of composition.

(a) At 800 K calculate the Gibbs free energy of formation of this periodic system from an initially uniform system for a cube 1 cm on a side without the gradient energy contribution.

(b) Make the same calculation with the gradient energy assuming $\kappa = 8 \times 10^{10}$ (J-cm^5)/(mol)^2.

(c) Compare the contribution of the gradient energy to the total energy of mixing.

REFERENCES

1. Haase, Rolf: *Thermodynamics of Irreversible Processes*, Addison-Wesley Publishing Co., Reading, Mass., pp. 64–69, 1969.
2. Cahn, J. W., and J. E. Hilliard: *J. Chem. Phys.*, vol. 28, p. 258, 1958.
3. Cahn, J. W., and J. E. Hilliard: *J. Chem. Phys.*, vol. 31, p. 688, 1959.
4. Hilliard, J. E.: "Spinodal Decomposition," in *Phase Transformations*, H. I. Aaronson, ed., ASM, Materials Park, Ohio, p. 497, 1970.
5. Philofsky, E. M., and J. E. Hilliard: *J. Appl. Phys.*, vol. 40, p. 2198, 1969.

CHAPTER
15

ELECTROCHEMISTRY

Many of the processes that sustain life combine chemical change with electrical exchanges. This complex combination of influences can degrade materials in corrosion, or can be used to refine them, as in the electrolysis of aluminum, titanium, copper, and other metals, and in the preparation of ceramic powders and polymer blends. The subset of science that deals with phenomena that combine chemical and electrical effects, *electrochemistry*, is exceedingly broad [15.1, 15.2, 15.3]. In this introductory text the focus is limited to the description of the conditions for equilibrium in such systems and applications derived from these conditions.

The *ionization energy* of an element is the difference in energy between an atom of that element in its vapor state and an ion of the same element. When that element is present as a component in a solution its ionization energy may be greatly reduced as a result of its interactions with the surrounding solvent molecules. A salt dissolved in water dissociates to a greater or lesser extent to form cations and anions. The resulting solution is called an *electrolyte* because it transports a charge when subjected to an electric field.[1]

Water is the most familiar and important solvent for electrolytes and life itself depends upon this property of water. However, many other so called *polar* solvents,

---

[1]The term *electrolyte* is also used to describe the compound that forms ions in solution. NaCl is an electrolyte; a solution of NaCl in water is also called an electrolyte.

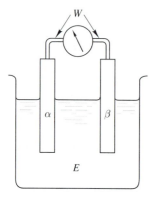

FIGURE 15.1
A galvanic cell consists of four phases: two electrodes (α) and (β), an electrolyte (E), and the external wire connecting them with measuring apparatus (W).

such as acetone or glycerol, also form electrolyte solutions. *Solid* electrolytes, in which a charge current is carried by diffusion of the ions in the system, also exist and find useful applications. The conditions for equilibrium in an electrolyte solution are derived in Sec. 15.1.

In addition to the intensive properties dealt with in earlier chapters, every phase has an *internal electric potential*. If no external electrostatic field is applied, this potential is uniform within the phase at equilibrium. In considering the equilibrium *between* phases, it is not necessary to consider their differing electric potentials unless charged particles are transferred between the phases in the process being considered. Section 15.2 presents the conditions for equilibrium between phases and electrolytes that may communicate electrical charges.

Absolute values of the electric potentials of phases cannot be measured nor can the potential difference between two phases. Because the measurement of an electrical potential requires a closed circuit, the minimum configuration of a system that can be devised in which an electric potential can be measured is the *galvanic cell*. Such a system consists of at least four phases; two electrodes inserted into an electrolyte and externally connected by a wire (which also provides connections to the electric measuring devices that monitor the system), Fig. 15.1. The properties of galvanic cells and associated conditions for equilibrium are presented in Sec. 15.3.

Practical visualization of the complexities of the conditions for equilibrium in cells is conveniently represented in *potential-pH* predominance diagrams, commonly called *Pourbaix* diagrams after their inventor. The strategy for deriving and interpreting Pourbaix diagrams is laid out in Sec. 15.4 together with applications to corrosion and related phenomena.

15.1 EQUILIBRIUM WITHIN AN ELECTROLYTE SOLUTION

The compound $M_u X_v$ is an electrolyte if it dissociates, either partially or completely, into its cations and anions when dissolved in water or other appropriate solvent. This compound may be an *acid* (in which case M is hydrogen), a *base* (X is the hydroxyl group OH) or a *salt* (M is usually a metal and X a nonmetal, as in NaCl).

Depending on the level of detail required in a given application, this solution can be considered to consist of two components: H_2O and M_uX_v; or alternatively, six components: undissociated solvent (H_2O) and its associated ions,[2] H^+ and OH^-; and undissociated compound (M_uX_v) and its ions, M^{z+} and X^{z-}. Complexes involving other combinations of these elements, for example, $MX_2O_4^z$, may also exist in the system and thereby further increase the number of components considered in its behavior.

It is awkward to designate the chemical potentials, number of moles, molar concentrations, activities, and activity coefficients of all these components with self-explanatory subscripts because these designations themselves have superscripts. For example, M^{z+} describes the cation obtained from the compound electrolyte and the number of moles of this ionic component could be written $n_{M^{z+}}$. In the descriptions that follow the notation can be simplified without loss of clarity with the designations shown in Table 15.1.

The solution is formed by adding n_C moles of the compound M_uX_v to n_W moles of water. In the solution the compound dissociates to form n_+ moles of the cation M^{z+} and n_- moles of the anion X^{z-}, leaving n_U moles of the compound undissociated. Introduce the concept of the *degree of dissociation* of the electrolyte α, such that αn_C is the number of moles of the electrolyte that dissociates into ions. Since each molecule of the compound M_uX_v contains u cations and v anions, $u(\alpha n_C) = n_+$ moles of the cation M^{z+} and $v(\alpha n_C) = n_-$ moles of the anion X^{z-} are formed. The number of moles of undissociated electrolyte in the solution n_U is $(1-\alpha)n_C$. This latter relation emphasizes the distinction between n_C, the total number of moles of the compound added to the system, and n_U, the number of moles

TABLE 15.1
Notation used to designate the number of moles of components in electrolytes

Component	Symbol
Preparation of the solution:	
Number of moles of water added	n_W
Number of moles of compound M_uX_v added	n_C
Components in the solution:	
Number of moles of undissociated M_uX_v	n_U
Number of moles of cations M^{z+} formed	n^+
Number of moles of anions X^{z-} formed	n^-

---

[2]Since the hydrogen atom consists of a single proton and an electron, the hydrogen ion H^+ is a proton. Experimental studies of kinetic phenomena demonstrate that free protons do not generally exist in aqueous solution but are associated with a water molecule, forming the *hydronium ion, H_3O^+*. However, this distinction, while critical for explaining kinetic phenomena, is not important in rationalizing thermodynamic behavior. Accordingly, in this chapter H^+ is used to designate the positive ion derived from the dissociation of H_2O.

of the compound that remain *undissociated* in the solution after it comes to equilibrium.

There exists a class of compounds called *strong electrolytes*, which dissociate completely into ions in the solution. For these compounds, $\alpha = 1$ so that n_U is zero. *Weak electrolytes* are partially dissociated, with α values ranging between 0 and 1. The behavior of strong electrolytes and weak electrolytes is discretely different and is treated separately in the thermodynamics of electrolytes.

15.1.1   Equilibrium in Weak Electrolytes

The change in Gibbs free energy associated with a change in composition at constant temperature and pressure of an electrolyte solution can be written in either of two forms:

$$dG'_{T,P} = \mu_W dn_W + \mu_C dn_C \tag{15.1}$$

$$dG'_{T,P} = \mu_{H_2O} dn_{H_2O} + \mu_{H^+} dn_{H^+} + \mu_{OH^-} dn_{OH^-} + \mu_U dn_U + \mu_+ dn_+ + \mu_- dn_- \tag{15.2}$$

A chemical potential can be defined for each of the components. For example,

$$\mu_+ \equiv \left(\frac{\partial G'}{\partial n_+} \right)_{T,P,n_{H_2O} n_{H^+} n_{OH^-} n_U n_-} \tag{15.3}$$

Guggenheim [15.4] has observed that the chemical potential of ionic components is of academic interest only since it is not possible to create a process in which one ionic component such as n_+ is added to a system while keeping the number of moles of the other ionic components constant. Such a process would violate the condition of charge neutrality. In practice, the composition of the electrolyte can be changed only by adding the compound $M_u X_v$, (increasing n_C) or the solvent H_2O (increasing n_W). Thus, the chemical potentials (and the related activities and activity coefficients introduced below) of the ionic components are conceptually useful, but cannot be measured.

The activity corresponding to the chemical potential defined in Eq. (15.3) is given by

$$\mu_+ = \mu_+^\circ + RT \ln a_+ \tag{15.4}$$

where μ_+° is the chemical potential of the cation in its standard state. An activity coefficient for the ionic component γ_+ can be defined in the usual way,

$$a_+ \equiv \gamma_+ X_+ \tag{15.5}$$

Similar definitions can be applied to the other ionic components.

In the description of the composition of electrolytes, the molar concentration or *molarity* of a component is normally used in the literature in place of the mole fraction. The molarity of a component is defined to be the *number of moles of the component per liter of solution*. The molarity of component k is usually written as $[k]$; it is identical to the molar concentration c_k defined and used in Chap. 14. The molarity of component k is simply related to its mole fraction by

$$c_k = [k] = \frac{X_k}{V} \tag{15.6}$$

where V is the molar volume of the solution. This quantity is a convenient measure of composition because it is easy to prepare a liquid solution of a desired molarity in the laboratory. A known number of moles of solute is weighed and added to the solvent and dissolved. Solvent is then added to bring the total volume of the solution to one liter, which then has the desired number of moles of solute per liter of solution.

In this context it is useful to introduce an alternate definition of the activity coefficient such that

$$a_k = f_k c_k = f_k[k] \tag{15.7}$$

Note that, since a_k is unitless, f_k must have units of (liters/mole of k). Evidently,

$$f_k = \gamma_k V \tag{15.8}$$

The strategy for finding conditions for equilibrium within an electrolyte yields the usual conditions for thermal and mechanical equilibrium. Applying the isolation constraints that conserve H, O, M, and X atoms yields combinations of the chemical potentials of the components that arrange themselves into affinities, each implying its corresponding reaction:

$$\mathscr{A}_W = (\mu_{H^+} + \mu_{OH^-}) - \mu_{H_2O} \tag{15.9}$$

$$H_2O = H^+ + OH^- \tag{15.1}$$

$$\mathscr{A}_C = (u\mu_+ + v\mu_-) - \mu_U \tag{15.10}$$

$$M_u X_v = uM^{z+} + vX^{z-} \tag{15.2}$$

Additional independent equations appear if complex ionic components are assumed to be present.

At equilibrium these affinities are each separately equal to zero. Substitution of the definitions of the corresponding activities and standard free energy changes leads to the familiar working equations for chemical equilibrium:

$$\Delta G_W^o = -RT \ln K_W \qquad \text{where} \qquad K_W = \frac{a_{H^+} a_{OH^-}}{a_{H_2O}} \tag{15.11}$$

$$\Delta G_C^o = -RT \ln K_C \qquad \text{where} \qquad K_C = \frac{a_+^u a_-^v}{a_U} \tag{15.12}$$

In this application the equilibrium constants K_W and K_C are called *dissociation constants*. The ΔG^o terms are differences in free energies between reactants and products when they are in their reference states.

The equilibrium constant for the dissociation reaction [15.2] can be expressed in terms of the activity coefficients and molarities:

$$K_C = \frac{a_+^u a_-^v}{a_U} = \frac{f_+^u f_-^v}{f_U} \cdot \frac{[M^{z+}]^u [X^{z-}]^v}{[M_u X_v]}$$

$$K_C = K_f \cdot \frac{[M^{z+}]^u [X^{z-}]^v}{[M_u X_v]} \tag{15.13}$$

Reference states for the components in an electrolyte solution are chosen so that in dilute solutions the activity coefficients f_k approach 1. Values of tabulated dissociation constants are based upon this choice of reference state. Thus the ratio of activity coefficients K_f is approximately 1 for dilute solutions and the dissociation constant can be expressed in terms of the concentrations:

$$K_C \cong \frac{[M^{z+}]^u [X^{z-}]^v}{[M_u X_v]} \tag{15.14}$$

It is assumed that $K_f = 1$ in most practical applications and that assumption is used in the applications presented in this chapter. However, use of this assumption in concentrated solutions may lead to significant errors.

The concentration ratio in equation (15.14) can be expressed in terms of c_C, the total concentration of electrolyte compound added to make the solution and the degree of dissociation α.

$$[M^{z+}] = u\alpha c_C \qquad [X^{z-}] = v\alpha c_C \qquad [M_u X_v] = (1-\alpha)c_C \tag{15.15}$$

$$K_C = \frac{(u\alpha c_C)^u (v\alpha c_C)^v}{(1-\alpha)c_C} = u^u v^v \cdot \frac{\alpha^{(u+v)} c_C^{(u+v-1)}}{(1-\alpha)} \tag{15.16}$$

Let $\omega = u + v$ be the total number of ions formed when a molecule of $M_u X_v$ dissociates. Equation (15.16) can be written

$$K_C = (u^u v^v) \frac{\alpha^{\omega} c_C^{\omega-1}}{(1-\alpha)} \tag{15.17}$$

For example, for the dissociation of $CaCl_2$, $u = 1$, $v = 2$ and $\omega = 3$

$$K_{CaCl_2} = (1^1 \cdot 2^2) \cdot \frac{\alpha^3}{1-\alpha} \cdot c_{CaCl_2}^{(3-1)} = 4 \cdot \frac{\alpha^3}{1-\alpha} \cdot c_{CaCl_2}^2$$

With a value of the dissociation constant, for a given total concentration of electrolyte, the value of α can be computed and the equilibrium concentrations of cations, anions, and undissociated electrolyte can be determined with Eq. (15.15).

For many dissociation reactions K_C is many orders of magnitude smaller than one and over a broad concentration range the resulting degree of dissociation $\alpha \ll 1$. In this case $(1 - \alpha)$ can be replaced safely by 1 and Eq. (15.17) takes on a simpler form:

$$K_C \cong (u^u v^v)\alpha^{\omega} c_C^{\omega-1} \tag{15.18}$$

This equation can be solved explicitly for the degree of dissociation in terms of the concentration of electrolyte added to the solution:

$$\alpha = \left(\frac{K_C}{u^u v^v}\right)^{\frac{1}{\omega}} \cdot c_C^{(\frac{1}{\omega}-1)} \tag{15.19}$$

Thus for most weak electrolytes the degree of dissociation increases as the electrolyte concentration decreases.

Example 15.1. The dissociation constant of pure water at 25°C is very close to $K_W = 10^{-14}$. Compute the hydrogen and hydroxyl concentrations in one liter of water at 25°C.

Solution. For the dissociation of water,

$$H_2O = H^+ + OH^- \qquad K_W = 10^{-14}$$

Since water is the only component, the reaction requires that equal numbers of H^+ and OH^- ions be formed; $[H^+] = [OH^-]$. The concentration of undissociated H_2O is negligibly different from 1. Assuming K_f for this reaction is 1, the dissociation constant can be written as

$$K_W = 10^{-14} = \frac{[H^+][OH^-]}{[H_2O]} = \frac{[H^+][H^+]}{1} = [H^+]^2$$

so that

$$[H^+] = 10^{-7} \left(\frac{\text{moles}}{\text{liter}}\right)$$

Thus in pure water the activities of hydrogen and hydroxyl ions are equal to their concentrations, which are each about 10^{-7} moles per liter.

One of the most important characteristics of aqueous solutions is the hydrogen ion concentration contained. Since in ordinary solutions, values of this quantity are found to vary over a range of several orders of magnitude, a convenient measure of this quantity has been defined as the *pH* of an electrolyte solution[3] as

$$\text{pH} \equiv -\log_{10}(a_{H^+}) = -\log_{10}[H^+] = -\log[H^+] \qquad (15.20)$$

The value of the pH for pure water is $-\log(10^{-7}) = -(-7) = +7$. The hydrogen ion activity is higher than 7 in acids, which contribute hydrogen ions to the solution, and lower than 7 in basic solutions, which contribute hydroxyl ions. (Since H_2O is the solvent, $a_{H_2O} = 1$; thus $a_{H^+}a_{OH^-} = K_W = 10^{-14}$. A high activity of OH^- implies a low H^+ activity and vice versa.) The scale of pH values normally obtained in the laboratory is illustrated in Fig. 15.2. The pH ranges from 2 for strong acids to 12 for

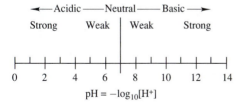

FIGURE 15.2
The scale of pH values in electrolyte solutions.

---

[3]Since the chemical potential of hydrogen is proportional to the logarithm of its activity at a fixed temperature, pH represents the *potential of hydrogen* in the electrolyte. Since the activity of H^+ is normally a small fraction, its logarithm is negative and inclusion of the minus sign in the definition makes the pH a positive number for the dilute solutions of H^+ ions ordinarily encountered.

strong base solutions. Since in most applications the concentrations of H^+ and OH^- are in the dilute range, the activities can be replaced by the molar concentrations.

Example 15.2. Compute the pH of a solution containing 0.05 molar concentration of acetic acid, HAc (where Ac is shorthand for CH_3COO^-). The dissociation constant for acetic acid at $25°C$ is 1.8×10^{-5}.

Solution. The dissociation reaction for acetic acid is

$$HAc = H^+ + Ac^- \qquad [15.3]$$

In writing the relation between the dissociation constant and the degree of dissociation, Eq. (15.16), set $u = v = 1$ and

$$K_{HAc} = (1^1 \cdot 1^1) \frac{\alpha^2}{1 - \alpha} c_{HAc}$$

Substitute for K_C and c_C from the information given:

$$1.8 \times 10^{-5} = \frac{\alpha^2}{1 - \alpha} \cdot (0.05)$$

Solve for the degree of dissociation:

$$\alpha = 0.019$$

The hydrogen ion concentration is

$$[H^+] = \alpha[HAc] = 0.019(0.05) = 9.4 \times 10^{-4}$$

Set this concentration equal to the activity in the definition of pH, Eq. (15.20):

$$pH = -\log(9.4 \times 10^{-4}) = 3.0$$

In a multicomponent electrolyte containing a variety of cations and anions, the conditions for equilibrium can be written

$$\Delta G_W^o = -RT \ln K_W \qquad (15.21)$$

$$\Delta G_r^o = -RT \ln K_r \qquad (r = 1, 2, \dots) \qquad (15.22)$$

where Eq. (15.21) applies to the solvent and Eqs. (15.22) to all the other components. Each of the latter equilibrium constants can be written

$$K_r = K_{f,r} \cdot K_{c,r} = K_{f,r} \cdot \frac{[+]_r^u [-]_r^v}{[U]_r} \qquad (r = 1, 2, \dots) \qquad (15.23)$$

where $K_{f,r}$ is the proper quotient of activity coefficients and $K_{c,r}$ is the corresponding ratio of the concentrations. Values for the dissociation constants are tabulated for many electrolyte compounds including salts, acids, and bases. Examples of some common compounds are given in Table 15.2.

The problem of determining the equilibrium concentrations of all the ions in the system is mathematically identical to that treated in Sec. 11.1.2 for chemical reactions in a multicomponent ideal gas mixture. Indeed, provided the dissociation constants are known and the $K_{f,r}$ terms are taken to be unity, the computer program

TABLE 15.2
Dissociation constants for common inorganic acids and bases

Compound	Dissociating into	K
NH_4OH	$NH_4^+ + OH^-$	1.79×10^{-5}
$Ca(OH)_2$	$Ca(OH)^+ + OH^-$	3.74×10^{-3}
$CaOH^+$	$Ca^{++} + OH^-$	4.0×10^{-2}
$AgOH$	$Ag^+ + OH^-$	1.1×10^{-4}
H_2CO_3	$H^+ + HCO_3^-$	4.3×10^{-7}
HCO_3^-	$H^+ + CO_3^-$	5.61×10^{-11}
HI	$H^+ + I^-$	1.69×10^{-1}
CH_2O_2	$H^+ + CHO_2^-$	1.77×10^{-4}

Source: *Handbook of Chemistry and Physics*, 57th Edition, R. C. Weast, Editor, CRC Press, Cleveland, OH (1976) p. D-149 and D-151.

SOLGASMIX, which provides solutions for a multivariate reacting mixture of gases, can be used to compute the concentrations of ions in an equilibrated multicomponent electrolyte solution.

Example 15.3. A quantity of 1.5 grams of phosphoric acid (H_3PO_4) is added to a beaker and water is added to fill the beaker to the 200 ml level. Compute the concentrations of all the components of this solution at equilibrium.

Solution. Like most acids that contain more than one hydrogen atom, phosphoric acid dissociates in stages, one hydrogen atom at a time. Thus the components existing in this system can be considered as: H^+, OH^-, H_2O, H_3PO_4, $H_2PO_3^-$, HPO_4^{2-}, and PO_4^{3-}. Conservation equations for the elements H, O, and P yield three equations among their concentrations. The four independent reaction equations required to provide seven relations among the concentrations of the seven components are presented with their dissociation constants.

$$H_2O = H^+ + OH^- \qquad K_W = 1 \times 10^{-14} = \frac{[H^+][OH^-]}{[H_2O]}$$

$$H_3PO_4 = H^+ + H_2PO_4^- \qquad K_{a1} = 7.1 \times 10^{-3} = \frac{[H^+][H_2PO_4^-]}{[H_3PO_4]}$$

$$H_2PO_4^- = H^+ + HPO_4^{2-} \qquad K_{a2} = 6.3 \times 10^{-8} = \frac{[H^+][HPO_4^{2-}]}{H_2PO_4^-]}$$

$$HPO_4^{2-} = H^+ + PO_4^{3-} \qquad K_{a3} = 4.7 \times 10^{-13} = \frac{[H^+][PO_4^{3-}]}{[HPO_4^{2-}]}$$

The atomic weights of the elements involved are: H = 1; O = 16; and P = 31 gm. The molecular weight of H_3PO_4 is thus $3(1)+31+4(16) = 98$ gm. The 1.5 grams of H_3PO_4 corresponds to 1.5/98 = 0.015 moles. The 0.015 moles dissolved to form 200 ml of solution corresponds to a concentration of 0.015 moles/0.2 liters = 0.077 moles per liter. Since this solution is very dilute, the 200 ml of solution corresponds closely to 200 gm of water. (Recall that the specific gravity of water is 1 gm/cc.) The molecular weight of H_2O is $2(1) + 16 = 18$ gm/mole. The number of moles of water in the solution is 200 gm/18 gm/mole = 11.11 moles. The number of gram atoms of each of the elements in the system can be computed from the initial compositions:

$$m_H = 2n_{H_2O} + 3n_{H_3PO_4} = 2(11.11) + 3(0.015) = 22.27 \text{ gm atoms of H}$$

$$m_O = n_{H_3PO_4} = 11.11 + 4(0.015) = 11.17 \text{ gm atoms of O}$$

$$m_P = n_{H_3PO_4} = 0.015 \text{ gm atoms of P}$$

The corresponding molar concentrations of each of the elements can be obtained by dividing by the volume of the solution. The conservation of the elements can be expressed in terms of molar concentrations:

$$\frac{m_H}{V} = \frac{22.27 \text{ gm-atoms}}{0.2 \text{ liters}} = 111.4 \left(\frac{\text{gm-atoms}}{\text{liter}} \right)$$
$$= 2[H_2O] + [H^+] + [OH^-] + 3[H_3PO_4] + 2[H_2PO_4^-] + [HPO_4^{2-}]$$

$$\frac{m_O}{V} = \frac{11.17 \text{ gm-atoms}}{0.2 \text{ liters}} = 55.85 \left(\frac{\text{gm-atoms}}{\text{liter}} \right)$$
$$= [H_2O] + [OH^-] + 4([H_3PO_4] + [H_2PO_4^-] + [HPO_4^{2-}] + [PO_4^{3-}])$$

$$\frac{m_P}{V} = \frac{0.015 \text{ gm-atoms}}{0.2 \text{ liters}} = 0.075 \left(\frac{\text{gm-atoms}}{\text{liter}} \right)$$
$$= [H_3PO_4] + [H_2PO_4^-] + [HPO_4^{2-}] + [PO_4^{3-}]$$

These three conservation equations combine with the four relations derived from the dissociations constants to provide seven equations in seven unknowns. The equilibrium composition is the solution to these equations, obtained by using a solver program in a mathematics applications software package.

15.1.2   Equilibrium in a Strong Electrolyte

By definition, a *strong electrolyte* is a compound that when added to a solvent, dissociates completely into its ionic components. Thus, the number of moles of undissociated molecules n_U in the solution is zero and the degree of dissociation α is equal to one for a strong electrolyte. Some compounds are essentially completely dissociated (are strong electrolytes) in dilute solutions but become weak electrolytes with increasing concentration. Indeed, this behavior is true of essentially all electrolytes at sufficiently high concentrations.

The thermodynamic behavior of weak electrolytes is based upon the condition for equilibrium expressed in terms of dissociation constants, Eq. (15.12) or (15.16). These equations can be used to compute the degree of dissociation and then the pH and other properties of the system. For strong electrolytes it is not necessary to invoke this calculation since, by definition, the degree of dissociation α is equal to 1. Thus, if a strong acid such as HCl is dissolved in a known quantity of water, the concentration of H^+ ions is evident by inspection, being equal to the concentration of HCl added. *Every* HCl molecule dissociates and contributes one hydrogen ion.

Example 15.4. Compute the pH of a 0.08 molar solution of HCl in water.

Solution. The number of moles of HCl added per liter of solution is 0.08. Since every HCl molecule dissociates forming one hydrogen ion, the concentration of H^+ resulting is 0.08 moles/liter. The corresponding pH of this solution is

$$pH = -\log(0.08) = 1.09$$

Example 15.5. Evaluate the pH of a solution formed by mixing equal quantities of 0.04 molar HCl with 0.06 molar NaOH, both of which are strong electrolytes.

Solution. The $[H^+]$ contributed from HCl is 0.04 moles/liter of HCl added; the sodium hydroxide contributes 0.06 moles of $[OH^-]$ per liter of NaOH added. Each of these is diluted by a factor of two when the two solutions are mixed so that the concentrations contributed are respectively 0.02 and 0.03 moles per liter. Essentially all 0.02 moles/liter of the hydrogen ions contributed reacts with 0.02 moles/liter of hydroxyl ions to form 0.02 moles/liter of water. The unreacted concentration of OH^- ions $(0.03 - 0.02 = 0.01$ moles/liter$)$ remain in solution. The corresponding hydrogen ion concentration can be determined from the dissociation constant for water:

$$K_W = 10^{-14} = [H^+][OH^-] = [H^+][0.01]$$

Solve for

$$[H^+] = 10^{-14}(0.01) = 10^{-12}$$

The corresponding pH is

$$pH = -\log(10^{-12}) = 12$$

which represents a strongly basic solution.

Manipulation of compositions of solutions involving strong electrolytes is evidently significantly simplified in comparison with the analysis of weak electrolytes.

15.2 EQUILIBRIUM IN TWO PHASE SYSTEMS INVOLVING AN ELECTROLYTE

The general strategy for finding conditions for equilibrium can be applied to a system consisting of two phases capable of exchanging electrical charge. A simple system of this class is shown in Fig. 15.3. A copper rod is immersed in an aqueous solution of copper chloride, $CuCl_2$. The copper rod is called an *electrode* and the solution an *electrolyte*.

If the concentration of $CuCl_2$ in the electrolyte is dilute, copper atoms dissolve from the rod and form cupric ions in the solution. Copper atoms are said to be *oxidized* to cupric ions. For each copper atom dissolved, two electrons are released and remain in the metal rod. The rod becomes negatively charged as electrons accumulate while the solution develops a positive charge. Both of these accumulations alter the total energy of the system by changing the electrostatic potential energy of the phases involved.

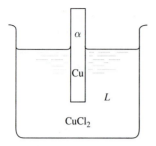

FIGURE 15.3
A copper rod (the α phase) is immersed in a solution of copper chloride in water (the ϵ phase).

On the other hand, if the electrolyte is sufficiently concentrated, cupric ions plate out on the copper rod, each ion capturing two electrons from the metal. Cupric ions in the solution are *reduced* to copper atoms on the metal surface. The number of free electrons in the copper rod is reduced and the rod develops a positive charge. In the solution the number of cupric ions is reduced relative to the negatively charged chloride ions and the solution develops a negative charge. Both effects alter the electrical potential energy of the two phase system.

These exchanges of atoms and charges can be succinctly represented by the reaction

$$Cu^\alpha = Cu^{2+L} + 2e^{-\alpha} \qquad [15.4]$$

If this reaction proceeds in the direction as written it is an *oxidation* reaction; the reverse is a *reduction* reaction.

Eventually, after sufficient charge is transferred, an equilibrium condition is attained in each phase. The general strategy for finding conditions for equilibrium can be applied to determine the relations that govern this equilibrium state.

Figure 15.4 shows a rod of the metal M (the α phase) immersed in an aqueous solution (the ϵ phase) containing the electrolyte $M_u X_v$. Let z^+ be the valence of the cations (M^{z+}) and z^- that of the anions (X^{z-}) in the electrolyte. The components in the α phase are atoms of the metal M and electrons e. Components in the electrolyte are M^{z+}, X^{z-}, $M_u X_v$, H^+, OH^- and H_2O. Let ϕ^α be the electrical potential of the metallic phase (defined as in Chap. 14 as the potential energy per unit charge) and

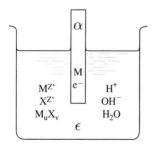

FIGURE 15.4
A metal rod *M* (the α phase) is immersed in a solution of the electrolyte $M_u X_v$ in water (the ϵ phase).

ϕ^ϵ that of the electrolyte solution.[4] The exchange of charged particles between the α and ϵ phases contributes to the change in the total energy of the system and must be incorporated into its thermodynamic description.

Equilibria established within the electrolyte are determined by application of the conditions for equilibrium within an electrolyte derived in Sec. 15.2. The dissociation constants of the electrolyte and of water establish the relationships between the electrolyte components. The following development focuses on the equilibrium established between the electrode and the electrolyte. The components involved in this exchange are the metal M and the electrons e$^-$ in the electrode and the metal ions M^{z+} in the electrolyte.

The change in entropy of the metallic phase α that constitutes the electrode is

$$dS'^{\alpha}_{\text{sys}} = \frac{1}{T^\alpha}du'^{\alpha} + \frac{P^\alpha}{T^\alpha}dV'^{\alpha} - \frac{1}{T^\alpha}[\mu^\alpha_M dn^\alpha_M + \mu^\alpha_e dn^\alpha_e] \tag{15.24}$$

In the electrolyte phase

$$dS'^{\epsilon}_{\text{sys}} = \frac{1}{T^\epsilon}dU'^{\epsilon} + \frac{P^\epsilon}{T^\epsilon}dV'^{\epsilon} - \frac{1}{T^\epsilon}\mu^\epsilon_{M^{z+}}dn^\epsilon_{M^{z+}} \tag{15.25}$$

The change in entropy of the system is the sum of these two expressions,

$$dS'_{\text{sys}} = dS'^{\alpha} + dS'^{\epsilon}$$

The isolation constraints must be carefully and completely formulated for this case. Conservation of the element M can be simply stated as

$$dm_M = 0 = dn^\alpha_M + dn^\alpha_{M^{z+}} \rightarrow dn^\alpha_{M^{z+}} = -dn^\alpha_M \tag{15.26}$$

Each electron that is added to the α phase adds a charge of $(-1)e$, where e is the magnitude of the charge on an electron. If dn^α_e moles of electrons are added to the α phase, the change in the charge contained in that phase is

$$dq^\alpha = (-1)\mathcal{F}dn^\alpha_e = -\mathcal{F}dn^\alpha_e \tag{15.27}$$

where $\mathcal{F} = N_o e$ is Faraday's constant (see Sec. 14.3.5). Each metal ion M^{z+} carries a charge of z^+. If $dn^\epsilon_{M^{z+}}$ moles of these ions are added to the electrolyte, the associated change in charge on the ϵ phase is

$$dq^\epsilon = (+z^+)\mathcal{F}dn^\epsilon_{M^{z+}} \tag{15.28}$$

The total charge accumulated in the system must be zero:

$$dq_{\text{tot}} = 0 = dq^\alpha + dq^\epsilon = -\mathcal{F}dn^\alpha_e + (z^+)\mathcal{F}dn^\epsilon_{M^{z+}} \tag{15.29}$$

So that

$$dn^\alpha_e = (z^+)dn^\epsilon_{M^{z+}} \tag{15.30}$$

---

[4] ϕ^α can be thought of as the change in potential energy when unit charge is transferred from infinity to the α phase. ϕ^ϵ can be similarly defined for the ϵ phase. It is not practical to determine the absolute values of these potentials and only differences in potentials appear in the working equations.

From Eq. (15.26),

$$dn_e^\alpha = z^+(-dn_M^\alpha) = -z^+ dn_M^\alpha \tag{15.31}$$

Equations (15.26) and (15.31) permit expression of the three compositional variables in Eqs. (15.24) and (15.25) in terms of dn_M^α.

Additional isolation constraints derive from the condition of constant volume in an isolated system,

$$dV'_{sys} = 0 = dV'^\alpha + dV'^\epsilon \rightarrow dV'^\epsilon = -dV'^\alpha \tag{15.32}$$

and the requirement that the total energy of an isolated system cannot change. In formulating the total energy constraint it is necessary to include changes in the electrical energies of the phases due to the exchange of charged particles.

$$dE'_{tot} = dU'^\alpha + dU'^\epsilon + \phi^\alpha dq^\alpha + \phi^\epsilon dq^\epsilon \tag{15.33}$$

Equations (15.27) and (15.28) can be used to evaluate dq^α and dq^ϵ:

$$dE'_{tot} = dU'^\alpha + dU'^\epsilon + \phi^\alpha(-\mathcal{F}dn_e^\alpha) + \phi^\epsilon(z^+\mathcal{F}dn_{M^{z+}}^\epsilon)$$

Both $dn_e{}^\alpha$ and $dn_{M+}{}^\epsilon$ can be expressed in terms of $dn_M{}^\alpha$, as in Eqs. (15.26) and (15.31):

$$dE'_{tot} = dU'^\alpha + dU'^\epsilon + z^+\mathcal{F}(\phi^\alpha - \phi^\alpha)dn_M^\alpha)$$

In an isolated system, $dE'_{tot} = 0$; this isolation constraint therefore requires that

$$dU'^\epsilon = -\left[dU'^\alpha + z^+\mathcal{F}(\phi^\alpha - \phi^\epsilon)dn_M^\alpha)\right] \tag{15.34}$$

The change in the total entropy of the system obtained by adding expressions for dS' in Eqs. (15.24) and (15.25) is expressed in terms of seven variables. The isolation constraints provide four relations among these seven variables. Use Eqs. (15.26) and (15.31) to replace $dn_e{}^\alpha$ and $dn_{M_{z+}}{}^\epsilon$ with $dn_M{}^\alpha$. Use Eq. (15.32) to eliminate dV'^ϵ and Eq. (15.34) to substitute for dU'^ϵ.

$$dS'_{sys,iso} = \frac{1}{T^\alpha}dU'^\alpha + \frac{P^\alpha}{T^\alpha}dV'^\alpha - \frac{1}{T^\alpha}\left[\mu_M^\alpha dn_M^\alpha + \mu_e^\alpha(-z^+ dn_M^\alpha)\right]$$

$$+ \frac{1}{T^\epsilon}(-1)\left[dU'^\alpha + z^+\mathcal{F}(\phi^\alpha - \phi^\epsilon)dn_M^\alpha\right]$$

$$\frac{P^\epsilon}{T^\epsilon}(-dV^\alpha) - \frac{1}{T^\epsilon}\mu_{M^{z+}}^\epsilon(-dn_M^\alpha)$$

Grouping like terms

$$dS'_{sys,iso} = \left(\frac{1}{T^\alpha} - \frac{1}{T^\epsilon}\right)dU'^\alpha + \left(\frac{P^\alpha}{T^\alpha} - \frac{P^\epsilon}{T^\epsilon}\right)dV'^\alpha$$

$$- \frac{1}{T^\alpha}\left[\mu_M^\alpha - z^+\mu_e^\alpha - \mu_{M^{z+}}^\epsilon + z^+\mathcal{F}(\phi^\alpha - \phi^\epsilon)\right]dn_M^\alpha \tag{15.35}$$

At equilibrium, all the coefficients in this equation are zero. This condition for the extremum in entropy in an isolated system yields the following *conditions for equilibrium*:

$$\frac{1}{T^\alpha} - \frac{1}{T^\epsilon} = 0 \rightarrow T^\alpha = T^\epsilon \quad \text{[Thermal Equilibrium]} \quad (15.36)$$

$$\frac{P^\alpha}{T^\alpha} - \frac{P^\epsilon}{P^\epsilon} = 0 \rightarrow P^\alpha = P^\epsilon \quad \text{[Mechanical Equilibrium]} \quad (15.37)$$

$$\mu_M^\alpha - \left(z^+ \mu_e^\alpha + \mu_{M^{z+}}^\epsilon\right) + z^+ \mathcal{F}(\phi^\alpha - \phi^\epsilon) = 0 \quad \text{[Electrochemical Equilibrium]} \quad (15.38)$$

These relations are familiar except Eq. (15.38), which is the condition for *electrochemical* equilibrium in the system. Rewrite this expression as

$$\mu_M^\alpha - \left(z^+ \mu_e^\alpha + \mu_{M^{z+}}^\epsilon\right) = -z^+ \mathcal{F}(\phi^\alpha - \phi^\epsilon) \quad (15.39)$$

The left side of this equation can be recognized as the affinity for the reduction reaction

$$(M^{z+})^\epsilon + z^+ e^- = M^\alpha \quad [15.5]$$

Evidently equilibrium is reached when the affinity for this reaction equals the change in potential energy of the system associated with transferring z^+ gram atoms of positive charge from the metal electrode to the electrolyte.

Unfortunately, the potential difference between two phases in equilibrium cannot be measured. A probe inserted into the electrolyte and connected through external instrumentation to the electrode to measure this potential would itself constitute an electrode-electrolyte system with its own equilibrium conditions. Such a system with two electrodes immersed in an electrolyte is called an *electrochemical cell*. It is the simplest configuration that can be assembled to explore the relationship between electrical and chemical effects in such systems.

15.3 EQUILIBRIUM IN AN ELECTROCHEMICAL CELL

Figure 15.5 shows an electrochemical cell consisting of two dissimilar electrodes, α and β, each immersed in an appropriate electrolyte solution ϵ and η respectively. To make the illustration concrete, suppose the α electrode is copper and immersed in a $CuCl_2$ solution, and the β electrode is zinc immersed in a $ZnCl_2$ solution. Each electrode forms divalent ions in its electrolyte solution. If each electrode reaches equilibrium with its electrolyte separately, as in Fig. 15.5a, the condition for equilibrium in each case is given by the appropriate adaptation of Eq. (15.39). For the copper electrode (α) the reaction is

$$(Cu^{2+})^\epsilon + 2(e^-)^\epsilon = Cu^\alpha \quad [15.6]$$

The condition for equilibrium for this electrode is:

$$\mu_{Cu}^\alpha - \left(2\mu_e^\alpha + \mu_{Cu^{2+}}^\epsilon\right) = -2\mathcal{F}(\phi^\alpha - \phi^\epsilon) \quad (15.40)$$

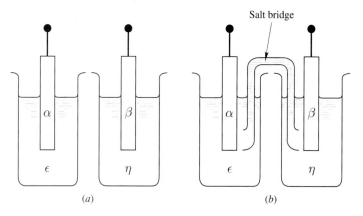

FIGURE 15.5

(a) Two separate single electrodes systems achieve their own internal equilibrium. (b) They are then connected in the electrolyte with a salt bridge. The electrical potential difference between the electrodes can now be measured.

At the zinc electrode (β) the reaction is

$$(Zn^{2+})^\eta + 2(e^-)^\beta = Zn^\beta \qquad [15.7]$$

and the condition for equilibrium gives

$$\mu_{Zn}^\beta - \left(2\mu_e^\beta + \mu_{Zn^{2+}}^\eta\right) = -2\mathcal{F}(\phi^\beta - \phi^\eta) \qquad (15.41)$$

Let the two electrolyte baths be connected by a *salt bridge*, Fig. 15.5b. The salt bridge is a tube filled with an electrolyte that permits transfer of charge between the electrolytes without disturbing the equilibrium between Zn and its electrolyte and Cu and its electrolyte. As a consequence, within the continuous electrolyte phase, ϵ the potential is uniform with the value ϕ^ϵ. The chemical potentials of the electrons in the metallic electrodes can also be taken to be the same. Subtraction of Eq. (15.40) from (15.41) yields an expression for the potential difference between the electrodes.

$$\left[\mu_{Zn}^\beta - \left(2\mu_e^\beta + \mu_{Zn^{2+}}^\eta\right)\right] - \left[\mu_{Cu}^\alpha - \left(2\mu_e^\alpha + \mu_{Cu^{2+}}^\epsilon\right)\right] = -2\mathcal{F}(\phi^\beta - \phi^\alpha) \qquad (15.42)$$

The first term in brackets on the left side of this equation corresponds to the affinity $\mathcal{A}_{[15.7]}$ for the electrode reaction [15.7] for the zinc (β) electrode, and the second term is the affinity $\mathcal{A}_{[15.6]}$ for the copper (α) reaction [15.6]. Equation (15.42) demonstrates that when a galvanic cell reaches equilibrium, the external potential, which can be measured experimentally with a potentiometer or sensitive voltmeter, is determined by the difference in affinities for the two electrode reactions

$$\mathcal{A}^\beta - \mathcal{A}^\alpha = -2\mathcal{F}(\phi^\beta - \phi^\alpha) \qquad (15.43)$$

Since $\mu_e^\alpha = \mu_e^\beta$, the left side of Eq. (15.43) can also be written

$$\left[\mu_{Zn}^\beta - \mu_{Zn^{2+}}^\eta\right] - \left[\mu_{Cu}^\alpha - \mu_{Cu^{2+}}^\epsilon\right] = 2\mathcal{F}(\phi^\beta - \phi^\alpha) \qquad (15.44)$$

The left side now corresponds to the affinity for the overall cell reaction

$$Zn^\beta + (Cu^{2+})^\epsilon = Cu^\alpha + (Zn^{2+})^\eta \qquad [15.8]$$

15.3.1 Conditions for Equilibrium in a General Galvanic Cell

Equation (15.44) can be written for a general galvanic cell. To avoid confusion with respect to the meaning of the sign of the cell potential, it is useful to establish the following conventions for the representation of the cell:

1. The affinities and the electrode reactions are written for *reduction* reactions, that is, so that electrons are consumed as reactants.
2. In visualizing the cell geometry, the emf reported for the cell is the potential for the right electrode minus that for the left.[5]

This convention (right minus left) can be conveniently remembered by recognizing that it is the same as the convention for interpreting chemical reactions: (products minus reactants) is (right side of reaction minus left side) by convention.

To help visualize the cell under consideration, it is useful to adopt a simple notation for describing its configuration. The generic cell representation is:

Left Electrode|Electrolyte 1||Electrolyte 2|Right Electrode

The single vertical line represents the interface between electrode and electrolyte on each side. The double vertical line represents an interface between the two electrolytes. The cell discussed in the last section can be written

$$Zn^\alpha |ZnCl_2||CuCl_2|Cu^\beta$$

More explicit information can be supplied in representing the system

$$Zn^\alpha(pure)|Zn^{++}(a_{Zn++} = 0.005)||Cu^{++}(a_{Cu++} = 0.002)||Cu(pure)$$

The electrode reactions written so far in this section are examples of oxidation reactions in which electrons are generated. The reverse of these reactions are *reduction reactions*. All subsequent reactions are written as reduction reactions according to the convention just adopted. For a set of examples of reduction reactions see Table 15.3.

Consider the cell shown in Fig. 15.6. The α electrode is visualized as on the left and β on the right. By the convention chosen, the emf of this cell would be reported as $\phi^\beta - \phi^\alpha$. Suppose the α phase is composed of metal M_1 which forms ions with

---

[5]In a significant subset of the literature the opposite convention is adopted, and the emf is written as that of the left electrode minus that of the right. As a consequence, for the same cell the reported emf would be the *negative* of that obtained for the convention adopted here.

TABLE 15.3
Selected values from the electromotive series

Reduction reaction	emf(volts)
$Ca^+ + e^- = Ca$	−3.80
$Na^+ + e^- = Na$	−2.71
$Al^{+++} + 3e^- = Al$	−1.662
$Fe^{++} + 2e^- = Fe$	−0.447
$Ni^{++} + 2e^- = Ni$	−0.257
$Fe^{+++} + 3e^- = Fe$	−0.037
$2H^+ + 2e^- = H_2$	0.00000
$Cu^{++} + e^- = Cu^+$	0.153
$Cu^+ + e^- = Cu$	0.521
$Fe^{+++} + e^- = Fe^{++}$	0.771
$O_2 + 4H^+ + 4e^- = 2H_2O$	1.229
$Au^+ + e^- = Au$	1.692
$Ag^{++} + e^- = Ag^+$	1.980

Source: *CRC Handbook of Chemistry and Physics, 71st Edition*, D. R. Lide, Editor, CRC Press, Boca Raton, FL, (1991).

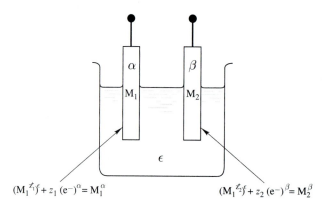

FIGURE 15.6
A general cell with two electrodes; the phase α on the left is the metal M_1; and the β phase on the right is M_2. The electrode reactions, written as reduction reactions, are indicated.

charge z_1 in the electrolyte solution, and electrode β is the metal M_2 with ions of charge z_2. The electrode reactions are:

$$\left(M_1^{z_1+}\right)^\epsilon + z_1(e^-)^\alpha = M_1^\alpha \qquad [15.9]$$

and

$$\left(M_2^{z_2+}\right)^\epsilon + z_2(e^-)^\beta = M_2^\beta \qquad [15.10]$$

Apply the condition for equilibrium, Eq. (15.39), to each of these electrode-electrolyte systems:

$$\mathscr{A}^\alpha = \mu^\alpha_{M_1} - \left(z_1\mu^\alpha_e + \mu^\epsilon_{M_1^{z_1+}}\right) = -z_1\mathcal{F}(\phi^\alpha - \phi^\epsilon) \qquad (15.45)$$

$$\mathscr{A}^\beta = \mu^\beta_{M_2} - \left(z_2\mu^\beta_e + \mu^\epsilon_{M_2^{z_2+}}\right) = -z_2\mathcal{F}(\phi^\beta - \phi^\epsilon) \qquad (15.46)$$

Multiply Eq. (15.45) by (z_2/z_1) and subtract the result from Eq. (15.46) to obtain

$$\left[\mu^\beta_{M_2} - \left(z_2\mu^\beta_e + \mu^\epsilon_{M_2^{z_2+}}\right)\right] - \frac{z_2}{z_1}\left[\mu^\alpha_{M_1} - \left(z_1\mu^\alpha_e + \mu^\epsilon_{M_2^{z_1+}}\right)\right] = -z_2\mathcal{F}(\phi^\beta - \phi^\alpha) \quad (15.47)$$

The left-hand side is again seen to correspond to the difference in the affinities for the two electrode reactions, balanced so that each transfers the same number of electrons.

$$\mathscr{A}^\beta - \frac{z_2}{z_1}\mathscr{A}^\alpha = -z_2\mathcal{F}(\phi^\beta - \phi^\alpha) \qquad (15.48)$$

Equation (15.48) demonstrates that in a galvanic cell at equilibrium the difference in affinity for the two electrode reactions is proportional to the externally measured difference in potential between the electrodes.

Because the chemical potentials of the electrons in the electrodes are equal, Eq. (15.47) can also be written

$$\mathscr{A}_{cell} = \left[\mu^\beta_{M_2} + \frac{z_2}{z_1}\mu^\epsilon_{M_1^{z_1+}}\right] - \left[\mu^\epsilon_{M_2^{z_2+}} + \frac{z_2}{z_1}\mu^\alpha_{M_1}\right] = -z_2\mathcal{F}(\phi^\beta - \phi^\alpha) \qquad (15.49)$$

The left side of this equation corresponds to the affinity for the overall cell reaction

$$\left(M_2^{z_2+}\right)^\epsilon + \frac{z_2}{z_1}M_1^\alpha = M_2^\beta + \frac{z_2}{z_1}\left(M_1^{z_1+}\right)^\epsilon \qquad [15.11]$$

To illustrate this general result, suppose α is zinc, for which z_1 is 2, and β is aluminum, for which z_2 is 3. This result then reads

$$\left[\mu^\beta_{Al} + \frac{3}{2}\mu^\epsilon_{Zn^{2+}}\right] - \left[\mu^\epsilon_{Al^{3+}} + \frac{3}{2}\mu^\alpha_{Zn}\right] = -3\mathcal{F}(\phi^\beta - \phi^\alpha) \qquad (15.50)$$

The left side is the affinity for the cell reaction

$$(Al^{3+})^\epsilon + \tfrac{3}{2}Zn^\alpha = Al^\beta + \tfrac{3}{2}(Zn^{2+})^\epsilon \qquad [15.12]$$

Recall from Sec. 11.1, Eq. (11.32), that the affinity for a reaction can be written in terms of the standard free energy change for the reaction and the proper quotient of activities:

$$\mathscr{A} = \Delta G^\circ + RT \ln Q \qquad (11.32)$$

Substituting this expression into Eq. (15.49),

$$\mathscr{A}_{cell} = \Delta G^\circ_{cell} + RT \ln Q_{cell} = -z\mathcal{F}\,\mathcal{E}_{cell} \qquad (15.51)$$

where $\mathcal{E}_{cell}$ is the potential difference between the electrodes, $\mathcal{E}_{cell} = (\phi^\beta - \phi^\alpha)$, reported, in accordance with the convention adopted, as the potential of the right electrode minus the left. Note that the order of the potentials in this definition $(\beta - \alpha)$

corresponds to the order of the electrode reactions, written as reduction reactions, implied by the affinity on the right hand side of Eq. (15.51).

If a cell is constructed so that the components in the electrode reactions are in their standard states, then for that cell, $Q = 1$ and $\ln Q = 0$. As an example, for the $Cu|CuCl_2||ZnCl_2|Zn$ cell considered previously, the electrodes would be pure copper and pure zinc and the two electrolytes would be prepared with a molality that yields an activity of 1. Equation (15.51) becomes

$$\Delta G^{\circ}_{cell} = -z\mathcal{F}\,\mathcal{E}^{\circ}_{cell} \tag{15.52}$$

The measured emf $\mathcal{E}$ is the *standard electrode potential* for the cell reaction. Substitution of this result into Eq. (15.51) yields, after minor rearrangement,

$$\mathcal{E} = \mathcal{E}^{\circ} - \frac{RT}{z\mathcal{F}}\ln Q \tag{15.53}$$

This relation is known as the *Nernst equation*. It is the working equation of electrochemistry, most often applied in the solution of practical problems.

Most applications of electrochemistry to aqueous solutions take place near room temperature, taken to be 25°C. It is also convenient in most applications to replace the natural logarithm in this expression with base ten logarithms where $\ln(x) = 2.303\log_{10}(x)$. With these constraints, the Nernst equation can be written

$$\mathcal{E} = \mathcal{E}^{\circ} - \frac{8{,}314\left(\frac{\text{J}}{\text{mol-K}}\right)(2.98.17\text{ K})(2.303)}{z \times 96{,}512\left(\frac{\text{J}}{\text{volt-mol}}\right)}\log Q$$

$$\mathcal{E} = \mathcal{E}^{\circ} - \frac{0.05915}{z}\log Q \tag{15.54}$$

where the units of $\mathcal{E}$ is volts. While this is the most widely used form of the Nernst equation, it is emphasized that the relationship in Eq. (15.53) is general and is also frequently applied to cells at elevated temperatures in which the electrolyte is a molten salt or even a solid ionic conductor.

15.3.2   Temperature Dependence of the Electromotive Force of a Cell

Electrochemical cells designed to study a given chemical reaction can be used to measure the standard entropy change for the reaction and the heat of reaction. This can be achieved by constructing a cell with electrodes and their solutions in their standard states and measuring their reversible emf at a series of temperatures. The computation of these properties is then straightforward.

Recall that the change in entropy for a reaction is the negative of the temperature derivative of its Gibbs free energy change. Apply this coefficient relation to Eq. (15.52):

$$\Delta S^{\circ} = -\left(\frac{\partial \Delta G^{\circ}}{\partial T}\right)_{P,n_k} = +z\mathcal{F}\left(\frac{\partial \mathcal{E}^{\circ}}{\partial T}\right)_{P,n_k} \tag{15.55}$$

The enthalpy change for the reaction can be computed from the definitional relationship as

$$\Delta H^\circ = \Delta G^\circ + T\,\Delta S^\circ = z\mathcal{F}\left(-\mathcal{E}^\circ + T\left(\frac{\partial \mathcal{E}^\circ}{\partial T}\right)_{P,n_k}\right) \tag{15.56}$$

If experiments are carried out carefully, the resulting values for ΔS° and ΔH° compare well with those obtained calorimetrically. The electrochemical cell is evidently an extremely useful tool in the experimental evaluation of thermodynamic properties.

15.3.3 The Standard Hydrogen Electrode

It would be very convenient if it were possible to measure and tabulate the electrical potentials of single electrodes, such as the left electrode in Fig. 15.5a. The equilibrium electrical potential of any galvanic cell could then be computed as the difference between two "single electrode potentials." A simple strategy has been devised that essentially achieves this level of convenience without the necessity of measuring true "single electrode potentials." Experimentalists have agreed to report equilibrium electromotive force (emf) measurements for cells in which one electrode is chosen because it bears upon the practical problem at hand and the other electrode is *always the same* half cell configuration. If all potential differences are recorded *relative to this standard electrode* as $(\phi^\alpha - \phi^{STD})$, $(\phi^\beta - \phi^{STD})$, etc., then potential differences for cells with electrodes α and β can be determined by subtraction as

$$\mathcal{E}^{\alpha\beta} = \phi^\beta - \phi^\alpha = (\phi^\beta - \phi^{STD}) - (\phi^\alpha - \phi^{STD}) \tag{15.57}$$

In practice it is not even necessary to make all measurements on cells that contain the standard electrode. It is only necessary to measure the emf of system in which one of the electrodes has previously been calibrated against the standard electrode. Suppose the emf of a cell has been measured with an electrode β and the standard electrode, $(\phi^\beta - \beta^{STD})$. Then the potential of a new electrode material α, against the standard electrode can be determined from a cell with α and β as electrodes $(\phi^\beta - \phi^\alpha)$, since

$$\phi^\alpha - \phi^{STD} = (\phi^\beta - \phi^{STD}) - (\phi^\beta - \phi^\alpha) \tag{15.58}$$

Since the potential of the standard electrode cancels in the computation of cell potentials, it is convenient to choose the potential ϕ^{STD} to be at zero volts. With this choice, any potential measured relative to the standard electrode configuration

$$\mathcal{E}^\beta = \phi^\beta - \phi^{STD} = \phi^\beta - 0 = \phi^\beta \tag{15.59}$$

can be thought of as a "half cell potential" for that electrode.

The choice of electrode configuration for the standard is based on considerations that are analogous to those that determined the choice of the 0° and 100°C benchmarks for the centigrade temperature scale. The ice point and boiling point of pure water were chosen because water is available in reasonable purity in every laboratory and the benchmarks (freezing and boiling at one atmosphere pressure) are reproducible.

The standard electrode to which all electrochemical emf values are referred is the "Standard Hydrogen Electrode" (SHE) shown in Fig. 15.7. An inert platinum electrode is immersed in an acid solution and purified hydrogen gas at one atmosphere pressure is bubbled over the platinum-electrolyte interface with the system maintained at 25°C. The reaction at the electrode is

$$2(H^+)^\epsilon + 2(e^-)^{Pt} = H_2^g \qquad [15.13]$$

A variety of electrodes have been designed that can be conveniently used as reference electrodes in cell measurements in the laboratory [15.2], p. 101–123. All have been carefully calibrated with respect to the standard hydrogen electrode so that potentials determined with their use can be computed relative to the agreed standard, as in Eq. (15.58), and results compared with compiled potential measurements.

Application of this strategy for determining "half cell potentials" or "single electrode potentials" to cells in which the components are in their standard states generates $\mathcal{E}°$ values, *standard single electrode potentials*, for the corresponding single electrode reactions. A collection of standard single electrode potentials (measured relative to the standard hydrogen electrode and reported for the reduction reactions), is presented in Table 15.3. Some compilations report these electrode potentials for the oxidation reactions. In this case all signs would be reversed and the order of the reactions in the table inverted. This ordering of the elements according to their single electrode cell emf is called the *electromotive series* of the elements.

The standard free energy change for a reaction is related to its equilibrium constant K. From Eq. (15.52),

$$\Delta G° = -RT \ln K = -z\mathcal{F}\,\mathcal{E}°$$

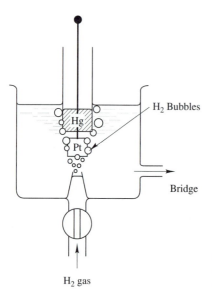

H$_2$ Bubbles

Hg

Pt

Bridge

H$_2$ gas

FIGURE 15.7
One form of the standard hydrogen electrode used as a reference electrode in reporting cell potentials.

or

$$\mathcal{E}° = \frac{RT}{z\mathcal{F}} \ln K \qquad (15.60)$$

Reactions near the top of Table 15.3, with relatively large positive values of $\mathcal{E}°$, correspond to the condition $K > 1$. Thus for these reactions at equilibrium the activity of the products (the metal) is large in comparison with the reactants (the ion in solution). Elements at this end of the series are said to be *noble* because they resist dissolution to form their ions. Near the bottom of the series in Table 15.3, $\mathcal{E}°$ values are negative, implying $K < 1$. The reactants (in this case the ion in solution) dominate at equilibrium. Elements at the bottom end of the series are said to be *reactive*. The behavior changes smoothly from noble through neutral to reactive as the electromotive series is traversed from top to bottom.

Example 15.6. The measured reversible emf for the cell

$$Pt/H_2|H^+||Cu^{++}|Cu$$

at 25°C is found to be +0.295 volts. Compute the activity of Cu^{++}.

Solution. The standard electrode potential for the reduction of cupric ion is found in Table 15.3 to be +0.337 volts. The electrode reaction is

$$Cu^{2+} + 2e^- = Cu$$

for which $Q = a_{Cu}/a_{Cu2+} = 1/a_{Cu2+}$. Applying the Nernst equation, Eq. (15.54)

$$0.295 \text{ volts} = 0.337 \text{ volts} - \frac{.05915 \text{ volts}}{2} \log \frac{1}{a_{Cu^{2+}}}$$

$$\log a_{Cu^{2+}} = (0.295 \text{ volts} - 0.337 \text{ volts}) \cdot \frac{2}{0.05915 \text{ volts}} = -1.42$$

$$a_{Cu^{2+}} = 0.038$$

15.4 POURBAIX DIAGRAMS

Predominance diagrams similar in concept to those devised in Sec. 11.4 can be constructed to represent electrochemical equilibria to describe the behavior of multivariate reacting systems. Variables chosen to describe the system are cell emf and electrolyte pH, which form the axes of Pourbaix diagrams. Construction of these plots is based upon repeated application of the Nernst equation, Eq. (15.54), to all the competing reactions that may occur within a given cell. Lines derived from these conditions divide the (pH, $\mathcal{E}$) space into regions of predominance for each of the components known to exist in the system. For a complete overview of the input information, calculation and application of these diagrams see Ref. [15.5].

15.4.1 The Stability of Water

The simplest Pourbaix diagram is for pure water. The components assumed to exist in such a system are H_2O, H^+, OH^-, $H_2(g)$, and $O_2(g)$. The activity of H^+, and implicitly that of OH^-, is represented on the pH axis of the diagram. Regions of predominance of the remaining components, H_2, O_2, and H_2O, can be computed by considering two electrode reactions.

For the formation of hydrogen consider the reaction

$$2H^+ + 2e^- = H_2(g) \qquad \mathcal{E}^\circ \equiv 0.000 \text{ volts} \qquad [15.14]$$

For this half cell reaction,

$$\log Q = \log \left[\frac{a_{H_2}^g}{a_{H^+}^2 a_{e^-}^2} \right] = \log \left[\frac{P_{H_2}}{a_{H^+}^2} \right] = \log P_{H_2} - 2 \log a_{H^+}$$

$$\log Q = \log P_{H_2} + \text{pH} \qquad (15.61)$$

Let $b = 2.303 RT / \mathcal{F} = 0.05915$ volts, since the constant appears repeatedly in these calculations. The Nernst equation for the given values of $\mathcal{E}^\circ$ and Q is

$$\mathcal{E} = 0.000 - \frac{b}{2} [\log P_{H_2} + 2\text{pH}]$$

$$\mathcal{E} = -b\text{pH} - \frac{b}{2} \log P_{H_2} \qquad (15.62)$$

For a given value of P_{H_2} this relationship plots as a straight line in (pH, $\mathcal{E}$) space. The slope of the line is $(-b) = -0.05915$ volts, and the intercept on the vertical line at pH $= 0$ is equal to $[-(b/2) \log P_{H_2}]$. Values of intercept for $P_{H_2} = 10^{-2}, 10^{-1}, 10^0, 10^1$, and 10^2 are respectively $+0.05915, +0.02958, 0.000, -0.02958$, and -0.05915 volts. This set of five lines is plotted in Fig. 15.8. As the emf of the cell decreases in the region of these lines, the equilibrium hydrogen pressure increases from 10^{-2} atm to 10^2 atm. Thus in the (pH, $\mathcal{E}$) domain below the line representing $P_{H_2} = 1$ atm, water decomposes to form H_2 gas. Above this line the equilibrium hydrogen pressure decreases very rapidly as the emf increases and H_2O is stable. The line represented by the equation

$$\mathcal{E} = -b\text{pH} = -0.05915 \text{ pH} \qquad (15.63)$$

can be termed the *limit of predominance* between H_2 gas and H_2O.

The competition between H_2O and O_2 can be examined by considering the electrode reaction

$$O_2 + 4H^+ + 4e^- = 2H_2O \qquad \mathcal{E}^\circ = +1.229 \text{ volts} \qquad [15.15]$$

For this reaction,

$$\log Q = \log \frac{a_{H_2O}^2}{a_{O_2}^g a_{H^+}^4 a_{e^-}^4} = \log \frac{1}{P_{O_2} a_{H^+}^4} = -\log P_{O_2} - 4 \log a_{H^+}$$

$$\log Q = -\log P_{O_2} + 4\text{pH} \qquad (15.64)$$

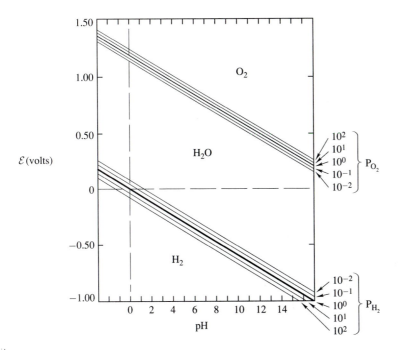

FIGURE 15.8
Region of the stability of water in $\mathcal{E}$–pH space.

The condition for equilibrium expressed in the Nernst equation is

$$\mathcal{E} = +1.229 - \frac{1}{4}b \left[-\log P_{O_2} + 4\text{pH} \right] \qquad (15.65)$$

$$\mathcal{E} = \left[1.229 + \frac{1}{4}b \log P_{O_2} \right] - b\text{pH} \qquad (15.66)$$

For a fixed value of P_{O_2}, this equation represents a straight line with slope equal to $-b = -0.05915$ and intercept $[1.229 + (b/4) \log P_{O_2}]$. As P_{O_2} varies from 10^{-2} to 10^2 the intercept varies from 1.199 to 1.259. These lines have the same slope as do the hydrogen reaction lines. They are closer together because the coefficient of log P_{O_2} is $(b/4)$ as compared to $(b/2)$ for the hydrogen case. Also, the order is reversed because the sign of the log P_{O_2} term is opposite to that for the H_2 lines. Thus as the cell emf *increases*, the oxygen pressure changes from 10^{-2} to 10^2 atm. The region of predominance of oxygen gas thus lies *above* the line corresponding to $P_{O_2} = 1$ atm; H_2O is the predominant component below that line.

Figure 15.8 shows that the Pourbaix diagram for pure water has three regions of predominance. At high cell emf values, water decomposes to form oxygen gas; at intermediate emf values, water is stable; and in a specific range of negative emf values, water decomposes to form hydrogen gas. From a strictly theoretical point of view the consideration of all electrochemical reactions that presume the presence of liquid water must be limited to the (pH, $\mathcal{E}$) range corresponding to the stability of H_2O in Fig. 15.8. However, from a practical point of view, the *rate* of evolution of hydrogen at low (negative) emf values and oxygen at high positive values may be very slow in comparison to other electrochemical reactions that may be of interest in the system. Accordingly, the practical application of these thermodynamic considerations span a range that is significantly broader than the region of stability of H_2O.

15.4.2   Pourbaix Diagram for Copper

The general strategy for computing Pourbaix diagrams is illustrated for the copper-water system in this section [15.5]. In addition to the components normally present in water, it is necessary to enumerate the ionic and nonionic components containing copper atoms that are known to exist in the system. The seven copper-containing components considered in calculating this diagram can be divided into two categories: three solid substances, Cu, CuO, and Cu_2O, and four dissolved ionic species, Cu^+, Cu^{++}, $HCuO_2^-$, and CuO_2^-. It is further assumed that the electrolyte contains additional components, anions or cations, necessary to vary the pH of the solution, which do not participate in the reactions involving copper.

When H^+, H_2O, and e^- are included, there are ten components and three elements (Cu, H, O) participating in this system. Thus, there are $r = (c-e) = (10-3) = 7$ independent reactions. Standard electrode potentials must be obtained for each of these reactions.

Each line on the diagram represents a zone of transition in predominance between two competing components. The equation describing a particular line is obtained by applying the condition for equilibrium, the Nernst equation, to a cell in which the competing pair of components is involved in the electrode reactions. The number of reactions that must be considered is the number of ways the seven copper-containing components can be placed in competition two at a time: $[7!/(2!)(5!)] = 21$. Thus, the Pourbaix diagram for copper consists at most of twenty-one lines (more precisely, twenty-one zones of transition). Some of these lines may not appear on the diagram. For example, the line representing the competition between components X and Y may lie within a region in which a third component Z predominates over both X and Y. In this case, the XY line is absent from the diagram.

The twenty-one reactions that must be considered are listed in Table 15.4. According to the convention established, reactions that involve the transfer of electrons are written as reduction reactions. Also, when a reaction can be written involving either H^+ or OH^- ions, the chosen reaction is expressed in terms of the H^+ ion. This choice is made because the abscissa on the Pourbaix diagram is expressed in terms of the activity of the hydrogen ion and OH^- and H^+ activities are related through the water reaction. Thus, for example, reaction [H] describing the competition between Cu and CuO could be written

$$Cu + 2OH^- = CuO + H_2O + 2e^-$$

TABLE 15.4
**Reactions considered in computing
the Pourbaix Diagram for pure
copper**

$Cu^{++} + 2H_2O = HCuO_2^- + 3H^+$	[A]
$Cu^{++} + 2H_2O = CuO_2^{--} + 4H^+$	[B]
$HCuO_2^- = CuO_2^{--} + H^+$	[C]
$Cu^+ = Cu^{++} + e^-$	[D]
$Cu^+ + 2H_2O = HCuO_2^- + 3H + e^-$	[E]
$Cu^+ + 2H_2O = CuO_2^{--} + 4H^+ + e^-$	[F]
$Cu + H_2O = Cu_2O + 2H^+ + 2e^-$	[G]
$Cu + H_2O = CuO + 2H^+ + 2e^-$	[H]
$Cu_2O + H_2O = 2CuO + 2H^+ + 2e^-$	[I]
$2Cu^+ + H_2O = Cu_2O + 2H^+$	[J]
$Cu^{++} + H_2O = CuO + 2H^+$	[K]
$CuO + H_2O = HCuO_2^- + H^+$	[L]
$CuO + H_2O = CuO_2^{--} + 2H^+$	[M]
$Cu = Cu^+ + e^-$	[N]
$Cu = Cu^{++} + 2e^-$	[O]
$Cu + 2H_2O = HCuO_2^- + 3H^+ + 2e^-$	[P]
$Cu + 2H_2O = CuO_2^{--} + 4H^+ + 2e^-$	[Q]
$Cu_2O + 2H^+ = 2Cu^{++} + H_2O + 2e^-$	[R]
$Cu_2O + 3H_2O = 2HCuO_2^- + 4H^+ + 2e^-$	[S]
$Cu_2O + 3H_2O = 2CuO_2^{--} + 6H^+ + 2e^-$	[T]
$Cu^+ + H_2O = CuO + 2H^+ + e^-$	[U]

However, the Nernst equation for this reaction involves $\log_{10} a_{OH^-}$, whereas the Pourbaix diagram describes this aspect of the behavior of the system with pH $= -\log_{10} a_{H^+}$. Subtraction of the water reaction from the above equation yields

$$Cu + 2OH^- + 2H_2O = CuO + H_2O + 2e^- + 2H^+ + 2OH^-$$

which simplifies to

$$Cu + H_2O = CuO + 2H^+ + 2e^- \qquad [H]$$

The inverse reaction is the proper reduction reaction for computation of the competition for the Pourbaix diagram.

This simple strategy permits any reaction involving OH^- ions to be rewritten in terms of H^+ ions so that the corresponding condition for equilibrium is expressed directly in terms of the pH of the solution.

The reactions summarized in Table 15.4 can be divided into three classes:

1. Reactions in which no electrons are transferred to the electrode include [A], [B], [C], [J], [K], [L], and [M]. These reactions rearrange charge over the ions in the

electrolyte. Because no charge is transferred to the electrode, the condition for equilibrium is independent of the external emf of the system. The corresponding limits of predominance plot as vertical lines at a fixed pH on the Pourbaix diagram.

2. Reaction [D], [N], and [O] do not depend upon the hydrogen ion concentration; the limit of predominance for this competition plots as a horizontal line on the diagram.

3. The remaining reactions involve both electrons and hydrogen ions. These are represented through the Nernst equation by sets of lines that have a slope that is a simple multiple of 0.05915. The multiplying factor depends upon the stoichiometric coefficients of H^+ and e^- in the reaction.

Sets of lines corresponding to these twenty-one reactions can be combined to generate the Pourbaix diagram for the copper-water system. It remains to determine which sets of lines do not appear on the diagram and to identify the regions of predominance outlined by each set. Automatic algorithms, based on linear programming theory, that make these decisions, have been adapted in the development of computer programs to calculate Pourbaix diagrams [15.6].

The completed Pourbaix diagram for the copper water system is presented in Fig. 15.9. From a practical point of view the variety of regions shown can be grouped into three classes:

1. A region of *immunity* in which the copper metal is the predominant component.
2. Regions of *corrosion* in which the predominant component is one of the copper-containing ionic species.
3. Regions of *passivation* in which the predominant component is one of the solid copper-containing species.

The latter condition is protective of the metal with respect to corrosion if the solid component reaction product forms a coherent layer on the metal, isolating it from the electrolyte solution and rendering it *passive* to corrosion. Figure 15.10 shows a simplified version of the Pourbaix diagram for copper in which only these three classes of behavior are identified.

Pourbaix diagrams can be computed for any metal or alloy for which the required thermodynamic information is available. The *Atlas of Electrochemical Equilibria in Aqueous Solutions* [15.5] provides a continuing compilation of this information. A computer program for generating such diagrams from the input thermodynamic data is available for use on personal computers [15.6]. With limited additional information it is possible to explore the effects of additional variables on these regions of predominance. For example, it is relatively straightforward to introduce a third variable such as chloride ion activity into the analysis and generate three dimensional ($\mathcal{E}$, pH, $-\log_{10} a_{Cl^-}$) diagrams or plot sections at selected chloride activities.

As is the case for phase diagrams, knowledge of Pourbaix potential-pH diagrams is just the starting point for developing an understanding of corrosion phenomena. These diagrams are generated by repeated application of the conditions for equilib-

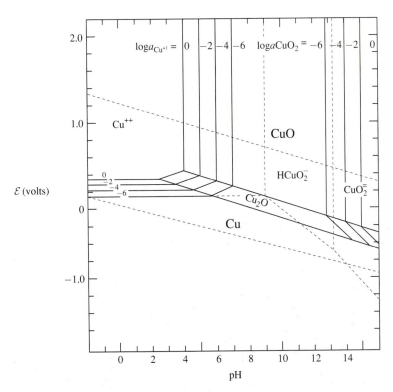

FIGURE 15.9
Complete Pourbaix diagram for copper in water [15.5].

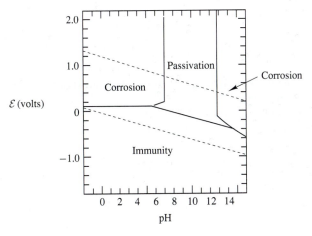

FIGURE 15.10
Simplified Pourbaix diagram for copper in water [15.5].

rium in electrochemical systems. Thus the domains of predominance they identify describe the predominant species for a given (pH, $\mathcal{E}$) *at equilibrium*. The diagram describes the ultimate condition of the system left indefinitely under fixed conditions. How the system approaches its ultimate destination may be of paramount importance in practical corrosion. Also, the systems treated in Pourbaix diagrams are generally greatly simplified in comparison to the electrolyte and electrode combinations that are of practical interest. Real-world corrosion problems are inherently dirty, experience changing conditions with time and may involve human factors that are unrelated to the physics, chemistry, and thermodynamics of the system. Pourbaix diagrams provide a point of departure in the solution of corrosion problems.

15.5 SUMMARY OF CHAPTER 15

When dissolved in an appropriate solvent an *electrolyte* dissociates into ionic components. Equilibrium within an electrolyte solution is determined by the *dissociation constant* through a condition for equilibrium that is analogous to that obtained for reacting systems. Weak electrolytes partially dissociate into ions with some of the compound remaining undissociated in solution. The degree of dissociation α for an electrolyte $M_u X_v$, is governed by the *dissociation constant K* according to the equation

$$K_c = (u^u v^v) \frac{\alpha^\omega}{1 - \alpha} [K]^{\omega - 1} \tag{15.17}$$

where $\omega = u + v$ and $[k]$ is the molarity (concentration in moles per liter) of the electrolyte in the solvent. Strong electrolytes, acids, bases, and salts dissociate completely into ions in solution.

Equilibrium between an electrode and its enclosing electrolyte solution gives rise to an electrostatic potential difference between the two phases. The system attains equilibrium when the affinity for the reaction that governs transfer of charge between the phases is proportional to the potential difference between electrode and electrolyte.

$$\mathcal{A} = \mu_M^\alpha - \left(z^+ \mu_{e-}^\alpha + \mu_{M^{z+}}^\epsilon \right) = -z^+ \mathcal{F}(\phi^\alpha - \phi^\epsilon) \tag{15.38}$$

Application of this condition for equilibrium to both electrodes in an electrochemical cell yields the Nernst equation

$$\mathcal{E} = \mathcal{E}^\circ - \frac{RT}{z\mathcal{F}} \ln Q \tag{15.53}$$

where $\mathcal{E}^\circ$ is the *standard electrode potential* for the cell and Q is the proper quotient of activities for the overall cell reaction.

A compilation of standard electrode potentials for a collection of reduction reactions involving elemental electrodes, called the *electromotive series*, provides a basis for ordering the elements in terms of their tendency to dissociate in water.

The conditions for electrochemical equilibrium of materials may be usefully visualized with potential-pH *predominance diagrams* devised by Pourbaix. Regions of immunity, corrosion, and passivation of materials can be mapped out in potential-pH space.

PROBLEMS

15.1. The dissociation constant for aluminum phosphate in aqueous solution at 298 K is 9.8×10^{-21}. Assume $K_f = 1$. Calculate the degree of dissociation of $AlPO_4$ as a function of concentration. Plot the result.

15.2. The dissociation constant of hydrocyanic acid (HCN) at $25°C$ is 4.93×10^{-10}. Compute the pH of a solution prepared with 0.05 molar concentration of HCN.

15.3. Beryllium hydroxide is a weak base with a dissociation constant given by 5×10^{-11} at $25°C$. How many grams of $Be(OH)_2$ must be added to a liter of water in order to obtain a pH of 8.5?

15.4. What concenteration of sodium hydroxide must be added to water to adjust the pH to 9.5?

15.5. Calcium hydroxide and nitric acid are strong electrolytes. Two stock solutions are available: one with 0.02 molar $Ca(OH)_2$ and the other with 0.05 molar HNO_3. How much of the nitric acid solution must be added to one liter of the calcium hydroxide solution to make a solution with a pH = 8.2?

15.6. Explain why the potential difference computed in Eq. (15.39) for a single electrode *cannot be measured.*

15.7. Describe *all* the subsystems in a galvanic cell, including the phases and the interfaces. Describe the conditions that must be met by each of these subsystems in order for the cell to operate *reversibly.*

15.8. Sketch a galvanic cell represented by the following notation.

$$Ni^{\alpha}(a_{Ni} = 0.23)|NiCl_2(a_{Ni} = 0.03)\|CuCl_2(a_{Cu} = 0.02)|Cu^{\alpha}(a_{Cu} = 1)$$

Label all the parts of the cell.

15.9. Given that it is not possible to measure the potential difference between an electrode and an electrolyte in which it is immersed, how is it possible to evaluate "half cell potentials" for the subsystems in an electrochemical cell?

15.10. Reversible emf values are measured for the following cell at 1073 K:

$$Mg \text{ pure liquid}|MgCl_2\text{–}ClCl_2 \text{ liquid eutectic}|Mg \text{ (in Al) liquid}$$

for a sequence of compositions:

X_{Mg}	0.0447	0.1130	0.1905	0.3400	0.5825	0.7490
$\mathcal{E}$ (mv)	152.840	109.868	82.501	54.195	30.446	20.871

Compute and plot the activity of magnesium in liquid Mg-Al alloys as a function of composition. (Source: Tiwari, Basant L.: *Metallurgical Transactions*, vol. 18A, p. 1645, 1987.)

15.11. Review the concept of the limit of predominance as it applies to the construction of Pourbaix pH-emf diagrams.

15.12. Construct a potential-pH diagram for the Ni-water system at $25°C$. Limit consideration to the following components: Ni, NiO, NiO_2, Ni^{++} and $HNiO_2^-$.

15.13. A sample of copper dissolves in water at $25°C$. Use the Pourbaix diagram in Fig. 15.10 to suggest three strategies to avoid the corrosion of copper in water.

REFERENCES

1. Robinson, R. A., and R. H. Stokes: *Electrolyte Solutions*, Butterworths, London, 1965.
2. Potter, E. C.: *Electrochemistry Principles & Applications* Cleaver-Hume Press Ltd, London, 1961.
3. Devereaux, O. F.: *Topics in Metallurgical Thermodynamics*, John Wiley & Sons, New York, Chapter 6, 1983.
4. Guggenheim, E. *Thermodynamics*, 5th edition, North-Holland, Amsterdam, 1967.
5. Pourbaix, M.: *Atlas of Electrochemical Equilibria in Aqueous Solutions*, second edition, National Association of Corrosion Engineers, Houston, Texas, 1974.
6. Froning, M. H., M. E. Shanley, and E. D. Verink, Jr.: *Corrosion Science,* vol. 16, p. 371, 1976.

APPENDIX
A

UNIVERSAL CONSTANTS AND CONVERSION FACTORS[1]

Universal Constants

Avogadro's number	N_0	6.0221367×10^{23}
Rest mass of the electron	m_e	9.1095×10^{-31} (kg)
Elementary charge	e	1.6022×10^{-19} (Coul)
Planck's constant	$\hbar$	1.0546×10^{-34} (J s)
Gas constant	R	8.314510 (J/mol-K)
	R	1.98722 (cal/mol-K)
	R	82.057 (cc-atm/mol-K)
	R	0.08057 (liter-atm/mol-K)
Boltzmann constant ($k = R/N_o$)	k	1.380658×10^{-23} (J/atom-K)

[1]*Source*: *CRC Handbook of Chemistry and Physics*, 71st edition, D. R. Lide, ed., CRC Press, Boca Raton, Fla., 1991.

Conversion Factors

Length	$1\ m = 100\ cm = 1000\ mm = 10^6\ microns = 10^9\ nm$
Volume	$1\ L = 1000\ cm^3 = 1000\ mL = 10^{-3}\ m^3$
Temperature	$T\ °C = T\ K - 273.15 = (5/9)(T\ °F - 32)$
Pressure	$1\ atm = 9.8692\ MPa = 0.98692\ bar = 0.05805\ psi$
Energy	$1\ J = 1\ N\text{-}m = 4.184\ cal = 9.869\ cc\text{-}atm = 0.009869\ liter\text{-}atm$

APPENDIX
B

ATOMIC NUMBERS AND ATOMIC WEIGHTS[1]

---

[1] *Source: CRC Handbook of Chemistry and Physics*, 71st edition, D. R. Lide, ed., CRC Press, Boca Raton, Fla., 1991.

AN = Atomic Number[1]

AW = Atomic Weight $\left(\dfrac{\text{gm}}{\text{gm-atom}}\right)^{1}$

	AN
	AW

1	2	3	4	5	6	7	8	9	10	11	12	13	14	15	16	17	18
Li 3 6.940	Be 4 9.013											B 5 10.82	C 6 12.011	N 7 14.008	O 8 16.00	F 9 19.00	Ne 10 20.183
Na 11 22.991	Mg 12 24.32											Al 13 26.98	Si 14 28.086	P 15 30.974	S 16 32.06	Cl 17 35.453	Ar 18 39.948
K 19 39.100	Ca 20 40.08	Sc 21 44.96	Ti 22 47.90	V 23 50.95	Cr 24 52.01	Mn 25 54.94	Fe 26 55.85	Co 27 58.94	Ni 28 58.71	Cu 29 63.546	Zn 30 65.38	Ga 31 69.72	Ge 32 72.59	As 33 74.922	Se 34 78.96	Br 35 79.904	Kr 36 83.80
Rb 37 85.48	Sr 38 87.63	Y 39 88.92	Zr 40 91.22	Nb 41 92.91	Mo 42 95.95	Tc 43 98	Ru 44 101.1	Rh 45 102.91	Pd 46 106.4	Ag 47 107.868	Cd 48 112.41	In 49 114.82	Sn 50 118.69	Sb 51 121.75	Te 52 127.60	I 53 126.905	Xe 54 131.30
Cs 55 132.91	Ba 56 137.36	La 57 138.92	Hf 72 178.58	Ta 73 180.95	W 74 183.86	Re 75 186.2	Os 76 190.2	Ir 77 192.22	Pt 78 195.09	Au 79 196.97	Hg 80 200.59	Tl 81 204.37	Pb 82 207.2	Bi 83 208.981	Po 84 (209)	At 85 (210)	Rn 86 (222)
Fr 87 (223)	Ra 88 226.025	Ac 89 (227)															Lu 71 174.97
																	Lw 103 (260)

Ce 58 140.12	Pr 59 140.91	Nd 60 144.24	Pm 61 (145)	Sm 62 150.4	Eu 63 151.96	Gd 64 157.25	Tb 65 158.93	Dy 66 162.50	Ho 67 164.93	Er 68 167.26	Tm 69 168.93	Yb 70 173.04
Th 90 232.04	Pa 91 231.04	U 92 238.03	Np 93 237.05	Pu 94 (244)	Am 95 (243)	Cm 96 (247)	Bk 97 (247)	Cf 98 (251)	Es 99 (254)	Fm 100 (257)	Md 101 (258)	No 102 (259)

Numbers in parentheses are mass numbers of the most stable isotope.

[1] Source: CRC Handbook of Chemistry and Physics, 71st Edition, D. R. Lide, ed., CRC Press, Boca Raton, FL (1991).

APPENDIX
C

VOLUMETRIC
PROPERTIES
OF THE
ELEMENTS

Legend

V = Atomic Volume $\left(\dfrac{cc}{\text{gm-atom}}\right)$ at 298 K

α = Volume Coefficient of Thermal Expansion $(K)^{-1}$ *

β = Isothermal Coefficient of Compressibility $(\text{bar})^{-1}$

V^L = Atomic Volume of Liquid $\left(\dfrac{cc}{\text{gm-atom}}\right)$

* Estimated from linear coefficient of expansion: $\alpha = 3\alpha_L$

Cell format:

V
$\alpha \times 10^6$
$\beta \times 10^7$
V^L

Element	V	$\alpha \times 10^6$	$\beta \times 10^7$	V^L
Li	13.00	168	—	13.21
Be	4.89	36	—	5.33
B	4.56	—	—	5.20
C	3.42	—	—	—
N	—	—	—	—
O	—	—	—	—
F	—	—	—	—
Ne	—	—	—	—
Na	23.70	213	—	24.80
Mg	13.97	78	—	15.29
Al	9.99	70.5	12	11.31
Si	12.00	23	—	11.19
P	—	—	—	—
S	—	—	—	26.96
Cl	—	—	—	54.57
Ar	—	—	—	—
K	45.46	249	—	47.28
Ca	26.03	66	53	29.36
Sc	15.03	36	15	—
Ti	10.64	27	—	11.65
V	8.35	25	54	8.94
Cr	7.32	20	—	8.28
Mn	7.42	69	—	9.59
Fe	7.10	36	5.9	7.96
Co	6.62	37	5.6	7.59
Ni	6.60	40	26	7.43
Cu	7.09	51	6.6	7.94
Zn	9.16	93	16	9.94
Ga	11.81	55	—	11.45
Ge	13.62	17	—	12.95
As	13.08	—	—	14.35
Se	16.49	111	—	19.80
Br	—	—	—	—
Kr	—	—	—	—
Rb	55.86	27	—	59.48
Sr	33.70	300	—	35.33
Y	19.85	32	22	—
Zr	14.06	18	—	(15.7)
Nb	10.84	22	—	(11.9)
Mo	9.40	15	4.0	(10.3)
Tc	—	—	—	—
Ru	8.29	29	—	9.28
Rh	8.30	26	—	9.53
Pd	8.87	33	—	10.14
Ag	10.27	57	9.0	11.54
Cd	13.01	93	21	14.01
In	15.73	74	—	16.35
Sn	16.26	71	17	16.96
Sb	18.21	27	—	18.77
Te	20.45	≠	—	22.35
I	—	—	—	—
Xe	—	—	—	—
Cs	71.07	231	—	71.69
Ba	39.22	54	—	41.34
La	22.50	15	36	23.17
Hf	13.61	18	89	16.07
Ta	10.90	20	43	(12.1)
W	9.52	14	2.9	(10.4)
Re	8.87	20	—	9.90
Os	8.45	14	—	(9.46)
Ir	8.58	20	—	9.6
Pt	9.10	27	3.5	10.3
Au	10.21	43	—	11.35
Hg	—	183	—	14.76
Tl	17.25	90	25	18.12
Pb	17.74	87	23	19.40
Bi	21.32	39	31	20.75
Po	22.35	—	—	—
At	51.48	—	—	—
Rn	—	—	—	—
Fr	—	—	—	(9.5)
Ra	—	45	—	—
Ac	—	—	—	—
Ce	20.76	24	54	20.96
Pr	20.50	14	41	21.31
Nd	20.59	20	34	21.56
Pm	—	—	—	—
Sm	19.96	—	26	—
Eu	28.98	—	87	—
Gd	19.91	19	25	(22)
Tb	19.21	21	—	—
Dy	19.01	26	23	—
Ho	18.74	29	—	—
Er	18.48	28	—	—
Tm	18.12	35	—	—
Yb	24.80	75	78	—
Lu	17.78	375	—	—
Th	20.18	34	—	22.10
Pa	15.04	—	—	—
U	12.49	≠	—	13.30
Np	11.59	—	—	—
Pu	12.3	165	—	14.7
Am	—	—	—	—
Cm	—	—	—	—
Bk	—	—	—	—
Cf	—	—	—	—
Es	—	—	—	—
Fm	—	—	—	—
Md	—	—	—	—
Lw	—	—	—	—

≠ Highly anisotropic

Sources: *Smithell's Metals Reference Book*, 6th edition, E. A. Brandes, ed., Butterworths, London (1983).

C. L. Reynolds, Jr., K. A. Faughnan, and R. E. Barker, Jr., *J. Chem. Phys.*, vol. 59 (1973) p. 2943.

C. L. Reynolds, Jr., and R. E. Barker, Jr., *J. Chem. Phys.*, vol. 61 (1974) p. 2564.

APPENDIX
D

ABSOLUTE ENTROPIES AND HEAT CAPACITIES OF SOLID ELEMENTS

S^0_{298} Absolute entropy at 298 K $\left(\dfrac{\text{Joules}}{\text{gm-atom-K}}\right)$

Heat capacity coefficients in the expression

$$Cp = a + bT + \dfrac{c}{T^2} \left(\dfrac{\text{Joules}}{\text{gm-atom-K}}\right)$$

Cell legend:

S^0_{298} / a	
$b \times 10^3$	
$c \times 10^{-5}$	

Element	S^0_{298} / a	$b \times 10^3$	$c \times 10^{-5}$
H (H₂)	130.57 / 27.3	3.2	0.50
Li	29.1 / 24.5	5.5	8.7
Be	9.50 / 19.0	8.9	-3.4
B (cryst)	5.85 / 19.8	5.8	-9.2
C (diam)	5.74 / 9.12	13.2	—
N (N₂)	191.50 / 27.9	4.2	—
O (O₂)	205.03 / 30.0	4.2	-1.7
F (F₂)	202.67 / 34.7	1.8	-3.3
He	126.04 / 20.79	—	—
Na	51.21 / 82.45	-369.3	—
Mg	32.7 / 22.3	10.3	-0.43
Al	28.3 / 20.7	12.3	—
Si	18.8 / 23.9	2.5	-4.1
P (yell.)	41.1 / 19.1	15.8	—
S (rhomb.)	31.8 / 15.0	26.1	—
Cl (Cl₂)	222.96 / 36.9	0.3	-2.8
Ne	146.22 / 20.79	—	—
K	64.18 / 25.3	13.1	—
Ca (α)	41.42 / 25.3	-7.26	—
Sc	34.6	—	—
Ti (α)	30.6 / 22.1	10.	—
V	28.9 / 20.5	10.8	0.8
Cr	23.8 / 24.4	9.87	-3.7
Mn (α)	32.0 / 23.8	14.1	-1.57
Fe (α)	27.3 / 37.12	6.17	—
Co (α)	30.0 / 21.4	14.3	-0.88
Ni (α)	29.9 / 17.0	29.5	—
Cu	33.14 / 22.6	5.6	—
Zn	41.6 / 22.4	10	—
Ga	40.9 / 26.2	—	—
Ge	31.1 / 21.6	5.6	—
As	35 / 23.2	5.52	—
Se	42.44 / 19.0	23.	—
Br (Br₂,l)	152.23 / 72.0	—	—
Kr	163.97 / 20.79	—	—
Rb	76.78 / 30.4	—	—
Sr	52.3	—	—
Y (α)	44.43 / 23.9	7.55	0.3
Zr (α)	39.0 / 22.0	11.6	-3.8
Nb	36.4 / 23.7	4.0	—
Mo	28.6 / 24.1	1.2	—
Tc	—	—	—
Ru (α)	28.5 / 22.1	4.6	—
Rh	31.5 / 23.0	8.6	—
Pd	37.6 / 24.3	5.8	—
Ag	42.6 / 21.3	8.5	1.5
Cd	51.76 / 22.2	12.3	—
In	57.82 / 24.3	10	—
Sn	51.55 / 21.6	18.2	—
Sb	45.69 / 23.1	7.4	—
Te	49.71 / 19.2	22.0	—
I (I₂,s)	180.68 / 40.1	49.8	—
Xe	169.57 / 20.79	—	—
Cs	85.22 / 31.0	—	—
Ba	62.8 / -5.69	80.3	—
La	56.9 / 25.8	6.69	—
Hf	43.56 / 23.5	7.6	—
Ta	41.5 / 27.8	-2.2	-1.9
W	32.6 / 24.0	3.2	—
Re	36.9 / 24.3	4.0	—
Os	33 / 23.80	3.7	—
Ir	35.5 / 23.3	5.9	—
Pt	41.6 / 24.3	5.4	—
Au	47.40 / 23.7	5.19	—
Hg	76.02 / 27.7	—	—
Tl	64.6 / 15.6	25.3	2.8
Pb	64.81 / 23.6	9.7	—
Bi	56.73 / 18.8	22.6	—
Po	—	—	—
At	—	—	—
Rn	176.1 / 20.79	—	—
Fr	95.4	—	—
Ra	71	—	—
Ac	56.5	—	—

Lanthanides:

Element	S^0_{298} / a	$b \times 10^3$	$c \times 10^{-5}$
Ce (α)	69.5 / 23.48	10.40	—
Pr (α)	73.2 / 26.0	—	—
Nd (α)	71.5 / 14.66	26.9	4.5
Pm	—	—	—
Sm (α)	69.58 / 25.1	24.4	-2.5
Eu	77.82	—	—
Gd	68.07	—	—
Tb	73.2	—	—
Dy	74.77	—	—
Ho	75.3	—	—
Er	73.18	—	—
Tm	74.01	—	—
Yb	59.87	—	—
Lu	50.96	—	—

Actinides:

Element	S^0_{298} / a	$b \times 10^3$	$c \times 10^{-5}$
Th	53.79 / 23.6	12.7	—
Pa	56	—	—
U (α)	50.20 / 10.9	37.4	4.9
Np	—	—	—
Pu (α)	— / 24.7	24	—
Am	—	—	—
Cm	—	—	—
Bk	—	—	—
Cf	—	—	—
Es	—	—	—
Fm	—	—	—
Md	—	—	—
Lw	—	—	—

Sources: CRC Handbook of Chemistry and Physics, 71st edition, D. R. Lide, ed., CRC Press, Boca Raton, FL (1991).

Smithell's Metals Reference Book, 6th edition, E. A. Brandes, ed., Butterworths, London (1983).

APPENDIX
E

PHASE TRANSITIONS: TEMPERATURE AND ENTROPIES OF MELTING AND VAPORIZATION

Legend:

T_m
ΔS_m^0
T_v
ΔS_v^0

Transition temperatures are in (K) at 1 atm

Transition entropies are in $\left(\dfrac{\text{Joules}}{\text{gm-atom-K}}\right)$

Element	T_m	ΔS_m^0	T_v	ΔS_v^0
H (H_2)	14.0	12.7	20.7	44.0
Li	454	6.45	1597	92.5
Be	1560	7.83	2740	107
B	2450	9.21	4100	141
C (graph)	4070	—	5300(s)	134.3
N (N_2)	63.2	11.39	77.4	72.1
O (O_2)	54.4	8.19	90.2	75.4
F (F_2)	53.6	29.8	85.2	76.69
Ne				
Na	371.0	7.12	1156	84.8
Mg	922	9.53	1360	93.7
Al	933.3	11.2	2793	104
Si	1685	30.1	3540	109
P (yellow)	317.3	8.32	553	—
S (rhomb.)	386.0	3.20	717.7	—
Cl (Cl_2)	172.2	37.2	239.1	85.4
K	336.4	7.10	1052	75.6
Ca	1116	7.49	1757	85.8
Sc	1811	—	(3140)	—
Ti	1940	(9.02)	3558	120
V	2175	7.70	3680	124
Cr	2130	(9.8)	2945	116
Mn	1517	(9.69)	2330	99.1
Fe	1809	8.40	3130	109
Co	1768	(8.8)	3203	(130)
Ni	1726	9.94	3180	118
Cu	1356.6	9.60	2830	108
Zn	692.7	10.51	1180	96.9
Ga	302.9	18.5	2690	100
Ge	1210	30.4	3100	106
As	1090	—	876	—
Se	493.7	12.7	958	100 (Se_2)
Br (Br_2)	265.9	39.7	331	92.3
Kr				
Rb	312.0	7.05	961	78.9
Sr	1043	(8.05)	1648	93.7
Y	1803	6.34	3600	103
Zr	2125	(9.08)	4700	123
Nb	2740	10.7	5010	136
Mo	2890	12.3	4880	121
Tc				
Ru	2520	—	4520	—
Rh	2239	(10.1)	4000	(120)
Pd	1825	9.15	3210	113
Ag	1234	8.99	2470	104.2
Cd	594.1	10.79	1040	95.8
In	429.6	7.61	2343	99.2
Sn	505.1	14.0	2898	102
Sb	903.7	22.0	1860	89.6 (Sb_2)
Te	723	24.3	1261	83.0 (Te_2)
I (I_2)	386.8	40.8	456	91.9
Xe				
Cs	303.0	6.90	973	68.4
Ba	1002	7.64	1970	90
La	1193	(7.01)	(3690)	(109)
Hf	2500	9.63	4870	117
Ta	3288	7.51	5640	—
W	3670	9.6	5828	(126)
Re	3450	9.7	5960	119
Os	3300	—	5300	—
Ir	2716	(9.6)	4700	130
Pt	2042	(9.65)	4400	107
Au	1336	9.56	3130	109.4
Hg	234.29	9.91	630	97.0
Tl	577	7.45	1746	95.2
Pb	600.6	8.01	2020	88.4
Bi	544	20.0	1837	97.5
Po	519	—	1238	81.5
At				
Rn				
Fr				
Ra	1000	—	1800	—
Ac				
Ce	1071	4.88	3700	101
Pr	1205	(9.4)	3780	—
Nd	1289	5.54	3340	—
Pm				
Sm	1345	6.63	2076	79.5
Eu	1099	—	1763	—
Gd	1585	—	3560	—
Tb	1633	—	3490	—
Dy	1682	—	2830	—
Ho				
Er	1795	—	3130	—
Tm				
Yb	1097	—	1467	—
Lu				
Th	2020	—	5060	(101)
Pa				
U	1405	8.90	4700	89
Np	910	—	—	—
Pu	913	3.18	3690	95.3
Am	—	—	2900	83
Cm				
Bk				
Cf				
Es				
Fm				
Md				
Lw				

Source: *Smithell's Metals Reference Book,* 6th edition, E. A. Brandes, ed., Butterworths, London (1983).

APPENDIX
F

SURFACE
TENSIONS
AND
INTERFACIAL
FREE
ENERGIES

Legend (box):

$$\begin{array}{|c|} \hline \sigma^{LV} \\ \gamma^{SV} \\ \gamma^{SL} \\ \hline \end{array}$$

σ^{LV} – surface tension of liquid-vacuum interface at the melting point $\left(\dfrac{mN}{m}\right)$

γ^{SV} – surface free energy of solid-vapor interface at the melting point $\left(\dfrac{mJ}{m^2} = \dfrac{ergs}{cm^2}\right)$

γ^{SL} – surface free energy of solid-liquid interface at the melting point $\left(\dfrac{mJ}{m^2} = \dfrac{ergs}{cm^2}\right)$

1	2	3	4	5	6	7	8	9	10	11	12	13	14	15	16	17	18
																	He 0.27
Li 408	Be — (2000)											B 1060	C	N 11.77	O	F	Ne
Na 200	Mg 525											Al 865 (910) 93	Si 720	P (white) 71.2	S 60.9	Cl	Ar 13.12
K 117	Ca 360	Sc 954	Ti 1650	V 1950	Cr 1700 (2100)	Mn 1100	Fe 1880 (1870)(δ) 204	Co 1880 (1900)	Ni 1822 (2040) 255	Cu 1360 (1700) 177	Zn 768	Ga 708	Ge 589	As	Se 95	Br (46.8)	Kr 16.34
Rb 77	Sr 303	Y 871	Zr 1480	Nb 1900 (2000)	Mo 2250 (2300)	Tc	Ru 2250	Rh 2700	Pd 1470	Ag 926 (1100)	Cd 640	In 560	Sn 540 685 55	Sb 368	Te 179	I 37.92	Xe 18.7
Cs 68.6	Ba 267	La 718	Hf 1490	Ta 2150	W 2500 (2500)	Re 2700	Os 2500	Ir 2250	Pt 1865 (1900) 240	Au (731) (1370) 132	Hg 490	Tl 451	Pb 462	Bi 376 (550) 61	Po	At	Rn
Fr	Ra	Ac															

Ce 706	Pr 707	Nd 688	Pm 680	Sm 431	Eu 264	Gd 664	Tb 669	Dy 648	Ho 650	Er 637	Tm –	Yb 320	Lu 940
Th 978	Pa	U 1500	Np	Pu 550	Am	Cm	Bk	Cf	Es	Fm	Md		Lw

Sources: *CRC Handbook of Chemistry and Physics,* 71st edition, D. R. Lide, ed., CRC Press, Boca Raton, FL (1990).

L. E. Murr, *Interfacial Phenomena in Metals and Alloys,* Addison-Wesley, Reading, MA (1975).

APPENDIX
G

THERMOCHEMISTRY
OF OXIDES

Legend box:

M_xO_v
S^0_{298}
ΔS^0_{298}
ΔH^0_{298}

S^0_{298} = Absolute entropy of the oxide at 298K $\left(\dfrac{J}{\text{mole-K}}\right)$

ΔS^0_{298} = Entropy of formation at 298K $\left(\dfrac{J}{\text{mole-K}}\right)$

ΔH^0_{298} = Enthalpy of formation at 298K $\left(\dfrac{KJ}{\text{mole}}\right)$

Element	Oxide	S^0_{298}	ΔS^0_{298}	ΔH^0_{298}
H	H_2O	188.8	−44.7	−242.0
Li	Li_2O	37.93	−121	−596.6
Be	BeO	14.1	−94.7	−608.4
B	B_2O_3	54.0	−265	−1272.5
C	CO_2	213.9	2.1	−393.77
Na	Na_2O	75.1	−130	−415.2
Mg	MgO	26.97	−102	−601.6
Al	Al_2O_3	51.1	−314	−1584.0
Si	$SiO_{2\,(q)}$	41.5	−371	−910.9
P	P_2O_5	229	—	−1493.0
S	$SO_{2\,(g)}$	248.07	−145.8	−297.05
K	K_2O	—	−142	−363.3
Ca	CaO	39.8	−102	−634.3
Sc	Sc_2O_3	77.0	−293	−1906.7
Ti	TiO_2	50.2	−273	−944.1
V	V_2O_5	131.0	−324	−1551.3
Cr	Cr_2O_3	81.2	−264	−1130.4
Mn	MnO	59.9	−74	−385.2
Fe	Fe_3O_4	151.6	−341	−1117.6
Co	CoO	52.96	−88	−239.1
Ni	NiO	38.1	−92	−240.7
Cu	Cu_2O	93.8	75.3	−167.5
Zn	ZnO	50.7	−197	−1101.3
Ga	Ga_2O_3	84.78	−304	−1083.5
Ge	GeO_2	39.7	−183	−580.2
As	As_2O_3	122.8	−255	−653.77
Se	SeO_2	66.7	—	−236.1
Rb	Rb_2O	—	—	−330.3
Sr	SrO	55.5	−99.6	−592.3
Y	Y_2O_3	99.2	−296	−1906.7
Zr	ZrO_2	50.7	−197	−1101.3
Nb	NbO_2	54.55	−187	−799.3
Mo	MoO_2	50.0	−184	−588.7
Tc	Tc_2O_7	184.2	−392	−1113.7
Ru	RuO_2	60.7	−174	−304.4
Rh	Rh_2O_3	92.1	—	−383.0
Pd	PdO	39.3	−101	−112.6
Ag	Ag_2O	121.8	−66.4	−30.6
Cd	CdO	54.8	−99	−259.4
In	In_2O_3	—	−311	−927.4
Sn	SnO_2	52.3	−204	−580.7
Sb	Sb_2O_3	123.1	—	−690.2
Te	TeO_2	74.1	−180	−322.4
Cs	Cs_2O	127.6	—	−317.8
Ba	BaO	70.3	−97.7	−553.8
La	La_2O_3	128.1	−293	−1794.1
Hf	HfO_2	59.5	−201	−1113.7
Ta	Ta_2O_5	143.2	−455	−2047.3
W	WO_2	50.6	—	−589.9
Re	ReO_2	62.8	−180	−432.9
Os	OsO_4	136.9	—	−393.9
Ir	IrO_2	56.5	−185	−241.5
Pt	$PtO_{2\,(g)}$	256.0	—	−168.7
Au				
Hg	HgO	70.3	−10	−90.9
Tl	Tl_2O_3	137.3	—	−390.6
Pb	PbO	66.3	−102	−219.4
Bi	Bi_2O_3	151.6	−247	−570.7
Ce	Ce_2O_3	150.7	−293	−1821.7
Pr	Pr_2O_3	158.6	—	−1828.8
Nd	Nd_2O_3	158.6	—	−1809.1
Sm	Sm_2O_3	151.1	−296	−1833
Gd	Gd_2O_3	150.7	−289	−1817.1
Tb	Tb_2O_3	—	—	−1828.8
Dy	Dy_2O_3	149.9	−309	−1866.5
Ho	Ho_2O_3	158.3	−300	−1882.4
Er	Er_2O_3	153.2	−301	−1889.1
Tm	Tm_2O_3	—	—	−1899.9
Yb	Yb_2O_3	133.1	—	−1815.3
Th	ThO_2	65.3	−180	−1227.6
U	UO_2	77.9	−177	−1085.2
Np	NpO_2	80.4	—	−1030.0
Pu	PuO_2	82.5	−18.3	−1058.4

Other element boxes (N, O, F, Ne, Cl, Ar, Br, Kr, I, Xe, At, Rn, Lu, Po, Au, Fr, Ra, Ac, Pm, Eu, Pa, Am, Cm, Bk, Cf, Es, Fm, Md, Lw) appear without data values.

Source: Smithell's Metals Reference Book, 6th edition, E. A. Brandes, ed., Butterworths, London (1985).

APPENDIX
H

THERMOCHEMISTRY
OF NITRIDES

Legend:

$$M_u N_v$$
$$S^0_{298}$$
$$\Delta S^0_{298}$$
$$\Delta H^0_{298}$$

S^0_{298} = Absolute entropy of the nitride at 298 K $\left(\dfrac{J}{mole\text{-}K}\right)$

ΔS^0_{298} = Entropy of formation at 298 K $\left(\dfrac{J}{mole\text{-}K}\right)$

ΔH^0_{298} = Enthalpy of formation at 298 K $\left(\dfrac{KJ}{mole}\right)$

Periodic table of nitrides (values listed per element as: nitride formula; S^0_{298}; ΔS^0_{298}; ΔH^0_{298}):

Element	Nitride	S^0_{298}	ΔS^0_{298}	ΔH^0_{298}
H	NH_3	192.7	−98	−46.1
Li	Li_3N	—	−142	−196.8
Be	Be_3N_2	34.3	−186	−589.9
B	BN	14.8	—	−254.1
Mg	Mg_3N_2	93.7	−208	−461.8
Al	AlN	20.2	−105	−318.6
Si	Si_3N_4	113.0	−410	−745.1
Ca	Ca_3N_2	108.0	−209	−439.6
Ti	TiN	30.1	−95	−336.6
V	VN	37.3	—	−217.3
Cr	CrN	—	−88	−123.1
Mn	Mn_5N_2	—	−152	−201.8
Fe	Fe_4N	(156.2)	−50	−10.9
Co	Co_3N	98.8	—	+8.4
Ni	Ni_3N	—	−88	+0.8
Cu	Cu_3N	—	—	−74.5
Zn	Zn_3N_2	140.2	—	−22.2
Ga	GaN	29.7	—	−109.7
Ge	Ge_3N_4	(167)	—	−65.3
Sr	Sr_3N_2	123.5	−225	−3910
Zr	ZrN	38.9	−98	−365.5
Nb	Nb_2N	67.0	—	−248.6
Mo	Mo_2N	(87.9)	4.3	−69.5
Cd	Cd_3N_2	—	—	+161.6
In	InN	43.5	—	−138.1
Ba	Ba_3N_2	(152.4)	−163	−341.1
La	LaN	44.4	−96	−299.4
Hf	HfN	50.6	—	−369.3
Ta	TaN	42.7	−125	−252.4
Ce	CeN	—	−105	−326.6
Th	Th_3N_4	—	−377	−1298.0
U	UN	62.7	−116	−294.7

Other periodic table cells shown without data: H group (Na, K, Rb, Cs, Fr); Sc, Y, Ac; Ra; Tc, Ru, Rh, Pd, Ag; Re, Os, Ir, Pt, Au, Hg; C, N, O, F, Ne; P, S, Cl, Ar; As, Se, Br, Kr; Sn, Sb, Te, I, Xe; Tl, Pb, Bi, Po, At, Rn; Pr, Nd, Pm, Sm, Eu, Gd, Tb, Dy, Ho, Er, Tm, Yb, Lu; Pa, Np, Pu, Am, Cm, Bk, Cf, Es, Fm, Md, Lw.

Source: Smithell's Metals Reference Book, 6th edition, E. A. Brandes, ed., Butterworths, London (1985).

APPENDIX
I

THERMOCHEMISTRY
OF CARBIDES

Legend box:

M_nC_v
S^0_{298}
ΔS^0_{298}
ΔH^0_{298}

S^0_{298} = Absolute entropy of the nitride at 298 K $\left(\dfrac{J}{\text{mole-K}}\right)$

ΔS^0_{298} = Entropy of formation at 298 K $\left(\dfrac{J}{\text{mole-K}}\right)$

ΔH^0_{298} = Enthalpy of formation at 298 K $\left(\dfrac{KJ}{\text{mole}}\right)$

Element	Compound M_nC_v	S^0_{298}	ΔS^0_{298}	ΔH^0_{298}
H	CH_4 (g)	186.3	−81	−74.9
Be	Be_2C	16.3	−8.0	−117.2
B	B_4C	27.09	−1.3	−71.6
Mg	Mg_2C_3	−62.8	—	+79.5
Al	Al_4C_3	88.7	−41	−215.8
Si	SiC	16.54	(−59)	−67.0
Ca	CaC_2	70.3	+18	−59
Ti	TiC	24.3	−12.7	−183.8
V	VC	27.6	—	−100.9
Cr	Cr_7C_3	200.9	−15.6	−228.1
Mn	Mn_7C_3	239.0	−2.7	−112.2
Fe	Fe_3C	104.7	+20.3	+25.1
Co	Co_3C	135.6	+5.3	−98.4
Zr	ZrC	33.1	−12	−202
Nb	NbC	35.1	—	−138.1
Ta	TaC	42.2	−4.7	−143.6
W	WC	41.8	−1.3	−37.7
Ce	Ce_2C_3	173.7	—	−176.6
Th	ThC_2	70.3	—	−117.2
U	U_2C_3	138.4	+0.3	−205.1

Other periodic table positions shown without carbide data:

Li; Na; K, Sc, Ni, Cu, Zn, Ga, Ge, As, Se, Br, Kr; Rb, Sr, Y, Mo, Tc, Ru, Rh, Pd, Ag, Cd, In, Sn, Sb, Te, I, Xe; Cs, La, Hf, Re, Os, Ir, Pt, Au, Hg, Tl, Pb, Bi, Po, At, Rn; Fr, Ra, Ac; C, N, O, F, Ne, P, S, Cl.

Lanthanide series: Pr, Nd, Pm, Sm, Eu, Gd, Tb, Dy, Ho, Er, Tm, Yb, Lu.

Actinide series: Pa, Np, Pu, Am, Cm, Bk, Cf, Es, Fm, Md, Lw.

Source: Smithell's Metals Reference Book, 6th edition, E. A. Brandes, ed., Butterworths, London (1985).

ELECTROCHEMICAL SERIES[1]

Electrode potentials relative to the standard hydrogen electrode arranged in descending order.

---

[1] *Source: CRC Handbook of Chemistry and Physics*, 71st edition, D. R. Lide, ed., CRC Press, Boca Raton, Fla., 1991.

Reduction Reaction	Emf (volts)	Reduction Reaction	Emf (volts)
$Sr^+ + e^- = Sr$	-4.10	$2H^+ + 2e^- = H_2$	0.00000
$Ca^+ + e^- = Ca$	-3.80	$Cu^{++} + e^- = Cu^+$	0.153
$Li^+ + e^- = Li$	-3.0401	$Re^{+++} + 3e^- = Re$	0.300
$Rb^+ + e^- = Rb$	-2.98	$Cu^{++} + 2e^- = Cu$	0.3419
$K^+ + e^- = K$	-2.931	$Ru^{++} + 2e^- = Ru$	0.455
$Cs^+ + e^- = Cs$	-2.92	$Cu^+ + e^- = Cu$	0.521
$Ba^{++} + 2e^- = Ba$	-2.912	$Rh^{++} + 2e^- = Rh$	0.600
$Sr^{++} + 2e^- = Sr$	-2.89	$Rh^+ + e^- = Rh$	0.600
$Ca^{++} + 2e^- = Ca$	-2.868	$Rh^{+++} + 3e^- = Rh$	0.758
$Na^+ + e^- = Na$	-2.71	$Fe^{+++} + e^- = Fe^{++}$	0.771
$Mg^+ + e^- = Mg$	-2.70	$Ag^+ + e^- = Ag$	0.7996
$Mg^{++} + 2e^- = Mg$	-2.372	$Hg^{++} + 2e^- = Hg$	0.851
$Sc^{+++} + 3e^- = Sc$	-2.077	$Pd^{++} + 2e^- = Pd$	0.951
$Be^{++} + 2e^- = Be$	-1.847	$Pt^{++} + 2e^- = Pt$	1.118
$Al^{+++} + 3e^- = Al$	-1.662	$Ir^{+++} + 3e^- = Ir$	1.156
$Ti^{++} + 2e^- = Ti$	-1.630	$O_2 + 4H^+ + 4e^- = 2H_2O$	1.229
$Ba^{++} + 2e^- = Ba \ (Hg)$	-1.570	$Au^{+++} + 2e^- = Au^{++}$	1.401
$Mn^{++} + 2e^- = Mn$	-1.185	$Au^{+++} + 3e^- = Au$	1.498
$V^{++} + 2e^- = V$	-1.175	$Mn^{+++} + e^- = Mn^{++}$	1.5415
$Nb^{+++} + 3e^- = Nb$	-1.099	$Au^+ + e^- = Au$	1.692
$Cr^{+++} + 3e^- = Cr$	-0.744	$Ag^{++} + e^- = Ag^+$	1.980
$Ga^{+++} + 3e^- = Ga$	-0.560		
$Fe^{++} + 2e^- = Fe$	-0.447		
$Cd^{++} + 2e^- = Cd$	-0.352		
$Co^{++} + 2e^- = Co$	-0.28		
$Ni^{++} + 2e^- = Ni$	-0.257		
$Mo^{+++} + 3e^- = Mo$	-0.200		
$Pb^{++} + 2e^- = Pb$	-0.1205		
$Fe^{+++} + 3e^- = Fe$	-0.037		
$2H^+ + 2e^- = H_2$	0.00000		

THE
CARNOT
CYCLE

The connection between the entropy change and the reversible heat absorbed for an arbitrary process can be traced to a development first presented by Sadi Carnot in the middle of the last century. Carnot visualized a thermodynamic system containing an ideal gas taken through a specific process sequence that returns it to its original state. This process is called the Carnot cycle and the sequence of states of the Carnot engine as its cycle is represented in Fig. K.1. A pictorial representation of the steps in the process is represented in (P, V) coordinates in Fig. K.2. All steps in the process are assumed to be carried out *reversibly*.

PROCESS I. The state of the system is initially given by (T_I, P_0), with a corresponding volume V_0. It is in contact with a heat reservoir at the same temperature T_I. Except for its area of contact with the heat reservoir, the boundary is thermally insulated. The pressure is reduced to P_1, allowing the gas to expand to volume V_1 *isothermally* at T_I. A quantity of heat Q_I flows from the reservoir into the system. When the system reaches the state (P_1, V_1) a thermally insulating barrier is inserted to separate it from the reservoir.

PROCESS II. The pressure is further reduced to P_2. Since the entire system is thermally insulated from its surroundings, process II is *adiabatic* and the heat absorbed Q_{II} is equal to 0. It is presently shown that the path in (P, V) space can be computed for this process so that the final volume (V_2) and temperature (T_{III}) can be calculated given P_2.

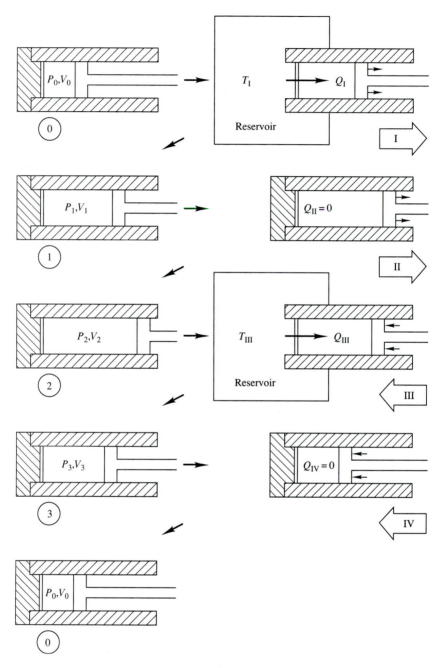

FIGURE K.1
States and processes that connect them in the Carnot cycle.

PROCESS III. The system is brought into contact with an external heat reservoir with a temperature T_{III} matching the value attained at the end of step II. The insulating partition is removed and the system is *isothermally* compressed at T_{III} to the pressure P_3. A quantity of heat Q_{III} is exchanged with the reservoir and since the heat flows from the system to the surroundings, Q_{III} is negative. The final state is represented by the condition (P_3, V_3). An insulating barrier is again inserted between the reservoir and the system in preparation for the next step.

PROCESS IV. The final pressure P_3 at the end of step III is not chosen arbitrarily. it is the unique pressure at T_{III} that takes the system along an adiabatic path back to the starting state at (P_0, V_0) completing the cycle. The heat absorbed in this process Q_{IV} is equal to 0. The changes associated with each of these four steps can be computed for one mole of an ideal gas without invoking the second law of thermodynamics. This observation is crucial, since the argument is intended to establish a key element of the second law, namely the relation between the change in a state function and the *reversible* heat absorbed during an arbitrary process. Later, the change in state function is identified with the entropy.

Process I is an isothermal expansion. Since the properties of an ideal gas depend only upon its temperature, $dU = 0$ for any isothermal process. From the first law,

$$dU = \delta Q + \delta W = 0 \rightarrow \delta Q = -\delta W \qquad \text{(K.1)}$$

Since, for any reversible process, $\delta W = -PdV$,

$$\delta Q = -(-PdV) = PdV = \frac{RT}{P}dP$$

For the isothermal process at T_1, integration gives

$$Q_I = RT_1 \ln \frac{P_1}{P_0} \qquad \text{(K.2)}$$

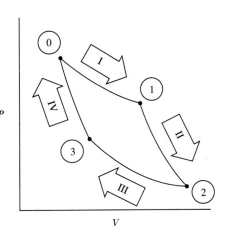

FIGURE K.2
Sequence of states in (V, P) space traversed by a system during a Carnot cycle.

Similarly, for process III carried out isothermally at T_{III} from P_2 to P_3,

$$Q_{III} = RT_{III} \ln \frac{P_3}{P_2} \tag{K.3}$$

For an adiabatic process, $\delta Q = 0$, and the first law becomes

$$dU = C_v dT = \delta W = -PdV = -\frac{RT}{V} dV$$

Separating the variables

$$\frac{dT}{T} = -\frac{R}{C_v} \frac{dV}{V}$$

In process II the temperature changes from T_I to T_{III}, and the volume changes from V_1 to V_2. Integration gives

$$\ln \frac{T_{III}}{T_I} = \ln \left(\frac{V_1}{V_2} \right)^{\frac{R}{C_v}}$$

Note that the minus sign on the right side of the previous equation causes the ratio of volumes to be inverted in the integrated expression. Equality of the logarithms implies equality of their arguments. Thus,

$$\frac{T_{III}}{T_I} = \left(\frac{V_1}{V_2} \right)^{\frac{R}{C_v}} \tag{K.4}$$

Since the states are described in terms of the pressures involved, the ideal gas law can be used to express V_2 and V_1 in terms of corresponding T and P values

$$\frac{T_{III}}{T_I} = \left(\frac{\frac{RT_I}{P_1}}{\frac{RT_{III}}{P_2}} \right)^{\frac{R}{C_v}}$$

Algebraic manipulation of this result gives the relation between T and P along an adiabatic path as

$$\frac{P_2}{P_1} = \left(\frac{T_{III}}{T_I} \right)^{\frac{C_v + R}{R}} \tag{K.5}$$

The analogous relation holds for process IV in which the system changes adiabatically from (T_{III}, P_3) to the initial condition (T_I, P_0),

$$\frac{P_0}{P_3} = \left(\frac{T_I}{T_{III}} \right)^{\frac{C_v + R}{R}} \tag{K.6}$$

Next, examine the hypothesis that, for any reversible process,

$$\oint \frac{\delta Q_{rev}}{T}$$

is a state function with a value independent of the path followed during the process, and depends only on the initial and final states. The crucial test of this hypothesis rests on the proof that for any cyclic process, that is, one that returns to the original state, this integral must be zero. This integral can be easily evaluted for the Carnot cycle:

Process I

$$\oint \frac{\delta Q_{rev}}{T} = \frac{1}{T_I} \oint \delta Q_{rev} = \frac{Q_I}{T_I} = \frac{RT_I}{T_I} \ln \frac{P_1}{P_0} = R \ln \frac{P_1}{P_0}$$

Process II

$$\oint \frac{\delta Q_{rev}}{T} = 0$$

Process III

$$\oint \frac{\delta Q_{rev}}{T} = \frac{1}{T_{III}} \oint \delta Q_{rev} = \frac{Q_{III}}{T_{III}} = \frac{RT_{III}}{T_{III}} \ln \frac{P_3}{P_2} = R \ln \frac{P_3}{P_2}$$

Process IV

$$\oint \frac{\delta Q_{rev}}{T} = 0$$

The result for the cycle is the sum of these four contributions:

$$\oint \frac{\delta Q_{rev}}{T} = R \ln \frac{P_1}{P_0} + 0 + R \ln \frac{P_3}{P_2} = 0$$

which can be written

$$\oint \frac{\delta Q_{rev}}{T} = R \ln \left(\frac{P_1 P_3}{P_0 P_2} \right) \tag{K.7}$$

The pressure ratios in this equation, (P_2/P_1) and (P_0/P_3), can be expressed simply in terms of the temperatures of the two isothermal steps by applying the adiabatic path equations, Eqs. (K.5) and (K.6)

$$\oint \frac{\delta Q_{rev}}{T} = R \ln \left(\frac{P_1}{P_2} \right) \left(\frac{P_3}{P_0} \right) = R \ln \left(\frac{T_I}{T_{III}} \right)^{\frac{C_v+R}{R}} \left(\frac{T_{III}}{T_I} \right)^{\frac{C_v+R}{R}}$$

Or

$$\oint \frac{\delta Q_{rev}}{T} = R \ln(1) = 0 \tag{K.8}$$

Thus, it is demonstrated that for any Carnot cycle using an ideal gas operating between any pair of temperatures for the isothermal steps, any choice of the independent pressures gives

$$\oint \frac{\delta Q_{rev}}{T} = 0$$

Accordingly, within these restrictions, this integral has the characteristics of a *state function*.

Carnot and others went on to demonstrate that *every* cyclic path that could be devised can be decomposed into an infinite series of infinitesimal Carnot cycles. Since the integral in question is zero for each Carnot cycle, it totals to zero for any sum of Carnot cycles. Thus, the integral has this characteristic of a state function for *every* process that can be imagined for an ideal gas.

This result was further generalized to include systems containing arbitrary working substances. The advancement was achieved by imagining the system with the real working substance coupled to one with an ideal gas as the working substance so that the energy, heat, and work exchanges accompanying the Carnot cycles for these two systems are equal and opposite at each step. It is concluded that

$$\oint \frac{\delta Q_{rev}}{T} = 0$$

for any real system taken along any arbitrary path. Thus, the result is completely general and this integral is demonstrated to be a state function of general application. The symbol widely adopted for this state function is S; the origins of its name, *entropy*, are obscure. For any arbitrary system taken along any reversible path,

$$\Delta S = \oint dS = \oint \frac{\delta Q_{rev}}{T} \tag{K.9}$$

It follows that for each infinitesimal step along the reversible path

$$dS = \frac{\delta Q_{rev}}{T} \tag{K.10}$$

and the general relation between entropy and the reversible heat absorbed is established.

The development of the concept of entropy did not end with the Carnot cycle argument. The statistical interpretation of entropy, introduced in Chap. 6, evolved half a century later. Notions of configurational entropy represented, for example, by the entropy of mixing solutions, expanded the concept. The quantification of the role of the entropy production in irreversible processes took another half a century, and continues to evolve. The Carnot cycle argument lies at the foundation of the development of thermodynamics; however, its role is largely hidden in the treatment of the complex multicomponent, multiphase systems with capillarity and electrochemical effects, which are the bread and butter of materials science.

ANSWERS TO
HOMEWORK
PROBLEMS

NUMERICAL ANSWERS TO HOMEWORK PROBLEMS

CHAPTER 2
2.5.

$$dz = 36u^2 v \cos(x) + 12u^3 \cos(x) - 12u^3 v \sin(x)$$

CHAPTER 3
3.16.

Reaction	ΔS°_{298} (J/mol K)
$C + O_2 = CO_2$	2.92
$2Al + 3/2\ O_2 = Al_2O_3$	-279.65
$Si + C = SiC$	-8.04
$C + 1/2\ O_2 = CO$	89.695
$Si + O_2 = SiO_2$	-181.77
$Si + CO_2 = SiC + O_2$	-10.96
$Al_2O_3 + 3/2\ Si = 3/2\ SiO_2 + 2\ Al$	7.00
$2\ Al + 3\ CO_2 = Al_2O_3 + 3\ CO$	-19.33
$SiO_2 + 2\ C = CO_2 + SiC$	176.65
$CO + 1/2\ O_2 = CO_2$	-86.775

Etc.

CHAPTER 4
4.3. After isothermal step, $V = 25.97$ (cc/mol)
After the second (compression) step, $V = 25.76$ (cc/mol)

513

4.4. (*a*) 20.63 (J/mol K)
 (*b*) −2.68 (J/mol K)
 (*c*) 44.62 (J/mol K)
 (*d*) −1.92 (J/mol K)
 (*e*) 21.25 (J/mol K)
 (*f*) 95.72 (J/mol K)

4.5. 14.87 (J/mol K),   −0.64 (J/mol K)

4.6. (*a*) $n = 0.536$ mol, $\Delta U = 162 - 90 = 72$ (J/mol);
 (*b*) $T_f = 1365K$, $\Delta U = 72$ (J/mol)

4.7.

$$\Delta S(T, P) = C_p \ln \left(\frac{T}{300} \right) - R \ln \left(\frac{P}{1} \right)$$

4.8. (*a*) $dU = \frac{C_v}{R}(V dP + P dV) = C_v dT$
 (*b*) $DH = \frac{C_p}{R}(V dP + P dV) = C_p dT$

4.9.

$$\left(\frac{\partial H}{\partial G} \right)_s = \frac{C_p}{C_p - T S \alpha}$$

4.10.

$$dF = - \left[S + P V \alpha - \frac{P \beta}{T \alpha} C_p \right] dT - \frac{P \beta}{\alpha} dS$$

4.12. (*a*) $T_i = 219.4$ K, $T_f = 292.5$ K
 (*b*) $Q = -33.77$ (J/mol)
 (*c*) $W = 945.62$ (J/mol)
 (*d*) $\Delta U = 911.45$ (J/mol), $\Delta H = 1520$ (J/mol), $\Delta S = 0.217$ (J/mol K), $\Delta F = -9205$
  (J/mol), $\Delta G = -9198$ (J/mol)

4.13. (*a*) For titanium, $P = 4307$ atm
 (*b*) For alumina, $P = 979$ atm

4.14.

$$\left(\frac{\partial F}{\partial S} \right)_V = -\frac{ST}{C_V}$$

Ideal gas: −3494 K; iron: −334 K.

4.16. Plot the surface given by:

$$G(T, P) - G(298, 1) = RT \ln P - 34.6T \ln T + 26.7T + 50,790$$

4.18. Assuming no heat losses, the temperature would rise 3556 K.

4.19. For potassium, $C_P - C_V = 0.154$ (J/mol K); for W, 0.194 (J/mol K).

CHAPTER 5

5.7. Unconstrained, $z_{max} = 4$ at $(x = 2, y = 2)$;
 (*a*), (*b*) Constrained by $y = 1 - x$, $z'_{max} = 8.5$ at $(x = 0.5, y = 0.5)$.

CHAPTER 6

6.2. List the states systematically, beginning with (10, 0, 0), (9, 0, 1), (9, 1, 0), (8, 0, 2), (8, 1, 1), (8, 2, 0), etc.

6.3. (a) $2^4 = 16$ microstates;
(b) Enumerate systematically; there are 10 macrostates.

6.4. (a) $4^3 = 64$;
(b) $4^{15} = 1.074 \times 10^9$;
(c) $15^4 = 50,625$
(d) $30^{50} = 7.18 \times 10^{73}$;
(e) $100^{1000} = (10^2)^{1000} = 10^{2000}$.

6.5. (a) $\Delta U = 0$;
(b) $\Omega_{II}/\Omega_{I} = 12$;
(c) II is more likely.

6.6. (a) $10! = 3.629 \times 10^6$, $30! = 2.653 \times 10^{32}$, $60! = 8.321 \times 10^{81}$;
(b) Stirling's formula: 4.54×10^5, 1.927×10^{31}, 4.28×10^{80};
(c) error in $\ln x!$: 0.138, 0.035, 0.016.

6.7. $\Delta S = -30.5$ (J/mol K)$= -5.07 \times 10^{-23}$ (J/atom K).

6.9. $\Delta U = 2,311$ (J/mol).

6.10. (a) $\Delta S = 2.162$ (J/mol K);
(b) $\Delta S = 2.162$ (J/mol K).

CHAPTER 7

7.3. Plot:

$$\mu(T, P) - \mu(298, 1) = -20.79T \ln T + 104.2T - 4,236 + RT \ln P$$

7.6. The error in log(P) is about 1.5% at the melting point.

7.7. $\Delta S_m = 22.00$ (J/mol K).

7.8. $T^{\epsilon L} = 1778$ K.

7.9. $\Delta H_v = 285,000$ (J/gm atom).

7.10. $P_{\alpha\gamma G} = 1.65 \times 10^{-10}$, $P_{\gamma\delta G} = 1.14 \times 10^{-5}$, $P_{\delta LG} = 8.29 \times 10^{-5}$

7.11. $(T, P)_{\beta LG} = (576, \ 6.38 \times 10^{-10})$; $(T, P)_{\epsilon\beta G} = (500, \ 4.47 \times 10^{-12})$.

7.12. $(T, P)_{\epsilon LG} = (569, \ 4.32 \times 10^{-10})$.

CHAPTER 8

8.1. (a) wt%$= 4.4$;
(b) $c_0 = 0.011$ (mol/cc);
(c) $\rho_0 = 0.18$ (gm/cc)

8.5. Plot the result of a Gibbs-Duhem integration:

$$\Delta \overline{H_{Cn}} = 12500X_{Pn}^2\left(X_{Pn} - \frac{1}{2}\right); \qquad \Delta H_{mix} = 6250X_{Cn}X_{Pn}^2$$

8.9. Plot the activity of Zn obtained from

$$RT \ln a_{Zn} = \left(1 - \frac{T}{4000}\right)X_{Al}^2\left[13200(X_{Al} - X_{Zn}) + 19200X_{Zn}\right] + RT \ln X_{Zn}$$

8.10. (*b*)

	Total Properties	PMP for A	PMP for B
G^{xs} (J/mol)	−903	1108	3821
S^{xs} (J/mol K)	6.80	0.764	2.64
V (cc/mol)	−0.407	−0.219	−0.756
H (J/mol)	2832	1528	5271
U (J/mol)	2842	1526	5263
F^{xs} (J/mol)	−898.9	1110	3829

8.13. Plot as a function of temperature

$$\gamma_A^o = \gamma_B^o = e^{-\frac{13500}{RT}}$$

8.14. $E_{AB} = -6.111 \times 10^{-20}$ (J/bond).

8.15. Plot $f_{AB} = [2X_A X_B](1.3)$, then $f_{AA} = X_A - f_{AB}/2$, $f_{BB} = X_B - f_{AB}/2$.

8.16. Plot

$$f_{AB} = \frac{\Delta H_{\text{mix}}}{\left[\frac{1}{2}N_0 z\left(e_{AB} - \frac{1}{2}(e_{AA} + e_{BB})\right)\right]}$$

$$f_{AA} = X_A - \frac{1}{2}f_{AB} \qquad f_{BB} = X_B - \frac{1}{2}f_{AB}$$

CHAPTER 9

No numerical problems.

CHAPTER 10

10.1. Given

$$\Delta G_{\text{mix}}^\alpha[\alpha; \alpha] = 8400 X_A^\alpha X_B^\alpha + RT(X_A^\alpha \ln X_A^\alpha + X_B^\alpha \ln X_B^\alpha)$$

$$\Delta G_{\text{mix}}^L[L; L] = 10500 X_A^L X_B^L + RT(X_A^L \ln X_A^L + X_B^L \ln X_B^L)$$

(*a*) Compute and plot:

$$\Delta G_{\text{mix}}^\alpha[L; L] = \Delta G_{\text{mix}}^\alpha[\alpha; \alpha] - X_A^\alpha \Delta G_A^{o\alpha \to L} - X_B^\alpha \Delta G_B^{o\alpha \to L}$$

$$\Delta G_{\text{mix}}^L[L; L] = \Delta G_{\text{mix}}^L[L; L]$$

(*b*) Compute and plot:

$$\Delta G_{\text{mix}}^\alpha[L; \alpha] = \Delta G_{\text{mix}}^\alpha[\alpha; \alpha] - X_A^\alpha \Delta G_A^{o\alpha \to L}$$

$$\Delta G_{\text{mix}}^L[L; \alpha] = \Delta G_{\text{mix}}^L[L; L] + X_B^L \Delta G_B^{o\alpha \to L}$$

(*c*) Compute and plot:

$$\Delta G_{\text{mix}}^\alpha[\alpha; \alpha] = \Delta G_{\text{mix}}^\alpha[\alpha; \alpha]$$

$$\Delta G_{\text{mix}}^L[\alpha; \alpha] = \Delta G_{\text{mix}}^L[L; L] + X_A^L \Delta G_A^{o\alpha \to L} + X_B^L \Delta G_B^{o\alpha \to L}$$

10.2. (*a*) Compute and plot:

$$\Delta G_{\text{mix}}^\beta[\beta; \beta] = -8200 X_A^\beta X_B^\beta + RT(X_A^\beta \ln X_A^\beta + X_B^\beta \ln X_B^\beta)$$

$$\Delta G_{\text{mix}}^L[L; L] = -10500 X_A^L X_B^L + RT(X_A^L \ln X_A^L + X_B^L \ln X_B^L)$$

(b) Compute and plot:

$$\Delta G^{\beta}_{mix}[\beta; L] = \Delta G^{\beta}_{mix}[\beta; \beta] - X^{\beta}_B \frac{6800}{660}(660 - T)$$

$$\Delta G^{L}_{mix}[\beta; L] = \Delta G^{\beta}_{mix}[L; L] + X^{L}_A \frac{8200}{1050}(1050 - T)$$

10.3. (a) Compute and plot: $a_B[L; L] = e^{\frac{8400}{RT}}$

(b) Compute and plot: $a_B[L; \alpha] = a_B[L; L]e^{-\frac{1200}{RT}}$

10.9. (a) $X_2 = 0.307$ to $X_2 = 0.693$;

(b) Compute and plot: $\Delta \mu^{\alpha}_2 = 12700(1 - X_2)^2 + RT \ln X_2$

(c) Compute and plot: $\left(\frac{d\Delta\mu^{\alpha}_2}{dX_2}\right) = \frac{RT}{X_2} - 2 \cdot 12700(1 - X_2)$

10.16. Solve the following equations simultaneously:

$$317 - 5600X_2^{L2} + 11400X_2^{\alpha2} + R(1300) \ln\left[\frac{1 - X_2^L}{1 - X_2^\alpha}\right] = 0$$

$$-1246 - 5600(1 - X_2^L)^2 + 11400(1 - X_2^\alpha)^2 + R(1300) \ln\left[\frac{X_2^L}{X_2^\alpha}\right] = 0$$

$$X_2^\alpha = 0.763, \quad X_2^L = 0.824.$$

10.17. $T_c = 637.5$ K

(a) Spinodal boundary $X_{2,spin} = \frac{1}{2}\left[-1 \pm \sqrt{1 - \frac{T}{T_c}}\right]$

(b) Phase boundary (minima): $T(X_2) = \frac{2T_c(1-2X_2)}{\ln\left[\frac{1-X_2}{X_2}\right]}$

10.18. $X_{2,max} = 0.649, \quad T_c = 652.3$ K.

10.19.

$$T(X^*) = \frac{-9380X^*(1 - X^*) + 1283(8.8)(1 - X^*) + 942(6.3)X^*}{8.8(1 - X^*) + 6.3X^*}$$

$$T(X^*) = \frac{10000X^*(1 - X^*) + 1283(8.8)(1 - X^*) + 942(6.3)X^*}{8.8(1 - X^*) + 6.3X^*}$$

10.20. $\gamma^{\circ}_0 = 370$

10.21. 7362 K.

CHAPTER 11

11.2. $\mathcal{A} = -160,600$ J; products form.

11.3.

$$H_2 + 1/2\,O_2 = H_2O; \qquad \mathcal{A} = -168,800 \text{ J}$$

$$CO + 1/2\,O_2 = CO_2; \qquad \mathcal{A} = -144,600 \text{ J}$$

$$CH_4 + O_2 = CO_2 + 2\,H_2 \qquad \mathcal{A} = -328,300 \text{ J}$$

Oxygen is consumed; oxides form.

11.4. $X_{CO} = 0.295$, $X_{CO2} = 0.705$, $X_{O2} = 1.8 \times 10^{-20}$

11.6. (a) $\Delta G_f(T) = -244,300 - T\,105.8$ J

    (b) $K(T) = \exp\{105.8/R\}\exp\{244,300/RT\}$

    (c) $P_{O_2} = 1.60 \times 10^{-19}$

    (d) $A = -333,700$ J

11.7. (a) $\Delta G_f(SiO_2) = -715,300$ J, $\Delta G_f(H_2O) = -194,500$ J

    (b) $\Delta G_r = -326,300$ J

    (c) $K = 7.68 \times 10^{15}$, $(H_2/H_2O) = 8.76 \times 10^7$

    (d) $P_{O_2} = 1.5 \times 10^{-35}$

    (e) $P_{O_2} = 1.5 \times 10^{-35}$

11.9. (a) $P_{O_2} = 2 \times 10^{-12}$

    (b) $K = 3 \times 10^{33}$

    (c) $H_2/H_2O = 4 \times 10^5$

    (d) Yes

    (e) $(CO/CO_2) = 6 \times 10^2$

    (f) $P_{O_2} = 10^{-22}$

11.10. (a) Yes

    (b) No

    (c) No

    (d) Yes

11.11.

$$Cu_2O : A = -14,510 \text{ J} \qquad CaO : A = -519,100 \text{ J}$$

$$NiS : A = -138,400 \text{ J} \qquad CaCl_2 : A = -571,200 \text{ J}$$

11.13.

$$Na_2O + O_2 + 1/2\ S_2 = Na_2SO_3$$

$$2/5\ Nb_2O_5 + O_2 + 4/5\ S_2 = 4/5\ NbS_2O_5$$

$$2/7\ Cu_2O + O_2 + 2/7\ S_2 = 4/7\ CuSO_4$$

$$1/4\ Cu_2S + O_2 + 1/8\ S_2 = 1/2\ CuSO_4$$

11.15. If total pressure is 1 atm, for $P_{CO} > 0.447$ oxidation will be prevented. Carburization requires $P_{CO} > 0.96$. Thus, if $P_{CO} > 0.96$, both conditions are satisfied.

11.16. Reactions: $2\ CO + O_2 = 2\ CO_2$ and $2H_2 + O_2 = 2\ H_2O$

    (a) Equilibrium mole fractions: $CO = 0.40$; $CO_2 = 0.09$; $H_2 = 0.376$; $H_2O = 0.134$; $O_2 = 1.5 \times 10^{-20}$.

    (b) Affinities: $A_{[2H2O]} = -378,800$ J; $A_{[2CO2]} = -374,500$ J.

CHAPTER 12

12.6. Equation of the surface:

$$\ln P(H, T) = \ln P(H = 0, T) - \frac{\Delta H_v}{RT} + \frac{2\gamma V^L}{RT}H$$

12.7. $P(H = 0) = 1.73 \times 10^{-6}$ atm; $P(H = -0.5\mu) = 6.8 \times 10^{-7}$ atm.

12.9. (a) $\lambda^\alpha = 1.51 \times 10^{-7}$ cm, $\lambda^\beta = 1.28 \times 10^{-7}$ cm

12.10. (a) $X_B^\epsilon = 0.109$, $X_B^\beta = 0.127$

(b) $\lambda^\beta = 1.16 \times 10^{-7}$ cm.

12.12. 0.866.

12.14. 549 ergs/cm^2.

12.15. Yes.

12.16. (a) $\phi = 163°$

(b) $s_c = 4.09\mu$.

CHAPTER 13

13.1. $\Delta S_v = 8.85$ (J/mol K); $\Delta H_v = 95,800$ (J/mol).

13.2. (a) $X_i = \exp\{7.6/R\} \exp\{-188,000/RT\}$

13.3. $X_{vv} = X_v^2 \exp\{-0.89/R\} \exp\{+9,500/RT\}$

13.4. High range: $X_f = \exp\{10/R\} \exp\{-350,000/RT\}$

Low range: $X_f = \exp\{1/R\} \exp\{-550,000/RT\}$

13.7. (a) $M_{2+0.00501}O_3$

(b) $M_2O_{3-0.00749}$

(c) $M_{2-0.0025}O_3$

(d) $M_2O_{3+0.00375}$

CHAPTER 14

14.1. $T = 398.38$ K at all x; (b) $\Delta S = 0.266$ (J/mol K).

14.6. Doubling the isotope ratio requires a centrifuge with 50 cm radius rotating at 20,000 rpm.

14.9. (a) $\Delta G = -219.3$ J

(b) $\Delta G_{grad} = +133.0$ J, $\Delta G_{tot} = -86.3$ J.

CHAPTER 15

15.1. Compute and plot:

$$\alpha(c) = \frac{-K \pm \sqrt{K^2 + 4Kc}}{2c} \qquad \text{Use } + \text{ sign.} \qquad \alpha(c) \approx \sqrt{\frac{K}{c}}$$

15.2. pH $= 5.30$

15.3. 1.4×10^{-5} gm.

15.4. $C_{NaOH} = 3.162 \times 10^{-5}$ (mol/liter).

15.5. 1.26×10^{-7} (liters) $= 0.126$ microliters.

15.10. For the compositions given,

$a_{Mg} = -0.006, \qquad 0.024, \qquad 0.061, \qquad 0.160, \qquad 0.357, \qquad 0.494$

15.12. Formation energies needed as input, Joules: NiO $= -215,900$; NiO$_2 = -215,100$; Ni$^{++} = -48,200$; HNiO$_2^- = -349,200$; H$_2$O $= -273,200$.

Triple points (E, pH) occur as follows: {Ni, Ni^{++}, NiO} at $(-0.250, 6.09)$; {Ni^{++}, NiO, NiO$_2$} at $(0.863, 6.09)$; {NiO, NiO$_2$, HNiO$_2^-$} at $(0.146, 18.21)$. {Ni, NiO, HNIO$_2^-$} at $(-0.967, 18.21)$.

INDEX

QUEEN MARY & WESTFIELD
COLLEGE LIBRARY
(MILE END)

WITHDRAWN
FROM STOCK
QMUL LIBRARY